ELEMENTS OF MATHEMATICS

Springer
Berlin
Heidelberg
New York
Hong Kong
London
Milan
Paris
Tokyo

NICOLAS BOURBAKI

Algebra II

Chapters 4-7

Translated by P.M. Cohn & J. Howie

Springer

Originally published as
ALGÈBRE, CHAPITRES 4 À 7
Masson, Paris, 1981

Mathematics Subject Classification (2000):
06-XX, 10XX, 20XX, 15XX, 18XX

Cataloging-in-Publication Data applied for
A catalog record for this book is available from the Library of Congress.
Bibliographic information published by Die Deutsche Bibliothek
Die Deutsche Bibliothek lists this publication in the Deutsche Nationalbibliografie; detailed bibliographic data is available in the Internet at http://dnb.ddb.de

Softcover printing of the 1st English edition of 1990

ISBN 3-540-00706-7 Springer-Verlag Berlin Heidelberg New York

ISBN 3-540-19375-8 1st edition Springer-Verlag Berlin Heidelberg New York

Springer-Verlag Berlin Heidelberg New York
a member of Springer Science+Business Media
http:/www.springer.de

Printed in Germany

Cover Design: *Design & Production* GmbH, Heidelberg
Printed on acid-free paper 41/3111 db 5 4 3 2 1 SPIN 11419846

To the reader

1. The Elements of Mathematics Series takes up mathematics at the beginning, and gives complete proofs. In principle, it requires no particular knowledge of mathematics on the readers' part, but only a certain familiarity with mathematical reasoning and a certain capacity for abstract thought. Nevertheless, it is directed especially to those who have a good knowledge of at least the content of the first year or two of a university mathematics course.

2. The method of exposition we have chosen is axiomatic, and normally proceeds from the general to the particular. The demands of proof impose a rigorously fixed order on the subject matter. It follows that the utility of certain considerations will not be immediately apparent to the reader unless he has already a fairly extended knowledge of mathematics.

3. The series is divided into Books and each Book into chapters. The Books already published, either in whole or in part, in the French edition, are listed below. When an English translation is available, the corresponding English title is mentioned between parentheses. Throughout the volume a reference indicates the English edition, when available, and the French edition otherwise.

Théorie des Ensembles (Theory of Sets)	designated by	E	(*Set Theory*)
Algèbre (Algebra[1])	—	A	(*Alg*)
Topologie Générale (General Topology)	—	TG	(*Gen. Top.*)
Fonctions d'une Variable Réelle	—	FVR	
Espaces Vectoriels Topologiques (Topological Vector Spaces)	—	EVT	(*Top. Vect. Sp.*)
Intégration	—	INT	
Algèbre Commutative (Commutative Algebra[2])	—	AC	(*Comm. Alg.*)
Variétés Différentielles et Analytiques	—	VAR	
Groupes et Algèbres de Lie (Lie Groups and Lie Algebras[3])	—	LIE	(*Lie*)
Théories Spectrales	—	TS	

(1) So far, chapters I to VII only have been translated.
(2) So far, chapters I to VII only have been translated.
(3) So far, chapters I to III only have been translated.

In the first six books (according to the above order), every statement in the text assumes as known only those results which have already been discussed in the same chapter, or in the previous chapters ordered as follows : E ; A, chapters I to III ; TG, chapters I to III ; A, from chapters IV on ; TG, from chapter IV on ; FVR ; EVT ; INT.

From the seventh Book on, the reader will usually find a precise indication of its logical relationship to the other Books (the first six Books being always assumed to be known).

4. However we have sometimes inserted examples in the text which refer to facts the reader may already know but which have not yet been discussed in the series. Such examples are placed between two asterisks : *...*. Most readers will undoubtedly find that these examples will help them to understand the text. In other cases, the passages between *...* refer to results which are discussed elsewhere in the Series. We hope the reader will be able to verify the absence of any vicious circle.

5. The logical framework of each chapter consists of the *definitions*, the *axioms*, and the *theorems* of the chapter. These are the parts that have mainly to be borne in mind for subsequent use. Less important results and those which can easily be deduced from the theorems are labelled as « propositions », « lemmas », « corollaries », « remarks », etc. Those which may be omitted at a first reading are printed in small type. A commentary on a particularly important theorem appears occasionally under the name of « scholium ».

To avoid tedious repetitions it is sometimes convenient to introduce notations or abbreviations which are in force only within a certain chapter or a certain section of a chapter (for example, in a chapter which is concerned only with commutative rings, the word « ring » would always signify « commutative ring »). Such conventions are always explicitly mentioned, generally at the beginning of the chapter in which they occur.

6. Some passages in the text are designed to forewarn the reader against serious errors. These passages are signposted in the margin with the sign ☡ (« dangerous bend »).

7. The Exercises are designed both to enable the reader to satisfy himself that he has digested the text and to bring to his notice results which have no place in the text but which are nonetheless of interest. The most difficult exercises bear the sign ¶.

8. In general, we have adhered to the commonly accepted terminology, *except where there appeared to be good reasons for deviating from it.*

9. We have made a particular effort always to use rigorously correct language, without sacrificing simplicity. As far as possible we have drawn attention in the text to *abuses of language*, without which any mathematical text runs the risk of pedantry, not to say unreadability.

10. Since in principle the text consists of the dogmatic exposition of a theory, it contains in general no references to the literature. Bibliographical references are

gathered together in *Historical Notes*. The bibliography which follows each historical note contains in general only those books and original memoirs which have been of the greatest importance in the evolution of the theory under discussion. It makes no sort of pretence to completeness.

As to the exercises, we have not thought it worthwhile in general to indicate their origins, since they have been taken from many different sources (original papers, textbooks, collections of exercises).

11. In the present Book, references to theorems, axioms, definitions, ... are given by quoting successively :

— the Book (using the abbreviation listed in Section 3), chapter and page, where they can be found ;

— the chapter and page only when referring to the present Book.

The *Summaries of Results* are quoted by the letter R ; thus *Set Theory*, R signifies « *Summary of Results of the Theory of Sets* ».

CHAPTER IV

Polynomials and rational fractions

Throughout this chapter A *denotes a commutative ring.*

§ 1. POLYNOMIALS

1. Definition of polynomials

Let I be a set. We recall (III, p. 452) that the free commutative algebra on I over A is denoted by $A[(X_i)_{i \in I}]$ or $A[X_i]_{i \in I}$. The elements of this algebra are called *polynomials* with respect to the indeterminates X_i (or in the indeterminates X_i) with coefficients in A. Let us recall that the indeterminate X_i is the canonical image of i in the free commutative algebra on I over A ; sometimes it is convenient to denote this image by another symbol such as X_i', Y_i, T_i, etc. This convention is often introduced by a phrase such as : « Let $\mathbf{Y} = (Y_i)_{i \in I}$ be a family of indeterminates » ; in this case the algebra of polynomials in question is denoted by $A[\mathbf{Y}]$. When $I = \{1, 2, \ldots, n\}$, one writes $A[X_1, X_2, \ldots, X_n]$ in place of $A[(X_i)_{i \in I}]$.

For $\nu \in \mathbf{N}^{(I)}$ we put

$$X^\nu = \prod_{i \in I} X_i^{\nu_i}.$$

Then $(X^\nu)_{\nu \in \mathbf{N}^{(I)}}$ is a basis of the A-module $A[(X_i)_{i \in I}]$. The X^ν are called *monomials* in the indeterminates X_i. For $\nu = 0$ we obtain the unit element of $A[(X_i)_{i \in I}]$. Every polynomial $u \in A[(X_i)_{i \in I}]$ can be written in exactly one way in the form

$$u = \sum_{\nu \in \mathbf{N}^{(I)}} \alpha_\nu X^\nu$$

where $\alpha_\nu \in A$ and the α_ν are zero except for a finite number ; the α_ν are called the *coefficients* of u ; the $\alpha_\nu X^\nu$ are called the *terms* of u (often the element $\alpha_\nu X^\nu$ is called the term in X^ν), in particular the term of $\alpha_0 X^0$, identified with α_0, is called the *constant term* of u. When $\alpha_\nu = 0$, we say by abuse of language that *u contains no element* in X^ν ; in particular when $\alpha_0 = 0$, we say that u is a polynomial *without constant term* (III, p. 453). Any scalar multiple of 1 is called a *constant polynomial.*

Let B be a commutative ring and $\rho : A \to B$ a ring homomorphism. We consider $B[(X_i)_{i \in I}]$ as an A-algebra by means of ρ. Thus the mapping σ of $A[(X_i)_{i \in I}]$ into $B[(X_i)_{i \in I}]$ which transforms $\sum \alpha_\nu X^\nu$ into $\sum \rho(\alpha_\nu) X^\nu$ is a homomorphism of A-algebras ; if $u \in A[(X_i)_{i \in I}]$, we sometimes denote by ${}^\rho u$ the image of u by this homomorphism. The homomorphism of $B \otimes_A A[(X_i)_{i \in I}]$ into $B[(X_i)_{i \in I}]$ canonically defined by σ transforms, for every $i \in I$, $1 \otimes X_i$ into X_i ; this is an isomorphism of B-algebras (III, p. 449).

Let M be a free A-module with basis $(e_i)_{i \in I}$. There exists precisely one unital homomorphism φ of the symmetric algebra $\mathbf{S}(M)$ into the algebra $A[(X_i)_{i \in I}]$ such that $\varphi(e_i) = X_i$ for each $i \in I$, and this homomorphism is an isomorphism (III, p. 506). This isomorphism is said to be *canonical*. It allows us to apply to polynomial algebras certain properties of symmetric algebras. For example, let $(I_\lambda)_{\lambda \in L}$ be a partition of I. Let φ_λ be the homomorphism of $P_\lambda = A[(X_i)_{i \in I_\lambda}]$ into $P = A[(X_i)_{i \in I}]$ which transforms X_i (qua element of P_λ) into X_i (qua element of P). Then the φ_λ define a homomorphism of the algebra $\bigotimes_{\lambda \in L} P_\lambda$ into the algebra P, and this homomorphism is an isomorphism (III, p. 503, Prop. 9).

Let E be an A-module, and put $E \otimes_A A[(X_i)_{i \in I}] = E[(X_i)_{i \in I}]$. The elements of the A-module $E[(X_i)_{i \in I}]$ are called polynomials in the indeterminates X_i with coefficients in E. Such a polynomial can be written in just one way as $\sum_{\nu \in \mathbf{N}^{(I)}} e_\nu \otimes X^\nu$, where $e_\nu \in E$ and the e_ν are zero for all but a finite number of suffixes ; we frequently write $e_\nu X^\nu$ instead of $e_\nu \otimes X^\nu$.

2. Degrees

Let $P = A[(X_i)_{i \in I}]$ be a polynomial algebra. For each integer $n \in \mathbf{N}$ let P_n be the submodule of P generated by the monomials X^ν such that $|\nu| = \sum_{i \in I} \nu_i$ equals n. Then $(P_n)_{n \in \mathbf{N}}$ is a graduation which turns $A[(X_i)_{i \in I}]$ into a graded algebra of type $\mathbf{N}$ (III, p. 459). The homogeneous elements of degree n in $A[(X_i)_{i \in I}]$ are sometimes called *forms of degree n* with respect to the indeterminates X_i.

When we are dealing with the degree of inhomogeneous polynomials, we shall agree to adjoin to the set $\mathbf{N}$ of natural numbers, an element written $-\infty$ and to extend the order relation and the addition of $\mathbf{N}$ to $\mathbf{N} \cup \{-\infty\}$ by the following conventions, where $n \in \mathbf{N}$,

$$-\infty < n , \quad (-\infty) + n = n + (-\infty) = -\infty , \quad (-\infty) + (-\infty) = -\infty .$$

Let $u = \sum_{\nu \in \mathbf{N}^{(I)}} \alpha_\nu X^\nu$ be a polynomial. The homogeneous component u_n of degree

n of u (for the graduation of type **N** defined above) is equal to $\sum_{|\nu|=n} \alpha_\nu X^\nu$, and we clearly have $u = \sum_{n \in \mathbf{N}} u_n$. If $u \neq 0$, the u_n are not all zero, and we define the *degree* (or *total degree*) of u, written $\deg u$, as the greatest of the numbers n such that $u_n \neq 0$; in other words (III, p. 453), the degree of u is the largest of the integers $|\nu|$ for the multi-indices ν such that $\alpha_\nu \neq 0$. When $u = 0$, the degree of u is $-\infty$ by convention. For every integer $p \in \mathbf{N}$, the relation $\deg u \leqslant p$ is thus equivalent to « $\alpha_\nu = 0$ for every multi-index ν with $|\nu| > p$ » ; the set of polynomials u such that $\deg u \leqslant p$ is thus an A-submodule of $A[(X_i)_{i \in I}]$, equal to $P_0 + P_1 + \cdots + P_p$ with the above notations.

Let E be an A-module. The family $(E \otimes P_n)_{n \in \mathbf{N}}$ is a graduation of type **N** of the module $E[(X_i)_{i \in I}] = E \otimes_A A[(X_i)_{i \in I}]$ of polynomials with coefficients in E. We extend the conventions adopted above for the degree of inhomogeneous polynomials to this case.

PROPOSITION 1. — *Let u and v be two polynomials.*

(i) *If* $\deg u \neq \deg v$, *we have*

$$u + v \neq 0 \quad \textit{and} \quad \deg(u+v) = \sup(\deg u, \deg v).$$

If $\deg u = \deg v$ *we have* $\deg(u+v) \leqslant \deg u$.

(ii) *We have* $\deg(uv) \leqslant \deg u + \deg v$.

The proof is immediate.

Let $J \subset I$ and $B = A[(X_i)_{i \in I-J}]$; we shall identify $A[(X_i)_{i \in I}]$ with $B[(X_i)_{i \in J}]$ (III, p. 453 f.). The degree of $u \in A[(X_i)_{i \in I}]$, qua element of $B[(X_i)_{i \in J}]$, is called the degree of u with respect to the X_i of index $i \in J$ (III, p. 454).

Let $u = \sum_{k=0}^{n} \alpha_k X^k \in A[X]$ be a non-zero polynomial of degree n in a single indeterminate. The coefficient α_n which is $\neq 0$ by hypothesis, is called the *leading coefficient* of u. A polynomial $u \neq 0$ whose leading coefficient is equal to 1 is called a *monic polynomial*.

In $A[X_1, X_2, \ldots, X_q]$ the number of monomials of total degree p is equal to the number of elements $(n_k)_{1 \leqslant k \leqslant q}$ of $\mathbf{N}^q$ such that $\sum_{k=1}^{q} n_k = p$, that is $\binom{q+p-1}{p}$ (Sets III, Prop. 15, p. 182).

More generally, let Δ be a commutative monoid and $(\delta_i)_{i \in I}$ a family of elements of Δ. There exists a unique graduation of type Δ of the algebra $A[(X_i)_{i \in I}]$ such that each monomial X^ν is of degree $\sum_{i \in I} \nu_i \delta_i$ (III, p. 458, example 3). The case considered above is that where $\Delta = \mathbf{N}$ and $\delta_i = 1$. In the general case, to avoid confusion, we shall use the word « weight » instead of « degree » and « isobaric » instead of « homogeneous ». For example, there exists a unique graduation of type **N** of the algebra $A[(X_i)_{i \geqslant 1}]$ such that X_i is of weight i

for each integer $i \geqslant 1$. The isobaric elements of weight n are the polynomials of the form $\sum_{\nu} a_\nu X^\nu$, where $a_\nu = 0$ for $\sum_{i \geqslant 1} i \cdot \nu_i \neq n$.

3. Substitutions

Let E be a unital associative algebra over A and $\mathbf{x} = (x_i)_{i \in I}$ a family of pairwise permutable elements of E. Let $\mathbf{X} = (X_i)_{i \in I}$ be a family of indeterminates. By III, Prop. 7, p. 449, there exists a unique unital homomorphism f of $A[\mathbf{X}]$ into E such that $f(X_i) = x_i$ for every $i \in I$. The image of an element u of $A[\mathbf{X}]$ under f is written $u(\mathbf{x})$ and is called *the element of* E *obtained by substituting* x_i *for* X_i *in* u, or also *the value of* u *for* $X_i = x_i$. In particular, $u = u((X_i)_{i \in I})$. If $I = \{1, \dots, n\}$ we write $u(x_1, \dots, x_n)$ in place of $u((x_i)_{i \in I})$. More generally, if M is an A-module and if v is an element of

$$M[(X_i)_{i \in I}] = M \otimes_A A[(X_i)_{i \in I}],$$

we denote the image of v in $M \otimes_A E = M_{(E)}$ under the mapping $1_M \otimes f$ by $v(\mathbf{x})$.

If the homomorphism $u \mapsto u(\mathbf{x})$ of $A[\mathbf{X}]$ into E is injective, we say that the family $\mathbf{x}$ is *algebraically free* over A, or that the x_i are *algebraically independent* over A. This also means that the monomials $\mathbf{x}^\nu$ ($\nu \in \mathbf{N}^{(I)}$) are linearly independent over A.

If λ is a unital homomorphism of E into a unital associative A-algebra E', we have

$$\lambda(u((x_i)_{i \in I})) = u((\lambda(x_i)_{i \in I})), \tag{1}$$

because $\lambda \circ f$ is a homomorphism of $A[\mathbf{X}]$ into E' which maps X_i to $\lambda(x_i)$.

Let $u \in A[\mathbf{X}]$. If E is commutative, the mapping $\mathbf{x} \mapsto u(\mathbf{x})$ of E^I into E is called the *polynomial function* defined by u (and the algebra E) ; we shall sometimes denote it by $\tilde{u}$ (or even just u).

Let $\mathbf{Y} = (Y_j)_{j \in J}$ be another family of indeterminates, and let us take for E the polynomial algebra $A[\mathbf{Y}]$. Given $u \in A[\mathbf{X}]$, take $g_i \in A[\mathbf{Y}]$ for $i \in I$ and put $\mathbf{g} = (g_i)_{i \in I}$; let $u(\mathbf{g}) \in A[\mathbf{Y}]$ be the polynomial obtained by substituting the polynomials g_i for X_i in u. Let $\mathbf{y} = (y_j)_{j \in J}$ be a family of pairwise permutable elements of a unital associative A-algebra E' ; applying (1) and taking for λ the homomorphism $g \mapsto g(\mathbf{y})$ of E into E', we obtain

$$(u(\mathbf{g}))(\mathbf{y}) = u((g_i(\mathbf{y}))). \tag{2}$$

If $\mathbf{f} = (f_i)_{i \in I} \in (A[(X_j)_{j \in J}])^I$ and $\mathbf{g} = (g_j)_{j \in J} \in (A[(Y_k)_{k \in K}])^J$, we denote by $\mathbf{f} \circ \mathbf{g}$ or $f(\mathbf{g})$, the family of polynomials $(f_i(\mathbf{g}))_{i \in I} \in (A[(Y_k)_{k \in K}])^I$. If we denote by $\tilde{\mathbf{f}}$ the mapping $\mathbf{x} \mapsto (f_i(\mathbf{x}))_{i \in \mathrm{I}}$ of E'^{J} into E'^{I} (where E′ is a unital associative and commutative A-algebra), then the relation (2) implies

$$(3) \qquad (\mathbf{f} \circ \mathbf{g})^{\sim} = \tilde{\mathbf{f}} \circ \tilde{\mathbf{g}} .$$

If $\mathbf{h} = (h_k)_{k \in \mathrm{K}} \in (\mathrm{A}[(\mathrm{Z}_l)_{l \in \mathrm{L}}])^{\mathrm{K}}$, it follows from (2) that :

$$(4) \qquad \mathbf{f} \circ (\mathbf{g} \circ \mathbf{h}) = (\mathbf{f} \circ \mathbf{g}) \circ \mathbf{h} .$$

PROPOSITION 2. — *Let $\mathbf{a} = (a_i)_{i \in \mathrm{I}}$ be a family of elements of* A *and let $u \in \mathrm{A}[\mathbf{X}]$. If v is the polynomial obtained by substituting $\mathrm{X}_i + a_i$ for X_i for each $i \in \mathrm{I}$, then the constant term of v is equal to $u(\mathbf{a})$.*

The constant term of v is obtained by substituting 0 for X_i in v for each $i \in \mathrm{I}$. The result therefore follows by (2).

COROLLARY 1. — *Let $\mathfrak{m}$ be the ideal of polynomials $u \in \mathrm{A}[\mathbf{X}]$ such that $u(\mathbf{a}) = 0$. Then $\mathfrak{m}$ is generated by the polynomials $\mathrm{X}_i - a_i$ (for $i \in \mathrm{I}$).*

It is clear that $\mathrm{X}_i - a_i \in \mathfrak{m}$ for each $i \in \mathrm{I}$. Let $u \in \mathfrak{m}$ and let v be as in Prop. 2. Since v has no constant term, there exists a family $(\mathrm{P}_i)_{i \in \mathrm{I}}$ of polynomials in $\mathrm{A}[\mathbf{X}]$ with finite support such that

$$v(\mathbf{X}) = \sum_{i \in \mathrm{I}} \mathrm{X}_i \,.\, \mathrm{P}_i(\mathbf{X}) .$$

If we replace X_i by $\mathrm{X}_i - a_i$ for each $i \in \mathrm{I}$ in the above equation, we obtain a relation of the form $u(\mathbf{X}) = \sum_{i \in \mathrm{I}} (\mathrm{X}_i - a_i) \,.\, \mathrm{P}'_i(\mathbf{X})$, whence the corollary.

COROLLARY 2. — *Let $\mathbf{X} = (\mathrm{X}_i)_{i \in \mathrm{I}}$ and $\mathbf{Y} = (\mathrm{Y}_i)_{i \in \mathrm{I}}$ be two families of indeterminates. The set of polynomials $u \in \mathrm{A}[\mathbf{X}, \mathbf{Y}]$ such that $u(\mathbf{X}, \mathbf{X}) = 0$ is the ideal of $\mathrm{A}[\mathbf{X}, \mathbf{Y}]$ generated by the polynomials $\mathrm{X}_i - \mathrm{Y}_i$ (for $i \in \mathrm{I}$).*

This corollary results directly from Cor. 1 on replacing A by $\mathrm{A}[\mathbf{Y}]$ and a_i by Y_i, interpreting $\mathrm{A}[\mathbf{X}, \mathbf{Y}]$ as polynomial ring in the X_i with coefficients in $\mathrm{A}[\mathbf{Y}]$.

PROPOSITION 3. — *Let $u \in \mathrm{A}[\mathbf{X}]$ and let $\mathbf{X} \,.\, \mathrm{Z}$ be the family $(\mathrm{X}_i\mathrm{Z})_{i \in \mathrm{I}}$ of elements of the polynomial ring $\mathrm{A}[\mathbf{X}][\mathrm{Z}]$. The coefficient of Z^k in $u(\mathbf{X} \,.\, \mathrm{Z})$ is the homogeneous component of degree k of u, for every positive integer k.*

It suffices to prove this Prop. in the case where u is a monomial, and in this case the result is clear.

COROLLARY. — *For a polynomial $u \in \mathrm{A}[\mathbf{X}]$ to be homogeneous of degree k it is necessary and sufficient that :*

$$u(\mathbf{X} \,.\, \mathrm{Z}) = u(\mathbf{X}) \,.\, \mathrm{Z}^k .$$

Remark. — Let $\mathbf{x} \in \mathrm{A}^{\mathrm{I}}$ and let f be the mapping $u \mapsto u(\mathbf{x})$ of $\mathrm{A}[\mathbf{X}]$ into A. Given an A-module M, we consider the homomorphism $1 \otimes f$ of $\mathrm{M}[\mathbf{X}] = \mathrm{M} \otimes_{\mathrm{A}} \mathrm{A}[\mathbf{X}]$

into $M \otimes_A A = M$. For each $v \in M[\mathbf{X}]$ we have $(1 \otimes f)(v) = v(\mathbf{x})$. If $v = \sum_{\nu \in \mathbf{N}^{(I)}} e_\nu X^\nu$, then $v(\mathbf{x}) = \sum_{\nu \in \mathbf{N}^{(I)}} \mathbf{x}^\nu e_\nu$.

4. Differentials and derivations

Let $B = A[(X_i)_{i \in I}]$, then by III, p. 569 there exists for each $i \in I$ one and only one A-derivation D_i of B such that

(5) $$D_i X_i = 1 , \quad D_i X_j = 0 \quad \text{for} \quad j \neq i .$$

The polynomial $D_i P$ is called the *partial derivative* of P *with respect to* X_i ; we shall also denote it by $D_{X_i} P$ or $\dfrac{\partial P}{\partial X_i}$ or P'_{X_i}. By III, p. 558, formula (21), we have, for $\nu = (\nu_j) \in \mathbf{N}^{(I)}$,

(6) $$D_i(X^\nu) = \begin{cases} \nu_i X_i^{\nu_i - 1} \prod_{j \in I - \{i\}} X_j^{\nu_j} & \text{if} \quad \nu_i > 0 \\ 0 & \text{if} \quad \nu_i = 0 . \end{cases}$$

It follows from (6) that $D_i D_j = D_j D_i$ for any $i, j \in I$. For $\nu = (\nu_i)_{i \in I} \in \mathbf{N}^{(I)}$ we shall put $D^\nu = \prod_{i \in I} D_i^{\nu_i}$ and $\nu ! = \prod_{i \in I} (\nu_i !)$. With the product ordering on $\mathbf{N}^{(I)}$ we have

$$D^\nu(X^\mu) = \begin{cases} \dfrac{\mu!}{(\mu - \nu)!} X^{\mu - \nu} & \text{if} \quad \nu \leqslant \mu , \\ 0 & \text{if not} . \end{cases}$$

When P is a polynomial in a single indeterminate X, the unique partial derivative of P is written DP or $\dfrac{dP}{dX}$ or P' and is called simply the *derivative* of P.

Again let $B = A[(X_i)_{i \in I}]$; by III, p. 569 the B-module of A-differentials of B, $\Omega_A(B)$, has the family $(dX_i)_{i \in I}$ of differentials of the X_i as basis. Let ∂_i be the coordinate form of index i relative to this basis over $\Omega_A(B)$. Then the mapping $u \mapsto \langle \partial_i, du \rangle$ of B into itself is a derivation of B which maps X_i to 1 and X_j to 0 for $j \neq i$, and hence is D_i ; in other words, we have

(7) $$du = \sum_{i \in I} (D_i u)\, dX_i$$

for each $u \in B$. If I is finite, $(D_i)_{i \in I}$ is a basis of the B-module of derivations of B.

PROPOSITION 4. — *Let* E *be an associative, commutative and unital* A*-algebra,* $\mathbf{x} = (x_i)_{i \in I}$ *a family of elements of* E, u *an element of* $A[(X_i)_{i \in I}]$ *and* $y = u(\mathbf{x})$. *Then for every derivation* D *of* E *into an* E*-module we have*

$$Dy = \sum_{i \in I} (D_i u)(\mathbf{x}) \,.\, Dx_i \,.$$

It suffices to prove the proposition when u is a monomial, and in that case it follows from III, p. 558, Prop. 6.

COROLLARY. — *Let $f \in A[X_1, \dots, X_p]$ and $g_i \in A[Y_1, \dots, Y_q]$ for $1 \leqslant i \leqslant p$, and write $h = f(g_1, \dots, g_p)$, then for $1 \leqslant j \leqslant q$ we have*

$$\frac{\partial h}{\partial Y_j} = \sum_{i=1}^{p} D_i f(g_1, \dots, g_p) \cdot \frac{\partial g_i}{\partial Y_j}. \tag{8}$$

This is the special case $E = A[Y_1, \dots, Y_q]$, $x_i = g_i$ and $D = \partial/\partial Y_j$ of Prop. 4.

Let $\mathbf{X} = (X_i)_{i \in I}$, $\mathbf{Y} = (Y_i)_{i \in I}$ be two disjoint families of indeterminates, and write $\mathbf{X} + \mathbf{Y}$ for the family $(X_i + Y_i)_{i \in I}$. Given $u \in A[\mathbf{X}]$, consider the element $u(\mathbf{X} + \mathbf{Y})$ of $A[\mathbf{X}, \mathbf{Y}]$. For $\nu \in \mathbf{N}^{(I)}$ we denote by $\Delta^\nu u$ the coefficient of $\mathbf{Y}^\nu$ in $u(\mathbf{X} + \mathbf{Y})$, considered as polynomial in the Y_i with coefficients in $A[\mathbf{X}]$. By definition we have $\Delta^\nu u \in A[\mathbf{X}]$ and

$$u(\mathbf{X} + \mathbf{Y}) = \sum_{\nu} (\Delta^\nu u)(\mathbf{X})\, \mathbf{Y}^\nu. \tag{9}$$

(Here and in the rest of this No., the summations are extended over the index set $\mathbf{N}^{(I)}$ unless the contrary is said.)

Let $\mathbf{a} \in A^I$, then on substituting $\mathbf{a}$ for $\mathbf{X}$ and $\mathbf{X} - \mathbf{a}$ for Y in (9) we obtain

$$u(\mathbf{X}) = \sum_{\nu} (\Delta^\nu u)(\mathbf{a})(\mathbf{X} - \mathbf{a})^\nu. \tag{10}$$

In particular we have

$$u(\mathbf{X}) = \sum_{\nu} (\Delta^\nu u)(0)\, \mathbf{X}^\nu. \tag{11}$$

If $u, v \in A[\mathbf{X}]$, we have

$$\begin{aligned}(uv)(\mathbf{X} + \mathbf{Y}) &= \left(\sum_{\nu} (\Delta^\nu u)(\mathbf{X})\, \mathbf{Y}^\nu\right) \left(\sum_{\rho} (\Delta^\rho v)(\mathbf{X})\, \mathbf{Y}^\rho\right) \\ &= \sum_{\sigma} \left[\sum_{\nu + \rho = \sigma} (\Delta^\nu u)(\mathbf{X})(\Delta^\rho v)(\mathbf{X}) \right] \mathbf{Y}^\sigma\end{aligned}$$

hence

$$\Delta^\sigma(uv) = \sum_{\nu + \rho = \sigma} (\Delta^\nu u)(\Delta^\rho v). \tag{12}$$

Let $\mathbf{Z} = (Z_i)_{i \in I}$ be another family of indeterminates.

We have :

$$\sum_{\nu} (\Delta^{\nu} u)(\mathbf{X})(\mathbf{Y} + \mathbf{Z})^{\nu} = u(\mathbf{X} + \mathbf{Y} + \mathbf{Z})$$

$$= \sum_{\sigma} (\Delta^{\sigma} u)(\mathbf{X} + \mathbf{Y}) \mathbf{Z}^{\sigma}$$

$$= \sum_{\rho, \sigma} (\Delta^{\rho} \Delta^{\sigma} u)(\mathbf{X}) \mathbf{Y}^{\rho} \mathbf{Z}^{\sigma} ,$$

hence by I, p. 99, Cor. 2 :

$$\Delta^{\rho} \Delta^{\sigma} u = \frac{(\rho + \sigma)!}{\rho! \, \sigma!} \Delta^{\rho + \sigma} u . \tag{13}$$

PROPOSITION 5. — *For any $u \in \mathrm{A}[\mathbf{X}]$ and $\nu \in \mathbf{N}^{(\mathrm{I})}$ we have*

$$\mathrm{D}^{\nu} u = \nu! \, \Delta^{\nu} u .$$

Suppose first that ν has length 1 ; then there exists an element i of I such that $\nu = \varepsilon_i$, that is, $\nu_i = 1$ and $\nu_j = 0$ for all $j \neq i$ in I. Formula (12) shows that Δ^{ε_i} is a derivation of the A-algebra $\mathrm{A}[\mathrm{X}]$ which clearly has the value zero on X_j for $j \neq i$ and the value 1 on X_i. We thus have $\Delta^{\varepsilon_i} = \mathrm{D}_i$ for each $i \in \mathrm{I}$.

By (13) we have

$$(\rho! \, \Delta^{\rho}) \, . \, (\sigma! \, \Delta^{\sigma}) = (\rho + \sigma)! \, \Delta^{\rho + \sigma} \tag{14}$$

in the endomorphism algebra of the A-module $\mathrm{A}[\mathrm{X}]$. It now follows by induction on the length of ν that $\nu! \, \Delta^{\nu} = \mathrm{D}^{\nu}$.

If A is a **Q**-algebra, the formulae (9), (10), (11) may thus be written

$$u(\mathbf{X} + \mathbf{Y}) = \sum_{\nu} \frac{1}{\nu!} (\mathrm{D}^{\nu} u)(\mathbf{X}) \mathbf{Y}^{\nu} \tag{15}$$

$$u(\mathbf{X}) = \sum_{\nu} \frac{1}{\nu!} (\mathrm{D}^{\nu} u)(\mathbf{a})(\mathbf{X} - \mathbf{a})^{\nu} \tag{16}$$

$$u(\mathbf{X}) = \sum_{\nu} \frac{1}{\nu!} (\mathrm{D}^{\nu} u)(0) \, \mathbf{X}^{\nu} . \tag{17}$$

The formulae (15), (16), (17) are all three called « Taylor's formula ».

PROPOSITION 6 (« Euler's identity »). — *Let $u \in \mathrm{A}[\mathbf{X}]$ be a homogeneous polynomial of degree r ; we have*

$$\sum_{i \in \mathrm{I}} \mathrm{X}_i \, . \, \mathrm{D}_i u = ru .$$

Let D be the A-linear mapping of A[**X**] into itself such that $D(v) = sv$ when v is homogeneous of degree s. We know (III, p. 554, Example 6) that D is a derivation of A[**X**]. Thus Prop. 6 is a Corollary of Prop. 4 (IV, p. 6).

5. Divisors of zero in a polynomial ring

PROPOSITION 7. — *Let $f \in A[X]$ be a non-zero polynomial in an indeterminate and α its leading coefficient. If α is cancellable in A (in particular if f is monic) then we have for any non-zero element g of $A[X]$,*

$$fg \neq 0 \quad \textit{and} \quad \deg(fg) = \deg f + \deg g .$$

Let $g \in A[X]$ be a non-zero polynomial, β its leading coefficient, $n = \deg f$ and $p = \deg g$. Then the coefficient of X^{n+p} in fg is $\alpha\beta$ which does not vanish, whence the proposition.

PROPOSITION 8. — *If A is an integral domain, then so is $A[(X_i)_{i \in I}]$.*

Let u, v be two non-zero elements of $A[(X_i)_{i \in I}]$; we have to show that $uv \neq 0$. Now u and v belong to a ring $A[(X_j)_{j \in J}]$ where J is a finite subset of I. Thus we can limit ourselves to the case when I is finite and equal to $\{1, 2, \ldots, p\}$. On the other hand, the ring $A[X_1, \ldots, X_p]$ is isomorphic to the polynomial ring in X_p with coefficients in $A[X_1, \ldots, X_{p-1}]$. By induction on p we are thus reduced to proving the proposition for $A[X]$, and now it suffices to apply Prop. 7.

COROLLARY 1. — *If A is an integral domain, and u, v are elements of $A[(X_i)_{i \in I}]$, then $\deg(uv) = \deg u + \deg v$.*

We can limit ourselves to the case where u and v are non-zero. Let $m = \deg u$, $n = \deg v$; we have

$$u = u_0 + u_1 + \cdots + u_m , \quad v = v_0 + v_1 + \cdots + v_n$$

where u_h (resp. v_h) is the homogeneous component of degree h of u (resp. v). Since $u_m \neq 0$ and $v_n \neq 0$, we have $u_m v_n \neq 0$ (Prop. 8). Now $uv = u_m v_n + w$ with $\deg w < m + n$, whence the result.

COROLLARY 2. — *If A is an integral domain, the invertible elements of $A[(X_i)_{i \in I}]$ are the invertible elements of A.*

This follows immediately from Cor. 1.

PROPOSITION 9. — *Let $u \in A[(X_i)_{i \in I}]$; then for u to be nilpotent in the ring $A[(X_i)_{i \in I}]$ it is necessary and sufficient for all its coefficients to be nilpotent in the ring A.*

As in the proof of Prop. 8 we can make a reduction to the case of polynomials in one variable X. If all the coefficients of u are nilpotent, then u is nilpotent (I,

p. 99, Cor. 1). Suppose that u is nilpotent but not zero and let n be its degree ; we shall argue by induction on n. Let α be the leading coefficient of u. There exists an integer $m > 0$ such that $u^m = 0$. The leading coefficient of u^m is α^m, hence $\alpha^m = 0$. Now $u - \alpha X^n$ is nilpotent (I, *loc. cit.*) and the induction hypothesis shows all the coefficients of $u - \alpha X^n$ to be nilpotent. Thus all the coefficients of u are nilpotent.

Remark. — Let u and v be elements of $A[(X_i)_{i \in I}]$, and suppose that A is an integral domain, v is a non-zero multiple of u and v is homogeneous ; then u is also homogeneous. For let $u' \in A[(X_i)_{i \in I}]$ be such that $v = uu'$; we have $u \neq 0$, $u' \neq 0$, and if

$$u = u_h + u_{h+1} + \cdots + u_k$$
$$u' = u'_{h'} + u'_{h'+1} + \cdots + u'_{k'}$$

are the decompositions of u and u' into homogeneous components, with $u_h \neq 0$, $u_k \neq 0$, $u'_{h'} \neq 0$, $u'_{k'} \neq 0$, then $v = u_h u'_{h'} + u_h u'_{h'+1} + \cdots + u_k u'_{k'}$ and $u_h u'_{h'}$ is non-zero homogeneous of degree $h + h'$ while $u_k u'_{k'}$ is non-zero homogeneous of degree $k + k'$ (Prop. 8). Since v is homogeneous, we have $h + h' = k + k'$ whence $h = k$, $h' = k'$.

6. Euclidean division of polynomials in one indeterminate

PROPOSITION 10. — *Let f and g be non-zero elements of* A[X] *of degrees m and n respectively. Let α_0 be the leading coefficient of f and $\mu = \sup(n - m + 1, 0)$. There exist u, $v \in A[X]$ such that*

$$\alpha_0^\mu g = uf + v, \quad \deg v < m .$$

If α_0 is cancellable in A, *then u and v are uniquely determined by these properties.*

The existence of u and v is clear when $n < m$, because then we can take $u = 0$ and $v = g$. For $n \geqslant m$ we shall use induction on n. Let β be the leading coefficient of g ; if $f = \sum_{k=0}^{m} \alpha_k X^{m-k}$, we can write $\alpha_0^\mu g = \alpha_0^{\mu-1} \beta X^{n-m} f + \alpha_0^{\mu-1} g_1$, where $g_1 \in A[X]$ and $\deg g_1 < n$. By the induction hypothesis there exist u_1, $v \in A[X]$ such that $\alpha_0^{\mu-1} g_1 = u_1 f + v$ and $\deg v < m$. Hence

$$\alpha_0^\mu g = (\alpha_0^{\mu-1} \beta X^{n-m} + u_1) f + v$$

and it suffices to put $u = \alpha_0^{\mu-1} \beta X^{n-m} + u_1$.

Assuming α_0 to be cancellable in A, let us now prove the uniqueness of u and v. Let u, v, u_1, $v_1 \in A[X]$ be such that

$$\alpha_0^\mu g = uf + v = u_1 f + v_1, \quad \deg v < m, \quad \deg v_1 < m .$$

We have $(u - u_1) f = v_1 - v$ and $\deg(v_1 - v) < m$, hence $u - u_1 = 0$ (IV, p. 9, Prop. 7) and therefore $v_1 - v = 0$.

COROLLARY (« *Euclidean division of polynomials* »). — *Let f be a non-zero element of* $A[X]$ *whose leading coefficient is invertible and* $m = \deg f$.

(i) *For every* $g \in A[X]$ *there exist* $u, v \in A[X]$ *such that*

$$g = uf + v , \quad \deg v < m .$$

Moreover, these conditions determine u and v uniquely.

(ii) *The sub-A-modules* $A + AX + \cdots + AX^{m-1}$ *and* $fA[X]$ *of* $A[X]$ *are supplementary in* $A[X]$.

(iii) *Assume f non-constant and consider* $A[X]$ *as an* $A[T]$*-module by means of the homomorphism* $u(T) \mapsto u(f(X))$ *of* $A[T]$ *into* $A[X]$. *Then* $A[X]$ *is a free* $A[T]$*-module with basis* $(1, X, \ldots, X^{m-1})$.

Assertions (i) and (ii) are immediate consequences of Prop. 10.

Let us prove (iii). Let ψ be the homomorphism $v \mapsto v(f(X), X)$ of $A[T, X]$ into $A[X]$. Consider $A[T, X]$ first as polynomial ring in T with coefficients in $A[X]$; Cor. 1 of IV, p. 5 shows that the kernel $\mathfrak{a}$ of ψ is the ideal $(T - f(X))$ of $A[T, X]$. Consider now $A[T, X]$ as polynomial ring in X with coefficients in $A[T]$; then ψ is an $A[T]$-linear mapping of $A[T][X]$ into $A[X]$. Assertion (ii) above (applied to the polynomial $f(X) - T$ in X with coefficients in $A[T]$) shows that $(1, X, \ldots, X^{m-1})$ is a basis of an $A[T]$-submodule of $A[T, X]$ supplementary to $\mathfrak{a}$. Since $\psi(X^i) = X^i$ for every integer $i \geqslant 0$, (iii) follows at once.

With the notations of (i) we shall say that u is the *quotient* and v the *remainder* in the Euclidean division of g by f; for the remainder to vanish it is necessary and sufficient that f should divide g.

7. Divisibility of polynomials in one indeterminate [1]

PROPOSITION 11. — *Let* K *be a commutative field.*

(i) *For every non-zero ideal* $\mathfrak{a}$ *of* $K[X]$ *there exists precisely one monic polynomial f in* $K[X]$ *such that* $\mathfrak{a} = (f)$.

(ii) *Let* f_1 *and* f_2 *be in* $K[X]$; *for* $(f_1) = (f_2)$ *to hold it is necessary and sufficient that there exist a non-zero element* λ *of* K *such that* $f_2 = \lambda f_1$.

Let us prove (ii), the sufficiency of the stated condition being clear. The case where f_1 and f_2 generate the zero ideal is trivial. Thus assume that the non-zero polynomials f_1 and f_2 generate the same ideal of $K[X]$. Then there exist polynomials u_1 and u_2 such that $f_1 = u_1 f_2$ and $f_2 = u_2 f_1$; it follows that

[1] The reader will note the analogy between the results of this No. and the next and the divisibility properties of the ring **Z** of integers (I, p. 112). They depend essentially on the fact that in the rings **Z** and $K[X]$ every ideal is principal, as we shall see in Chapter VII, § 1.

$u_1u_2 = 1$, whence $\deg u_1 + \deg u_2 = 0$ and so $\deg u_2 = 0$. We have thus shown that u_2 is a non-zero element of K.

To prove (i), let f be a monic polynomial in $\mathfrak{a}$ of least possible degree. Given g in $\mathfrak{a}$, let u and v be the quotient and remainder of the Euclidean division of g by f; then $v = g - uf$ belongs to $\mathfrak{a}$ and we have $\deg v < \deg f$; if v were non-zero, there would be a non-zero element λ of K such that λv is monic, and since $\lambda v \in \mathfrak{a}$, this would contradict the definition of f. We thus have $\mathfrak{a} = (f)$; the uniqueness of the monic polynomial f such that $\mathfrak{a} = (f)$ now follows from (ii).

PROPOSITION 12. — *Let* K *be a commutative field and f, g two elements of* K[X]. *For every polynomial d in* K[X] *the following properties are equivalent:*

(i) *The polynomial d divides f and g and every polynomial which divides both f and g divides d.*

(ii) *The polynomial d divides f and g and there exist two polynomials u and v such that $d = uf + vg$.*

(iii) *The relation $(d) = (f) + (g)$ holds between ideals in* K[X].

The polynomial d is determined up to multiplication by a non-zero element of K *by these properties. If f and g are not both zero, then $d \neq 0$ and the degree of d majorizes the degree of every polynomial dividing both f and g.*

When f and g are zero, each of the properties (i) to (iii) is satisfied only for $d = 0$, hence they are then equivalent. Henceforth we assume that f, g are not both 0 and we denote by $\mathfrak{a}$ the ideal $(f) + (g)$ of K[X].

We remark that for any polynomials u and v in K[X] the properties $(u) \supset (v)$ and « u divides v » are equivalent. The assertion (ii) is thus equivalent to « $(d) \supset (f)$ and $(d) \supset (g)$ and $d \in (f) + (g)$ », that is (iii). It is clear that (ii) implies (i). Finally suppose that (i) holds; we have $(d) \supset (f)$ and $(d) \supset (g)$, whence $(d) \supset \mathfrak{a}$; on the other hand, by Prop. 11 (IV, p. 12) there exists a polynomial d_1 such that $\mathfrak{a} = (d_1)$; since d_1 divides both f and g, it divides d by hypothesis, whence $(d) \subset \mathfrak{a}$, and finally we have $(d) = \mathfrak{a}$, that is, (iii).

The other assertions of Prop. 12 are immediate consequences of Prop. 11 applied to the ideal $\mathfrak{a} = (f) + (g)$.

DEFINITION 1. — *With the notation of Prop.* 12 *we say that d is a greatest common divisor* (*gcd* for short) *of f and g. We say that f and g are relatively prime or that f is prime to g if* 1 *is a gcd of f and g.*

To say that f and g are relatively prime thus means that there exist polynomials u and v in K[X] such that $uf + vg = 1$.

COROLLARY 1. — *Let d be a gcd of f and g and* K' *a commutative field containing* K *as subfield. Then d is a gcd of f and g considered as elements of* K'[X].

This follows from Prop. 12, (iii).

COROLLARY 2. — *Let d be a gcd of f and g.*

(i) *If $u \in K[X]$, du is a gcd of fu and gu.*

(ii) *If $v \in K[X]$ is a divisor ($\neq 0$) of f and g, then d/v is a gcd of f/v and g/v.*

This follows from Prop. 12, (ii).

COROLLARY 3. — *Let w be a common factor of f and g. For w to be a gcd of f and g it is necessary and sufficient that f/w and g/w are relatively prime.*

This follows from Cor. 2.

COROLLARY 4. — *Let $f, g, h \in K[X]$. If f divides gh and is prime to g, then f divides h.*

For f divides gh and fh, hence f divides every gcd of gh and fh, in particular h (Cor. 2, (i)).

COROLLARY 5. — *Let $f, g \in K[X]$. For f and g to be relatively prime it is necessary and sufficient that the canonical image of g in $K[X]/(f)$ should be invertible.*

For this condition means that there exist $u, v \in K[X]$ such that $uf + vg = 1$.

COROLLARY 6. — *Let $f, g_1, g_2, \ldots, g_n \in K[X]$. If f is prime to $g_1, g_2, \ldots, g_n$, then f is prime to $g_1 g_2 \ldots g_n$.*

* COROLLARY 7. — *For f and g to be relatively prime it is necessary and sufficient that they have no common roots in any extension of* K.

For if d is a gcd of f, g then the roots common to f and g in an extension K′ of K are the roots of d in K′. Now the corollary follows from V, p. 21, Prop. 4. *

8. Irreducible polynomials

DEFINITION 2. — *Let* K *be a commutative field. We say that $f \in K[X]$ is irreducible if* $\deg f \geq 1$ *and f is not divisible by any polynomial g such that* $0 < \deg g < \deg f$.

It comes to the same to say that $\deg f \geq 1$ and the only divisors of f in K[X] are the scalars $\neq 0$ and the products of f by scalars $\neq 0$. Since the relation $(f) \subset (g)$ means that g divides f, we see that the irreducible polynomials of K[X] may also be defined as the polynomials f such that the ideal (f) is *maximal* (I, p. 104).

Let $f, g \in K[X]$. If f is irreducible, it is clear that either f and g are relatively prime or that f divides g. If f and g are irreducible, either f and g are relatively prime or each is a product of the other by a scalar $\neq 0$. In particular, two distinct irreducible monic polynomials are relatively prime.

PROPOSITION 13. — *Let $\mathscr{I}$ be the set of irreducible monic polynomials in* K[X]. *Let f be a non-zero element of* K[X] *and α its leading coefficient; then there* *exists precisely one family of positive integers* $(v_p)_{p \in \mathcal{I}}$ *with finite support, such that we have a decomposition*

$$f = \alpha \prod_{p \in \mathcal{I}} p^{v_p} . \tag{18}$$

It suffices to prove the proposition when f is monic, that is when $\alpha = 1$. We shall argue by induction on the degree n of f, the case $n = 0$ being trivial. Suppose then that $n \geqslant 1$ and that the proposition has been established for all polynomials of degree $< n$.

Let E be the set of monic polynomials $\neq 1$ which divide f; we have $f \in \mathrm{E}$ hence E is not empty and there exists in E a polynomial g of least degree. It is clear that g is irreducible and there exists a monic polynomial h of degree $< n$ such that $f = gh$; by the induction hypothesis h is the product of a finite family of irreducible monic polynomials, hence f has the same property. This proves the *existence* of the decomposition (18).

Let us now prove the *uniqueness* of the decomposition (18). Let $(w_p)_{p \in \mathcal{I}}$ be a family of positive integers with finite support, such that $f = \prod_{p \in \mathcal{I}} p^{w_p}$. Since f is of degree $n \geqslant 1$, there exists $p \in \mathcal{I}$ such that $w_p > 0$; if we had $v_p = 0$, f would be the product of a family of elements of $\mathcal{I}$ distinct from p, hence it would be prime to p (IV, p. 13, Cor. 6), contrary to the fact that p divides f. By the induction hypothesis the polynomial f/p admits a unique decomposition of type (18); hence we conclude the equality $w_q = v_q$ for every $q \in \mathcal{I}$.

Let f be a non-zero polynomial in K[X]. We shall say that f has *no multiple factors* if the exponents v_p in the decomposition (18) are all $\leqslant 1$; it comes to the same to say that f is the product of a finite sequence of pairwise distinct irreducible polynomials, or also that f is not divisible by the square of any non-constant polynomial of K[X].

§ 2. ZEROS OF POLYNOMIALS

1. Roots of a polynomial in one indeterminate. Multiplicity

Let $g \in \mathrm{A}[(\mathrm{X}_i)_{i \in \mathrm{I}}]$ and let E be a unital associative A-algebra. Let $\mathbf{x} = (x_i)_{i \in \mathrm{I}}$ be a family of pairwise permutable elements of E. We shall say that $\mathbf{x}$ is a *zero* of g in E^{I} if $g(\mathbf{x}) = 0$. If f is a polynomial in a single indeterminate, a zero of f in E is also called a *root* of f in E.

PROPOSITION 1. — *Let* $f \in \mathrm{A}[\mathrm{X}]$ *and* $\alpha \in \mathrm{A}$. *The remainder of the division of* f *by* $\mathrm{X} - \alpha$ *is* $f(\alpha)$. *For* α *to be a root of* f *it is necessary and sufficient that* $\mathrm{X} - \alpha$ *should be a divisor of* f *in* A[X].

For if $u, v \in A[X]$ are such that $f = (X - \alpha)u + v$, $\deg v < 1$, then v is a scalar and $f(\alpha) = (\alpha - \alpha)u(\alpha) + v = v$. This proves the first assertion and the second follows from it.

PROPOSITION 2. — *Let $f \in A[X]$, $\alpha \in A$, and let h be an integer ≥ 0. The following conditions are equivalent:*

(i) *f is divisible by $(X - \alpha)^h$ but not by $(X - \alpha)^{h+1}$;*

(ii) *there exists $g \in A[X]$ such that $f = (X - \alpha)^h g$ and $g(\alpha) \neq 0$.*

(i) $\Rightarrow$ (ii) follows at once from Prop. 1.

(ii) $\Rightarrow$ (i): Suppose that $f = (X - \alpha)^h g$, where g does not admit α as a root. Then f is divisible by $(X - \alpha)^h$; if $g_1 \in A[X]$ existed such that $f = (X - \alpha)^{h+1} g_1$, then since $(X - \alpha)^n$ is not a divisor of zero in $A[X]$ (IV, p. 9, Prop. 7), we have $g = (X - \alpha) g_1$ and so $g(\alpha) = 0$, which is absurd.

PROPOSITION 3. — *Let f be a non-zero element of $A[X]$ and $\alpha \in A$. There exists just one integer $h \geq 0$ satisfying the conditions* (i) *and* (ii) *of Prop.* 2.

This is clear for condition (i), bearing in mind the fact that if f is divisible by $(X - \alpha)^h$, then $\deg f \geq h$ (IV, p. 9, Prop. 7).

DEFINITION 1. — *With the above notation we say that α is of order h, or multiplicity h relative to f.*

If $h > 0$ we also say that α is a root of order h or multiplicity h of f. A root of order 1 is called a simple root, a root of order 2 a double root,... A root of order > 1 is said to be multiple.

Remarks. — 1) If $f = 0$ we agree to say that α has order $\geq h$ relative to f, whatever $\alpha \in A$ and the integer $h \geq 0$. For any $f \in A[X]$ and $\alpha \in A$, to say that α has order $\geq h$ relative to f means that $(X - \alpha)^h$ divides f.

2) Let B be a commutative ring containing A as subring. Let $f \in A[X]$ be non-zero and $\alpha \in A$. The order of α relative to f is the same, whether we consider f as element of $B[X]$ or as element of $A[X]$. This is clear from condition (ii) of Prop. 2.

PROPOSITION 4. — *Let f and g be non-zero elements of $A[X]$. Let $\alpha \in A$, and let the orders of α relative to f and g be p and q respectively.*

(i) *The order of α relative to $f + g$ is $\geq \inf(p, q)$. It is equal to $\inf(p, q)$ if $p \neq q$.*

(ii) *The order of α relative to fg is $\geq p + q$. It is equal to $p + q$ if A is an integral domain.*

For we have $f(X) = (X - \alpha)^p f_1(X)$, $g(X) = (X - \alpha)^q g_1(X)$ with $f_1(\alpha) \neq 0$, $g_1(\alpha) \neq 0$. Suppose for example that $p \leq q$; then we have

$$f(X) + g(X) = (X - \alpha)^p (f_1(X) + (X - \alpha)^{q-p} g_1(X)),$$

and if $p < q$, α is not a root of $f_1(X) + (X - \alpha)^{q-p} g_1(X)$; this proves (i). On the other hand, we have $f(X)\, g(X) = (X - \alpha)^{p+q} f_1(X)\, g_1(X)$ and $f_1(\alpha)\, g_1(\alpha) \neq 0$ if A is an integral domain ; this proves (ii).

PROPOSITION 5. — *Suppose that* A *is an integral domain. Let* f *be a non-zero element of* A[X], *and* $\alpha_1, \ldots, \alpha_p$ *pairwise distinct roots of* f *in* A, *of orders* $k_1, \ldots, k_p$. *We have*

$$f(X) = (X - \alpha_1)^{k_1} (X - \alpha_2)^{k_2} \ldots (X - \alpha_p)^{k_p} g(X)$$

where $g \in A[X]$ *and* $\alpha_1, \ldots, \alpha_p$ *are not roots of* g.

We proceed by induction on p, the proposition being evident for $p = 1$, by Def. 1. Suppose then that $f(X) = g_1(X)\, g_2(X)$, where

$$g_1(X) = (X - \alpha_1)^{k_1} \ldots (X - \alpha_{p-1})^{k_{p-1}}, \quad g_2(X) \in A[X].$$

Since A is an integral domain and α_p is distinct from $\alpha_1, \ldots, \alpha_{p-1}$ it follows that α_p is not a root of $g_1(X)$, hence α_p is a root of order k_p of $g_2(X)$ (Prop. 4, (ii)). It follows that $g_2(X)$ is divisible by $(X - \alpha_p)^{k_p}$, and so

$$f(X) = (X - \alpha_1)^{k_1} \ldots (X - \alpha_p)^{k_p} g(X)$$

where $g(X) \in A[X]$. Clearly $\alpha_1, \ldots, \alpha_p$ are not roots of g.

THEOREM 1. — *Let* A *be an integral domain. Given a non-zero element* f *of* A[X], *of degree* n, *then the sum of the orders of all the roots of* f *in* A *is* $\leq n$.

This follows immediately from Prop. 5.

COROLLARY. — *Assume that* A *is an integral domain and let* $f, g \in A[X]$, *of degrees* $\leq n$. *If there exist* $n + 1$ *pairwise distinct elements* $x_1, \ldots, x_{n+1}$ *of* A *such that* $f(x_i) = g(x_i)$ *for* $1 \leq i \leq n + 1$, *then* $f = g$.

It suffices to apply Th. 1 to $f - g$.

PROPOSITION 6 (Lagrange interpolation formula). — *Let* K *be a commutative field,* $\alpha_1, \alpha_2, \ldots, \alpha_n$ *distinct elements of* K *and* $\beta_1, \beta_2, \ldots, \beta_n$ *any elements of* K. *For* $i = 1, 2, \ldots, n$ *we put*

$$f_i(X) = \prod_{j \in U(i)} (X - \alpha_j)/(\alpha_i - \alpha_j),$$

where $U(i)$ *is the set of integers* j *such that* $j \neq i$ *and* $1 \leq j \leq n$. *Then* $\beta_1 f_1 + \cdots + \beta_n f_n$ *is the unique element* f *of* K[X] *such that* $\deg f < n$ *and* $f(\alpha_i) = \beta_i$ *for* $1 \leq i \leq n$.

The uniqueness of f follows from the Cor. to Th. 1. Let $f = \beta_1 f_1 + \cdots + \beta_n f_n$, then since f_i has degree $n - 1$, we have $\deg f < n$. On the other hand, $f_i(\alpha_j) = 0$ for $j \neq i$ and $f_i(\alpha_i) = 1$, hence $f(\alpha_i) = \beta_i$ for $1 \leq i \leq n$.

COROLLARY. — *Suppose that* A *is an integral domain. Let* $f \in A[X]$, *of degree* $< n$, *and let* K *be a subring of* A *which is a field. If there exist n distinct elements* $\alpha_1, \ldots, \alpha_n$ *of* A *such that* $\alpha_i \in K$ *and* $f(\alpha_i) \in K$ *for* $i = 1, \ldots, n$, *then* $f \in K[X]$.

2. Differential criterion for the multiplicity of a root

PROPOSITION 7. — *Let* $f \in A[X]$ *and let* $\alpha \in A$ *be a root of f. For* α *to be a simple root of f it is necessary and sufficient that* α *should not be a root of the derivative* Df *of f.*

By hypothesis we have $f = (X - \alpha) g$, where $g \in A[X]$. For α to be a simple root of f it is necessary and sufficient that $g(\alpha) \neq 0$. Now we have $Df = g + (X - \alpha) Dg$, whence $(Df)(\alpha) = g(\alpha)$.

More generally :

PROPOSITION 8. — *Let* $f \in A[X]$ *and* $\alpha \in A$, *and suppose that* α *has order* $k \geqslant 1$ *relative to f. Then* α *has order* $\geqslant k - 1$ *relative to* Df. *If* $k \,.\, 1$ *is cancellable in* A, *then* α *has order* $k - 1$ *relative to* Df.

By hypothesis there exists $g \in A[X]$ such that $f = (X - \alpha)^k g$ and $g(\alpha) \neq 0$. Hence $Df = k(X - \alpha)^{k-1} g + (X - \alpha)^k Dg = (X - \alpha)^{k-1}(kg + (X - \alpha) Dg)$, which establishes the first part of the proposition. The value of $kg + (X - \alpha) Dg$ for $X = \alpha$ is $kg(\alpha)$, and this is non-zero if $k \,.\, 1$ is cancellable in A ; this proves the second part of the proposition.

Let k be an integer > 0 such that $k \,.\, 1 = 0$ in A. If $f(X) = X^k$, then 0 is a root of order k of f, and a root of arbitrarily high order of Df.

COROLLARY. — *Let* $f \in A[X]$, $\alpha \in A$ *and p an integer* $\geqslant 0$, *further, suppose that* $p! \,.\, 1$ *is cancellable in* A. *Then for* α *to be a root of order p of f, it is necessary and sufficient that* α *should be a root of* $f, Df, \ldots, D^{p-1} f$ *and not a root of* $D^p f$.

This follows from Prop. 8 by induction on p.

3. Polynomial functions on an infinite integral domain

PROPOSITION 9. — *Assume that* A *is an integral domain. Let* I *be a set,* $(H_i)_{i \in I}$ *a family of infinite subsets of* A *and* $H = \prod_{i \in I} H_i \subset A^I$. *If f is a non-zero element of* $A[(X_i)_{i \in I}]$, *and* H_f *the set of all* $\mathbf{x} \in H$ *such that* $f(\mathbf{x}) \neq 0$, *then* H *and* H_f *are equipotent.*

a) First suppose that I is finite and put $n = \operatorname{Card} I$. The proposition is clear for $n = 0$; we shall prove it by induction on n. Choose an element i_0 of I and put $J = I - \{i_0\}$, $B = A[(X_i)_{i \in J}]$. Since $f \neq 0$, we can write $f = \sum_{k=0}^{m} g_k X_{i_0}^k$, where

$g_0, \ldots, g_m \in \mathrm{B}$ and $g_m \neq 0$. By the induction hypothesis the set K of all $\mathbf{x} \in \prod_{i \in \mathrm{J}} \mathrm{H}_i$ such that $g_m(\mathbf{x}) \neq 0$ is equipotent with $\prod_{i \in \mathrm{J}} \mathrm{H}_i$. For $\mathbf{x} \in \mathrm{K}$ the polynomial

$$h(\mathrm{X}_{i_0}) = \sum_{k=0}^{m} g_k(\mathbf{x}) \mathrm{X}_{i_0}^k \in \mathrm{A}[\mathrm{X}_{i_0}]$$

is non-zero. By Th. 1 (IV, p. 16) the set of $\alpha \in \mathrm{H}_{i_0}$ such that $h(\alpha) \neq 0$ is equipotent with H_{i_0} whence

$$\mathrm{Card}\,\mathrm{H} \geqslant \mathrm{Card}\,\mathrm{H}_f \geqslant (\mathrm{Card}\,\mathrm{K}) \,.\, (\mathrm{Card}\,\mathrm{H}_{i_0}) = \mathrm{Card}\,\mathrm{H}\,,$$

and so $\mathrm{Card}\,\mathrm{H} = \mathrm{Card}\,\mathrm{H}_f$.

b) In the general case there is a finite subset I′ of I such that $f \in \mathrm{A}[(\mathrm{X}_i)_{i \in \mathrm{I}'}]$. Let H'_f be the set of all $\mathbf{x} \in \prod_{i \in \mathrm{I}'} \mathrm{H}_i$ such that $f(\mathbf{x}) \neq 0$. Then $\mathrm{H}_f = \mathrm{H}'_f \times \left(\prod_{i \in \mathrm{I} - \mathrm{I}'} \mathrm{H}_i \right)$, and it suffices to apply the first part of the proof to H'_f.

COROLLARY 1. — *We keep the hypothesis and notation of Prop.* 9. *If* I *is non-empty, then* H_f *is infinite.*

COROLLARY 2. — *Suppose that* A *is an infinite integral domain or that* A *is an algebra over an infinite field. For every* $f \in \mathrm{A}[(\mathrm{X}_i)_{i \in \mathrm{I}}]$, *let* $\tilde{f} : \mathrm{A}^{\mathrm{I}} \to \mathrm{A}$ *be the polynomial function defined by* f (IV, p. 4). *Then the mapping* $f \mapsto \tilde{f}$ *is injective.*

When A is an infinite integral domain, the corollary follows at once from Prop. 9. Suppose that A is an algebra over an infinite field k. Let $f = \sum_{\nu \in \mathbf{N}^{(\mathrm{I})}} \alpha_\nu \mathrm{X}^\nu$ be a non-zero element of $\mathrm{A}[(\mathrm{X}_i)_{i \in \mathrm{I}}]$; then there exists $\nu_0 \in \mathbf{N}^{(\mathrm{I})}$ such that $\alpha_{\nu_0} \neq 0$, and a k-linear form φ on A such that $\varphi(\alpha_{\nu_0}) \neq 0$. Let $g = \sum_{\nu \in \mathbf{N}^{(\mathrm{I})}} \varphi(\alpha_\nu) \mathrm{X}^\nu \in k[(\mathrm{X}_i)_{i \in \mathrm{I}}]$; we have $g \neq 0$, hence there exists $\mathbf{x} \in k^{\mathrm{I}}$ such that $g(\mathbf{x}) \neq 0$. Then $\varphi(f(\mathbf{x})) = g(\mathbf{x}) \neq 0$, and so $f(\mathbf{x}) \neq 0$.

When A is an infinite integral domain or when A is an algebra over an infinite field, we shall usually identify f with $\tilde{f}$.

Suppose that A is finite and let $f(\mathrm{X}) = \prod_{\alpha \in \mathrm{A}} (\mathrm{X} - \alpha)$, then $f \neq 0$, but $\tilde{f} = 0$. For other examples see IV, p. 88, exercises 7 and 8.

THEOREM 2 (Principle of extension of algebraic identities). — *Suppose that* A *is an infinite integral domain. Let* $g_1, \ldots, g_m$, f *be elements of* $\mathrm{A}[(\mathrm{X}_i)_{i \in \mathrm{I}}]$ *and assume the following hypotheses :*

a) $g_1 \neq 0, \ldots, g_m \neq 0$;

b) for all $\mathbf{x} \in A^I$ *such that* $g_1(\mathbf{x}) \neq 0, \ldots, g_m(\mathbf{x}) \neq 0$, *we have* $f(\mathbf{x}) = 0$. *Then* $f = 0$.

For if $f \neq 0$, we have $f g_1 \ldots g_m \neq 0$ (IV, p. 9, Prop. 8), hence there exists $\mathbf{x} \in A^I$ such that $f(\mathbf{x}) g_1(\mathbf{x}) \ldots g_m(\mathbf{x}) \neq 0$ (IV, p. 18, Cor. 2), which contradicts the hypothesis.

Scholium. — Let A be an integral domain and $f \in A[(X_i)_{i \in I}]$. Th. 2 provides a convenient means of proving that $f = 0$. It suffices to consider an infinite integral domain E containing A as subring ; if we can show that $f((x_i)) = 0$ for all $(x_i) \in E^I$ (or even for those $(x_i) \in E^I$ at which a finite number of given non-zero polynomials do not vanish) then it follows that $f = 0$. If A itself is not infinite, we can for example take E to be the ring $A[X]$ or its field of fractions.

Once we have proved the relation $f = 0$, we can clearly deduce that $f((y_i)) = 0$ for all $(y_i) \in F^I$ where F is any unital associative and commutative A-algebra whatsoever ; in particular F may be finite or non-integral.

In other words, the proof of the identity $f((x_i)) = 0$ when the x_i run over an infinite integral domain containing A as subring (with the possible restriction that $g_k((x_i)) \neq 0$ for $1 \leqslant k \leqslant m$, where the g_k are non-zero polynomials) implies the same identity when the x_i run over any unital associative and commutative A-algebra.

In particular let $f \in \mathbf{Z}[(X_i)_{i \in I}]$. If $f((x_i)) = 0$ when the x_i run over $\mathbf{Z}$ (with the possible restriction that $g_k((x_i)) \neq 0$ for $1 \leqslant k \leqslant m$, where the g_k are non-zero elements of $\mathbf{Z}[(X_i)]$), then we have the same identity when the x_i run over an arbitrary commutative ring.

§ 3. RATIONAL FRACTIONS

1. Definition of rational fractions

DEFINITION 1. — *Let* K *be a commutative field and* I *a set. The field of fractions* (I, p. 116) *of the integral domain* $K[(X_i)_{i \in I}]$ *is denoted by* $K((X_i)_{i \in I})$ *or* $K(X_i)_{i \in I}$. *Its elements are called rational fractions in the indeterminates* X_i *with coefficients in* K.

For $I = \{1, 2, \ldots, n\}$ we write $K(X_1, X_2, \ldots, X_n)$ in place of $K((X_i)_{i \in I})$.

Let A be an integral domain and K its field of fractions. The ring $A[(X_i)_{i \in I}]$ may be identified with a subring of $K[(X_i)_{i \in I}]$, hence also of $K((X_i)_{i \in I})$. For each $f \in K[(X_i)_{i \in I}]$ there exists a non-zero element α of A such that $\alpha f \in A[(X_i)_{i \in I}]$. Hence every element of $K((X_i)_{i \in I})$ can be written as u/v where $u, v \in A[(X_i)_{i \in I}]$, $v \neq 0$. Thus $K((X_i)_{i \in I})$ may be identified with the field of fractions of $A[(X_i)_{i \in I}]$.

Now let K be a commutative field, I a set and $J \subset I$. Put $B = K[(X_i)_{i \in J}]$, then $K[(X_i)_{i \in I}] = B[(X_i)_{i \in I - J}]$, and by what has been said, $K((X_i)_{i \in I})$ may be identified with $K'((X_i)_{i \in I - J})$, where $K' = K((X_i)_{i \in J})$.

2. Degrees

Let K be a commutative field. For every element r of $K((X_i)_{i \in I})$ there exist u, $v \in K[(X_i)_{i \in I}]$ such that $v \neq 0$ and $r = \frac{u}{v}$. The relation $\frac{u}{v} = \frac{u_1}{v_1}$, where $v \neq 0$, $v_1 \neq 0$ is equivalent to $uv_1 = vu_1$; if $r \neq 0$, we have $u \neq 0$ and $u_1 \neq 0$, and so $\deg u + \deg v_1 = \deg v + \deg u_1$ (IV, p. 9), or also $\deg u - \deg v = \deg u_1 - \deg v_1$. The rational integer $\deg u - \deg v$ thus depends only on r; we call it the *degree*, or the *total degree* of r, and denote it by $\deg r$. We shall agree to write $\deg 0 = -\infty$. If $J \subset I$, we can likewise define the degree with respect to the X_j of index $j \in J$. When r is a polynomial, these notions coincide with those defined in IV, p. 2.

PROPOSITION 1. — *Let r, s be two rational fractions.*

(i) *If* $\deg r \neq \deg s$, *we have*

$$r + s \neq 0 \quad \textit{and} \quad \deg(r+s) = \sup(\deg r, \deg s).$$

If $\deg r = \deg s$, *we have* $\deg(r+s) \leqslant \deg r$.

(ii) *We have* $\deg(rs) = \deg r + \deg s$.

We can limit ourselves to the case where r and s are non-zero.

Let us write $r = \frac{u}{v}$, $s = \frac{w}{z}$, where u, v, w, z are non-zero polynomials. We have $rs = \frac{uw}{vz}$, and so

$$\deg(rs) = \deg(uw) - \deg(vz) = \deg u - \deg v + \deg w - \deg z = \deg r + \deg s.$$

On the other hand, we have $r + s = \frac{uz + vw}{vz}$. Suppose that $\deg r \neq \deg s$, in other words $\deg u + \deg z \neq \deg w + \deg v$. Then $uz + wv \neq 0$, and

$$\begin{aligned} \deg(r+s) &= \deg(uz + wv) - \deg(vz) \\ &= \sup(\deg(uz), \deg(wv)) - \deg(vz) \\ &= \sup(\deg(uz) - \deg(vz), \deg(wv) - \deg(vz)) \\ &= \sup(\deg r, \deg s). \end{aligned}$$

Suppose that $\deg r = \deg s$, that is $\deg u + \deg z = \deg w + \deg v$. If $r + s \neq 0$, then we have

$$\begin{aligned} \deg(r+s) &= \deg(uz + wv) - \deg(vz) \\ &\leqslant \deg(uz) - \deg(vz) = \deg r. \end{aligned}$$

* The mapping $r \mapsto -\deg r$ is thus a discrete valuation on the field $K((X_i)_{i \in I})$. *

3. Substitutions

Let K be a commutative field, E a unital associative K-algebra, $\mathbf{x} = (x_i)_{i \in I}$ a family of pairwise permutable elements of E. Let $B = K[(X_i)_{i \in I}]$ and $S_{\mathbf{x}}$ the set of all non-zero $v \in B$ such that $v(\mathbf{x})$ is invertible in E. Let $u \in B$, $v \in S_{\mathbf{x}}$ and $f = \frac{u}{v} \in K((X_i)_{i \in I})$. The element $u(\mathbf{x})v(\mathbf{x})^{-1} = v(\mathbf{x})^{-1}u(\mathbf{x})$ is defined in E; moreover, if u_1, v_1 are two polynomials such that $f = \frac{u_1}{v_1}$ and $v_1 \in S_{\mathbf{x}}$, then $uv_1 = vu_1$, hence $u(\mathbf{x})v_1(\mathbf{x}) = v(\mathbf{x})u_1(\mathbf{x})$ and so

$$u(\mathbf{x})v(\mathbf{x})^{-1} = u_1(\mathbf{x})v_1(\mathbf{x})^{-1}.$$

Let $f \in K((X_i)_{i \in I})$. If there exists *at least one couple* (u, v) such that $f = \frac{u}{v}$ and $v \in S_{\mathbf{x}}$, we shall say that $\mathbf{x}$ is *substitutable* in f; the element $u(\mathbf{x})v(\mathbf{x})^{-1}$ which only depends on f and $\mathbf{x}$ is then denoted by $f(\mathbf{x})$ or $f((x_i))$ or $f((x_i)_{i \in I})$.

PROPOSITION 2. — *Let* K *be a commutative field,* E *a unital associative* K*-algebra and* $\mathbf{x} = (x_i)_{i \in I}$ *a family of pairwise permutable elements of* E. *The set* $S_{\mathbf{x}}^{-1}B$ *of* $f \in K((X_i)_{i \in I})$ *such that* $\mathbf{x}$ *is substitutable in* f *is a* K*-subalgebra of* $K((X_i)_{i \in I})$. *The mapping* $f \mapsto f(\mathbf{x})$ *is a unital homomorphism* φ *of* $S_{\mathbf{x}}^{-1}B$ *into* E. *The image* $\varphi(S_{\mathbf{x}}^{-1}B)$ *is the set of all* yz^{-1}, *where* y *runs over the unital subalgebra* $K[\mathbf{x}]_E$ *of* E *generated by the family* $\mathbf{x}$ *and where* z *runs over the set of all invertible elements of* $K[\mathbf{x}]_E$.

Let $f_1 = \frac{u_1}{v_1}$, $f_2 = \frac{u_2}{v_2}$ be two elements of $K((X_i)_{i \in I})$ such that $v_1, v_2 \in S_{\mathbf{x}}$. We have $f_1 + f_2 = \frac{u_1v_2 + u_2v_1}{v_1v_2}$, $f_1f_2 = \frac{u_1u_2}{v_1v_2}$ and $v_1, v_2 \in S_{\mathbf{x}}$. Hence $S_{\mathbf{x}}^{-1}B$ is a K-subalgebra of $K((X_i)_{i \in I})$. The rest of the proposition is clear.

COROLLARY. — *Let* L *be a commutative field,* K *a subfield of* L, $\mathbf{x} = (x_i)_{i \in I}$ *a family of elements of* L, M *the set consisting of the* x_i, U *the set of all* $f \in K((X_i)_{i \in I})$ *such that* $\mathbf{x}$ *is substitutable in* f *and* φ *the homomorphism* $f \mapsto f(\mathbf{x})$ *of* U *into* L. *Then* $\varphi(U)$ *is the subfield of* L *generated by* $K \cup M$.

Let L′ be the subfield of L generated by $K \cup M$. We have

$$K \cup M \subset \varphi(U) \subset L'$$

and $\varphi(U)$ is a subring of L. Now Prop. 2 implies that $\varphi(U)$ is a subfield of L, whence $\varphi(U) = L'$.

Let $f \in K((X_i)_{i \in I})$ and let $(g_i)_{i \in I}$ be a family of elements of $K((Y_l)_{l \in L})$. If

(g_i) is substitutable in f, then $f((g_i))$ is an element of $K((Y_l)_{l \in L})$. In particular, $(X_i)_{i \in I}$ is substitutable in f and $f = f((X_i)_{i \in I})$.

PROPOSITION 3. — *Let* E *be an algebra over* K *which is associative, commutative, unital and non-zero. Let* $f \in K((X_i)_{i \in I})$ *and for each* $i \in I$, *let* $g_i \in K((Y_l)_{l \in L})$. *Given a family* $\mathbf{y} = (y_l)_{l \in L}$ *of elements of* E, *suppose that* $\mathbf{y}$ *is substitutable into each* g_i *and* $(g_i(\mathbf{y}))_{i \in I}$ *is substitutable in* f. *Then :*

(i) $(g_i)_{i \in I}$ *is substitutable in* f;

(ii) *if we denote by* h *the element* $f((g_i))$ *of* $K((Y_l)_{l \in L})$, *then* $\mathbf{y}$ *is substitutable in* h *and* $h(\mathbf{y}) = f((g_i(\mathbf{y})))$.

We may take I to be finite. By hypothesis, for each $i \in I$, g_i can be put in the form p_i/q_i where $p_i, q_i \in K[(Y_l)_{l \in L}]$ and $q_i(\mathbf{y})$ is invertible in E. Likewise f can be written as u/v, where $u, v \in K[(X_i)_{i \in I}]$ and $v((g_i(\mathbf{y})))$ is invertible. Let $m = \sup(\deg u, \deg v)$, and let $w = \prod_{i \in I} q_i \in K[(Y_l)_{l \in L}]$, $u_1 = u((g_i))\, w^m$, $v_1 = v((g_i))\, w^m$. The polynomial u is a K-linear combination of monomials $\prod_{i \in I} X_i^{\nu_i}$ such that $\sum_{i \in I} \nu_i \leqslant m$. We have $w^m \prod_{i \in I} g_i^{\nu_i} = w^m \left(\prod_{i \in I} p_i^{\nu_i}\right)\left(\prod_{i \in I} q_i^{\nu_i}\right)^{-1}$ $\in K[(Y_l)_{l \in L}]$ by the choice of m. Hence $u_1 \in K[(Y_l)_{l \in L}]$ and similarly $v_1 \in K[(Y_l)_{l \in L}]$. Moreover, $v_1(\mathbf{y}) = (w(\mathbf{y}))^m v((g_i(\mathbf{y})))$ is invertible. Hence $v_1 \neq 0$, because $E \neq 0$, and so $v((g_i)) \neq 0$. The family (g_i) is thus substitutable in f. Besides we have $f((g_i)) = u_1/v_1$, hence $\mathbf{y}$ is substitutable in $h = f((g_i))$, and $h(\mathbf{y}) = u_1(\mathbf{y})/v_1(\mathbf{y}) = u((g_i(\mathbf{y})))/v((g_i(\mathbf{y}))) = f((g_i(\mathbf{y})))$.

Let K be a commutative field, E a commutative associative and unital K-algebra, and let $f \in K((X_i)_{i \in I})$. Let T_f be the set of all $\mathbf{x} = (x_i)_{i \in I} \in E^I$ which are substitutable in f. The mapping $\mathbf{x} \mapsto f(\mathbf{x})$ of T_f into E is called the *rational function* associated with f (and E) ; we sometimes denote it by $\tilde{f}$. If $g \in K((X_i)_{i \in I})$ we have $T_f \cap T_g \subset T_{f+g}$, $T_f \cap T_g \subset T_{fg}$, hence the rational function associated with $f + g$ (resp. fg) is defined on $T_f \cap T_g$ and has the same value on this set as $\tilde{f} + \tilde{g}$ (resp. $\tilde{f}\tilde{g}$). Let T'_f be the set of $\mathbf{x} \in T_f$ such that $f(\mathbf{x})$ is invertible ; if $\mathbf{x} \in T'_f$, $\mathbf{x}$ is substitutable in $1/f$ and the rational function associated with $1/f$ takes at $\mathbf{x}$ the value $f(\mathbf{x})^{-1}$.

If K is an *infinite* commutative field, $f \in K((X_i)_{i \in I})$, $g \in K((X_i)_{i \in I})$ and $\tilde{f}, \tilde{g}$ are the rational functions associated with f, g (and K), and if $\tilde{f}(\mathbf{x}) = \tilde{g}(\mathbf{x})$ for all $\mathbf{x} \in T_f \cap T_g$ then $f = g$. For if $f = u/v$ and $g = u_1/v_1$, where u, v, u_1, v_1 are polynomials, we have $u(\mathbf{x})\, v_1(\mathbf{x}) = u_1(\mathbf{x})\, v(\mathbf{x})$ for all $\mathbf{x}$ such that $v(x) v_1(x) \neq 0$, hence $uv_1 = u_1 v$ (IV, p. 18, Th. 2). Therefore the mapping $f \mapsto \tilde{f}$ is injective and we shall often identify f and $\tilde{f}$.

* Using the factoriality of $K[(X_i)_{i \in I}]$ (Comm. Alg., VII, § 3, No. 2 p. 502 and Cor. of Th. 2 p. 506), one easily shows the following: for every $f \in K((X_i)_{i \in I})$ there exist u, $v \in K[(X_i)_{i \in I}]$ such that:

1) $f = u/v$;

2) for $\mathbf{x} \in K^I$ to be substitutable in f it is necessary and sufficient that $v(\mathbf{x}) \neq 0$. *

4. Differentials and derivations

Let K be a commutative field. By III, p. 558, Prop. 5, every derivation D of $K[(X_i)_{i \in I}]$ extends in a unique fashion to a derivation $\bar{D}$ of $K((X_i)_{i \in I})$. If D, D′ are permutable derivations of $K[(X_i)_{i \in I}]$, then the bracket $[D, D'] = DD' - D'D$ is zero, hence $[\bar{D}, \bar{D}']$ which is a derivation of $K((X_i)_{i \in I})$ extending $[D, D']$ is zero; in other words, $\bar{D}$ and $\bar{D}'$ are permutable. In particular the derivations D_i (IV, p. 6) extend to derivations of $K((X_i)_{i \in I})$ again denoted by D_i and which are pairwise permutable. If $f \in K((X_i)_{i \in I})$, $D_i f$ is also written $D_{X_i} f$ or $\frac{\partial f}{\partial X_i}$ or f'_{X_i}. When there is only a single indeterminate X one uses the notation Df, $\frac{df}{dX}$, f'.

Let $B = K[(X_i)_{i \in I}]$, $C = K((X_i)_{i \in I})$. By III, p. 574, Prop. 23, the canonical mapping

$$\Omega_K(B) \otimes_B C \to \Omega_K(C)$$

is an isomorphism of vector C-spaces. Bearing in mind III, p. 570, we see that the vector C-space $\Omega_K(C)$ admits as a basis the family $(dX_i)_{i \in I}$ of the differentials of the X_i. Let ∂_i be the coordinate form of index i on $\Omega_K(C)$ relative to that basis. Then the mapping $u \mapsto \langle \partial_i, du \rangle$ of C into itself is a derivation of C which maps X_i to 1 and X_j to 0 for $j \neq i$, and so is equal to D_i; in other words, we have

$$du = \sum_{i \in I} (D_i u)\, dX_i \qquad (1)$$

for every $u \in C$. If I is finite, $(D_i)_{i \in I}$ is a basis of the vector C-space of derivations of C.

PROPOSITION 4. — *Let* E *be an associative, commutative and unital* K*-algebra,* $\mathbf{x} = (x_i)_{i \in I}$ *a family of elements of* E *and* $f \in K((X_i)_{i \in I})$. *Suppose that* $\mathbf{x}$ *is substitutable in* f *and* $y = f(\mathbf{x})$.

(i) *For every derivation* Δ *of* $K((X_i)_{i \in I})$ *which maps* $K[(X_i)_{i \in I}]$ *into itself,* $\mathbf{x}$ *is substitutable in* Δf.

(ii) *For every derivation* D *of* E *into an* E*-module we have*

$$Dy = \sum_{i \in I} (D_i f)(\mathbf{x}) \,.\, Dx_i \,.$$

Let $f = \frac{u}{v}$ with $u, v \in K[(X_i)_{i \in I}]$ and $v(\mathbf{x})$ invertible in E. Let Δ be a derivation of $K((X_i)_{i \in I})$ mapping $K[(X_i)_{i \in I}]$ into itself. We have

$$\Delta f = \frac{(\Delta u)\, v - u(\Delta v)}{v^2}$$

and $v^2(\mathbf{x})$ is invertible, hence $\mathbf{x}$ is substitutable in Δf. Secondly put $r = u(\mathbf{x})$, $s = v(\mathbf{x})$; we have $y = s^{-1}r$, hence for every derivation D of E into an E-module we have

$$\begin{aligned} \mathrm{D}y &= s^{-2}(s(\mathrm{D}r) - r(\mathrm{D}s)) \\ &= s^{-2}\Big(s \sum_{i \in I} (\mathrm{D}_i u)(\mathbf{x}) \,.\, \mathrm{D}x_i - r \sum_{i \in I} (\mathrm{D}_i v)(\mathbf{x}) \,.\, \mathrm{D}x_i\Big) \end{aligned}$$

by Prop. 4 of IV, p. 6. Thus $\mathrm{D}y = \sum_{i \in I} w_i \,.\, Dx_i$ with

$$w_i = v(\mathbf{x})^{-2}(v(\mathbf{x})(\mathrm{D}_i u)(\mathbf{x}) - u(\mathbf{x})(\mathrm{D}_i v)(\mathbf{x})) = (\mathrm{D}_i f)(\mathbf{x})\,.$$

§ 4. FORMAL POWER SERIES

1. Definition of formal power series. Order

Let I be a set. We recall (III, p. 454 and 456) that the total algebra of the monoid $\mathbf{N}^{(I)}$ over A is called *the algebra of formal power series with respect to the indeterminates* X_i $(i \in I)$ (or in the indeterminates X_i) *with coefficients in* A. It is denoted by $A[[X_i]]_{i \in I}$ or $A[[(X_i)_{i \in I}]]$ or also $A[[\mathbf{X}]]$, on denoting by $\mathbf{X}$ the family $(X_i)_{i \in I}$: in this paragraph we shall mainly use the notation $A[[I]]$. Sometimes it is convenient to designate the canonical image in $A[[I]]$ of the element i of I by a symbol other than X_i, for example Y_i, Z_i, T_i, ... ; the conventions used in this case are analogous to those for polynomials (IV, p. 1). The algebra $A[[I]]$ is then designated by $A[[Y_i]]_{i \in I}$, or $A[[\mathbf{Y}]]$ etc.

When I is a finite set of p elements, we also say that $A[[I]]$ is an algebra of formal power series in p indeterminates. These algebras are all isomorphic, for fixed p. An algebra of formal power series in 1, 2, ... indeterminates will also be denoted by $A[[X]]$, $A[[U, V]]$, ..., the set I of indices not being specified.

A formal power series u is conventionally written $u = \sum_{\nu \in \mathbf{N}^{(I)}} \alpha_\nu X^\nu$ (cf. IV, p. 1).

The α_ν are the *coefficients* of u ; there may be infinitely many of them $\neq 0$. The $\alpha_\nu X^\nu$ are called *terms* of u ; for u to be a polynomial it is necessary and sufficient that u should have only a finite number of terms $\neq 0$. The terms $\alpha_\nu X^\nu$ such that

$|\nu| = p$ are called the terms of total degree p. The formal power series $u_p = \sum_{|\nu| = p} \alpha_\nu X^\nu$ is called the *homogeneous component of degree p* of u (it is a polynomial when I is finite) ; u_0 is identified with an element of A called also the *constant term* of u. We say that u is homogeneous of degree p if $u = u_p$. If u, $v \in A[[I]]$ and $w = uv$, we have

$$w_p = \sum_{q+r=p} u_q v_r \tag{1}$$

for every integer $p \geqslant 0$.

We recall (III, p. 456), that the *order* $\omega(u)$ of a formal power series $u \neq 0$ is the least integer p such that $u_p \neq 0$. We shall agree to adjoin to $\mathbf{Z}$ an element written ∞ and extend the order relation and addition from $\mathbf{Z}$ to $\mathbf{Z} \cup \{\infty\}$ by the conventions

$$n < \infty\,, \quad \infty + \infty = \infty\,, \quad \infty + n = n + \infty = \infty$$

for every $n \in \mathbf{Z}$; we also put $\omega(0) = \infty$. With these conventions we have the relations

$$\begin{aligned} \omega(u+v) &\geqslant \inf(\omega(u), \omega(v))\,, \\ \omega(u+v) &= \inf(\omega(u), \omega(v)) \quad \text{if} \quad \omega(u) \neq \omega(v)\,, \\ \omega(uv) &\geqslant \omega(u) + \omega(v)\,, \end{aligned}$$

for any formal power series u and v in A[[I]].

We recall (III, p. 457) that for any subset J of I we identify A[[I]] with A[[I − J]] [[J]], which allows us to define the order $\omega_J(u)$ of a formal power series with respect to the X_j ($j \in J$), the homogeneous component of u with respect to the X_j ($j \in J$) etc.

Let φ be a homomorphism of A into a ring B. We extend φ to a homomorphism $\bar{\varphi}$ of A[[I]] into B[[I]] by letting each formal power series $u = \sum_\nu \alpha_\nu X^\nu$ correspond to the formal power series $\sum_\nu \varphi(\alpha_\nu) X^\nu$; we say that the latter is obtained by *applying φ to the coefficients of the formal power series u*. We shall sometimes write ${}^\varphi u$ for $\bar{\varphi}(u)$.

In particular if A is a subring of B and φ the canonical injection of A into B, then the homomorphism $\bar{\varphi}$ of A[[I]] into B[[I]] is injective ; we shall in general identify A[[I]] with a subring of B[[I]] by means of $\bar{\varphi}$.

2. Topology on the set of formal power series. Summable families

By definition A[[I]] is nothing other than the product set $A^{\mathbf{N}^{(I)}}$. Except for express mention to the contrary we shall equip A with the discrete topology and A[[I]] with the product topology (Gen. Top. I, p. 31 f.) which we shall call the *canonical topology*. Equipped with addition and the discrete topology, A is a separated and complete topological group ; hence for addition A[[I]] is a *separated and complete* topological group (Gen. Top., III, p. 238 and 242 and Gen. Top., II, p. 187). Morever the algebra $A[(X_i)_{i \in I}]$ of polynomials is dense in A[[I]] (Gen. Top., III p. 238, Prop. 25) and we may thus consider A[[I]] as the *completion* of $A[(X_i)_{i \in I}]$.

For each $\beta \in \mathbf{N}^{(I)}$ let S_β be the set of multi-indices ν such that $\nu \leqslant \beta$ and let $\mathfrak{a}_\beta$ be the set of formal power series $u = \sum_\nu \alpha_\nu X^\nu$ such that $\alpha_\nu = 0$ for $\nu \in S_\beta$. Clearly S_β is a finite subset of $\mathbf{N}^{(I)}$, and every finite subset of $\mathbf{N}^{(I)}$ is contained in a set of the form S_β. It follows that the family $(\mathfrak{a}_\beta)_{\beta \in \mathbf{N}^{(I)}}$ is a fundamental system of neighbourhoods of 0 in A[[I]]. The sets $\mathfrak{a}_\beta$ are ideals in A[[I]], hence (Gen. Top., III, p. 275) A[[I]] is a *topological ring*.

Lemma 1. — *Let* L *be an infinite set and* $(u_\lambda)_{\lambda \in L}$ *a family of elements of* A[[I]], *and put* $u_\lambda = \sum_\nu \alpha_{\lambda,\nu} X^\nu$ *for* $\lambda \in L$. *Then the following conditions are equivalent :*

(i) *The family* $(u_\lambda)_{\lambda \in L}$ *is summable* (Gen. Top., III, p. 262) *in* A[[I]].

(ii) *We have* $\lim u_\lambda = 0$, *taken along the filter of complements of finite subsets of* L.

(iii) *For every* $\nu \in \mathbf{N}^{(I)}$ *we have* $\alpha_{\lambda,\nu} = 0$ *except for a finite number of indices* $\lambda \in L$.

When these conditions hold, the series $u = \sum_{\lambda \in L} u_\lambda$ *is equal to* $\sum_\nu \alpha_\nu X^\nu$ *with* $\alpha_\nu = \sum_{\lambda \in L} \alpha_{\lambda,\nu}$ *for each* $\nu \in \mathbf{N}^{(I)}$.

The equivalence of (i) and (ii) follows from Cor. 2 of Gen. Top., III, p. 263.

The equivalence of (ii) and (iii) follows from the properties of limits in a product space (Gen. Top., I, p. 55, Cor. 1).

The last assertion follows from Prop. 4 of Gen. Top., III, p. 266.

Let us give some examples of summable families.

a) Let $u \in A[[I]]$ and let α_ν be the coefficient of X^ν in u. The family $(\alpha_\nu X^\nu)_{\nu \in \mathbf{N}^{(I)}}$ is then summable, with sum u (which justifies writing $u = \sum_\nu \alpha_\nu X^\nu$).

b) Let $u \in A[[I]]$; for every integer $p \geqslant 0$ let u_p be the homogeneous component of degree p of u. Then the family $(u_p)_{p \in \mathbf{N}}$ is summable and we have $u = \sum_{p \geqslant 0} u_p$.

c) Let $(u_\lambda)_{\lambda \in L}$ be a family of elements of A[[I]] and suppose that for every integer $n \geqslant 0$ the set of $\lambda \in L$ such that $\omega(u_\lambda) < n$ is finite. Then the family $(u_\lambda)_{\lambda \in L}$ is summable.

Remark. — Suppose that I is *finite*. For every integer $n \geqslant 0$ let $\mathfrak{b}_n$ be the set of formal power series $u \in A[[I]]$ such that $\omega(u) \geqslant n$. The sequence $(\mathfrak{b}_n)_{n \geqslant 0}$ is a fundamental system of neighbourhoods of 0 in A[[I]]. Therefore a family of elements u_λ of A[[I]] ($\lambda \in L$) is summable if and only if for every $n \in N$ the set of $\lambda \in L$ such that $\omega(u_\lambda) < n$ is finite.

PROPOSITION 1. — *Let $(u_\lambda)_{\lambda \in L}$ and $(v_\mu)_{\mu \in M}$ be two summable families of elements of* A[[I]]. *Then the family $(u_\lambda v_\mu)_{(\lambda,\mu) \in L \times M}$ is summable and we have*

$$\sum_{(\lambda,\mu) \in L \times M} u_\lambda v_\mu = \left(\sum_{\lambda \in L} u_\lambda\right)\left(\sum_{\mu \in M} v_\mu\right). \tag{2}$$

Let $(\alpha_{\lambda,\nu})_{\nu \in N^{(I)}}$ (resp. $(\beta_{\mu,\nu})_{\nu \in N^{(I)}}$) be the family of coefficients of u_λ (resp. v_μ). For each $\nu \in N^{(I)}$ there exists only a finite number of pairs $(\nu_1, \nu_2) \in N^{(I)} \times N^{(I)}$ such that $\nu_1 + \nu_2 = \nu$, hence only a finite number of pairs $(\lambda, \mu) \in L \times M$ such that the coefficient of X^ν in $u_\lambda v_\mu$ is $\neq 0$. Hence the family $(u_\lambda v_\mu)_{(\lambda,\mu) \in L \times M}$ is summable. Now the formula (2) follows from the associativity of the sum (Gen. Top., III, p. 265, formula (2)).

In A[[I]] the product is an associative and commutative composition law. We may therefore speak of a *multipliable family* of elements of A[[I]] and of the *product* of a multipliable family (Gen. Top., III, p. 262, remark 3).

PROPOSITION 2. — *Let $(u_\lambda)_{\lambda \in L}$ be a summable family of elements of* A[[I]].

(i) *The family $(1 + u_\lambda)_{\lambda \in L}$ is multipliable.*

(ii) *Let $\mathfrak{F}$ be the set of all finite subsets of* L. *For any $M \in \mathfrak{F}$ put $u_M = \prod_{\lambda \in M} u_\lambda$. Then the family $(u_M)_{M \in \mathfrak{F}}$ is summable and we have*

$$\sum_{M \in \mathfrak{F}} u_M = \prod_{\lambda \in L} (1 + u_\lambda).$$

Let us define the ideals $\mathfrak{a}_\beta$ as at the beginning of this No., and let $\beta \in N^{(I)}$. There exists a finite subset L_0 of L such that $u_\lambda \in \mathfrak{a}_\beta$ for $\lambda \notin L_0$. Then for every $M \in \mathfrak{F}$ such that $M \not\subset L_0$ we have $u_M \in \mathfrak{a}_\beta$. It follows that the family $(u_M)_{M \in \mathfrak{F}}$ is summable. On the other hand, for any finite subset M_0 of L we have

$$\sum_{M \subset M_0} u_M = \prod_{\lambda \in M_0} (1 + u_\lambda).$$

Taken along the filtered ordered set $\mathfrak{F}$, the left-hand side has as limit $\sum_{M \in \mathfrak{F}} u_M$. Hence the right-hand side has as limit $\sum_{M \in \mathfrak{F}} u_M$, which proves (i) and (ii) at the same time.

PROPOSITION 3. — *Let $u = \sum_\nu \alpha_\nu X^\nu \in$ A[[I]] and m an integer > 0. For every $n \in N$ let $(\alpha_{\nu,n})_{\nu \in N^{(I)}}$ be the family of coefficients of u^n. If $\alpha_0^m = 0$, then $\alpha_{\nu,n} = 0$ for $n \geqslant |\nu| + m$.*

Let $\nu \in \mathbf{N}^{(\mathrm{I})}$ and $n \in \mathbf{N}$. We have

$$\alpha_{\nu, n} = \sum_{\nu(1) + \cdots + \nu(n) = \nu} \alpha_{\nu(1)} \ldots \alpha_{\nu(n)} .$$

If $n \geqslant |\nu| + m$ and $\nu(1) + \cdots + \nu(n) = \nu$, we have $|\nu(1)| + \cdots + |\nu(n)| \leqslant n - m$. We thus have $\nu(r) = 0$ and so $\alpha_{\nu(r)} = \alpha_0$ for at least m distinct values of r; it follows that $\alpha_{\nu(1)} \ldots \alpha_{\nu(n)} = 0$, whence the result.

COROLLARY. — *Let $u \in \mathrm{A}[[\mathrm{I}]]$; then for $\lim_{n \to \infty} u^n = 0$ to hold it is necessary and sufficient that the constant term of u should be nilpotent.*

Let α_0 be the constant term of u. The constant term of u^n is α_0^n, hence the stated condition is necessary; it is sufficient by Prop. 3.

3. Substitutions

Let E be an A-algebra. A topology on E is said to be *linear* if it is invariant under translation and if there exists a fundamental system of neighbourhoods of 0 consisting of ideals of E (Gen. Top., III, p. 223). The topology on E is then compatible with its A-algebra structure (when A carries the discrete topology). An A-algebra with a linear topology is called a *linearly topologized* A-algebra.

PROPOSITION 4. — *Let* I *be a set and* E *an associative, commutative, unital, linearly topologized separated complete* A*-algebra.*

(i) *Let φ be a continuous homomorphism of* $\mathrm{A}[[\mathrm{I}]]$ *into* E *and* $x_i = \varphi(\mathrm{X}_i)$. *Then:*

(*a*) *for all $i \in \mathrm{I}$, x_i^n tends to* 0 *as n tends to $+\infty$;*

(*b*) *if* I *is infinite, x_i tends to* 0 *along the filter of complements of finite subsets of* I.

(ii) *Let $\mathbf{x} = (x_i)_{i \in \mathrm{I}}$ be a family of elements of* E *satisfying a) and b) of* (i). *Then there exists one and only one unital continuous homomorphism φ of* $\mathrm{A}[[\mathrm{I}]]$ *into* E *such that $\varphi(\mathrm{X}_i) = x_i$ for all $i \in \mathrm{I}$.*

For all $i \in \mathrm{I}$, X_i^n clearly tends to 0 in $\mathrm{A}[[\mathrm{I}]]$ as n tends to $+\infty$; on the other hand when I is infinite, X_i tends to 0 along the filter of complements of finite subsets of I. This proves (i).

Let $(x_i)_{i \in \mathrm{I}}$ be a family of elements of E satisfying conditions *a*) and *b*) of (i), let ψ be the homomorphism $u \mapsto u((x_i)_{i \in \mathrm{I}})$ of $\mathrm{A}[(\mathrm{X}_i)_{i \in \mathrm{I}}]$ into E, and let V be a neighbourhood of 0 of E which is an ideal of E. By *b*) there exists a finite subset J of I such that $x_i \in \mathrm{V}$ for all $i \in \mathrm{I} - \mathrm{J}$. Next there exists by *a*) an integer $n \geqslant 0$ such that $x_i^n \in \mathrm{V}$ for all $i \in \mathrm{J}$. Let β be the element of $\mathbf{N}^{(\mathrm{I})}$ such that $\beta_i = n - 1$ for $i \in \mathrm{J}$ and $\beta_i = 0$ for $i \in \mathrm{I} - \mathrm{J}$. If we define the ideal $\mathfrak{a}_\beta$ of $\mathrm{A}[[\mathrm{I}]]$ as at the beginning of No. 2 (IV, p. 26), then

$$u \in \mathrm{A}[(\mathrm{X}_i)_{i \in \mathrm{I}}] \cap \mathfrak{a}_\beta \Rightarrow \psi(u) \in \mathrm{V} .$$

This shows that ψ is continuous if we equip $A[(X_i)_{i \in I}]$ with the topology induced by that of $A[[I]]$. Since E is separated and complete, ψ extends to a continuous unital homomorphism φ of $A[[I]]$ into E. We have $\varphi(X_i) = \psi(X_i) = x_i$ for all $i \in I$. Finally let φ' be a continuous unital homomorphism of $A[[I]]$ into E such that $\varphi'(X_i) = x_i$. We have $\varphi'(u) = \varphi(u)$ for all $u \in A[(X_i)_{i \in I}]$, hence $\varphi' = \varphi$ because $A[(X_i)_{i \in I}]$ is dense in $A[[I]]$.

Let us keep the previous notation. If $u \in A[[I]]$, the image of u by φ is denoted by $u(\mathbf{x})$ or $u((x_i)_{i \in I})$ (or also $u(x_1, \ldots, x_n)$ if $I = \{1, 2, \ldots, n\}$) and is called *the element of* E *obtained by substitution of* x_i *for* X_i *in* u, or *the value of* u *for the values* x_i *of the* X_i or also *the value of* u *for* $X_i = x_i$. In particular we have $u = u((X_i)_{i \in I})$.

Let E′ be an associative commutative and unital linearly topologized separated and complete A-algebra. Let λ be a continuous unital homomorphism of E into E′, and $(x_i)_{i \in I}$ a family of elements of E satisfying conditions *a*) and *b*) of Prop. 4 (IV, p. 28). The family $(\lambda(x_i))_{i \in I}$ satisfies the same conditions *a*) and *b*). For every $u \in A[[I]]$ we have

$$\lambda(u((x_i)_{i \in I})) = u((\lambda(x_i))_{i \in I}), \tag{3}$$

for the mapping $u \mapsto \lambda(u((x_i)_{i \in I}))$ is a continuous unital homomorphism of $A[[I]]$ into E′ which transforms X_i into $\lambda(x_i)$ for all $i \in I$.

If J and K are two sets, we denote by $A_{J,K}$ the set of all families $(g_j)_{j \in J}$ satisfying the following conditions:

(i) *for all* $j \in J$, g_j *is an element of* $A[[K]]$ *whose constant term is nilpotent*;

(ii) *if* J *is infinite*, g_j *tends to* 0 *along the filter of complements of finite subsets of* J.

We note that if J is finite, every family of formal power series $(g_j)_{j \in J}$ without constant term in $A[[K]]$ belongs to $A_{J,K}$.

Let $(g_j)_{j \in J}$ be in $A_{J,K}$. By the Cor. of Prop. 3 (IV, p. 28) we have $\lim_{n \to \infty} g_j^n = 0$ for all $j \in J$. Let $f \in A[[J]]$; we can substitute g_j for the variable of index j in f and obtain a formal power series $f((g_j)_{j \in J})$ belonging to $A[[K]]$. Moreover, the mapping $f \mapsto f((g_j)_{j \in J})$ is a *continuous homomorphism* of A-algebras $A[[J]]$ into $A[[K]]$.

In particular if $J = \{1, \ldots, p\}$ and $f \in A[[X_1, \ldots, X_p]]$, we can substitute for each X_j a formal power series $g_j \in A[[K]]$ without constant term; the result of this substitution is written $f(g_1, \ldots, g_p)$.

Let $\mathbf{x} = (x_k)_{k \in K}$ be a family of elements of E satisfying conditions *a*) and *b*) of Prop. 4 (IV, p. 28). Let us apply (3), taking for λ the homomorphism $u \mapsto u(\mathbf{x})$ of $A[[K]]$ into E; we obtain

$$f((g_j)_{j \in J})(\mathbf{x}) = f((g_j(\mathbf{x}))_{j \in J}). \tag{4}$$

Let $\mathbf{f} = (f_i)_{i \in I} \in (A[[J]])^I$ and $g = (g_j)_{j \in J} \in A_{J,K}$. We denote by $\mathbf{f}(\mathbf{g})$ or $\mathbf{f} \circ \mathbf{g}$ the element $(f_i((g_j)_{j \in J}))_{i \in I}$ of $(A[[K]])^I$. If $\mathbf{f} \in A_{I,J}$, we have $\mathbf{f} \circ \mathbf{g} \in A_{I,K}$ because the mapping $f \mapsto f((g_j)_{j \in J})$ of $A[[I]]$ into $A[[K]]$ is continuous.

Let $\mathbf{f} \in (A[[J]])^I$, $\mathbf{g} \in A_{J,K}$, $\mathbf{h} \in A_{K,L}$. Then $\mathbf{g} \circ \mathbf{h} \in A_{J,L}$ and by (4), we have

$$(5) \qquad (\mathbf{f} \circ \mathbf{g}) \circ \mathbf{h} = \mathbf{f} \circ (\mathbf{g} \circ \mathbf{h}).$$

4. Invertible formal power series

PROPOSITION 5. — *In the ring* $A[[T]]$ *of formal power series in one indeterminate the polynomial* $1 - T$ *is invertible, and we have* $(1 - T)^{-1} = \sum_{n=0}^{\infty} T^n$.

For

$$(1 - T)\left(\sum_{n=0}^{\infty} T^n\right) = \sum_{n=0}^{\infty} T^n - \sum_{n=0}^{\infty} T^{n+1} = 1.$$

PROPOSITION 6. — *Let* $u \in A[[I]]$; *then for* u *to be invertible in* $A[[T]]$ *it is necessary and sufficient that its constant term should be invertible in* A.

Suppose that there exists $v \in A[[I]]$ such that $uv = 1$. Let α, β be the constant terms of u and v, then $\alpha\beta = 1$, so α is invertible.

Conversely, suppose that the constant term α of u is invertible. Then there exists a formal power series $t \in A[[I]]$ such that $u = \alpha(1 - t)$ and $\omega(t) > 0$. Now there is a ring homomorphism $\varphi : A[[T]] \to A[[I]]$ such that $\varphi(T) = t$, and $1 - T$ is invertible in $A[[T]]$ (Prop. 5) ; consequently $1 - t$ is invertible in $A[[I]]$, and hence so is u.

Remark. — Let $\mathcal{M}$ be the set of all formal power series with constant term 1. By Prop. 6, $\mathcal{M}$ is a commutative group under multiplication ; the multiplicative group of $A[[I]]$ is thus the direct product of $\mathcal{M}$ and the multiplicative group of A. We shall equip $\mathcal{M}$ with the topology induced from that of $A[[I]]$. For each $\beta \in \mathbf{N}^{(I)}$ we have in IV, p. 26 defined the ideal $\mathfrak{a}_\beta$ of $A[[I]]$; then $1 + \mathfrak{a}_\beta$ is a subgroup of $\mathcal{M}$ and the family $(1 + \mathfrak{a}_\beta)$ is a fundamental system of neighbourhoods of 1 in $\mathcal{M}$. Since the multiplication in $\mathcal{M}$ is continuous, we see that $\mathcal{M}$ is a topological group (Gen. Top., III, p. 223) ; in other words, the *mapping* $f \mapsto f^{-1}$ *is continuous in* $\mathcal{M}$.

Let K be a commutative field and $\mathfrak{O}$ the subring of the field of rational fractions $K((X_i)_{i \in I})$ formed of rational fractions in which the element 0 of K^I is substitutable. If $f \in \mathfrak{O}$, we have $f = \frac{u}{v}$, where u and v are polynomials such that the constant term of v is $\neq 0$, hence v is invertible in $K[[I]]$. We can verify at once that the element uv^{-1} of $K[[I]]$ depends only on f; we say that the formal power series uv^{-1} is the *expansion at the origin of the rational fraction* $\frac{u}{v}$. The mapping $f \mapsto uv^{-1}$ is an injective homomorphism of $\mathfrak{O}$ into $K[[I]]$; we shall often identify $\mathfrak{O}$ with its image under this mapping.

5. Taylor's formula for formal power series

Let $\mathbf{X} = (X_i)_{i \in I}$ and $\mathbf{Y} = (Y_i)_{i \in I}$ be two families of indeterminates relative to the same index set I. We denote by $\mathbf{X} + \mathbf{Y}$ the family $(X_i + Y_i)_{i \in I}$ of formal power series in $A[[\mathbf{X}, \mathbf{Y}]]$. It is clear that we can substitute $X_i + Y_i$ for X_i in a formal power series $u \in A[[\mathbf{X}]]$, the result being written $u(\mathbf{X} + \mathbf{Y})$. For each $\nu \in \mathbf{N}^{(I)}$ we denote by $\Delta^\nu u$ the coefficient of $\mathbf{Y}^\nu$ in the formal power series $u(\mathbf{X} + \mathbf{Y})$ considered as belonging to $A[[\mathbf{X}]][[\mathbf{Y}]]$ (III, p. 456). In other words, we have

$$u(\mathbf{X} + \mathbf{Y}) = \sum_\nu \Delta^\nu u(\mathbf{X}) \,.\, \mathbf{Y}^\nu \quad (u \in A[[\mathbf{X}]])\,. \tag{6}$$

Substituting $(0, \mathbf{X})$ for $(\mathbf{X}, \mathbf{Y})$ we obtain

$$u(\mathbf{X}) = \sum_\nu \Delta^\nu u(0) \,.\, \mathbf{X}^\nu\,; \tag{7}$$

In other words, the constant term of $\Delta^\nu u$ is the coefficient of $\mathbf{X}^\nu$ in u. Since the mapping $u \mapsto u(\mathbf{X} + \mathbf{Y})$ of $A[[\mathbf{X}]]$ into $A[[\mathbf{X}, \mathbf{Y}]]$ is continuous, the mappings $u \mapsto \Delta^\nu u$ of $A[[\mathbf{X}]]$ into itself are again continuous.

As in the case of polynomials (IV, p. 7) we can prove the formulae

$$\Delta^\sigma(uv) = \sum_{\nu + \rho = \sigma} \Delta^\nu(u)\,\Delta^\rho(v)\,, \tag{8}$$

$$\Delta^\rho \Delta^\sigma u = \frac{(\rho + \sigma)!}{\rho!\,\sigma!}\,\Delta^{\rho + \sigma} u\,. \tag{9}$$

The binomial formula (I, p. 99, Cor. 2) gives the following value for $\Delta^\nu u$ when $u = \sum_\lambda \alpha_\lambda \mathbf{X}^\lambda$

$$\Delta^\nu u = \sum_\lambda \alpha_{\lambda + \nu} \frac{(\lambda + \nu)!}{\lambda!\,\nu!} \mathbf{X}^\lambda\,. \tag{10}$$

Consider in particular the case $\nu = \varepsilon_i$, that is $\nu_i = 1$, $\nu_j = 0$ for $j \neq i$. We shall put $D_i u = \Delta^{\varepsilon_i} u$; put differently, $D_i u$ is the coefficient of Y_i in $u(\mathbf{X} + \mathbf{Y})$. By (10) we thus have

$$D_i u = \sum_\lambda (\lambda_i + 1)\,\alpha_{\lambda + \varepsilon_i} \mathbf{X}^\lambda\,; \tag{11}$$

in particular we have $D_i(X_i) = 1$ and $D_i(X_j) = 0$ for $j \neq i$. The formula (8) shows that D_i is a derivation of $A[[\mathbf{X}]]$, and from (9) we deduce the relation

$$D^\nu u = \nu!\,\Delta^\nu u \tag{12}$$

as in the case of polynomials (IV, p. 8) (we have put $D^\nu = \prod_{i \in I} D_i^{\nu_i}$ for $\nu = (\nu_i)_{i \in I}$ in $\mathbf{N}^{(I)}$). When A is a **Q**-algebra, the formulae (6), (7) and (12) imply the « Taylor formulae » :

$$u(\mathbf{X} + \mathbf{Y}) = \sum_{\nu} \frac{1}{\nu!} D^\nu u(\mathbf{X}) \,.\, \mathbf{Y}^\nu \,, \tag{13}$$

$$u(\mathbf{X}) = \sum_{\nu} \frac{1}{\nu!} D^\nu u(0) \,.\, \mathbf{X}^\nu \,. \tag{14}$$

Remarks. — 1) We often say that $D_i u$ is *the partial derivative of u with respect to* X_i ; we also use the notation $D_{X_i} u$, $\frac{\partial u}{\partial X_i}$ and u'_{X_i}. For a single indeterminate X the unique partial derivative Du (also written $\frac{du}{dX}$ or u') is called the *derivative* of u.

2) The formula (9) shows that the endomorphisms Δ^p of the A-module $A[[\mathbf{X}]]$ commute pairwise. Hence the same holds of the endomorphisms D_i.

3) If $u \in A[(X_i)_{i \in I}]$ is a polynomial, the polynomials $\Delta^p u$ and $D_i u$ defined in IV, p. 6 and 7 coincide with the formal power series denoted by the same symbols.

6. Derivations in the algebra of formal power series

Let I be a set, E an associative, commutative and unital linearly topologized, separated and complete A-algebra, and $\mathbf{x} = (x_i)_{i \in I}$ a family of elements of E satisfying conditions *a*) and *b*) of Prop. 4 (IV, p. 28). Let φ be the continuous homomorphism $u \mapsto u(x)$ of $A[[I]]$ into E ; it equips E with an $A[[I]]$-module structure. By III, p. 552, an A-derivation D of $A[[I]]$ into the $A[[I]]$-module E is thus an A-linear mapping $D : A[[I]] \to E$ satisfying the relation

$$D(uv) = u(\mathbf{x}) \,.\, D(v) + D(u) \,.\, v(\mathbf{x}) \tag{15}$$

for u, v in $A[[I]]$.

PROPOSITION 7. — *Let* $(y_i)_{i \in I}$ *be a family of elements of* E. *When* I *is infinite, we assume that* y_i *tends to* 0 *along the filter of complements of finite subsets of* I. *There exists then a unique continuous* A-*derivation* D *of* $A[[I]]$ *into the* $A[[I]]$-*module* E *such that* $D(X_i) = y_i$ *for all* $i \in I$. *We have*

$$D(u) = \sum_{i \in I} (D_i u)(\mathbf{x}) \,.\, y_i \quad (u \in A[[I]]) \,. \tag{16}$$

Since 0 admits in E a fundamental system of neighbourhoods consisting of ideals, the family $((D_i u)(\mathbf{x}) \,.\, y_i)_{i \in I}$ is summable in E for all $u \in A[[I]]$ (Gen. Top., III, p. 263, Cor. 2). Formula (16) thus defines an A-linear mapping $D : A[[I]] \to E$. We leave the reader to verify that D is a continuous derivation.

Let D_1 be a continuous A-derivation of $A[[I]]$ into E, such that $D_1(X_i) = y_i$ for all $i \in I$. The kernel of the continuous derivation $D - D_1$ is a closed subalgebra B of A[[I]] containing 1 and the indeterminates X_i. Since the polynomial algebra $A[(X_i)_{i \in I}]$ is dense in A[[I]], we have $B = A[[I]]$ and so $D_1 = D$.

COROLLARY 1. — *Let* Δ *be a continuous derivation of the* A-*algebra* E. *For every formal power series* $u \in A[[I]]$ *the family* $((D_i u)(\mathbf{x}) \,.\, \Delta x_i)_{i \in I}$ *is summable and we have*

$$\Delta(u(\mathbf{x})) = \sum_{i \in I} (D_i u)(\mathbf{x}) \,.\, \Delta x_i \,. \tag{17}$$

This follows from Prop. 7 because the mapping $u \mapsto \Delta(u(\mathbf{x}))$ is a continuous derivation of A[[I]] into the A[[I]]-module E.

COROLLARY 2. — *The derivation* D_i *is the unique continuous derivation of the* A-*algebra* A[[I]] *such that*

$$D_i(X_i) = 1 \,, \quad D_i(X_j) = 0 \quad \textit{for} \quad j \neq i \,. \tag{18}$$

This follows from Cor. 1.

COROLLARY 3. — *Let* $f \in A[[X_1, \dots, X_p]]$ *and* $g_i \in A[[Y_1, \dots, Y_q]]$ *for* $1 \leqslant i \leqslant p$. *Suppose that for* $1 \leqslant i \leqslant p$ *the constant term of* g_i *is zero, and put* $h = f(g_1, \dots, g_p)$. *Then for* $1 \leqslant j \leqslant q$ *we have*

$$\frac{\partial h}{\partial Y_j} = \sum_{i=1}^{p} D_i f(g_1, \dots, g_p) \cdot \frac{\partial g_i}{\partial Y_j} \,. \tag{19}$$

This is the special case $E = A[[Y_1, \dots, Y_q]]$, $x_i = g_i$ and $\Delta = \partial/\partial Y_j$ of Cor. 1.

PROPOSITION 8. — *Let* $\mathbf{X} = (X_i)_{i \in I}$ *be a* finite *family of indeterminates.*

(i) *Every derivation of the ring of formal power series* A[[**X**]] *is continuous.*

(ii) *Every derivation of the polynomial ring* A[**X**] *into the ring of formal power series* A[[**X**]] *extends in a unique fashion to a derivation of the ring* A[[**X**]].

(iii) *The family* $(D_i)_{i \in I}$ *is a basis of the* A[[**X**]]-*module of* A-*derivations of* A[[**X**]] *into itself.*

Let $\mathfrak{b}_n$ be the set of all formal power series of order $\geqslant n$. It is clear that $\mathfrak{b}_n$ is an ideal in the ring A[[**X**]], generated by the monomials of degree n. Hence $\mathfrak{b}_n$ consists of finite sums of products of n formal power series without constant terms ; if D is a derivation of A[[**X**]], we have

$$D(f_1 \dots f_n) = \sum_{i=1}^{n} f_1 \dots f_{i-1} D(f_i) f_{i+1} \dots f_n \,,$$

whence it follows at once that $D\mathfrak{b}_n \subset \mathfrak{b}_{n-1}$ for $n \geqslant 1$. Since the sequence $(\mathfrak{b}_n)_{n \geqslant 0}$ is a fundamental system of neighbourhoods of 0 in A[[**X**]] (IV, p. 26 remark), D is continuous and (i) is proved.

Let Δ be a derivation of $A[\mathbf{X}]$ into $A[[\mathbf{X}]]$. Arguing as before, we can show that $\Delta(h)$ belongs to $\mathfrak{b}_{n-1}$ for every homogeneous polynomial h of degree $n \geqslant 1$. Now let $u \in A[[\mathbf{X}]]$ and let u_n be the homogeneous component of degree n of u. Since $\Delta(u_n) \in \mathfrak{b}_{n-1}$ for $n \geqslant 1$, the family $(\Delta(u_n))_{n \geqslant 0}$ is summable in $A[[\mathbf{X}]]$ and we can define a derivation D of $A[[\mathbf{X}]]$ into itself by

$$D(u) = \sum_{n \geqslant 0} \Delta(u_n).$$

We have $D(\mathfrak{b}_n) \subset \mathfrak{b}_{n-1}$, hence D is a continuous endomorphism of the additive group of $A[[\mathbf{X}]]$. The mapping $\Phi : (u, v) \mapsto D(uv) - uD(v) - D(u)v$ of $A[[\mathbf{X}]] \times A[[\mathbf{X}]]$ into $A[[\mathbf{X}]]$ is continuous and zero on $A[\mathbf{X}] \times A[\mathbf{X}]$. Since $A[\mathbf{X}]$ is dense in $A[[\mathbf{X}]]$, we have $\Phi = 0$; in other words, D is a derivation of $A[[\mathbf{X}]]$ into itself, extending Δ.

Finally, $A[\mathbf{X}]$ is dense in $A[[\mathbf{X}]]$ and every derivation of $A[[\mathbf{X}]]$ is continuous by (i) ; hence there exists a unique extension of Δ to a derivation of $A[[\mathbf{X}]]$. This proves (ii).

It remains to prove (iii). Formula (18) (IV, p. 33) shows that the family $(D_i)_{i \in I}$ is linearly independent over $A[[\mathbf{X}]]$, and formula (16) (IV, p. 32), applied in the case $E = A[[\mathbf{X}]]$, shows that every A-derivation is a linear combination of the D_i with coefficients in $A[[\mathbf{X}]]$.

PROPOSITION 9. — *Let $(u_\lambda)_{\lambda \in L}$ be a summable family of elements of* $A[[I]]$ *without constant term and* D *a continuous derivation of the* A*-algebra* $A[[I]]$. *If* $f = \prod_{\lambda \in L} (1 + u_\lambda)$ (IV, p. 27, Prop. 2), *then the family* $(Du_\lambda/(1 + u_\lambda))_{\lambda \in L}$ *is summable and we have*

$$D(f)/f = \sum_{\lambda \in L} D(u_\lambda)/(1 + u_\lambda). \tag{20}$$

If g and h are two invertible elements of $A[[I]]$, then

$$D(gh) = h \,.\, Dg + g \,.\, Dh$$

whence on division by gh,

$$D(gh)/gh = D(g)/g + D(h)/h. \tag{21}$$

For every finite subset M of L put $f_M = \prod_{\lambda \in M} (1 + u_\lambda)$. From (21) we deduce by induction on Card M the relation

$$D(f_M)/f_M = \sum_{\lambda \in M} D(u_\lambda)/(1 + u_\lambda). \tag{22}$$

This proves Prop. 9 when L is finite. Now suppose that L is infinite and write $\mathfrak{F}$ for the filtered ordered set of finite subsets of L. We have $\lim_{\mathfrak{F}} f_M = f$, and hence (IV, p. 30 remark)

$$D(f)/f = \lim_{\mathfrak{F}} D(f_M)/f_M .$$

Now Prop. 9 follows by passage to the limit in (22).

7. The solution of equations in a formal power series ring

Lemma 2. — *Let* $(g_i)_{i \in I}$ *be a family of elements of order* $\geqslant 2$ *in* $A[[I]]$. *When* I *is infinite, assume that* g_i *tends to* 0 *along the filter of complements of finite subsets of* I. *There exists one and only one automorphism* T *of the topological* A-*algebra* $A[[I]]$ *such that* $T(X_i) = X_i + g_i$ *for all* $i \in I$. *Further,*

$$\omega(T(u) - u) \geqslant \omega(u) + 1 \tag{23}$$

for each $u \in A[[I]]$.

The series $f_i = X_i + g_i$ has no constant term and when I is infinite, f_i tends to 0 along the filter of complements of finite subsets of I. Therefore (IV, p. 28, Prop. 4) there exists precisely one continuous endomorphism T of the A-algebra $A[[I]]$ such that $T(X_i) = f_i$ for all $i \in I$. For each $\nu \in \mathbf{N}^{(I)}$ we put

$$v_\nu = T(X^\nu) - X^\nu = \prod_{i \in I} (X_i + g_i)^{\nu(i)} - \prod_{i \in I} X_i^{\nu(i)} ;$$

the relations $\omega(g_i) \geqslant 2$ imply $\omega(v_\nu) \geqslant |\nu| + 1$, and the relation (23) follows at once from this.

Let us show that T is *injective*. Given $u \in A[[I]]$ such that $T(u) = 0$, by (23) we have $\omega(u) \geqslant \omega(u) + 1$, which is impossible if $u \neq 0$ because $\omega(u)$ would then be a positive integer.

For every formal series v in $A[[I]]$ we denote by $H_n(v)$ its homogeneous component of degree n. Let us put $S_0(v) = H_0(v)$ and define the continuous mappings $S_n : A[[I]] \to A[[I]]$ by the recursion equations

$$S_n(v) = H_n\left(v - T\left(\sum_{k=0}^{n-1} S_k(v)\right)\right) \quad \text{for} \quad n \geqslant 1 . \tag{24}$$

Put $S(v) = \sum_{n \geqslant 0} S_n(v)$; if $\nu \in \mathbf{N}^{(I)}$ and $n = |\nu|$, then the coefficient $S^\nu(v)$ of X^ν in $S(v)$ is equal to that of X^ν in $S_n(v)$; since S_n is a continuous mapping, the mapping $S^\nu : A[[I]] \to A$ is continuous. Hence, by the definition of the product topology on $A[[I]] = A^{\mathbf{N}^{(I)}}$, the mapping $S : A[[I]] \to A[[I]]$ is continuous.

We shall prove the relation $T(S(v)) = v$ for all $v \in A[[I]]$ which will complete the proof of the lemma. Let $v \in A[[I]]$, $u_n = S_n(v)$ and $u = S(v)$. Let n be a positive integer such that

$$\omega(v - T(u)) \geqslant n . \tag{25$_n$}$$

We have $\omega(u - (u_0 + \cdots + u_{n-1})) \geqslant n$, whence

$$\omega(\mathrm{T}(u) - \mathrm{T}(u_0 + \cdots + u_{n-1}) - u_n) \geqslant n + 1 \tag{26}$$

by (23). Now the recursion equation (24) shows that

$$u_n = \mathrm{H}_n(v - \mathrm{T}(u_0 + \cdots + u_{n-1}))\,. \tag{27}$$

By (26) the formal power series $v - \mathrm{T}(u)$ and $v - \mathrm{T}(u_0 + \cdots + u_{n-1}) - u_n$ have the same homogeneous component of degree n, and this component is zero, by (27). We thus have $\omega(v - \mathrm{T}(u)) \geqslant n + 1$, that is $(25)_n$ implies $(25)_{n+1}$. Since $(25)_0$ clearly holds, we thus have $\omega(v - \mathrm{T}(u)) \geqslant n$ for every integer $n \geqslant 0$, whence $v = \mathrm{T}(u) = \mathrm{T}(\mathrm{S}(v))$, as was to be proved.

For the rest of this No. we shall, for any set I, denote by $\mathrm{A}\{\mathrm{I}\}$ the set of families $(f_i)_{i \in \mathrm{I}}$ satisfying the following conditions :

(i) for each $i \in \mathrm{I}$, f_i is an element of $\mathrm{A}[[\mathrm{I}]]$ without constant term ;

(ii) if I is infinite, f_i tends to 0 along the filter of complements of finite subsets of I.

The set $\mathrm{A}\{\mathrm{I}\}$ is a monoid for the composition law $(\mathbf{f}, \mathbf{g}) \mapsto \mathbf{f} \circ \mathbf{g}$, with $\{\mathrm{X}_i\}_{i \in \mathrm{I}}$ as unit element. The set of invertible elements of $\mathrm{A}\{\mathrm{I}\}$ is thus a group.

On the other hand, let E be the monoid of all continuous unital endomorphisms of the A-algebra $\mathrm{A}[[\mathrm{I}]]$ leaving the ideal of all formal power series without constant term invariant. If $\mathbf{f} \in \mathrm{A}\{\mathrm{I}\}$ and $g \in \mathrm{A}[[\mathrm{I}]]$, then the element $g(\mathbf{f})$ is defined. For fixed $\mathbf{f}$, the mapping $g \mapsto g(\mathbf{f})$ of $\mathrm{A}[[\mathrm{I}]]$ into itself is an element $\mathrm{W}_{\mathbf{f}}$ of E. If $\mathbf{f}_1$, $\mathbf{f}_2 \in \mathrm{A}\{\mathrm{I}\}$ and $g \in \mathrm{A}[[\mathrm{I}]]$, we have, by formula (5) (IV, p. 30)

$$\mathrm{W}_{\mathbf{f}_1 \circ \mathbf{f}_2}(g) = g(\mathbf{f}_1 \circ \mathbf{f}_2) = g(\mathbf{f}_1) \circ \mathbf{f}_2 = \mathrm{W}_{\mathbf{f}_2}(\mathrm{W}_{\mathbf{f}_1}(g))$$

hence $\mathbf{f} \mapsto \mathrm{W}_{\mathbf{f}}$ is a homomorphism of the monoid opposite to $\mathrm{A}\{\mathrm{I}\}$ into E. By Prop. 4 (IV, p. 28) this homomorphism is bijective.

Let $\mathbf{f} = (f_i)_{i \in \mathrm{I}} \in \mathrm{A}\{\mathrm{I}\}$ and let $\sum_{j \in \mathrm{I}} \alpha_{ij}\mathrm{X}_j$ be the homogeneous component of degree 1 of f_i. For any fixed j in I we have $\alpha_{ij} = 0$ except for a finite number of suffixes i, by hypothesis (ii) above. If $(\lambda_i) \in \mathrm{A}^{(\mathrm{I})}$, we thus have $\left(\sum_{j \in \mathrm{I}} \alpha_{ij}\lambda_j\right) \in \mathrm{A}^{(\mathrm{I})}$.

We denote by $\mathrm{T}_{\mathbf{f}}$ the A-linear mapping [1]

$$(\lambda_i) \mapsto \left(\sum_{j \in \mathrm{I}} \alpha_{ij}\lambda_j\right)$$

of $\mathrm{A}^{(\mathrm{I})}$ into $\mathrm{A}^{(\mathrm{I})}$. If $\mathbf{g} \in \mathrm{A}\{\mathrm{I}\}$, it is easily verified that

$$\mathrm{T}_{\mathbf{f} \circ \mathbf{g}} = \mathrm{T}_{\mathbf{f}} \circ \mathrm{T}_{\mathbf{g}}\,. \tag{28}$$

[1] Sometimes $\mathrm{T}_{\mathbf{f}}$ is called the linear mapping tangent to $\mathbf{f}$.

PROPOSITION 10. — *Let* $\mathbf{f} \in A\{I\}$; *then the following conditions are equivalent :*
(i) $\mathbf{f}$ *is invertible in* $A\{I\}$ *for the law* $\circ$;
(ii) $T_{\mathbf{f}}$ *is invertible in the ring* $\mathrm{End}(A^{(I)})$.

The implication (i) $\Rightarrow$ (ii) is immediate from (28). Suppose now that $T_{\mathbf{f}}$ is invertible in $\mathrm{End}(A^{(I)})$. There exists $\mathbf{g} = (g_i)_{i \in I} \in A\{I\}$ such that each g_i is homogeneous of degree 1 and $T_{\mathbf{g}} \circ T_{\mathbf{f}}$ is the identity mapping of $A^{(I)}$. Write $\mathbf{h} = \mathbf{g} \circ \mathbf{f}$; then (28) shows that $T_{\mathbf{h}}$ is the identity mapping of $A^{(I)}$, which is equivalent to the assertion $\omega(h_i - X_i) \geqslant 2$. By Lemma 2 of IV, p. 35 $\mathbf{h}$ is therefore invertible in $A\{I\}$. It is clear that $\mathbf{g}$ is invertible in $A\{I\}$, hence $\mathbf{f}$ is invertible in $A\{I\}$.

COROLLARY. — *Let* $f_i(Y_1, Y_2, \ldots, Y_q, X_1, X_2, \ldots, X_p)$ $(1 \leqslant i \leqslant q)$ *be* q *formal power series without constant term in* $A[[Y_1, \ldots, Y_q, X_1, \ldots, X_p]]$. *If the constant term of the formal power series* $D = \det\left(\dfrac{\partial f_i}{\partial Y_j}\right)$ *is invertible in* A, *then there exists precisely one system of* q *formal power series* $u_1(X_1, \ldots, X_p), \ldots, u_q(X_1, \ldots, X_p)$ *such that*

$$f_i(u_1, \ldots, u_q, X_1, \ldots, X_p) = 0 \quad (1 \leqslant i \leqslant q). \tag{29}$$

Put $f_{q+1} = X_1, \ldots, f_{q+p} = X_p$, $\mathbf{f} = (f_1, \ldots, f_{p+q})$, then $\det T_{\mathbf{f}}$ is equal to the constant term of D, hence invertible in A ; therefore $T_{\mathbf{f}}$ is invertible. By Prop. 10 there exist formal power series without constant term

$$g_1, \ldots, g_{q+p} \in A[[Y_1, \ldots, Y_q, X_1, \ldots, X_p]]$$

such that on writing

$$\mathbf{g} = (g_1, \ldots, g_{p+q}), \quad \mathbf{1}_{p+q} = (Y_1, \ldots, Y_q, X_1, \ldots, X_p)$$

we have $\mathbf{f} \circ \mathbf{g} = \mathbf{g} \circ \mathbf{f} = \mathbf{1}_{p+q}$. The relation $\mathbf{f} \circ \mathbf{g} = \mathbf{1}_{p+q}$ in particular gives

$$g_{q+1} = X_1, \ldots, g_{q+p} = X_p .$$

Hence

$$f_i(g_1, \ldots, g_q, X_1, \ldots, X_p) = Y_i \quad (1 \leqslant i \leqslant q). \tag{30}$$

Now put

$$u_i(X_1, \ldots, X_p) = g_i(0, \ldots, 0, X_1, \ldots, X_p) \quad (1 \leqslant i \leqslant q); \tag{31}$$

substituting 0 for each Y_i in (30) we obtain the desired relation (29).

Conversely, suppose that the formal power series $u_1, \ldots, u_q$ in the ring $A[[X_1, \ldots, X_p]]$ satisfy the relation (29). The relation $\mathbf{g} \circ \mathbf{f} = \mathbf{1}_{p+q}$ implies

$$g_i(f_1, \ldots, f_q, X_1, \ldots, X_p) = Y_i \quad (1 \leqslant i \leqslant q); \tag{32}$$

and substituting u_i for Y_i for $1 \leqslant i \leqslant q$ in (32), we obtain (31), whence the uniqueness of the solution of the system (29).

8. Formal power series over an integral domain

PROPOSITION 11. — *Suppose that* A *is an integral domain.*

(i) *The ring* A[[I]] *is again an integral domain.*

(ii) *If u, v are non-zero elements of* A[[I]], *then* $\omega(uv) = \omega(u) + \omega(v)$.

For each $J \subset I$ let φ_J be the homomorphism of A[[I]] into A[[J]] obtained by substituting in each element of A[[I]], X_i for X_i when $i \in J$ and 0 for X_i when $i \in I - J$. Let u, v be non-zero elements of A[[I]], $p = \omega(u)$, $q = \omega(v)$; there exists a finite subset J of I such that

$$\varphi_J(u) \neq 0\,, \quad \varphi_J(v) \neq 0\,, \quad \omega(\varphi_J(u)) = p\,, \quad \omega(\varphi_J(v)) = q\,.$$

Let a (resp. b) be the homogeneous component of degree p (resp. q) of $\varphi_J(u)$ (resp. $\varphi_J(v)$). Since J is finite, a and b are polynomials. We have $a \neq 0$, $b \neq 0$, hence $ab \neq 0$ (IV, p. 9, Prop. 8). Hence $\varphi_J(u)\,\varphi_J(v)$ is non-zero, of order $p+q$. It follows that $uv \neq 0$ and $\omega(uv) \leqslant p+q$; but clearly $\omega(uv) \geqslant p+q$.

9. The field of fractions of the ring of formal power series in one indeterminate over a field

If K is a commutative field, we shall denote by K((X)) the field of fractions of the integral domain K[[X]].

PROPOSITION 12. — *Every non-zero element u of* K((X)) *may be written in a unique way as $u = X^k v$, where $k \in \mathbf{Z}$ and v is a formal power series in* X *of order* 0.

Let $u = w/t$, where w, t are non-zero elements of K[[X]]. We have $w = X^r w_1$, $t = X^s t_1$, where $r, s \in \mathbf{N}$ and w_1, t_1 are formal power series of order 0, hence invertible in K[[X]] (IV, p. 30, Prop. 6). Then $u = X^{r-s} w_1 t_1^{-1}$ and $w_1 t_1^{-1}$ is a formal power series of order 0.

Let us prove the uniqueness. Suppose that $u = X^{k_1} v_1 = X^{k_2} v_2$ where k_1, $k_2 \in \mathbf{Z}$ and v_1, v_2 are formal power series of order 0. Since $X^{k_1 - k_2} = v_2 v_1^{-1}$ is a formal power series of order 0, we have $k_1 = k_2$ whence $v_1 = v_2$ and this proves the uniqueness assertion.

We shall say that the elements of K((X)) are *generalized formal power series* in X with coefficients in K, or simply formal power series when no confusion can arise (the elements of K[[X]] are then called *formal power series with positive exponents*) ; if $u \neq 0$, the integer k defined in Prop. 12 is also called the *order* of u and is written $\omega(u)$, even if it is < 0 ; we also put $\omega(0) = \infty$. It may be verified at once that

$$\begin{aligned} &\omega(u+v) \geqslant \inf(\omega(u), \omega(v)) \\ &\omega(u+v) = \inf(\omega(u), \omega(v)) \quad \text{if} \quad \omega(u) \neq \omega(v) \\ &\omega(uv) = \omega(u) + \omega(v) \end{aligned}$$

still hold for generalized formal power series. In particular, if $u \neq 0$, then $\omega(u^{-1}) = -\omega(u)$. * In other words (Comm. Alg., VI, § 3, No. 6, p. 392, Def. 3), ω is a *normalized discrete valuation* of the field $K((X))$. *

For each integer $n \in \mathbf{Z}$ let $\mathfrak{p}_n$ be the set of all $u \in K((X))$ such that $\omega(u) \geqslant n$. Then $(\mathfrak{p}_n)_{n \in \mathbf{Z}}$ is a decreasing sequence of subgroups of the additive group $K((X))$, with intersection 0; there exists thus a topology on $K((X))$, invariant under translation, for which $(\mathfrak{p}_n)_{n \in \mathbf{Z}}$ is a fundamental system of neighbourhoods of 0 (Gen. Top., III, p. 223). We can easily verify that $K((X))$ is a topological field (Gen. Top., III, p. 281) and that $K[[X]]$ is an open and closed subspace of $K((X))$.

Let $(\alpha_n)_{n \in \mathbf{Z}}$ be a family of elements of K, and suppose that there exists an integer N such that $\alpha_n = 0$ for all $n < N$. Then the family $(\alpha_n X^n)_{n \in \mathbf{Z}}$ is summable in $K((X))$ (Gen. Top., III, p. 263, Cor.); put $u = \sum_{n \in \mathbf{Z}} \alpha_n X^n$, then $u = 0$ if and only if $\alpha_n = 0$ for all n; otherwise the order of u is the least integer k such that $\alpha_k \neq 0$. Finally every element of $K((X))$ may be written in a *unique* fashion in the form $\sum_{n \in \mathbf{Z}} \alpha_n X^n$, where the sequence (α_n) satisfies $\alpha_{-n} = 0$ for all sufficiently large n.

Since the ring $K[X]$ is a subring of $K[[X]]$, *every* rational fraction $u/v \in K(X)$ (u, v being polynomials in X) may be identified with the (generalized) formal power series uv^{-1} of $K((X))$, which we shall call its *expansion at the origin*; the field $K(X)$ is thus identified with a subfield of $K((X))$.

10. Exponential and logarithm

By the exponential power series we shall understand the element $\sum_{n \geqslant 0} \frac{X^n}{n!}$ of $\mathbf{Q}[[X]]$; it will be denoted by $\exp X$ or e^X.

PROPOSITION 13. — *In* $\mathbf{Q}[[X, Y]]$ *we have* $e^{X+Y} = e^X e^Y$.

For the binomial formula gives

$$\frac{(X+Y)^n}{n!} = \sum_{i+j=n} \frac{X^i}{i!} \frac{Y^j}{j!}.$$

Hence

$$e^X e^Y = \left(\sum_{i \geqslant 0} \frac{X^i}{i!}\right)\left(\sum_{j \geqslant 0} \frac{Y^j}{j!}\right) = \sum_{i,j \geqslant 0} \frac{X^i}{i!} \frac{Y^j}{j!} = \sum_{n \geqslant 0} \sum_{i+j=n} \frac{X^i}{i!} \frac{Y^j}{j!}$$
$$= \sum_{n \geqslant 0} \frac{(X+Y)^n}{n!} = e^{X+Y}.$$

We shall define two elements $e(X)$, $l(X)$ of $\mathbf{Q}[[X]]$ by

$$e(X) = e^X - 1 = \sum_{n \geqslant 1} \frac{X^n}{n!} \tag{33}$$

(34) $$l(X) = \sum_{n \geq 1} (-1)^{n-1} \frac{X^n}{n}.$$

We have

(35) $$e(X + Y) = e(X) + e(Y) + e(X)e(Y)$$

(36) $$D(e^X) = D(e(X)) = e^X$$

(37) $$D(l(X)) = \sum_{n \geq 0} (-X)^n = (1 + X)^{-1}.$$

PROPOSITION 14. — *We have* $l(e(X)) = e(l(X)) = X$.

The series l and e have no constant term and their terms of degree 1 are equal to X. By Prop. 10 of IV, p. 37 it suffices to prove the formula $l(e(X)) = X$. By the formulae (36) and (37) and Cor. 3 of IV, p. 33 we have

$$D(l(e(X))) = (1 + e(X))^{-1}D(e(X)) = (e^X)^{-1}e^X = 1$$

whence $l(e(X)) = X$.

Let K be a **Q**-algebra, then the elements of K[[I]] without constant term form a commutative group $\mathscr{E}$ under addition. The elements of K[[I]] with constant term 1 form a commutative group $\mathscr{M}$ under multiplication (IV, p. 30). For each $f \in \mathscr{E}$, we can define the elements $e \circ f$ and $l \circ f$ of $\mathscr{E}$, and by Prop. 14 above, the mappings $f \mapsto l \circ f$ and $f \mapsto e \circ f$ are mutually inverse permutations of $\mathscr{E}$; clearly they are continuous. Since $\exp X = e(X) + 1$, we see that the exponential mapping $f \mapsto \exp f = e \circ f + 1$ is a continuous bijection of $\mathscr{E}$ onto $\mathscr{M}$. By formula (4) of IV, p. 29 and Prop. 13, we have $\exp(f + g) = (\exp f)(\exp g)$ for $f, g \in \mathscr{E}$. Thus *the exponential is an isomorphism of the topological group* $\mathscr{E}$ *onto the topological group* $\mathscr{M}$.

The inverse isomorphism of $\mathscr{M}$ onto $\mathscr{E}$ is called the *logarithm* and is written $g \mapsto \log g$. We thus have $\log g = l(g - 1)$ for g in $\mathscr{M}$, and in particular,

(38) $$\log(1 + X) = l(X).$$

Since the logarithm is a homomorphism of $\mathscr{M}$ into $\mathscr{E}$, the formula $(1 + X)(1 + Y) = 1 + (X + Y + XY)$ implies

(39) $$l(X) + l(Y) = l(X + Y + XY).$$

Let $(u_\lambda)_{\lambda \in L}$ be a summable family of elements of $\mathscr{E}$, then the family $(\exp u_\lambda)_{\lambda \in L}$ is multipliable and we have

(40) $$\exp\left(\sum_{\lambda \in L} u_\lambda\right) = \prod_{\lambda \in L} \exp u_\lambda.$$

Similarly, if $(f_\lambda)_{\lambda \in L}$ is a multipliable family of elements of $\mathscr{M}$, the family $(\log f_\lambda)_{\lambda \in L}$ is summable and we have

$$\log\left(\prod_{\lambda \in L} f_\lambda\right) = \sum_{\lambda \in L} \log f_\lambda . \tag{41}$$

Let $g \in \mathscr{M}$, and let D be a continuous derivation of K[[I]]. We have $\log g = l(g-1)$, hence by Cor. 3 of IV, p. 33 and (37) we have

$$\mathrm{D} \log g = \mathrm{D}(g)/g . \tag{42}$$

The expression $\mathrm{D}(g)/g$ is called the *logarithmic derivative* of g (relative to D).

§ 5. SYMMETRIC TENSORS AND POLYNOMIAL MAPPINGS

1. Relative traces

Let H be a group and M a left A[H]-module [1]. We shall denote by M^H the set of all $m \in \mathrm{M}$ such that $hm = m$ for all $h \in \mathrm{H}$ [2]; this is a sub-A-module of M.

Let G be a subgroup of H, then M^G is a sub-A-module of M containing M^H.

Given $m \in \mathrm{M}^\mathrm{G}$, $h \in \mathrm{H}$, if $x = h\mathrm{G}$ is the left coset of h modulo G, then we have $xm = h\mathrm{G}m = \{hm\}$. By abuse of notation the element hm of M will be written xm. If $h' \in \mathrm{H}$, we have

$$h'(xm) = (h'x)\, m . \tag{1}$$

Suppose from now on that G is of finite index in H. Then

$$\sum_{x \in \mathrm{H}/\mathrm{G}} xm \in \mathrm{M}^\mathrm{H} . \tag{2}$$

For, given $h' \in \mathrm{H}$, we have by virtue of (1),

$$h'\left(\sum_{x \in \mathrm{H}/\mathrm{G}} xm\right) = \sum_{x \in \mathrm{H}/\mathrm{G}} (h'x)\, m = \sum_{y \in \mathrm{H}/\mathrm{G}} ym .$$

[1] We denote by A[H] the group algebra of H (III, p. 446).

[2] Care should be taken not to confuse this notation with that introduced in the study of products of sets (*Set Theory*, II, p. 102).

DEFINITION 1. — *If* G *is of finite index in* H, *we denote by* $\mathrm{Tr}_{H/G}$ *the mapping of* M^G *into* M^H *defined by*

$$\mathrm{Tr}_{H/G}\, m = \sum_{x \in H/G} xm . \tag{3}$$

This mapping is a homomorphism of the A-module M^G into the A-module M^H.

PROPOSITION 1. — (i) *Let* $m \in M^G$ *and* $h \in H$. *Then* $hm \in M^{hGh^{-1}}$ *and*

$$\mathrm{Tr}_{H/hGh^{-1}}(hm) = \mathrm{Tr}_{H/G}\, m .$$

(ii) *Let* F *be a subgroup of* G *of finite index in* G, *and let* $m \in M^F$, *then*

$$\mathrm{Tr}_{H/G}(\mathrm{Tr}_{G/F}\, m) = \mathrm{Tr}_{H/F}\, m .$$

(iii) *If* $m \in M^H$, *then* $\mathrm{Tr}_{H/G}\, m = (H:G)\, m$.

(i) Let $h \in H$. For $h' \in H$ and $m \in M$ let us put $\varphi(h') = hh'h^{-1}$ and $\psi(m) = hm$. We have $\varphi(h')\,\psi(m) = \psi(h'm)$; by transfer of structure we deduce that if $m \in M^G$, then $hm \in M^{hGh^{-1}}$ and

$$\mathrm{Tr}_{H/hGh^{-1}}(hm) = \psi(\mathrm{Tr}_{H/G}(m)) .$$

Since $\mathrm{Tr}_{H/G}(m) \in M^H$, this proves (i).

(ii) Let $m \in M^F$ and let $(g_\alpha)_{\alpha \in A}$ be a system of representatives of the left cosets of G mod F, and $(h_\beta)_{\beta \in B}$ a system of representatives of the left cosets of H mod G. Then $(h_\beta g_\alpha)_{(\beta,\alpha) \in B \times A}$ is a system of representatives of the left cosets of H mod F, so

$$\begin{aligned}\mathrm{Tr}_{H/G}(\mathrm{Tr}_{G/F}\, m) &= \sum_{\beta \in B} h_\beta \Big(\sum_{\alpha \in A} g_\alpha m \Big) \\ &= \sum_{(\beta,\alpha) \in B \times A} (h_\beta g_\alpha)\, m = \mathrm{Tr}_{H/F}\, m .\end{aligned}$$

(iii) This assertion is evident.

2. Definition of symmetric tensors

Let M be an A-module. We recall (III, p. 501) that $\mathfrak{S}_n$ operates on the left of the A-module $\mathbf{T}^n(M)$, in such a way that

$$\sigma(x_1 \otimes x_2 \otimes \dots \otimes x_n) = x_{\sigma^{-1}(1)} \otimes x_{\sigma^{-1}(2)} \otimes \dots \otimes x_{\sigma^{-1}(n)}$$

for any $x_1, \dots, x_n \in M$ and $\sigma \in \mathfrak{S}_n$. The elements $z \in \mathbf{T}^n(M)$ such that $\sigma . z = z$ for all $\sigma \in \mathfrak{S}_n$ are called *symmetric tensors of order n*; they form a sub-A-module of $\mathbf{T}^n(M)$ denoted by $\mathbf{TS}^n(M)$; we have $\mathbf{TS}^0(M) = A$, $\mathbf{TS}^1(M) = M$. We shall put

$\mathbf{TS}(M) = \bigoplus_{n=0}^{\infty} \mathbf{TS}^n(M)$; this is a graded sub-A-module of $\mathbf{T}(M)$. For every $z \in \mathbf{T}^n(M)$ the element $\sum_{\sigma \in \mathfrak{S}_n} \sigma . z$ belongs to $\mathbf{TS}^n(M)$; we denote it by $\boldsymbol{s} . z$ and call it the *symmetrization* of z. The mapping $\boldsymbol{s} : z \mapsto \boldsymbol{s} . z$ is a homomorphism of the A-module $\mathbf{T}^n(M)$ into the A-module $\mathbf{TS}^n(M)$. If $z \in \mathbf{TS}^n(M)$, then $\boldsymbol{s} . z = n!\, z$.

3. Product for symmetric tensors

Let p, $q \in \mathbf{N}$ and let $\mathfrak{S}_{p|q}$ be the subgroup of $\mathfrak{S}_{p+q}$ consisting of all permutations $\sigma \in \mathfrak{S}_{p+q}$ which leave the intervals $[1, p]$ and $[p+1, p+q]$ of $\mathbf{N}$ stable. If $\sigma \in \mathfrak{S}_p$ and $\sigma' \in \mathfrak{S}_q$, we can define an element σ'' of $\mathfrak{S}_{p|q}$ by putting $\sigma''(n) = \sigma(n)$ for $1 \leqslant n \leqslant p$ and $\sigma''(p+n) = p + \sigma'(n)$ for $1 \leqslant n \leqslant q$; the mapping $(\sigma, \sigma') \mapsto \sigma''$ is an isomorphism of $\mathfrak{S}_p \times \mathfrak{S}_q$ onto $\mathfrak{S}_{p|q}$.

Let $z \in \mathbf{TS}^p(M)$, $z' \in \mathbf{TS}^q(M)$, then the element $z \otimes z'$ of $\mathbf{T}^{p+q}(M)$ is invariant under $\mathfrak{S}_{p|q}$; we can therefore define the element $\mathrm{Tr}_{\mathfrak{S}_{p+q}/\mathfrak{S}_{p|q}}(z \otimes z')$ of $\mathbf{TS}^{p+q}(M)$. We shall equip $\mathbf{TS}(M)$ with the A-bilinear multiplication $(y, y') \mapsto yy'$ such that for p, $q \in \mathbf{N}$, $z \in \mathbf{TS}^p(M)$, $z' \in \mathbf{TS}^q(M)$ we have

$$zz' = \mathrm{Tr}_{\mathfrak{S}_{p+q}/\mathfrak{S}_{p|q}}(z \otimes z') . \tag{4}$$

If $y \in \mathbf{TS}(M)$ and $y' \in \mathbf{TS}(M)$, we shall call yy' the *symmetric product* of y and y'. The family $(\mathbf{TS}^p(M))_{p \in \mathbf{N}}$ is a graduation of type $\mathbf{N}$ of the algebra $\mathbf{TS}(M)$, and the unit element of $\mathbf{T}(M)$ is a unit element of $\mathbf{TS}(M)$.

Let $\mathfrak{S}_{p,q}$ be the set of those $\sigma \in \mathfrak{S}_{p+q}$ such that

$$\sigma(1) < \sigma(2) < \dots < \sigma(p)$$
$$\sigma(p+1) < \sigma(p+2) < \dots < \sigma(p+q) .$$

The mapping $(\sigma, \tau) \mapsto \sigma\tau$ of $\mathfrak{S}_{p,q} \times \mathfrak{S}_{p|q}$ into $\mathfrak{S}_{p+q}$ is bijective (I, p. 60, Example 2) ; hence if $z \in \mathbf{TS}^p(M)$ and $z' \in \mathbf{TS}^q(M)$, we have

$$zz' = \sum_{\sigma \in \mathfrak{S}_{p,q}} \sigma(z \otimes z') . \tag{5}$$

PROPOSITION 2. — (i) *The* A-*algebra* $\mathbf{TS}(M)$ *is associative, commutative and unital.*

(ii) *Let* $p_1, \dots, p_n$ *be integers* > 0, *and let* $\mathfrak{S}_{p_1|\dots|p_n}$ *be the set of all* $\sigma \in \mathfrak{S}_{p_1+\dots+p_n}$ *leaving the following intervals of* $\mathbf{N}$ *stable :*

$$[1, p_1], [p_1+1, p_1+p_2], \dots, [p_1+\dots+p_{n-1}+1, p_1+\dots+p_n] .$$

Let $z_1 \in \mathbf{TS}^{p_1}(\mathrm{M}), \ldots, z_n \in \mathbf{TS}^{p_n}(\mathrm{M})$, *then*

$$z_1 z_2 \ldots z_n = \mathrm{Tr}_{\mathfrak{S}_{p_1+\cdots+p_n}/\mathfrak{S}_{p_1|\cdots|p_n}}(z_1 \otimes z_2 \otimes \ldots \otimes z_n)\,.$$

In particular if $x_1, \ldots, x_n \in \mathrm{M}$, *we have* $x_1 \ldots x_n = s(x_1 \otimes \ldots \otimes x_n)$.

The assertion (ii) is clear for $n = 1$. Assume that the relation

$$z_2 \ldots z_n = \mathrm{Tr}_{\mathfrak{S}_{p_2+\cdots+p_n}/\mathfrak{S}_{p_2|\cdots|p_n}}(z_2 \otimes \ldots \otimes z_n)$$

has been proved, and let us identify $\mathfrak{S}_{p_2+\cdots+p_n}$ with the subgroup of $\mathfrak{S}_{p_1+\cdots+p_n}$ consisting of all permutations whose restriction to $(1, p_1)$ is the identity. Then

$$\mathrm{Tr}_{\mathfrak{S}_{p_1|p_2+\cdots+p_n}/\mathfrak{S}_{p_1|p_2|\cdots|p_n}}(z_1 \otimes z_2 \otimes \ldots \otimes z_n) =$$
$$= z_1 \otimes \mathrm{Tr}\,_{\mathfrak{S}_{p_2+\cdots+p_n}/\mathfrak{S}_{p_2|\cdots|p_n}}(z_2 \otimes \ldots \otimes z_n) = z_1 \otimes (z_2 \ldots z_n)\,.$$

Hence we have

$$z_1 z_2 \ldots z_n = z_1(z_2 \ldots z_n) =$$
$$= \mathrm{Tr}_{\mathfrak{S}_{p_1+p_2+\cdots+p_n}/\mathfrak{S}_{p_1|p_2+\cdots+p_n}}(z_1 \otimes (z_2 \ldots z_n))$$
$$= \mathrm{Tr}_{\mathfrak{S}_{p_1+\cdots+p_n}/\mathfrak{S}_{p_1|p_2+\cdots+p_n}}(\mathrm{Tr}_{\mathfrak{S}_{p_1|p_2+\cdots+p_n}/\mathfrak{S}_{p_1|p_2|\cdots|p_n}}(z_1 \otimes z_2 \otimes \ldots \otimes z_n))$$
$$= \mathrm{Tr}_{\mathfrak{S}_{p_1+\cdots+p_n}/\mathfrak{S}_{p_1|\cdots|p_n}}(z_1 \otimes z_2 \otimes \ldots \otimes z_n)$$

by Prop. 1, (ii) of IV, p. 42. Thus (ii) is established.

In particular,

$$z_1(z_2 z_3) = \mathrm{Tr}_{\mathfrak{S}_{p_1+p_2+p_3}/\mathfrak{S}_{p_1|p_2|p_3}}(z_1 \otimes z_2 \otimes z_3)\,,$$

and in the same way we show that

$$(z_1 z_2) z_3 = \mathrm{Tr}_{\mathfrak{S}_{p_1+p_2+p_3}/\mathfrak{S}_{p_1|p_2|p_3}}(z_1 \otimes z_2 \otimes z_3)\,.$$

Hence the algebra $\mathbf{TS}(\mathrm{M})$ is associative.

Let σ be the element of $\mathfrak{S}_{p_1+p_2}$ such that

$$\sigma(1) = p_2 + 1, \sigma(2) = p_2 + 2, \ldots, \sigma(p_1) = p_2 + p_1\,,$$
$$\sigma(p_1 + 1) = 1, \sigma(p_1 + 2) = 2, \ldots, \sigma(p_1 + p_2) = p_2\,.$$

Then

$$\begin{aligned} z_2 z_1 &= \mathrm{Tr}_{\mathfrak{S}_{p_1+p_2}/\mathfrak{S}_{p_2|p_1}}(z_2 \otimes z_1) \\ &= \mathrm{Tr}_{\mathfrak{S}_{p_1+p_2}/\sigma\mathfrak{S}_{p_1|p_2}\sigma^{-1}}\sigma(z_1 \otimes z_2) \\ &= \mathrm{Tr}_{\mathfrak{S}_{p_1+p_2}/\mathfrak{S}_{p_1|p_2}}(z_1 \otimes z_2) \quad \text{by Prop 1, (i)} \\ &= z_1 z_2\,. \end{aligned}$$

Hence the algebra **TS**(M) is commutative.

It is to be noted that the canonical injection of **TS**(M) in **T**(M) is not in general an algebra homomorphism. Worse still, **TS**(M) is not in general stable under the multiplication of **T**(M).

4. Divided powers

Let $x \in \mathrm{M}$ and $k \in \mathbf{N}$. It is clear that $x_1 \otimes x_2 \otimes \ldots \otimes x_k$, where

$$x_1 = x_2 = \cdots = x_k = x \,,$$

is an element of $\mathbf{TS}^k(\mathrm{M})$.

DEFINITION 2. — *If $x \in \mathrm{M}$, the element $x \otimes x \otimes \ldots \otimes x$ of $\mathbf{TS}^k(\mathrm{M})$ is denoted by $\gamma_k(x)$.*

PROPOSITION 3. — (i) *If $x \in \mathrm{M}$, the* pth *power of x, calculated in* **TS**(M), *is equal to $p!\,\gamma_p(x)$.*

(ii) *Let $x_1, \ldots, x_n \in \mathrm{M}$; then*

$$\gamma_p(x_1 + x_2 + \cdots + x_n) = \sum_{p_1 + p_2 + \cdots + p_n = p} \gamma_{p_1}(x_1)\,\gamma_{p_2}(x_2) \ldots \gamma_{p_n}(x_n)\,.$$

(iii) *Let $x_1, \ldots, x_n \in \mathrm{M}$, let $p_1, \ldots, p_n$ be integers $\geqslant 0$ and $p = p_1 + \cdots + p_n$. Let* E *be the set of mappings φ of $\{1, \ldots, p\}$ into $\{1, \ldots, n\}$ such that*

$$\mathrm{Card}\,\varphi^{-1}(1) = p_1, \ldots, \mathrm{Card}\,\varphi^{-1}(n) = p_n\,.$$

Then

$$\gamma_{p_1}(x_1)\,\gamma_{p_2}(x_2) \ldots \gamma_{p_n}(x_n) = \sum_{\varphi \in \mathrm{E}} x_{\varphi(1)} \otimes x_{\varphi(2)} \otimes \cdots \otimes x_{\varphi(p)}\,.$$

(iv) *Let $x \in \mathrm{M}$ and let q, r be integers $\geqslant 0$. Then*

$$\gamma_q(x)\,\gamma_r(x) = \frac{(q+r)!}{q!\,r!}\,\gamma_{q+r}(x)\,.$$

(v) *Let $x_1, \ldots, x_n \in \mathrm{M}$, and for $\mathrm{H} \subset \{1, \ldots, n\}$ put $x_{\mathrm{H}} = \sum_{i \in \mathrm{H}} x_i$, then*

$$(-1)^n x_1 x_2 \ldots x_n = \sum_{\mathrm{H} \subset \{1, \ldots, n\}} (-1)^{\mathrm{Card}\,\mathrm{H}}\,\gamma_n(x_{\mathrm{H}})\,.$$

The assertion (i) follows at once from Prop. 2 (ii).

Let us prove (ii) ; by an induction on n we see that it is enough to consider the case $n = 2$. Then we have

$$\begin{aligned}\gamma_p(x_1 + x_2) &= (x_1 + x_2) \otimes (x_1 + x_2) \otimes \dots \otimes (x_1 + x_2) \quad (p \text{ factors})\\ &= \sum_{p_1 + p_2 = p} \sum_{\sigma \in \mathfrak{S}_{p_1, p_2}} \sigma(x_1 \otimes x_1 \otimes \dots \otimes x_1 \otimes x_2 \otimes x_2 \otimes \dots \otimes x_2)\\ &= \sum_{p_1 + p_2 = p} \sum_{\sigma \in \mathfrak{S}_{p_1, p_2}} \sigma(\gamma_{p_1}(x_1) \otimes \gamma_{p_2}(x_2))\\ &= \sum_{p_1 + p_2 = p} \gamma_{p_1}(x_1)\, \gamma_{p_2}(x_2)\,.\end{aligned}$$

To prove (iii), let $\mathfrak{S}_{p_1, \dots, p_n}$ be the set of permutations of $[1, p_1 + \dots + p_n]$ whose restrictions to the intervals

$$[1, p_1], [p_1 + 1, p_1 + p_2], \dots, [p_1 + \dots + p_{n-1} + 1, p_1 + \dots + p_n]$$

are increasing. By I, p. 60, example 2 and Prop. 2, (ii) we have

$$\gamma_{p_1}(x_1)\, \gamma_{p_2}(x_2) \dots \gamma_{p_n}(x_n) = \sum_{\rho \in \mathfrak{S}_{p_1, p_2, \dots, p_n}} \rho(x_1 \otimes x_1 \otimes \dots \otimes x_1 \otimes x_2 \otimes x_2 \otimes \dots \otimes x_2 \otimes \dots \otimes x_n \otimes x_n \otimes \dots \otimes x_n)$$

(with p_i factors x_i) and this sum is equal to

$$\sum_{\varphi \in \mathrm{E}} x_{\varphi(1)} \otimes x_{\varphi(2)} \otimes \dots \otimes x_{\varphi(p)}\,.$$

In (iii) let us put $n = 2$, $x_1 = x_2 = x$, $p_1 = q$ and $p_2 = r$, then we obtain (iv) (I, *loc. cit.*).

Finally (v) follows from Prop. 2, (ii) and Prop. 2 of I, p. 100, applied to the elements x_i of the ring $\mathbf{T}(\mathrm{M})$.

Remarks. — 1) Let $(x_i)_{i \in \mathrm{I}}$ be a family of elements of M. For each $\nu \in \mathbf{N}^{(\mathrm{I})}$ put

$$x_\nu = \prod_{i \in \mathrm{I}} \gamma_{\nu_i}(x_i).$$

If $(\lambda_i) \in \mathrm{A}^{(\mathrm{I})}$ and $p \in \mathbf{N}$, then we have by Prop. 3 (ii),

$$\gamma_p\Big(\sum_{i \in \mathrm{I}} \lambda_i x_i\Big) = \sum_{\nu \in \mathbf{N}^{(\mathrm{I})},\ |\nu| = p} \lambda^\nu x_\nu\,. \tag{6}$$

2) Let $\mathscr{M}$ be the set of mappings of $[1, p]$ into I. We define a mapping $\rho \mapsto \rho^*$ of $\mathscr{M}$ into $\mathbf{N}^{(\mathrm{I})}$ by putting

$$\rho^*(i) = \operatorname{Card} \rho^{-1}(i)\,.$$

For two elements ρ_1, ρ_2 of $\mathscr{M}$ to satisfy $\rho_1^* = \rho_2^*$ it is necessary and sufficient that there should exist $\sigma \in \mathfrak{S}_p$ such that $\rho_2 = \rho_1 \circ \sigma$ (I, p. 95). By Prop. 3 (iii) we have, for $|\nu| = p$,

(7) $$x_\nu = \sum_{\rho \in \mathscr{M}, \rho^* = \nu} x_{\rho(1)} \otimes x_{\rho(2)} \otimes \dots \otimes x_{\rho(p)} .$$

5. Symmetric tensors over a free module

PROPOSITION 4. — *Let* M *be free and* $(e_i)_{i \in I}$ *a basis of* M.

(i) *For* $\nu \in \mathbf{N}^{(I)}$ *let* $e_\nu = \prod_{i \in I} \gamma_{\nu_i}(e_i)$. *Then* $(e_\nu)_{\nu \in \mathbf{N}^{(I)}}$ *is a basis of the* A*-module* $\mathbf{TS}(M)$. *In particular the algebra* $\mathbf{TS}(M)$ *is generated by the family of elements* $\gamma_k(x)$ *for* $k \in \mathbf{N}$ *and* $x \in M$.

(ii) *For each* $p \in \mathbf{N}$, $\mathbf{TS}^p(M)$ *is a direct factor of the* A*-module* $\mathbf{T}^p(M)$.

Let us use the notation of the Remark 2 above. The family $(e_{\rho(1)} \otimes \dots \otimes e_{\rho(p)})_{\rho \in \mathscr{M}}$ is a basis of $\mathbf{T}^p(M)$. Hence Prop. 4 follows from formula (7) and the following lemma, applied with $H = \mathfrak{S}_p$ and $U = \mathbf{T}^p(M)$.

Lemma 1. — *Let* H *be a finite group and* U *a left* A[H]*-module. Suppose that the* A*-module* U *has a basis* B *which is stable under the operations of* H *in* U, *and put* $\Omega = B/H$. *For each* $\omega \in \Omega$ *let* $u_\omega = \sum_{b \in \omega} b$; *then*

(i) $(u_\omega)_{\omega \in \Omega}$ *is a basis of the* A*-module* U^H.

(ii) *For each* $\omega \in \Omega$ *let* v_ω *be a point of* ω; *put* $\omega' = \omega - \{v_\omega\}$ *and* $B' = \bigcup_{\omega \in \Omega} \omega'$, *then* B′ *is a basis of a supplementary subspace for* U^H *in* U.

The union of the set of all u_ω (for $\omega \in \Omega$) and of B′ is a basis of U. If $U' = \sum_{\omega \in \Omega} Au_\omega$ and $U'' = \sum_{b \in B'} Ab$, we therefore have $U = U' \oplus U''$. On the other hand, we have $u_\omega \in U^H$ for all $\omega \in \Omega$, hence $U' \subset U^H$. Finally, let $(\alpha_b)_{b \in B}$ be a family of elements of A with finite support and let $x = \sum_{b \in B} \alpha_b b$. If $x \in U^H$, then $\alpha_{hb} = \alpha_b$ for all $b \in B$ and all $h \in H$, hence $x \in U'$, and it follows that $U' = U^H$.

PROPOSITION 5. — *Let* M *be a free* A*-module*, k *an integer* $\geqslant 0$, P *the sub-*A*-module of* $\mathbf{TS}^k(M)$ *generated by* $\gamma_k(M)$. *Assume* A *to be an infinite integral domain. Then for each* $z \in \mathbf{TS}^k(M)$ *there exists* $\alpha \in A - \{0\}$ *such that* $\alpha z \in P$.

Let K be the field of fractions of A. We identify $\mathbf{TS}^k(M)$ with a sub-A-module of the vector K-space $V = \mathbf{TS}^k(M) \otimes_A K$ (Prop. 4, and II, p. 314). We have to show that this vector K-space is generated by $\gamma_k(M)$, that is, every K-linear form f

on V satisfying $f(\gamma_k(\mathrm{M})) = 0$ is zero. Let $(e_i)_{i \in \mathrm{I}}$ be a basis of M, and define the e_ν as in Prop. 4. For any $(\alpha_i) \in \mathrm{A}^{(\mathrm{I})}$ we have, on taking (6) into account,

$$0 = f\left(\gamma_k\left(\sum_{i \in \mathrm{I}} \alpha_i e_I\right)\right) = \sum_{\nu \in \mathbf{N}^{(\mathrm{I})}, |\nu| = k} \alpha^\nu f(e_\nu) .$$

By Cor. 2 of IV, p. 18 it follows that $f(e_\nu) = 0$ for all $\nu \in \mathbf{N}^{(\mathrm{I})}$, whence $f = 0$.

6. The functor TS

Let M, N be A-modules and u a homomorphism of M into N ; it is clear that $\mathbf{T}(u)(\mathbf{TS}(\mathrm{M})) \subset \mathbf{TS}(\mathrm{N})$. The mapping of $\mathbf{TS}(\mathrm{M})$ into $\mathbf{TS}(\mathrm{N})$ obtained from $\mathbf{T}(u)$ is denoted by $\mathbf{TS}(u)$. It is easily verified that this is a unital homomorphism of graded algebras and we have $\mathbf{TS}(u)(\gamma_p(x)) = \gamma_p(u(x))$ for all $x \in \mathrm{M}$ and each integer $p \geqslant 0$. If $v : \mathrm{N} \to \mathrm{P}$ is a homomorphism of A-modules, we have

$$\mathbf{TS}(v \circ u) = \mathbf{TS}(v) \circ \mathbf{TS}(u) .$$

By definition of $\mathbf{TS}(u)$, the diagram

$$\begin{array}{ccccc} \mathrm{M} & \longrightarrow & \mathbf{TS}(\mathrm{M}) & \longrightarrow & \mathbf{T}(\mathrm{M}) \\ \downarrow{\scriptstyle u} & & \downarrow{\scriptstyle \mathbf{TS}(u)} & & \downarrow{\scriptstyle \mathbf{T}(u)} \\ \mathrm{N} & \longrightarrow & \mathbf{TS}(\mathrm{N}) & \longrightarrow & \mathbf{T}(\mathrm{N}) \end{array}$$

is commutative, where the horizontal arrows denote canonical injections.

If M is a direct factor of N and $i : \mathrm{M} \to \mathrm{N}$ is the canonical injection, then $\mathbf{TS}(i)$ is an injective homomorphism of $\mathbf{TS}(\mathrm{M})$ onto a direct factor R of $\mathbf{TS}(\mathrm{N})$, by which we normally identify $\mathbf{TS}(\mathrm{M})$ and R. That is proved as for the tensor algebra (III, p. 487).

Suppose that M is a direct sum of a family $(\mathrm{M}_\lambda)_{\lambda \in \mathrm{L}}$ of submodules. The canonical injections $\mathbf{TS}(\mathrm{M}_\lambda) \to \mathbf{TS}(\mathrm{M})$ define a unital homomorphism h of graded algebras, called canonical :

$$\bigotimes_{\lambda \in \mathrm{L}} \mathbf{TS}(\mathrm{M}_\lambda) \to \mathbf{TS}(\mathrm{M}) .$$

Let $\lambda_1, \lambda_2, \ldots, \lambda_n$ be pairwise distinct elements of L and let $x_1 \in \mathrm{M}_{\lambda_1}, \ldots, x_n \in \mathrm{M}_{\lambda_n}$. By Prop. 3, (ii) of IV, p. 45 we have

$$h\left(\sum_{p_1 + \cdots + p_n = p} \gamma_{p_1}(x_1) \otimes \ldots \otimes \gamma_{p_n}(x_n)\right) = \gamma_p(x_1 + \cdots + x_n) . \tag{8}$$

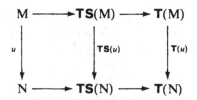

Let N be an A-module which is a direct sum of a family $(N_\lambda)_{\lambda \in L}$ of submodules. For every $\lambda \in L$ let u_λ be a homomorphism of M_λ into N_λ. Let u be the homomorphism of M into N defined by the u_λ. Then the diagram

$$\begin{array}{ccc} \bigotimes \mathbf{TS}(M_\lambda) & \xrightarrow{h} & \mathbf{TS}(M) \\ \Big\downarrow {\scriptstyle \otimes \mathbf{TS}(u_\lambda)} & & \Big\downarrow {\scriptstyle \mathbf{TS}(u)} \\ \bigotimes \mathbf{TS}(N_\lambda) & \xrightarrow{h'} & \mathbf{TS}(N) \end{array}$$

commutes, where h and h' are canonical homomorphisms. For if $z \in \mathbf{TS}(M_\lambda)$ and if i_λ (resp. j_λ) denotes the canonical injection of M_λ into M (resp. N_λ into N), then we have

$$\mathbf{TS}(u)\,(h(z)) = \mathbf{TS}(u) \circ \mathbf{TS}(i_\lambda)(z) = \mathbf{TS}(u \circ i_\lambda)(z) =$$
$$= \mathbf{TS}(j_\lambda \circ u_\lambda)(z) = \mathbf{TS}(j_\lambda) \circ \mathbf{TS}(u_\lambda)(z) = h'(\mathbf{TS}(u_\lambda)(z))\,.$$

PROPOSITION 6. — *Let* M *be an* A-*module which is the direct sum of a family* $(M_\lambda)_{\lambda \in L}$ *of submodules. If each* M_λ *is a free module, then the canonical homomorphism of* $\bigotimes_{\lambda \in L} \mathbf{TS}(M_\lambda)$ *into* $\mathbf{TS}(M)$ *is an isomorphism.*

Let $(e_{i,\lambda})_{i \in I_\lambda}$ be a basis of M_λ. For $\nu \in \mathbf{N}^{(I_\lambda)}$ put $e_{\nu,\lambda} = \prod_{i \in I_\lambda} \gamma_{\nu(i)}(e_{i,\lambda})$. The $e_{\nu,\lambda}$ for $\nu \in \mathbf{N}^{(I_\lambda)}$ form a basis of $\mathbf{TS}(M_\lambda)$ (IV, p. 47, Prop. 4 (i)) and $e_{0,\lambda}$ is the unit-element of $\mathbf{TS}(M_\lambda)$. Therefore the elements

$$\bigotimes_{\lambda \in L} e_{\nu_\lambda,\lambda} \tag{9}$$

where $\nu_\lambda \in \mathbf{N}^{(I_\lambda)}$ and $\nu_\lambda = 0$ except for a finite number of indices, form a basis of $\bigotimes_\lambda \mathbf{TS}(M_\lambda)$. The image of the element (9) under the canonical homomorphism of the proposition is $\prod_{\lambda \in L} e_{\nu_\lambda,\lambda}$. If we denote by $(e_i)_{i \in I}$ the disjoint union of the families $(e_{i,\lambda})_{i \in I_\lambda}$, then the above elements are precisely $\prod_{i \in I} \gamma_{\nu(i)}(e_i)$, where $\nu \in \mathbf{N}^{(I)}$, and so they constitute a basis of $\mathbf{TS}(M)$. This proves the proposition.

Under the conditions of Prop. 6, the inverse isomorphism $\mathbf{TS}(M) \to \bigotimes_\lambda \mathbf{TS}(M_\lambda)$ is also called *canonical.* Frequently $\mathbf{TS}(M)$ is identified with $\bigotimes_\lambda \mathbf{TS}(M_\lambda)$ by means of this isomorphism. It should be noted that if $z \in \mathbf{TS}(M_\lambda)$

and $z' \in \mathbf{TS}(\mathrm{M}_\mu)$ with $\lambda \neq \mu$, then the element of $\mathbf{TS}(\mathrm{M})$ which we are thus led to denote by $z \otimes z'$ is not the tensor product of z and z' in $\mathbf{T}(\mathrm{M})$ but the symmetric product of z and z'.

PROPOSITION 7. — *Let* M *be an* A*-module,* u *the mapping* $(x, y) \mapsto x + y$ *of* $\mathrm{M} \oplus \mathrm{M}$ *into* M *and* f *the composite mapping*

$$\mathbf{TS}(\mathrm{M}) \otimes \mathbf{TS}(\mathrm{M}) \xrightarrow{h} \mathbf{TS}(\mathrm{M} \oplus \mathrm{M}) \xrightarrow{\mathbf{TS}(u)} \mathbf{TS}(\mathrm{M})$$

where h *is the canonical homomorphism. If* $z, z' \in \mathbf{TS}(\mathrm{M})$, *then* $f(z \otimes z') = zz'$.

For let i be the mapping $x \mapsto (x, 0)$ of M into $\mathrm{M} \oplus \mathrm{M}$. We have $u \circ i = \mathrm{Id}_\mathrm{M}$, hence $\mathbf{TS}(u) \circ \mathbf{TS}(i) = \mathrm{Id}_{\mathbf{TS}(\mathrm{M})}$; therefore

$$f(z \otimes 1) = \mathbf{TS}(u)\,(h(z \otimes 1)) = \mathbf{TS}(u)\,(\mathbf{TS}(i)(z)) = z\,.$$

Likewise $f(1 \otimes z') = z'$, whence $f(z \otimes z') = f(z \otimes 1)\, f(1 \otimes z') = zz'$.

7. Coproduct for symmetric tensors

Let M be a *free* A-module, and $\Delta_\mathrm{M} = \Delta$ the diagonal homomorphism $x \mapsto (x, x)$ of M into $\mathrm{M} \oplus \mathrm{M}$. Let $c_\mathrm{M} = c$ be the unital homomorphism of graded A-algebras composed of the homomorphisms :

$$\mathbf{TS}(\mathrm{M}) \xrightarrow{\mathbf{TS}(\Delta)} \mathbf{TS}(\mathrm{M} \oplus \mathrm{M}) \xrightarrow{\sigma} \mathbf{TS}(\mathrm{M}) \otimes \mathbf{TS}(\mathrm{M})$$

where σ is the canonical isomorphism. Equipped with c, $\mathbf{TS}(\mathrm{M})$ is a graded A-cogebra.

For all $x \in \mathrm{M}$ and each integer $p \geqslant 0$ we have $\mathbf{TS}(\Delta)\,(\gamma_p(x)) = \gamma_p((x, x))$, hence by (8),

(10) $$c(\gamma_p(x)) = \sum_{r+s=p} \gamma_r(x) \otimes \gamma_s(x)\,.$$

In particular

(11) $$c(x) = x \otimes 1 + 1 \otimes x\,.$$

Let $(x_i)_{i \in \mathrm{I}}$ be a family of elements of M, and for $\nu \in \mathbf{N}^{(\mathrm{I})}$ put $x_\nu = \prod_{i \in \mathrm{I}} \gamma_{\nu_i}(x_i)$.

Then

(12) $$c(x_\nu) = \sum_{\rho + \sigma = \nu} x_\rho \otimes x_\sigma\,.$$

This follows from (10) because c is an algebra homomorphism.

PROPOSITION 8. — *Let* M *be a free* A*-module, then with its algebra and cogebra structures,* $\mathbf{TS}(\mathrm{M})$ *is a graded commutative and cocommutative bigebra. The counit is the* A*-linear mapping* $\varepsilon : \mathbf{TS}(\mathrm{M}) \to \mathbf{TS}^0(\mathrm{M}) = \mathrm{A}$ *which is zero on* $\mathbf{TS}^p(\mathrm{M})$ *for* $p > 0$ *and such that* $\varepsilon(1) = 1$.

We know that the A-algebra $\mathbf{TS}(\mathrm{M})$ is associative, commutative and unital. On the other hand, the coproduct is by construction a homomorphism of graded algebras ; now the fact that the cogebra $\mathbf{TS}(\mathrm{M})$ is coassociative and cocommutative follows by an easy calculation from the Formula (10). The mapping ε of $\mathbf{TS}(\mathrm{M})$ into A is a homomorphism of graded algebras such that $\varepsilon(1) = 1$. Finally for every $x \in \mathrm{M}$ we have $\varepsilon(\gamma_p(x)) = 0$ if $p > 0$, $\varepsilon(\gamma_0(x)) = 1$; if we bear in mind (10), this shows that $(\varepsilon \otimes 1) \circ c = (1 \otimes \varepsilon) \circ c = \mathrm{Id}_{\mathbf{TS}(\mathrm{M})}$; thus ε is the counit of $\mathbf{TS}(\mathrm{M})$.

PROPOSITION 9. — *Let* M *and* N *be free* A*-modules and* u *an* A*-homomorphism of* M *into* N ; *then* $\mathbf{TS}(u)$ *is a bigebra homomorphism.*

For we have $\Delta_{\mathrm{N}} \circ u = (u, u) \circ \Delta_{\mathrm{M}}$, hence the diagram

$$\begin{array}{ccccc}
\mathbf{TS}(\mathrm{M}) & \xrightarrow{\mathbf{TS}(\Delta_{\mathrm{M}})} & \mathbf{TS}(\mathrm{M} \oplus \mathrm{M}) & \xrightarrow{\sigma} & \mathbf{TS}(\mathrm{M}) \otimes \mathbf{TS}(\mathrm{M}) \\
\downarrow{\scriptstyle \mathbf{TS}(u)} & & \downarrow{\scriptstyle \mathbf{TS}(u,u)} & & \downarrow{\scriptstyle \mathbf{TS}(u) \otimes \mathbf{TS}(u)} \\
\mathbf{TS}(\mathrm{N}) & \xrightarrow[\mathbf{TS}(\Delta_{\mathrm{N}})]{} & \mathbf{TS}(\mathrm{N} \oplus \mathrm{N}) & \xrightarrow[\tau]{} & \mathbf{TS}(\mathrm{N}) \otimes \mathbf{TS}(\mathrm{N}) ,
\end{array}$$

where σ and τ are canonical isomorphisms, is commutative (IV, p. 49). Thus $c_{\mathrm{N}} \circ \mathbf{TS}(u) = (\mathbf{TS}(u) \otimes \mathbf{TS}(u)) \circ c_{\mathrm{M}}$.

PROPOSITION 10. — *Let* M *be a free* A*-module, then the primitive elements* (III, p. 602) *of the bigebra* $\mathbf{TS}(\mathrm{M})$ *are the elements of* M.

Let $(e_i)_{i \in \mathrm{I}}$ be a basis of M, and for $\nu \in \mathbf{N}^{(\mathrm{I})}$ put $e_\nu = \prod_{i \in \mathrm{I}} \gamma_{\nu_i}(e_i)$. Let $z = \sum_{\nu \in \mathbf{N}^{(\mathrm{I})}} \lambda_\nu e_\nu$ be an element of $\mathbf{TS}(\mathrm{M})$, then by (12) we have

$$c(z) = \sum_\nu \lambda_\nu \sum_{\rho, \sigma \in \mathbf{N}^{(\mathrm{I})}, \rho + \sigma = \nu} e_\rho \otimes e_\sigma = \sum_{\rho, \sigma} \lambda_{\rho + \sigma} e_\rho \otimes e_\sigma$$

hence

$$c(z) - 1 \otimes z - z \otimes 1 = \sum_{\rho \neq 0, \sigma \neq 0} \lambda_{\rho + \sigma} e_\rho \otimes e_\sigma - \lambda_0 e_0 \otimes e_0 .$$

and so

$$\begin{aligned}
z \text{ primitive} &\Leftrightarrow \lambda_{\rho + \sigma} = 0 \ \text{ when } \rho \neq 0 \text{ and } \sigma \neq 0 \text{ and } \lambda_0 = 0 \\
&\Leftrightarrow \lambda_\nu = 0 \quad \text{when } |\nu| \neq 1 \\
&\Leftrightarrow z \in \mathrm{M} .
\end{aligned}$$

8. Relations between **TS**(M) and **S**(M)

The canonical injection of M in **TS**(M) extends in a unique fashion to an algebra homomorphism of **T**(M) into **TS**(M) (III, p. 485, Prop. 1). By Prop. 2, (ii) of IV, p. 43, this homomorphism is the operator s of symmetrization. Since the algebra **TS**(M) is commutative there exists (III, p. 497) one and only one *algebra homomorphism* φ_M, called *canonical,* of the algebra **S**(M) into the algebra **TS**(M) such that the diagram

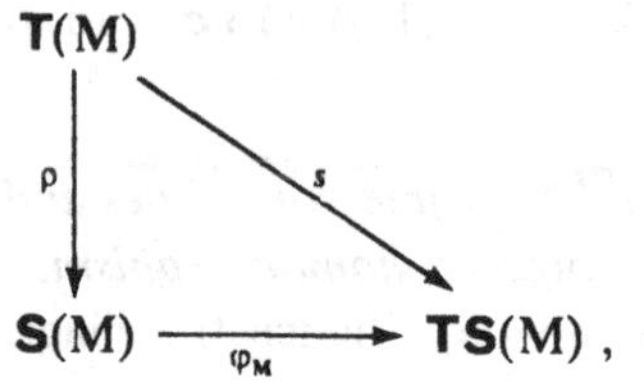

where ρ denotes the canonical homomorphism of **T**(M) onto **S**(M), is commutative. We have $\varphi_M(\mathbf{S}^p(M)) \subset \mathbf{TS}^p(M)$ for all $p \in \mathbf{N}$.

On the other hand, by composing the canonical injection i of **TS**(M) into **T**(M) with the canonical homomorphism ρ of **T**(M) onto **S**(M) we obtain a homomorphism ψ_M of *graded* A-*modules,* called *canonical.* The diagram

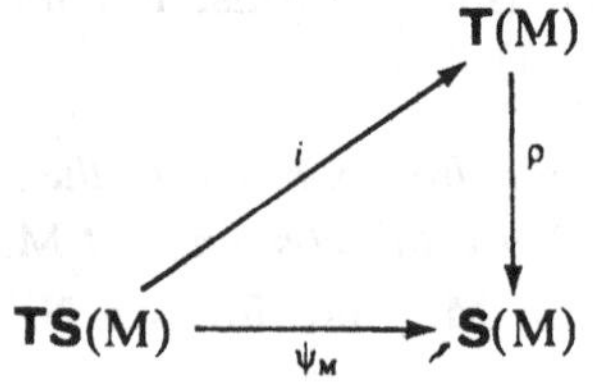

is commutative.

If $u : M \to N$ is a homomorphism of A-modules, then the diagram

(13)

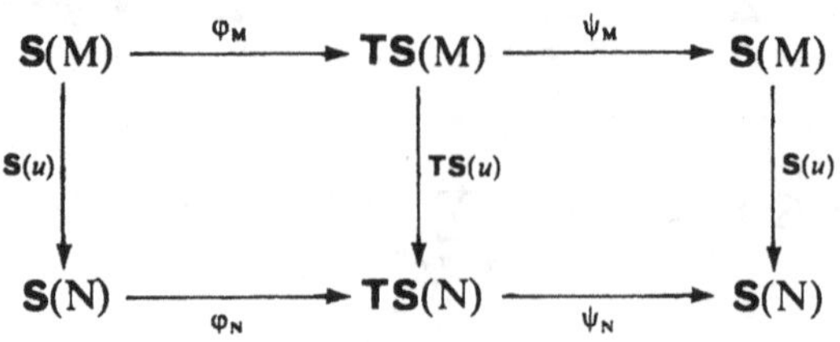

is commutative, as is easily verified.

If M is the direct sum of the modules M_λ, the diagram

$$\begin{array}{ccc} \otimes\, \mathbf{S}(\mathrm{M}_\lambda) & \xrightarrow{\otimes\, \varphi_{\mathrm{M}_\lambda}} & \otimes\, \mathbf{TS}(\mathrm{M}_\lambda) \\ \Big\downarrow f & & \Big\downarrow g \\ \mathbf{S}(\mathrm{M}) & \xrightarrow[\varphi_\mathrm{M}]{} & \mathbf{TS}(\mathrm{M}), \end{array} \tag{14}$$

where f and g are the canonical homomorphisms, is commutative. For $g \circ \bigotimes_\lambda \varphi_{\mathrm{M}_\lambda}$ and $\varphi_\mathrm{M} \circ f$ are algebra homomorphisms which coincide on M_λ for every λ.

PROPOSITION 11. — *If* M *is free, then* φ_M *is a morphism of graded bigebras.*

Using the commutativity of the diagrams (13) and (14), we obtain the commutative diagram

$$\begin{array}{ccccc} \mathbf{S}(\mathrm{M}) & \xrightarrow{\mathbf{S}(\Delta)} & \mathbf{S}(\mathrm{M} \oplus \mathrm{M}) & \xrightarrow{h} & \mathbf{S}(\mathrm{M}) \otimes \mathbf{S}(\mathrm{M}) \\ \Big\downarrow \varphi_\mathrm{M} & & \Big\downarrow \varphi_{\mathrm{M} \oplus \mathrm{M}} & & \Big\downarrow \varphi_\mathrm{M} \otimes \varphi_\mathrm{M} \\ \mathbf{TS}(\mathrm{M}) & \xrightarrow[\mathbf{TS}(\Delta)]{} & \mathbf{TS}(\mathrm{M} \oplus \mathrm{M}) & \xrightarrow[k]{} & \mathbf{TS}(\mathrm{M}) \otimes \mathbf{TS}(\mathrm{M}), \end{array}$$

where Δ is the diagonal homomorphism and h, k are canonical homomorphisms. The proposition follows from this.

PROPOSITION 12. — (i) *If* $u \in \mathbf{S}^n(\mathrm{M})$, *then* $\psi_\mathrm{M}(\varphi_\mathrm{M}(u)) = n!u$.

(ii) *If* $v \in \mathbf{TS}^n(\mathrm{M})$ *then* $\varphi_\mathrm{M}(\psi_\mathrm{M}(v)) = n!v$.

Let $x_1, \dots, x_n \in \mathrm{M}$ and let u be the product $x_1 \dots x_n$ calculated in $\mathbf{S}(\mathrm{M})$. Then $\varphi_\mathrm{M}(u)$ is the product $x_1 \dots x_n$ calculated in $\mathbf{TS}(\mathrm{M})$, that is

$$\sum_{\sigma \in \mathfrak{S}_n} x_{\sigma(1)} \otimes \dots \otimes x_{\sigma(n)} .$$

Hence $\psi_\mathrm{M}(\varphi_\mathrm{M}(u))$ equals $\sum_{\sigma \in \mathfrak{S}_n} x_{\sigma(1)} \dots x_{\sigma(n)}$ calculated in $\mathbf{S}(\mathrm{M})$, that is

$$n!x_1 \dots x_n = n!u .$$

Let $v = \sum_{i=1}^{p} x_1^i \otimes x_2^i \otimes \dots \otimes x_n^i$ be an element of $\mathbf{TS}^n(\mathrm{M})$, where the x_j^i belong to M ; then $\psi_\mathrm{M}(v)$ is equal to $\sum_{i=1}^{p} x_1^i x_2^i \dots x_n^i$ calculated in $\mathbf{S}(\mathrm{M})$, whence

$$\varphi_\mathrm{M}(\psi_\mathrm{M}(v)) = \sum_{i=1}^{p} \mathrm{s}(x_1^i \otimes x_2^i \otimes \dots \otimes x_n^i) = s(v) = n!v .$$

COROLLARY 1. — *If* A *is a* **Q**-*algebra, then the canonical homomorphism of* **S**(M) *into* **TS**(M) *is an algebra isomorphism. If moreover* M *is free, then it is an isomorphism of graded bigebras.*

COROLLARY 2. — *If* A *is a* **Q**-*algebra then the module* $\mathbf{TS}^n(\mathrm{M})$ *is generated by the* n-th *powers of elements of* M *in* **TS**(M).

This follows from Cor. 1 and the corresponding property of **S**(M) (III, p. 498).

9. Homogeneous polynomial mappings

PROPOSITION 13. — *Let* M *and* N *be* A-*modules,* q *an integer* $\geqslant 0$, *and* f *a mapping of* M *into* N. *Suppose that* M *is free, then the following conditions are equivalent :*

(i) *There exists a* q-*linear mapping* g *of* M^q *into* N *such that* $f(x) = g(x, x, \dots, x)$ *for all* $x \in \mathrm{M}$.

(ii) *There exists a linear mapping* h *of* $\mathbf{TS}^q(\mathrm{M})$ *into* N *such that* $f(x) = h(\gamma_q(x))$ *for all* $x \in \mathrm{M}$.

(iii) *There exists a basis* $(e_i)_{i \in \mathrm{I}}$ *of* M *and a family* $(u_\nu)_{\nu \in \mathbf{N}^{(\mathrm{I})}, |\nu| = q}$ *of elements of* N *such that*

$$f\left(\sum_{i \in \mathrm{I}} \lambda_i e_i\right) = \sum_{\nu \in \mathbf{N}^{(\mathrm{I})}, |\nu| = q} \lambda^\nu u_\nu \tag{15}$$

for all $(\lambda_i) \in \mathrm{A}^{(\mathrm{I})}$.

(iv) *For each basis* $(e_i)_{i \in \mathrm{I}}$ *of* M *there exists a family* $(u_\nu)_{\nu \in \mathbf{N}^{(\mathrm{I})}, |\nu| = q}$ *of elements of* N *such that*

$$f\left(\sum_{i \in \mathrm{I}} \lambda_i e_i\right) = \sum_{\nu \in \mathbf{N}^{(\mathrm{I})}, |\nu| = q} \lambda^\nu u_\nu$$

for all $(\lambda_i) \in \mathrm{A}^{(\mathrm{I})}$.

(i) $\Rightarrow$ (ii) : let g satisfy (i), then there exists a linear mapping g' of $\mathbf{T}^q(\mathrm{M})$ into N such that $g(x_1, x_2, \dots, x_q) = g'(x_1 \otimes x_2 \otimes \dots \otimes x_q)$ for any $x_1, \dots, x_q \in \mathrm{M}$. Then

$$f(x) = g(x, x, \dots, x) = g'(x \otimes x \otimes \dots \otimes x) = g'(\gamma_q(x)) ;$$

and on writing $h = g' | \mathbf{TS}^q(\mathrm{M})$ we see that condition (ii) holds.

(ii) $\Rightarrow$ (i) and (iv) : let h satisfy the conditions of (ii). By Prop. 4, (ii) (IV, p. 47) there exists a linear mapping g' of $\mathbf{T}^q(\mathrm{M})$ into N such that $h = g' | \mathbf{TS}^q(\mathrm{M})$. Let g be the q-linear mapping of M into N associated with g', then for any $x \in \mathrm{M}$ we have

$$f(x) = h(\gamma_q(x)) = g'(x \otimes x \otimes \dots \otimes x) = g(x, x, \dots, x) ,$$

whence (i). On the other hand, if $(e_i)_{i \in I}$ is a basis of M, we have, by formula (6) (IV, p. 46)

$$f\left(\sum_i \lambda_i e_i\right) = h\left(\gamma_q\left(\sum_i \lambda_i e_i\right)\right) = h\left(\sum_{|\nu| = q} \lambda^\nu e_\nu\right)$$

on writing $e_\nu = \prod_{i \in I} \gamma_{\nu_i}(e_i)$; we thus have

$$f\left(\sum_i \lambda_i e_i\right) = \sum_{|\nu| = q} \lambda^\nu h(e_\nu) .$$

(iv) $\Rightarrow$ (iii) is clear.

(iii) $\Rightarrow$ (ii) : given (e_i), (u_ν) satisfying the conditions of (iii), let us write $e_\nu = \prod_{i \in I} \gamma_{\nu_i}(e_i)$ and recall that $(e_\nu)_{|\nu| = q}$ is a basis of $\mathbf{TS}^q(\mathrm{M})$. Let h be the homomorphism of $\mathbf{TS}^q(\mathrm{M})$ into N defined by $h(e_\nu) = u_\nu$; then for each $x = \sum_{i \in I} \lambda_i e_i$ in M we have

$$f(x) = f\left(\sum_i \lambda_i e_i\right) = \sum_{|\nu| = q} \lambda^\nu u_\nu = h\left(\sum_{|\nu| = q} \lambda^\nu e_\nu\right) = h(\gamma_q(x)) .$$

DEFINITION 3. — *Let* M *and* N *be* A*-modules and* q *an integer* $\geqslant 0$. *Suppose that* M *is free, and denote by* $\mathrm{Pol}^q_{\mathrm{A}}(\mathrm{M}, \mathrm{N})$ *or simply* $\mathrm{Pol}^q(\mathrm{M}, \mathrm{N})$ *the set of mappings of* M *into* N *satisfying the conditions of Prop.* 13. *The elements of* $\mathrm{Pol}^q(\mathrm{M}, \mathrm{N})$ *are called homogeneous polynomial mappings of degree* q *of* M *into* N.

Prop. 13 (i) defines a homomorphism of A-modules :

$$\mathscr{L}_q(\mathrm{M}, \ldots, \mathrm{M} ; \mathrm{N}) \to \mathrm{Pol}^q(\mathrm{M}, \mathrm{N}) .$$

Prop. 13 (ii) defines a homomorphism of A-modules :

$$\mathrm{Hom}_{\mathrm{A}}(\mathbf{TS}^q(\mathrm{M}), \mathrm{N}) \to \mathrm{Pol}^q(\mathrm{M}, \mathrm{N}) .$$

These homomorphisms are called *canonical.* They are surjective.

Examples. — 1) The homogeneous polynomial mappings of degree 1 of M into N are the A-linear mappings of M into N.

2) Let $(\mathrm{N}_i)_{i \in I}$ be a family of A-modules, f_i a mapping of M into N_i, $i \in I$, and $f : \mathrm{M} \to \prod_{i \in I} \mathrm{N}_i$ the mapping with components f_i. For f to be a homogeneous polynomial mapping of degree q it is necessary and sufficient that each f_i be a homogeneous polynomial mapping of degree q.

3) Let $(\mathrm{M}_j)_{j \in J}$ be a finite family of free A-modules and $u : \prod_{j \in J} \mathrm{M}_j \to \mathrm{N}$ a multilinear mapping ; then u is polynomial of degree Card(J).

4) Let $(X_i)_{i \in I}$ be a family of indeterminates, N an A-module and $u \in N[(X_i)_{i \in I}]$ a homogeneous polynomial of degree q. The mapping $(x_i)_{i \in I} \mapsto u((x_i)_{i \in I})$ of $A^{(I)}$ into N is a homogeneous polynomial mapping of degree q : this is seen at once by condition (iii) of Prop. 13. If I is finite, every homogeneous polynomial mapping of degree q of $A^{(I)} = A^I$ into N is of that form.

5) The mapping $(x_i)_{i \in N} \mapsto x_0^2 + x_1^2 + \cdots + x_n^2 + \cdots$ of $A^{(N)}$ into A is a homogeneous polynomial mapping of degree 2. If $A = \mathbf{Z}/2\mathbf{Z}$, it coincides with the linear mapping $(x_i)_{i \in I} \mapsto x_0 + x_1 + \cdots + x_n + \cdots$

6) Let $f \in \mathrm{Pol}_A^q(M, N)$, let B be a commutative ring, ρ a homomorphism of B into A and M′ and N′ the B-modules derived from M and N by means of ρ. Assume that M′ is free ; then $f \in \mathrm{Pol}_B^q(M', N')$: this follows at once from condition (i) of Prop. 13.

PROPOSITION 14. — *Let* M, N, P *be* A-*modules*, q *and* r *integers* $\geqslant 0$, *and assume that* M *and* N *are free. If* $f \in \mathrm{Pol}^q(M, N)$, $f' \in \mathrm{Pol}^r(N, P)$, *then* $f' \circ f \in \mathrm{Pol}^{qr}(M, P)$.

There exists a q-linear mapping g of M^q into N and an r-linear mapping g' of N^r into P such that

$$f(x) = g(x, x, \ldots, x) \quad \text{for all } x \in M,$$
$$f'(y) = g'(y, y, \ldots, y) \quad \text{for all } y \in N.$$

Hence for every $x \in M$ we have

$$f'(f(x)) = g'(f(x), f(x), \ldots, f(x)) = g'(g(x, x, \ldots, x), \ldots, g(x, x, \ldots, x))$$

and the mapping $(x_1, \ldots, x_{qr}) \mapsto g'(g(x_1, \ldots, x_q), \ldots, g(x_{q(r-1)+1}, \ldots, x_{qr}))$ of M^{qr} into P is qr-linear.

PROPOSITION 15. — *Let* M *be a free* A-*module*, N *an* A-*module and* q *an integer* $\geqslant 0$. *We suppose that the mapping* $y \mapsto q!y$ *is an automorphism of* N. *Let* $f \in \mathrm{Pol}^q(M, N)$, *then there exists one and only one symmetric* q-*linear mapping* h *of* M^q *into* M *such that* $f(x) = h(x, x, \ldots, x)$ *for all* $x \in M$. *For any* $x_1, \ldots, x_q \in M$ *we have*

$$h(x_1, x_2, \ldots, x_q) = \frac{(-1)^q}{q!} \sum_{H \subset \{1,2,\ldots,q\}} (-1)^{\mathrm{Card}\, H} f\Big(\sum_{i \in H} x_i\Big). \tag{16}$$

a) There exists a q-linear mapping g of M^q into N such that $f(x) = g(x, x, \ldots, x)$ for all $x \in M$. Let us define a q-linear mapping h of M into N by

$$h(x_1, x_2, \ldots, x_q) = \frac{1}{q!} \sum_{\sigma \in \mathfrak{S}_q} g(x_{\sigma(1)}, x_{\sigma(2)}, \ldots, x_{\sigma(q)}).$$

Then h is symmetric and $f(x) = h(x, x, \ldots, x)$ for all $x \in M$.

b) Let h be a symmetric q-linear mapping of M^q into N such that

$f(x) = g(x, x, \ldots, x)$. Let l be the linear mapping of $\mathbf{T}^q(\mathrm{M})$ into N such that $h(x_1, \ldots, x_q) = l(x_1 \otimes \ldots \otimes x_q)$ for any $x_1, \ldots, x_q \in \mathrm{M}$. We have

$$(-1)^q q! h(x_1, \ldots, x_q) = (-1)^q \sum_{\sigma \in \mathfrak{S}_q} h(x_{\sigma(1)}, \ldots, x_{\sigma(q)}) =$$

$$= (-1)^q l(\mathbf{s}(x_1 \otimes \ldots \otimes x_q)) = \sum_{\mathrm{H} \subset \{1, \ldots, q\}} (-1)^{\mathrm{Card}\, \mathrm{H}} l\left(\gamma_q\left(\sum_{i \in \mathrm{H}} x_i\right)\right)$$

by Prop. 3, (v) (IV, p. 45). Now

$$l\left(\gamma_q\left(\sum_{i \in \mathrm{H}} x_i\right)\right) = h\left(\sum_{i \in \mathrm{H}} x_i, \ldots, \sum_{i \in \mathrm{H}} x_i\right) = f\left(\sum_{i \in \mathrm{H}} x_i\right)$$

and this proves formula (16) and the uniqueness of h.

PROPOSITION 16. — *Let* M *be a free* A*-module,* N *an* A*-module,* q *a positive integer and* u *the canonical homomorphism of* $\mathrm{Hom}(\mathbf{TS}^q(\mathrm{M}), \mathrm{N})$ *into* $\mathrm{Pol}^q(\mathrm{M}, \mathrm{N})$.

(i) *If* A *is an infinite integral domain and* N *is torsion-free, then* u *is an isomorphism.*

(ii) *If the mapping* $y \mapsto q!y$ *in* N *is injective,* u *is an isomorphism.*

In the two cases of the proposition we have to prove that u is injective, that is, every linear mapping f of $\mathbf{TS}^q(\mathrm{M})$ into N which is zero on $\gamma_q(\mathrm{M})$, vanishes.

Assume A to be an infinite integral domain and N torsion-free. For every $z \in \mathbf{TS}^q(\mathrm{M})$ there exists $\alpha \in \mathrm{A} - \{0\}$ such that αz is an A-linear combination of elements of $\gamma_q(\mathrm{M})$ (IV, p. 47, Prop. 5). Hence $\alpha f(z) = f(\alpha z) = 0$, and so $f(z) = 0$.

Suppose next that the mapping $y \mapsto q!y$ in N is injective, then by IV, p. 45, Prop. 3, (v), f vanishes on $\mathbf{s}\,.\,\mathbf{T}^q(\mathrm{M})$. Hence if $z \in \mathbf{TS}^q(\mathrm{M})$, we have $q!f(z) = f(\mathbf{s}z) = 0$, and so $f(z) = 0$.

COROLLARY. — *Let* M *be a free* A*-module,* N *an* A*-module,* q *a positive integer,* $h \in \mathrm{Pol}^q(\mathrm{M}, \mathrm{N})$ *and* $(e_i)_{i \in \mathrm{I}}$ *a basis of* M. *In the two cases of Prop.* 16 *there exists a unique family* $(u_\nu)_{\nu \in \mathbf{N}^{(\mathrm{I})}, |\nu| = q}$ *of elements of* N *such that* $h\left(\sum_{i \in \mathrm{I}} \lambda_i e_i\right) = \sum_{|\nu| = q} \lambda^\nu u_\nu$ *for all* $(\lambda_i) \in \mathrm{A}^{(\mathrm{I})}$.

10. Polynomial mappings

DEFINITION 4. — *Let* M *and* N *be* A*-modules and assume that* M *is free. Let* $\mathrm{Map}(\mathrm{M}, \mathrm{N})$ *be the* A*-module of all mappings of* M *into* N. *The submodule* $\sum_{q > 0} \mathrm{Pol}^q_{\mathrm{A}}(\mathrm{M}, \mathrm{N})$ *of* $\mathrm{Map}(\mathrm{M}, \mathrm{N})$ *is denoted by* $\mathrm{Pol}_{\mathrm{A}}(\mathrm{M}, \mathrm{N})$ *or simply* $\mathrm{Pol}(\mathrm{M}, \mathrm{N})$; *its elements are called polynomial mappings of* M *into* N.

Let $(e_i)_{i \in \mathrm{I}}$ be a basis of M and suppose that I is finite ; by Prop. 13 (IV, p. 54) a mapping f of M into N is polynomial if and only if there exists a polynomial F in the indeterminates X_i with coefficients in N such that

$$f\left(\sum_{i \in \mathrm{I}} x_i e_i\right) = \mathrm{F}(\mathbf{x})$$

for every family $\mathbf{x} = (x_i)_{i \in \mathrm{I}}$ in $\mathrm{A}^{(\mathrm{I})}$. This property is independent of the basis chosen for M and it justifies the terminology « polynomial mapping ».

PROPOSITION 17. — *Let* M *be a free* A*-module and* B *an associative, commutative and unital* A*-algebra. Then* $\mathrm{Pol}_{\mathrm{A}}(\mathrm{M}, \mathrm{B})$ *is a sub-*B*-algebra of the algebra* $\mathrm{Map}(\mathrm{M}, \mathrm{B})$.

This follows from Def. 4 and Prop. 13, (iv) (IV, p. 54).

PROPOSITION 18. — *Let* M, N, P *be* A*-modules, and assume that* M *and* N *are free. If* $f \in \mathrm{Pol}(\mathrm{M}, \mathrm{N})$, $g \in \mathrm{Pol}(\mathrm{N}, \mathrm{P})$, *then* $g \circ f \in \mathrm{Pol}(\mathrm{M}, \mathrm{P})$.

We can reduce at once to the case where there exists an integer q such that $g \in \mathrm{Pol}^q(\mathrm{N}, \mathrm{P})$; then there exists a q-linear mapping h of N^q into P such that $g(y) = h(y, y, \ldots, y)$ for all $y \in \mathrm{N}$. Writing f as a sum of homogeneous polynomial mappings we are thus reduced to proving that the mapping

$$x \mapsto h(f_1(x), f_2(x), \ldots, f_q(x))$$

of M into P, where $f_i \in \mathrm{Pol}^{q_i}(\mathrm{M}, \mathrm{N})$, is polynomial. For $i = 1, \ldots, q$ there exists a q_i-linear mapping l_i of M^{q_i} into N such that $f_i(x) = l_i(x, x, \ldots, x)$ for all $x \in \mathrm{M}$. Hence

$$h(f_1(x), f_2(x), \ldots, f_q(x)) = h(l_1(x, \ldots, x), \ldots, l_q(x, \ldots, x)),$$

from which our assertion follows.

Lemma 2. — *Let* N *be an* A*-module,* n *an integer* $\geqslant 0$ *and*

$$f = m_0 + m_1\mathrm{X} + \cdots + m_n\mathrm{X}^n \in \mathrm{N}[\mathrm{X}].$$

Suppose that there exist $\alpha_0, \alpha_1, \ldots, \alpha_n \in \mathrm{A}$ *such that* $f(\alpha_0) = \cdots = f(\alpha_n) = 0$, *and such that for* $i \neq j$ *the homothety of ratio* $\alpha_i - \alpha_j$ *in* N *is injective, then* $f = 0$.

(This lemma generalizes the Cor. of IV, p. 16.)

The lemma clearly holds for $n = 0$; we shall prove it by induction on n. We have

$$f(\mathrm{X}) = f(\mathrm{X}) - f(\alpha_0) = \sum_{i=1}^{n} m_i(\mathrm{X}^i - \alpha_0^i) = (\mathrm{X} - \alpha_0)\, g(\mathrm{X})$$

where g is an element of $N[X]$ of the form $m'_0 + m'_1 X + \cdots + m'_{n-1} X^{n-1}$. The hypotheses of the lemma imply that $g(\alpha_1) = \cdots = g(\alpha_n) = 0$, hence $g = 0$ by the induction hypothesis, and so $f = 0$.

PROPOSITION 19. — *Let* M *be a free* A*-module,* N *an* A*-module,* G *an infinite additive subgroup of* A*, and suppose that the homotheties of* N *defined by the non-zero elements of* G *are injective. Then* $\mathrm{Pol}(M, N)$ *is the direct sum of the* $\mathrm{Pol}^q(M, N)$.

Let $f_0, f_1, \ldots, f_n$ be such that $f_i \in \mathrm{Pol}^i(M, N)$ and suppose that we have the relation $f_0 + \cdots + f_n = 0$. Let $x \in M$, then for all $\lambda \in G$ we have

$$0 = \sum_{i=0}^{n} f_i(\lambda x) = \sum_{i=0}^{n} \lambda^i f_i(x).$$

By Lemma 2, applied to the polynomial $\sum_{i=0}^{n} f_i(x) X^i$ we have

$$f_0(x) = \cdots = f_n(x) = 0.$$

COROLLARY. — *Assume that* A *is an infinite integral domain ; let* M *be a free* A*-module and* N *a torsion-free* A*-module.*

(i) *We have* $\mathrm{Pol}(M, N) = \bigoplus_{q \geqslant 0} \mathrm{Pol}^q(M, N)$ *and each* $\mathrm{Pol}^q(M, N)$ *may be identified canonically with* $\mathrm{Hom}(\mathbf{TS}^q(M), N)$.

(ii) *Let* $f \in \mathrm{Pol}(M, N)$ *and* $(e_i)_{i \in I}$ *a basis of* M*. There exists one and only one family* $(u_\nu)_{\nu \in \mathbf{N}^{(I)}}$ *of elements of* N *such that* $f\left(\sum_{i \in I} \lambda_i e_i\right) = \sum_{\nu \in \mathbf{N}^{(I)}} \lambda^\nu u_\nu$ *for all* $(\lambda_i) \in A^{(I)}$.

The assertion (i) follows from Prop. 16 and 19, and (ii) follows from (i) and the Cor. of Prop. 16.

11. Relations between $\mathbf{S}(M^*)$, $\mathbf{TS}(M)^{*gr}$ and $\mathrm{Pol}(M, A)$

Let M be a free A-module, we shall equip the graded dual $\mathbf{TS}(M)^{*gr}$ with the structure of a commutative, associative and unital graded [1] algebra, derived from the graded cogebra structure of $\mathbf{TS}(M)$ (III, p. 580). By III, p. 497 there exists a unique homomorphism of graded A-algebras

$$\theta : \mathbf{S}(M^*) \to \mathbf{TS}(M)^{*gr}$$

inducing in degree 1 the identity mapping of M^*.

[1] A graded homomorphism of degree $-k$ of $\mathbf{TS}(M)$ into A is here considered as an element of degree k of $\mathbf{TS}(M)^{*gr}$ (II, p. 377).

PROPOSITION 20. — *If the A-module M is free and finitely generated, then θ is an isomorphism of graded algebras.*

Let $(e_i)_{i \in I}$ be a basis of M and $(e_i^*)_{i \in I}$ the dual basis of M^*. For $\nu \in \mathbf{N}^{I}$ put

$$e_\nu = \prod_{i \in I} \gamma_{\nu_i}(e_i) \in \mathbf{TS}(M).$$

By Prop. 4 (IV, p. 47) the family $(e_\nu)_{\nu \in \mathbf{N}^I}$ is a basis of $\mathbf{TS}(M)$; let (e_ν^*) be the basis of $\mathbf{TS}(M)^{*gr}$ dual to (e_ν). In the light of III, p. 505, Th. 1 it is enough to show that for any $\nu \in \mathbf{N}^I$ we have

$$e_\nu^* = \prod_{i \in I} (e_i^*)^{\nu_i},$$

or also that for $\rho, \sigma \in \mathbf{N}^I$ we have $e_\rho^* \cdot e_\sigma^* = e_{\rho+\sigma}^*$; but this last assertion follows from IV, p. 50, Formula (12).

Remark 1. — In the same way we see that if M is a finitely generated free A-module, then the graded algebra $\mathbf{S}(M)^{*gr}$ defined in III, p. 593, may be identified with $\mathbf{TS}(M^*)$.

PROPOSITION 21. — *The canonical homomorphism of* A*-modules* (IV, p. 55)

$$u : \mathbf{TS}(M)^{*gr} \to \mathrm{Pol}_A(M, A)$$

is an algebra homomorphism.

Let $a \in \mathbf{TS}^q(M)^*$, $b \in \mathbf{TS}^r(M)^*$, $x \in M$; we have

$$u(ab)(x) = \langle ab, \gamma_{q+r}(x)\rangle = \langle a \otimes b, c(\gamma_{q+r}(x))\rangle =$$
$$= \langle a \otimes b, \gamma_q(x) \otimes \gamma_r(x)\rangle = \langle a, \gamma_q(x)\rangle \langle b, \gamma_r(x)\rangle = u(a)(x) \cdot u(b)(x),$$

whence the result.

Remarks. — 2) The composite homomorphism $\lambda_m = u \circ \theta : \mathbf{S}(M^*) \to \mathrm{Pol}_A(M, A)$ is the unique unital homomorphism of algebras inducing the inclusion of

$$M^* = \mathrm{Pol}^1(M, A)$$

in $\mathrm{Pol}(M, A)$. If M is finitely generated free and A is an infinite integral domain, then λ_M is bijective (Prop. 20 and Cor. of Prop. 19). In particular if A is an infinite integral domain, then the canonical homomorphism $f \mapsto \tilde{f}$ of $A[X_1, \dots, X_n]$ into $\mathrm{Pol}(A^n, A)$ (IV, p. 4) is an isomorphism.

3) Consider the coproduct $c_{\mathbf{S}} : \mathbf{S}(M^*) \to \mathbf{S}(M^* \times M^*)$ (III, p. 575, *Example* 6). For any $v \in \mathbf{S}(M^*)$, $x, y \in M$, the polynomial mapping $\lambda_{M \times M}(c_{\mathbf{S}}(v))$: $M \times M \to A$ maps (x, y) to $\lambda_M(v)(x + y)$. For the two algebra homomorphisms

of $\mathbf{S}(M^*)$ into $\operatorname{Map}(M \times M, A)$ defined in this way agree on M^*, in virtue of the relation

$$(v \otimes 1 + 1 \otimes v)(x, y) = v(x + y) \quad (v \in M^*) .$$

§ 6. SYMMETRIC FUNCTIONS

1. Symmetric polynomials

Let n be a positive integer. For every permutation $\sigma \in \mathfrak{S}_n$ let φ_σ be the automorphism of the A-algebra $A[X_1, \ldots, X_n]$ which maps X_i to $X_{\sigma(i)}$ for $1 \leqslant i \leqslant n$. It is clear that $\sigma \mapsto \varphi_\sigma$ is a homomorphism of $\mathfrak{S}_n$ into the group of automorphisms of $A[X_1, \ldots, X_n]$. We shall put $\sigma f = \varphi_\sigma(f)$ for $\sigma \in \mathfrak{S}_n$ and $f \in A[X_1, \ldots, X_n]$. The polynomial f is said to be *symmetric* if $\sigma f = f$ for all $\sigma \in \mathfrak{S}_n$; the symmetric polynomials form a unital graded subalgebra of $A[X_1, \ldots, X_n]$; we shall denote it by $A[X_1, \ldots, X_n]^{\text{sym}}$ in the rest of this paragraph.

For every positive integer k we denote by $\mathfrak{P}_k$ the set of k-element subsets of the set $\{1, 2, \ldots, n\}$ and put

$$s_k = \sum_{H \in \mathfrak{P}_k} \prod_{i \in H} X_i \,. \tag{1}$$

When we wish to specify the integer n we shall write $s_{k,n}$ instead of s_k. We have in particular

$$\begin{aligned} s_0 &= 1 \\ s_1 &= \sum_{1 \leqslant i \leqslant n} X_i \\ s_2 &= \sum_{1 \leqslant i < j \leqslant n} X_i X_j \\ &\cdots\cdots\cdots\cdots\cdots \\ s_n &= X_1 \ldots X_n \end{aligned}$$

and $s_k = 0$ for $k > n$. It is clear that s_k is a homogeneous symmetric polynomial of degree k; we shall call it the *elementary symmetric polynomial of degree k*.

In the ring $A[X_1, \ldots, X_n, U, V]$ we have the relation

$$\prod_{i=1}^{n} (U + VX_i) = \sum_{k=0}^{n} U^{n-k} V^k s_k \,; \tag{2}$$

by appropriate substitutions we deduce the relations

$$\prod_{i=1}^{n} (1 + TX_i) = \sum_{k=0}^{n} s_k T^k \,, \tag{3}$$

$$(4) \qquad \prod_{i=1}^{n} (X - X_i) = \sum_{k=0}^{n} (-1)^{n-k} s_{n-k} X^k .$$

THEOREM 1. — *Let us put* $E = A[X_1, \dots, X_n]$ *and* $S = A[X_1, \dots, X_n]^{sym}$.

a) The A*-algebra* S *of symmetric polynomials is generated by* $s_1, \dots, s_n$.

b) The elements $s_1, \dots, s_n$ *of* E *are algebraically independent over* A (IV, p. 4).

c) The family of monomials $X^\nu = X_1^{\nu(1)} \dots X_n^{\nu(n)}$ *such that* $0 \leqslant \nu(i) < i$ *for* $1 \leqslant i \leqslant n$ *is a basis of the* S*-module* E. *In particular,* E *is a free* S*-module of rank* $n!$.

We shall prove the theorem by induction on n, the case $n = 0$ being trivial. Let us write $B = A[X_n]$ and denote by s'_k the elementary symmetric polynomial of degree k in $X_1, \dots, X_{n-1}$; we thus have $B[X_1, \dots, X_{n-1}] = A[X_1, \dots, X_n]$. If we replace n by $n - 1$ and A by B in the statement of Th. 1, we can formulate the induction hypothesis as follows:

(A) *The* B*-algebra* S′ *of polynomials* $f \in A[X_1, \dots, X_n]$ *invariant under all permutations of* $X_1, \dots, X_{n-1}$ *is generated by* $s'_1, \dots, s'_{n-1}$.

(B) *The elements* $s'_1, \dots, s'_{n-1}$ *of* E *are algebraically independent over* B.

(C) *The family of monomials* $X_1^{\nu(1)} \dots X_{n-1}^{\nu(n-1)}$ *such that* $0 \leqslant \nu(i) < i$ *for* $1 \leqslant i \leqslant n - 1$ *is a basis of the* S′*-module* E.

We have the obvious relation

$$(5) \qquad s_k = s'_k + s'_{k-1} X_n \quad (1 \leqslant k \leqslant n - 1),$$

from which we obtain by induction on k

$$(6) \qquad s'_k = (-1)^k X_n^k + \sum_{i=1}^{k} (-1)^{k-i} s_i X_n^{k-i} \quad (1 \leqslant k \leqslant n - 1).$$

We have $S \subset S'$, hence $s_1, \dots, s_n$ belong to S′; by (A) and formula (6) the B-algebra S′ is therefore generated by $s_1, \dots, s_{n-1}$.

By (B) there exists an endomorphism μ of the B-algebra S′ such that

$$(7) \qquad u(s'_k) = (-1)^k X_n^k + \sum_{i=1}^{k} (-1)^{k-i} s'_i X_n^{k-i} \quad (1 \leqslant k \leqslant n - 1).$$

By (5), we have $u(s_k) = u(s'_k) + u(s'_{k-1}) X_n$, whence $u(s_k) = s'_k$ by an easy calculation. Let $P \in B[Y_1, \dots, Y_{n-1}]$; then from $P(s_1, \dots, s_{n-1}) = 0$ we deduce

$$0 = u(P(s_1, \dots, s_{n-1})) = P(s'_1, \dots, s'_{n-1}),$$

whence $P = 0$ by (B). It follows that the B-algebra S′ is generated by the algebraically independent elements $s_1, \dots, s_{n-1}$. This property may be reformulated as follows:

(D) *The* A*-algebra* S′ *is generated by the algebraically independent elements* $s_1, \dots, s_{n-1}, X_n$.

We can thus identify S′ with the polynomial ring $C[X_n]$, where C is the sub-A-algebra of E generated by $s_1, \ldots, s_{n-1}$.

To prove a), let $f \in S$ be a homogeneous symmetric polynomial of degree m. We have $f \in S' = C[X_n]$, so there exists an element $g = P(s_1, \ldots, s_{n-1})$ of C, homogeneous of degree m in $X_1, \ldots, X_n$, such that $f - g$ is divisible by X_n. Since $f - g$ is symmetric, each of its terms is also divisible by $X_1, \ldots, X_{n-1}$, hence $f - g$ is divisible by $s_n = X_1 \ldots X_n$. In other words, there exists $h \in S$ such that $f = g + hs_n$, whence $\deg h < m$. By induction on m it follows that f belongs to

$$C[s_n]_E = A[s_1, \ldots, s_{n-1}, s_n]_E .$$

Thus we have shown that the A-algebra S is generated by $s_1, \ldots, s_n$.

Next we prove b). If we substitute X_n for X in (4), we find

$$(-1)^{n+1}s_n = X_n^n + \sum_{k=1}^{n-1} (-1)^{n-k}s_{n-k}X_n^k ;$$

in other words, $(-1)^{n+1}s_n$ is a monic polynomial in X_n of degree n and with coefficients in C. By IV, p. 11 we thus have the following property :

(E) *The C-algebra homomorphism* φ of $C[T]$ *into* $C[X_n] = S'$ *which maps* T *to* s_n *is injective, and* S′ *is a free module over the image of* φ, *with basis* $(1, X_n, \ldots, X_n^{n-1})$.

The elements $s_1, \ldots, s_{n-1}$ of C are algebraically independent over A, by (D) ; thus the injectivity of φ means that $s_1, \ldots, s_{n-1}, s_n$ are algebraically independent over A, whence *b*).

To prove c), the image of φ equals $C[s_n]_E = S$, hence by (E), S′ is a free module over S on the basis $(1, X_n, \ldots, X_n^{n-1})$. Now the assertion *c*) follows from the induction hypothesis (C) and Prop. 25 of II, p. 222. This completes the proof.

Let f be a symmetric polynomial in $X_1, \ldots, X_n$, homogeneous of degree m. By Th. 1 (IV, p. 62) there exists a polynomial $Q \in A[Y_1, \ldots, Y_n]$ such that $f = Q(s_1, \ldots, s_n)$. The preceding proof provides a *means of explicit calculation* of Q, by a double induction on n and m. For as we have seen, there exists a polynomial $P \in A[Y_1, \ldots, Y_{n-1}]$ and a polynomial h symmetric in $X_1, \ldots, X_n$ which is homogeneous of degree $m - n$, such that

$$f = P(s_1, \ldots, s_{n-1}) + s_n h . \tag{8}$$

For every polynomial $u \in A[X_1, \ldots, X_n]$ we put

$$u'(X_1, \ldots, X_{n-1}) = u(X_1, \ldots, X_{n-1}, 0) .$$

Then $s_1', \ldots, s_{n-1}'$ are the elementary symmetric polynomials in $X_1, \ldots, X_{n-1}$ and formula (8) implies

$$f' = P(s_1', \ldots, s_{n-1}') .$$

The determination of P is thus reduced to a calculation of symmetric polynomials in $n-1$ indeterminates, and h is obtained from (8).

We illustrate the method by two examples.

Examples. — 1) Let $n = 3$ and

$$f = X_1^2(X_2 + X_3) + X_2^2(X_3 + X_1) + X_3^2(X_1 + X_2) .$$

We have

$$f' = X_1^2 X_2 + X_1 X_2^2 = X_1 X_2 (X_1 + X_2) = s_1' s_2' .$$

Let us write $g = f - s_1 s_2$; we have

$$g = f - (X_1 + X_2 + X_3)(X_1X_2 + X_1X_3 + X_2X_3) = -3X_1X_2X_3 ,$$

hence finally

$$f = s_1 s_2 - 3s_3 .$$

2) Again let $n = 3$ and put $p = X_1^3 + X_2^3 + X_3^3$.

We have $p(X_1, 0, 0) = X_1^3 = s_1(X_1, 0, 0)^3$. Writing $q = p - s_1^3$, we obtain

$$q = -3f - 6X_1X_2X_3 = -3s_1s_2 + 3s_3$$

and finally

$$p = s_1^3 - 3s_1s_2 + 3s_3 .$$

Let $S_1, \ldots, S_n$ be indeterminates. We shall equip the polynomial algebra $A[S_1, \ldots, S_n]$ with the graduation of type $\mathbf{N}$ for which S_k is of *weight* k for $1 \leqslant k \leqslant n$ (IV, p. 3), and we equip $A[X_1, \ldots, X_n]$ with the usual graduation. For $1 \leqslant k \leqslant n$ the elementary symmetric polynomial $s_{k,n}$ in $X_1, \ldots, X_n$ is homogeneous of degree k. By Th. 1 (IV, p. 62) the mapping $g \mapsto g(s_{1,n}, \ldots, s_{n,n})$ is thus an isomorphism of graded algebras

$$\varphi_n : A[S_1, \ldots, S_n] \to A[X_1, \ldots, X_n]^{\text{sym}} .$$

Let m be an integer such that $0 \leqslant m \leqslant n$. For every integer k such that $1 \leqslant k \leqslant m$ we have

(9) $$s_{k,m}(X_1, \ldots, X_m) = s_{k,n}(X_1, \ldots, X_m, 0, \ldots, 0)$$

by definition (IV, p. 61, formula (1)) of s_k. Hence the diagram

(10) $$\begin{array}{ccc} A[S_1, \ldots, S_m] & \xrightarrow{\ j\ } & A[S_1, \ldots, S_n] \\ \Big\downarrow{\scriptstyle \varphi_m} & & \Big\downarrow{\scriptstyle \varphi_n} \\ A[X_1, \ldots, X_m]^{\text{sym}} & \xleftarrow[\ \rho\ ]{} & A[X_1, \ldots, X_n]^{\text{sym}} \end{array}$$

(where j denotes the canonical inclusion and p the homomorphism

$$g \mapsto g(X_1, \ldots, X_m, 0, \ldots, 0))$$

is commutative.

PROPOSITION 1. — *For every pair of positive integers k, n, let* $S_k^{(n)}$ *be the A-module consisting of all symmetric polynomials in* $X_1, \ldots, X_n$, *which are homogeneous of degree k. If the integer m satisfies* $0 \leqslant k \leqslant m \leqslant n$, *then the mapping* $f \mapsto f(X_1, \ldots, X_m, 0, \ldots, 0)$ *is an isomorphism of* $S_k^{(n)}$ *onto* $S_k^{(m)}$.

By the commutativity of the diagram (10) it is enough to show that every isobaric polynomial of weight k in $S_1, \ldots, S_n$ depends only on $S_1, \ldots, S_m$ under the hypotheses $0 \leqslant k \leqslant m \leqslant n$. Now the weight of a monomial $S_1^{\alpha(1)} \ldots S_n^{\alpha(n)}$ is equal to the integer $\alpha(1) + 2\alpha(2) + \cdots + n\alpha(n)$; since the integers $\alpha(1), \ldots, \alpha(n)$ are positive, the relation

$$\alpha(1) + 2\alpha(2) + \cdots + n\alpha(n) = k \leqslant n$$

implies $\alpha(j) = 0$ for $k < j \leqslant n$, whence the assertion.

Example 3. — By example 2 of IV, p. 64 and Prop. 1 above we thus have

$$\sum_{i=1}^{n} X_i^3 = s_{1,n}^3 - 3s_{1,n}s_{2,n} + 3s_{3,n}$$

for every integer $n \geqslant 3$. The commutativity of the diagram (10) moreover gives the formulae

$$X_1^3 + X_2^3 = s_{1,2}^3 - 3s_{1,2}s_{2,2}\,,$$
$$X_1^3 = s_{1,1}^3\,.$$

Remark. — Let n and k be two positive integers. We denote by $\Delta_{k,n}$ the set of elements of length k in $\mathbf{N}^n$ and further equip $\Delta_{k,n}$ with the order relation, written $\alpha \preccurlyeq \beta$, induced by the lexicographic order on $\mathbf{N}^n$ (*Set Theory*, III, p. 157), and define an action of the group $\mathfrak{S}_n$ on $\mathbf{N}^n$ by $(\sigma\alpha)\ (i) = \alpha(\sigma^{-1}(i))$ for $\sigma \in \mathfrak{S}_n$, $\alpha \in \mathbf{N}^n$ and $1 \leqslant i \leqslant n$. Moreover, we denote by D_k the set of elements $\alpha = (\alpha(1), \ldots, \alpha(k))$ of $\mathbf{N}^k$ such that

$$\alpha(1) \geqslant \alpha(2) \geqslant \cdots \geqslant \alpha(k)\,, \quad \alpha(1) + \cdots + \alpha(k) = k\,.$$

Suppose that $k \leqslant n$ and let us identify $\mathbf{N}^k$ with a subset of $\mathbf{N}^n$ by the mapping $(\alpha(1), \ldots, \alpha(k)) \mapsto (\alpha(1), \ldots, \alpha(k), 0, \ldots, 0)$. Then D_k consists of the elements α of $\Delta_{k,n}$ such that $\sigma\alpha \preccurlyeq \alpha$ for all $\sigma \in \mathfrak{S}_n$. It follows that every orbit of the group $\mathfrak{S}_n$ in $\Delta_{k,n}$ contains a unique element of D_k. For every $\alpha \in D_k$ let $0(\alpha)$ be the orbit of α in $\Delta_{k,n}$ under the operation of $\mathfrak{S}_n$; put

(11) $$M(\alpha) = \sum_{\beta \in 0(\alpha)} X^\beta\,.$$

It follows from Lemma 1 of IV, p. 47, that the family $(\mathrm{M}(\alpha))_{\alpha \in \mathrm{D}_k}$ is a basis of the A-module $\mathrm{S}_k^{(n)}$.

For each $\alpha \in \mathrm{D}_k$ put

$$(12) \quad \mathrm{S}(\alpha) = \prod_{i=1}^{k} s_i^{\alpha(i) - \alpha(i+1)} \quad \text{(where it is understood that } \alpha(k+1) = 0) ;$$

since we have $\sum_{i=1}^{k} i \cdot (\alpha(i) - \alpha(i+1)) = \sum_{i=1}^{k} \alpha(i) = k$, the symmetric polynomial $\mathrm{S}(\alpha)$ is homogeneous of degree k. It follows directly from Th. 1 (IV, p. 62) that the family $(\mathrm{S}(\alpha))_{\alpha \in \mathrm{D}_k}$ is a basis of the A-module $\mathrm{S}_k^{(n)}$.

Let α, β be in D_k, and let $c_{\alpha\beta}$ be the coefficient of the monomial X^β in the polynomial $\mathrm{S}(\alpha)$ defined by (12) in the case $\mathrm{A} = \mathbf{Z}$ and $k = n$. This is then a positive integer, *independent of the ring* A *and of the integer n*. By formula (9) (IV, p. 64) we then have

$$(13) \qquad \mathrm{S}(\alpha) = \sum_{\beta \in \mathrm{D}_k} c_{\alpha\beta} \cdot \mathrm{M}(\beta) \quad (\alpha \in \mathrm{D}_k) .$$

It may be shown (cf. IV, p. 101, Exercise 13) that the matrix $\mathrm{C} = (c_{\alpha\beta})_{\alpha,\beta \in \mathrm{D}_k}$ is such that $c_{\alpha\alpha} = 1$ and $c_{\alpha\beta} = 0$ for $\alpha < \beta$; generalizing the terminology introduced in II, p. 351, we shall say that C belongs to the lower total triangular group. The same is true of the matrix D inverse to C. The values of the matrices C and D when $2 \leqslant k \leqslant 5$, are given in the table (IV, p. 103-105).

Suppose now that $n < k$ and let us identify $\mathbf{N}^n$ with a subset of $\mathbf{N}^k$ by the mapping $(\alpha(1), \dots, \alpha(n)) \mapsto (\alpha(1), \dots, \alpha(n), 0, \dots, 0)$. For every α in $\mathrm{D}_k \cap \mathbf{N}^n$ we again denote by $0(\alpha)$ the orbit of α in $\Delta_{k,n}$ under the operation of $\mathfrak{S}_n$ and define $\mathrm{M}(\alpha)$ by (11). Further we define $\mathrm{S}(\alpha)$ by (12), then the families $(\mathrm{M}(\alpha))_{\alpha \in \mathrm{D}_k \cap \mathbf{N}^n}$ and $(\mathrm{S}(\alpha))_{\alpha \in \mathrm{D}_k \cap \mathbf{N}^n}$ are bases of the A-module $\mathrm{S}_k^{(n)}$. By formula (9) of IV, p. 64, we have a formula analogous to (13) where D_k is replaced by $\mathrm{D}_k \cap \mathbf{N}^n$, with the same integers $c_{\alpha\beta}$.

Example 4. — By Example 3 of IV, p. 65, we have

$$\mathrm{M}(3,0,0) = \mathrm{S}(3,0,0) - 3\mathrm{S}(2,1,0) + 3\mathrm{S}(1,1,1)$$

for every integer $n \geqslant 3$, and so

$$\mathrm{M}(3,0) = \mathrm{S}(3,0) - 3\mathrm{S}(2,1)$$

for $n = 2$.

2. Symmetric rational fractions

Let K be a commutative field and $X_1, X_2, \ldots, X_n$ indeterminates. For every $\sigma \in \mathfrak{S}_n$ we have defined in No. 1 (IV, p. 61) an automorphism φ_σ of $K[X_1, X_2, \ldots, X_n]$. This automorphism has a unique extension ψ_σ to the field $K(X_1, \ldots, X_n)$, and $\sigma \mapsto \psi_\sigma$ is an injective homomorphism of $\mathfrak{S}_n$ into the automorphism group of $K(X_1, \ldots, X_n)$. For each $f \in K(X_1, \ldots, X_n)$ we have $(\psi_\sigma f)\ (X_1, \ldots, X_n) = f(X_{\sigma(1)}, \ldots, X_{\sigma(n)})$. The rational fractions f such that $\psi_\sigma(f) = f$ for all $\sigma \in \mathfrak{S}_n$ are called *symmetric rational fractions*. The set of all symmetric rational fractions in $X_1, \ldots, X_n$ is a subfield of $K(X_1, \ldots, X_n)$.

PROPOSITION 2. — *The field of symmetric rational fractions in* $X_1, \ldots, X_n$ *is the field of fractions of the ring of symmetric polynomials in* $X_1, \ldots, X_n$.

Let $f \in K(X_1, \ldots, X_n)$ be a symmetric rational fraction, and let u_1, v_1 be two elements of $K[X_1, \ldots, X_n]$ such that $f = \dfrac{u_1}{v_1}$. We put $v = \prod_{\sigma \in \mathfrak{S}_n} \psi_\sigma(v_1) \in K[X_1, \ldots, X_n]$ and $u = fv \in K[X_1, \ldots, X_n]$; then v is symmetric, hence u is symmetric, because f is, and $f = \dfrac{u}{v}$, whence the result.

COROLLARY. — *Let* $s_1, s_2, \ldots, s_n$ *be the elementary symmetric polynomials in* $X_1, \ldots, X_n$. *For every rational fraction* $g \in K(S_1, S_2, \ldots, S_n)$ *the sequence* $(s_1, s_2, \ldots, s_n)$ *is substitutable in* g *and the mapping* $g \mapsto g(s_1, s_2, \ldots, s_n)$ *is an isomorphism of* $K(S_1, S_2, \ldots, S_n)$ *onto the field of symmetric rational fractions in* $X_1, \ldots, X_n$.

This follows from Prop. 2 and Th. 1 of IV, p. 62.

3. Symmetric formal power series

Let I be a set, $\mathbf{X} = (X_i)_{i \in I}$ a family of indeterminates and $A[[\mathbf{X}]]$ the algebra of formal power series in the X_i. For every permutation $\sigma \in \mathfrak{S}_I$ there exists a unique continuous automorphism φ_σ of the algebra $A[[\mathbf{X}]]$ which maps X_i to $X_{\sigma(i)}$ for each $i \in I$ (IV, p. 28, Prop. 4); it is clear that $\sigma \mapsto \varphi_\sigma$ is a homomorphism of $\mathfrak{S}_I$ into the group of continuous automorphisms of the algebra $A[[\mathbf{X}]]$. Let $f \in A[[\mathbf{X}]]$ be a formal power series; we put $\sigma f = \varphi_\sigma(f)$ and we shall say that the formal power series f is *symmetric* if $\sigma f = f$ for each $\sigma \in \mathfrak{S}_I$. The symmetric formal power series form a closed subalgebra of $A[[\mathbf{X}]]$ which is denoted by $A[[\mathbf{X}]]^{\text{sym}}$ and is equipped with the topology induced by that of $A[[\mathbf{X}]]$.

Let T be an indeterminate. In the ring of formal power series $A[[\mathbf{X}, T]]$ the family $(X_iT)_{i \in I}$ is summable, hence the family $(1 + X_iT)_{i \in I}$ is multipliable (IV, p. 27, Prop. 2) ; moreover we have

$$\prod_{i \in I} (1 + X_iT) = 1 + \sum_{k \geq 1} s_k T^k , \tag{14}$$

where the formal power series $s_k \in A[[\mathbf{X}]]$ is defined by

$$s_k = \sum_{H \in \mathfrak{P}_k} \left(\prod_{i \in H} X_i \right) \quad (k \geq 1) \tag{15}$$

(we denote by $\mathfrak{P}_k$ the set of all k-element subsets of I). In particular we have $s_1 = \sum_{i \in I} X_i$. When I is finite with n elements we have $s_k = 0$ for $k > n$; more precisely, when $I = \{1, \dots, n\}$ then the formal power series s_k is nothing other than the elementary symmetric polynomial of degree k in $X_1, \dots, X_n$.

Let $\mathbf{S} = (S_k)_{k \geq 1}$ be a sequence of indeterminates. Since the formal power series s_k is of order $\geq k$, and belongs to $A[[\mathbf{X}]]^{\mathrm{sym}}$, conditions a) and b) of Prop. 4 of IV, p. 28 are satisfied with $E = A[[\mathbf{X}]]^{\mathrm{sym}}$; there exists thus a unique continuous A-algebra homomorphism

$$\varphi_I : A[[\mathbf{S}]] \to A[[\mathbf{X}]]^{\mathrm{sym}}$$

such that $\varphi_I(S_k) = s_k$ for each integer $k \geq 1$.

THEOREM 2. — *a) If* I *is a finite set of n elements, then* φ_I *induces a bicontinuous isomorphism of* $A[[S_1, \dots, S_n]]$ *onto* $A[[\mathbf{X}]]^{\mathrm{sym}}$.

b) If I *is infinite,* φ_I *is a bicontinuous isomorphism of* $A[[\mathbf{S}]]$ *onto* $A[[\mathbf{X}]]^{\mathrm{sym}}$.

In case *a)* we put $B = A[[S_1, \dots, S_n]]$ and equip this algebra with the topology induced by that of $A[[\mathbf{S}]]$; we also denote by ψ_I the restriction of φ_I to B. In case *b)* we put $B = A[[\mathbf{S}]]$ and $\psi_I = \varphi_I$. We shall equip the polynomial algebra $A[\mathbf{S}]$ with the graduation of type $\mathbf{N}$ for which S_k is of weight k for every integer $k \geq 1$.

Lemma 1. — *Let* J *be a finite subset of* I, *r an integer such that* Card J $\geq r$ *and f a symmetric formal power series which is homogeneous of degree r in the* X_i $(i \in I)$. *Let* $\bar{f}$ *be the formal power series obtained by substituting* 0 *for* X_i *for each i in* I – J. *If* $\bar{f} = 0$, *then we have* $f = 0$.

Let us put $f = \sum_{|\alpha| = r} a_\alpha X^\alpha$ (where $|\alpha|$ is the length $\sum_{i \in I} \alpha_i$ of the multiindex $\alpha = (\alpha_i)_{i \in I}$). If α is a multiindex of length r, and J′ is the support of α (the set of $i \in I$ such that $\alpha_i \neq 0$), then Card J′ $\leq r$, hence there exists a permutation $\sigma \in \mathfrak{S}_I$ such that $\sigma(J') \subset J$. Put $\beta_i = \alpha_{\sigma^{-1}(i)}$ for $i \in I$, then the monomial $X^\beta = \prod_{i \in I} X_{\sigma(i)}^{\alpha_i}$ depends only on the indeterminates X_j $(j \in J)$, whence $a_\beta = 0$ by the hypothesis $\bar{f} = 0$. Since f is symmetric, we have $a_\alpha = a_\beta$ and since α was arbitrary, we conclude that $f = 0$.

Lemma 2. — *Let f be a symmetric formal power series of degree r in the X_i $(i \in I)$. There exists a unique polynomial* $P \in B \cap A[S]$, *isobaric of weight r, such that* $f = \psi_I(P)$.

The case when I is finite follows from Th. 1 (IV, p. 62).

Assume that I is infinite and choose a finite subset J of I, containing r elements. We keep the notation of Lemma 1 ; we remark that every isobaric polynomial of weight r in the S_n $(n \geqslant 1)$ depends only on $S_1, \ldots, S_r$ and that $\bar{s}_1, \ldots, \bar{s}_r$ are the elementary symmetric polynomials in the r indeterminates X_j $(j \in J)$. If P is an isobaric polynomial of weight r in the S_n and $h = f - \psi_I(P)$, then $\bar{h} = \bar{f} - P(\bar{s}_1, \ldots, \bar{s}_r)$ and Lemma 1 shows the relation $f = \psi_I(P)$ to be equivalent to $\bar{f} = P(\bar{s}_1, \ldots, \bar{s}_r)$. By Th. 1 (IV, p. 62) there exists a unique polynomial $P \in A[S]$, isobaric of weight r and such that $\bar{f} = P(\bar{s}_1, \ldots, \bar{s}_r)$, whence the result.

Lemma 3. — *For every integer $m \geqslant 0$ let $\mathfrak{c}_m$ be the ideal of the algebra* $A[[X]]^{\text{sym}}$ *consisting of all symmetric formal power series of order $\geqslant m$. The sequence $(\mathfrak{c}_m)_{m \geqslant 0}$ is a fundamental system of neighbourhoods of* 0 *in* $A[[X]]^{\text{sym}}$.

The lemma clearly holds when I is finite, so let us assume that I is infinite. For every finite subset J of I, consisting of m elements, denote by $\tilde{J}$ the set of elements of $\mathbf{N}^{(I)}$ of length $< m$ and with support in J. Further denote by $\mathfrak{a}_J'$ the set of formal power series containing no term aX^α with $\alpha \in \tilde{J}$. Since $\tilde{J}$ is a finite subset of $\mathbf{N}^{(I)}$ and every finite subset of $\mathbf{N}^{(I)}$ is contained in a set of the form $\tilde{J}$, the family $(\mathfrak{a}_J')$ is a base of neighbourhoods of 0 in $A[[X]]$ (IV, p. 26). Now Lemma 1 implies the relation $\mathfrak{a}_J' \cap A[[X]]^{\text{sym}} = \mathfrak{c}_m$ for every subset J with m elements, and this proves Lemma 3.

Since there are only a finite number of monomials of a given weight in the S_k, every formal power series $f \in B$ may be written in a unique way as $f = \sum_{r \geqslant 0} P_r$, where P_r is an isobaric polynomial of weight r in $B \cap A[S]$. For every integer $m \geqslant 0$ let $\mathfrak{b}_m$ be the ideal of B consisting of all formal power series of the above type such that $P_r = 0$ for $0 \leqslant r < m$. The sequence $(\mathfrak{b}_m)_{m \geqslant 0}$ is a base of neighbourhoods of 0 in B.

With the above notation $\psi_I(P_r)$ is a symmetric formal power series in the X_i, homogeneous of degree r, and so this is the homogeneous component of degree r of $\psi_I(f)$. Lemma 2 shows that ψ_I is an algebra homomorphism of B onto $A[[X]]^{\text{sym}}$, mapping $\mathfrak{b}_m$ to $\mathfrak{c}_m$ for every integer $m \geqslant 0$; now Lemma 3 shows ψ_I to be bicontinuous, and this completes the proof of Theorem 2.

4. Sums of powers

We again write $\mathbf{X} = (X_i)_{i \in I}$ for a family of indeterminates. The symmetric formal power series s_k are defined as before by

(16) $$s_k = \sum_{H \in \mathfrak{P}_k} \prod_{i \in H} X_i \quad (k \geqslant 1),$$

where $\mathfrak{P}_k$ is the set of all k-element subsets of I. We also write

(17) $$p_k = \sum_{i \in I} X_i^k \quad (k \geqslant 1).$$

This is a symmetric formal power series which is homogeneous of degree k.

Lemma 4 (« Newton's relations »). — *For every integer $d \geqslant 1$ we have*

(18) $$p_d = \sum_{k=1}^{d-1} (-1)^{k-1} s_k p_{d-k} + (-1)^{d+1} d s_d .$$

Let us define a continuous derivation Δ in $A[[\mathbf{X}]]$ by $\Delta(u) = \sum_{n \geqslant 0} n u_n$, where for every u in $A[[\mathbf{X}]]$, u_n is the homogeneous component of degree n of u. By (16) and Prop. 2 of IV, p. 27 we have

(19) $$1 + \sum_{k \geqslant 1} s_k = \prod_{i \in I} (1 + X_i).$$

By Prop. 9 of IV, p. 34 we thus have

(20) $$\left(\sum_{k \geqslant 1} k s_k\right) \cdot \left(1 + \sum_{k \geqslant 1} s_k\right)^{-1} = \sum_{i \in I} \Delta(X_i)/(1 + X_i).$$

We have $\Delta(X_i) = X_i$ and $X_i/(1 + X_i) = \sum_{k \geqslant 1} (-1)^{k-1} X_i^k$. The right-hand side of (20) is thus equal to $\sum_{k \geqslant 1} (-1)^{k-1} p_k$. Hence by (20) we have

$$\sum_{k \geqslant 1} k s_k = \left(1 + \sum_{k \geqslant 1} s_k\right) \cdot \left(\sum_{k \geqslant 1} (-1)^{k-1} p_k\right)$$

and now Lemma 4 follows by a comparison of homogeneous components of degree d.

Remark. — With the notation of the above proof we have

$$\Delta u = \sum_{i \in I} X_i \, . \, D_i(u)$$

(IV, p. 33, Cor. 1). In other words, *Euler's relation* (IV, p. 8, Prop 6) extends to formal power series : if $u \in A[\mathbf{X}]$ is homogeneous of degree n, then

$$n \cdot u = \sum_{i \in I} X_i \cdot D_i(u) . \tag{21}$$

When I is finite, consisting of n elements, we have $s_k = 0$ for $k > n$. Then Newton's relations may be written

$$p_2 = s_1 p_1 - 2 s_2$$
$$p_3 = s_1 p_2 - s_2 p_1 + 3 s_3$$
$$\cdots\cdots\cdots\cdots\cdots\cdots\cdots\cdots\cdots\cdots$$
$$p_{n-1} = s_1 p_{n-2} - s_2 p_{n-3} + \cdots + (-1)^{n-1} s_{n-2} p_1 + (-1)^n (n-1) s_{n-1}$$
$$p_n = s_1 p_{n-1} - s_2 p_{n-2} + \cdots + (-1)^n s_{n-1} p_1 + (-1)^{n+1} n s_n$$

and

$$p_k = s_1 p_{k-1} - s_2 p_{k-2} + \cdots + (-1)^{n+1} s_n p_{k-n} \quad (\text{for } k > n) . \tag{22}$$

The first n of the above relations hold for any I ; we find for example

$$p_1 = s_1 , \quad p_2 = s_1^2 - 2 s_2 , \quad p_3 = s_1^3 - 3 s_1 s_2 + 3 s_3 .$$

More generally, let $\mathbf{S} = (S_n)_{n \geqslant 1}$ be a family of indeterminates. Let us define the polynomials $P_d \in \mathbf{Z}[S_1, \ldots, S_d]$ recursively by $P_1 = S_1$ and

$$P_d = \sum_{k=1}^{d-1} (-1)^{k-1} S_k P_{d-k} + (-1)^{d+1} d S_d \quad (d \geqslant 2) .$$

Then we have the « *universal formulae* » $p_d = P_d(s_1, \ldots, s_d)$ holding over any ring A and family $\mathbf{X}$ of indeterminates.

Let $\mathbf{P} = (P_k)_{k \geqslant 1}$ be a sequence of indeterminates. Since p_k is homogeneous of degree k in $A[[\mathbf{X}]]$, there exists precisely one continuous A-algebra homomorphism

$$\lambda_I : A[[\mathbf{P}]] \to A[[\mathbf{X}]]^{\mathrm{sym}}$$

such that $\lambda_I(P_k) = p_k$ for every integer $k \geqslant 1$ (IV, p. 28). If we assign to P_k the weight k, then λ_I transforms an isobaric polynomial of weight n in the P_k into a formal power series which is homogeneous of degree n in the X_i.

PROPOSITION 3. — *a) If* I *is a finite set with n elements and* $n! \cdot 1$ *is invertible in* A, *then* λ_I *induces a bicontinuous isomorphism of* $A[[P_1, \ldots, P_n]]$ *onto* $A[[\mathbf{X}]]^{\mathrm{sym}}$.

b) If I *is infinite and* A *is a* **Q**-*algebra, then* λ_I *is a bicontinuous isomorphism of* $A[[\mathbf{P}]]$ *onto* $A[[\mathbf{X}]]^{\mathrm{sym}}$.

We shall treat the case *a*) only, the case *b*) being quite similar.

By Th. 2 (IV, p. 68) we can identify $A[[\mathbf{X}]]^{\text{sym}}$ with the algebra of formal power series $A[[S_1, \ldots, S_n]]$, with S_k corresponding to s_k. By Lemma 4 of IV, p. 70, there exist formal power series $g_1, \ldots, g_n$ of order $\geqslant 2$ in the indeterminates $s_1, \ldots, s_n$ such that

$$p_k = (-1)^{k+1} k s_k + g_k(s_1, \ldots, s_n) \quad (1 \leqslant k \leqslant n).$$

Since $k! \,.\, 1$ is invertible in A, Lemma 2 of IV, p. 35 proves the existence of an automorphism T of the topological A-algebra $A[[\mathbf{X}]]^{\text{sym}}$ which maps s_k to p_k for $1 \leqslant k \leqslant n$. Now Prop. 3, *a*) is an immediate consequence.

COROLLARY. — *Let $\xi_1, \ldots, \xi_n, \eta_1, \ldots, \eta_n$ be elements of* A *and suppose that* A *is an integral domain.*

a) *If $s_k(\xi_1, \ldots, \xi_n) = s_k(\eta_1, \ldots, \eta_n)$ for $1 \leqslant k \leqslant n$, then there exists a permutation $\sigma \in \mathfrak{S}_n$ such that $\eta_i = \xi_{\sigma(i)}$ for $1 \leqslant i \leqslant n$.*

b) *Suppose that $n! \,.\, 1 \neq 0$ in* A *and*

$$\xi_1^k + \cdots + \xi_n^k = \eta_1^k + \cdots + \eta_n^k \tag{23}$$

for $1 \leqslant k \leqslant n$. Then there exists a permutation $\sigma \in \mathfrak{S}_n$ such that $\eta_i = \xi_{\sigma(i)}$ for $1 \leqslant i \leqslant n$.

Under the hypotheses *a*) we have $\prod_{i=1}^{n} (X - \xi_i) = \prod_{i=1}^{n} (X - \eta_i)$. If we substitute η_n for X, we find that $\prod_{i=1}^{n} (\eta_n - \xi_i) = 0$ and since A is an integral domain, there is an integer $\sigma(n)$ such that $1 \leqslant \sigma(n) \leqslant n$ and $\eta_n = \xi_{\sigma(n)}$. Now the assertion *a*) follows easily by induction because A[X] is an integral domain.

Under the hypotheses of *b*) there exist by Prop. 3 polynomials $\Pi_1, \ldots, \Pi_n$ in n indeterminates, with coefficients in the field of fractions of A such that $s_k = \Pi_k(p_1, \ldots, p_n)$ for $1 \leqslant k \leqslant n$. Now the relation (23) implies

$$s_k(\xi_1, \ldots, \xi_n) = s_k(\eta_1, \ldots, \eta_n)$$

for $1 \leqslant k \leqslant n$ and so *b*) follows from *a*).

5. Symmetric functions in the roots of a polynomial

Consider a monic polynomial of degree n, with coefficients in A :

$$f = X^n + a_1 X^{n-1} + \cdots + a_{n-1} X + a_n .$$

We define an associative, commutative and unital A-algebra E_f with the generators $x_1, \ldots, x_n$ and relations

$$\sum_{i_1 < \ldots < i_k} x_{i_1} \ldots x_{i_k} = (-1)^k a_k \quad (1 \leqslant k \leqslant n). \tag{24}$$

More precisely, we have

$$E_f = A[X_1, \ldots, X_n]/\mathfrak{a}$$

where the ideal $\mathfrak{a}$ is generated by the polynomials $s_k + (-1)^{k+1}a_k$ for $1 \leqslant k \leqslant n$, and x_i is the residue class of X_i mod $\mathfrak{a}$ for $1 \leqslant i \in n$. The relation (24) is also equivalent to $f(X) = \prod_{i=1}^{n} (X - x_i)$. When there is any risk of ambiguity, we write $x_{1,f}, \ldots, x_{n,f}$ in place of $x_1, \ldots, x_n$.

PROPOSITION 4. — *Let* B *be a commutative ring,* ρ *a homomorphism of* A *into* B *and* $\xi_1, \ldots, \xi_n$ *elements of* B. *Suppose that the relation* ${}^{\rho}f(X) = \prod_{i=1}^{n} (X - \xi_i)$ *holds in* B[X]. *Then there exists one and only one ring homomorphism* $u : E_f \to B$ *such that* $\rho(a) = u(a \,.\, 1)$ *for all* $a \in A$ *and* $u(x_i) = \xi_i$ *for* $1 \leqslant i \leqslant n$.

We consider B as associative, commutative and unital A-algebra by means of ρ. Then the relation ${}^{\rho}f(X) = \prod_{i=1}^{n} (X - \xi_i)$ may also be written in the form

$$\sum_{i_1 < \cdots < i_k} \xi_{i_1} \cdots \xi_{i_k} = (-1)^k a_k \,.\, 1 \quad (1 \leqslant k \leqslant n)$$

in B. Since the relations (24) define a presentation of E_f, Prop. 4 follows.

Prop. 4 justifies the name « *universal decomposition algebra of* f » for E_f. The relation $f(X) = \prod_{i=1}^{n} (X - x_{i,f})$ is called the « *universal decomposition of* f ». Let $\sigma \in \mathfrak{S}_n$ be a permutation ; since $f(X) = \prod_{i=1}^{n} (X - x_{\sigma(i),f})$, there exists an automorphism t_σ of the A-algebra E_f characterized by $t_\sigma(x_{i,f}) = x_{\sigma(i),f}$ for $1 \leqslant i \leqslant n$. We have $t_{\sigma\tau} = t_\sigma \circ t_\tau$ for σ, τ in $\mathfrak{S}_n$, and hence obtain an action of the group $\mathfrak{S}_n$ an the A-algebra E_f.

PROPOSITION 5. — *In the universal decomposition algebra* E_f *the family of monomials* $x_1^{\nu(1)} \ldots x_n^{\nu(n)}$ *such that* $0 \leqslant \nu(i) < i$ *for* $1 \leqslant i \leqslant n$ *is a basis of the* A-*module* E_f. *In particular* E_f *is a free* A-*module of rank* $n!$.

Write $B = A[X_1, \ldots, X_n]$ and $C = A[X_1, \ldots, X_n]^{\text{sym}}$. By Th. 1 (IV, p. 62) we have $C = A[s_1, \ldots, s_n]$ and $s_1, \ldots, s_n$ are algebraically independent over A. The polynomials without constant term in $s_1, \ldots, s_n$ form an ideal C^+ in C, supplementary to A and generated by $s_1, \ldots, s_n$. Let $\mathfrak{c}$ be the ideal of C generated by $s_1 + a_1, s_2 - a_2, \ldots, s_n + (-1)^{n+1}a_n$. There exists an A-algebra automorphism of C which maps s_k to $s_k + (-1)^{k+1}a_k$ for $1 \leqslant k \leqslant n$, and hence maps C^+ onto $\mathfrak{c}$; therefore we have $C = A \oplus \mathfrak{c}$. Moreover Th. 1, *c*) of IV, p. 62 shows that

$$B = \bigoplus_{\nu \in S} CX^{\nu}$$

where S is the set of all $\nu \in \mathbf{N}^n$ such that $0 \leqslant \nu(i) < i$ for $1 \leqslant i \leqslant n$. The ideal $\mathfrak{a}$ of B is generated by $\mathfrak{c}$, whence $\mathfrak{a} = \mathrm{B}\mathfrak{c} = \bigoplus_{\nu \in \mathrm{S}} \mathfrak{c} \,.\, \mathrm{X}^\nu$. Since $\mathrm{C} = \mathrm{A} \oplus \mathfrak{c}$, we find

$$\mathrm{B} = \mathfrak{a} \oplus \bigoplus_{\nu \in \mathrm{S}} \mathrm{AX}^\nu ,$$

whence Prop. 5, because $\mathrm{E}_f = \mathrm{B}/\mathfrak{a}$.

COROLLARY. — *The canonical homomorphism of* A *into the universal decomposition algebra of the monic polynomial* $f \in \mathrm{A}[\mathrm{X}]$ *is injective.*

For the unit element of E_f forms part of a basis of the A-module E_f.

PROPOSITION 6. — *Let* $f \in \mathrm{A}[\mathrm{X}]$ *be a monic polynomial of degree n, and let* P *be a symmetric polynomial in* $\mathrm{X}_1, \ldots, \mathrm{X}_n$ *with coefficients in* A. *Then there exists precisely one element a of* A *with the following property :*
(FS) *For any ring homomorphism* $\rho : \mathrm{A} \to \mathrm{B}$ *and decomposition* ${}^\rho f(\mathrm{X}) = \prod_{i=1}^{n} (\mathrm{X} - \xi_i)$ *in* $\mathrm{B}[\mathrm{X}]$ *we have* $\rho(a) = \mathrm{P}(\xi_1, \ldots, \xi_n)$.

Write $f = \mathrm{X}^n + \sum_{k=1}^{n} a_k \mathrm{X}^{n-k}$, then by Th. 1, IV, p. 62 there exists a polynomial Π in n indeterminates with coefficients in A, such that $\mathrm{P} = \Pi(s_1, \ldots, s_n)$. Put $a = \Pi(-a_1, a_2, \ldots, (-1)^n a_n)$. Under the hypothesis (FS) we have

$$s_k(\xi_1, \ldots, \xi_n) = (-1)^k \rho(a_k)$$

whence

$$\begin{aligned} \rho(a) &= \Pi(-\rho(a_1), \rho(a_2), \ldots, (-1)^n \rho(a_n)) \\ &= \Pi(s_1(\xi_1, \ldots, \xi_n), \ldots, s_n(\xi_1, \ldots, \xi_n)) \\ &= \mathrm{P}(\xi_1, \ldots, \xi_n) . \end{aligned}$$

This proves the existence of an element a satisfying (FS). The uniqueness of a follows from the Cor. to Prop. 5, because we have $a \,.\, 1 = \mathrm{P}(x_{1,f}, \ldots, x_{n,f})$ in the universal decomposition algebra E_f.

With the notation of Prop. 6 one sometimes writes $a = \mathrm{P}^*(f)$. Here are some examples.

Examples. — 1) If $\mathrm{P} = s_k$, then $\mathrm{P}^*(f) = (-1)^k a_k$.

* 2) Let g be a polynomial in $\mathrm{A}[\mathrm{X}]$ and put

$$\mathrm{P}(\mathrm{X}_1, \ldots, \mathrm{X}_n) = g(\mathrm{X}_1) \ldots g(\mathrm{X}_n) .$$

Then $\mathrm{P}^*(f)$ is just the resultant $\operatorname{res}(f, g)$, by Cor. 1 of IV, p. 80.

3) Put $\Delta(\mathrm{X}_1, \ldots, \mathrm{X}_n) = \prod_{i<j} (\mathrm{X}_i - \mathrm{X}_j)^2$, then $\Delta^*(f)$ is just the discriminant of the monic polynomial f (IV, p. 82, formula (46)). *

4) Put $P(X_1, \ldots, X_n) = X_1^k + \cdots + X_n^k$; moreover, define the algebra $E = A[X]/(f)$ and write x for the image of X in E. We recall that the A-module E is free, with basis $(1, x, \ldots, x^{n-1})$ (IV, p. 11, Cor.). Let us show

$$\mathrm{Tr}_{E/A}(x^k) = P^*(f)\,. \tag{25}$$

Put $\pi_k = \mathrm{Tr}_{E/A}(x^k)$ for every integer $k \geqslant 1$. Bearing in mind Newton's relations (IV, p. 70) we need only establish the relations

$$\pi_k + a_1\pi_{k-1} + \cdots + a_{k-1}\pi_1 + ka_k = 0 \quad \text{for} \quad 1 \leqslant k \leqslant n \tag{26}$$

$$\pi_k + a_1\pi_{k-1} + \cdots + a_{n-1}\pi_{k-n+1} + a_n\pi_{k-n} = 0 \quad \text{for} \quad k > n \tag{27}$$

(which we shall also call « Newton's relations »). The relation (27) is clear, because the left-hand side is the trace of

$$x^{k-n}(x^n + a_1x^{n-1} + \cdots + a_{n-1}x + a_n) = 0\,.$$

Suppose that $1 \leqslant k \leqslant n$ and put

$$y = x^k + a_1x^{k-1} + \cdots + a_{k-1}x + a_k \cdot 1\,;$$

let $M = (m_{ij})$ be the matrix of the linear mapping $u \mapsto yu$ in E relative to the basis $(x^i)_{0 \leqslant i \leqslant n}$. We easily obtain the relations

$$m_{ii} = a_k \quad \text{for} \quad 0 \leqslant i < n - k$$
$$m_{ii} = 0 \quad \text{for} \quad n - k \leqslant i < n\,,$$

whence

$$\mathrm{Tr}_{E/A}(y) = \sum_{i=0}^{n-1} m_{ii} = (n-k)\,a_k\,.$$

Moreover we have

$$\mathrm{Tr}_{E/A}(y) = \pi_k + a_1\pi_{k-1} + \cdots + a_{k-1}\pi_1 + na_k\,,$$

whence formula (26) follows.

6. The resultant

In this No. we assume as given two positive integers p, q and two polynomials f, g in A[X] of the form

$$f = t_pX^p + t_{p-1}X^{p-1} + \cdots + t_0$$
$$g = u_qX^q + u_{q-1}X^{q-1} + \cdots + u_0$$

such that $\deg f \leqslant p$, $\deg g \leqslant q$. For every integer $n \geqslant 0$ we denote by S_n the sub-A-module of A[X] consisting of all polynomials of degree $< n$; it has the family $(X^i)_{0 \leqslant i < n}$ as basis, and hence is of rank n.

We provide $S_q \times S_p$ with the basis

$$B_1 = ((X^{q-1}, 0), \ldots, (X, 0), (1, 0), (0, X^{p-1}), \ldots, (0, X), (0, 1))$$

and S_{p+q} with the basis

$$B_2 = (X^{p+2-1}, \ldots, X, 1).$$

We define a linear mapping $\varphi : S_q \times S_p \to S_{p+q}$ by

$$\varphi(u, v) = uf + vg$$

and we denote by $M(f, g, p, q)$ the matrix of φ with respect to the bases B_1 and B_2. This is a square matrix of order $p + q$, indexed by the set $\{0, 1, \ldots, p + q - 1\}$. Its elements a_{ij} are given by the rules :

a) $a_{ij} = t_{p-i+j}$ for $0 \leqslant j \leqslant q - 1$,
b) $a_{ij} = u_{j-i}$ for $q \leqslant j \leqslant p + q - 1$,

where t_k is taken to be 0 if $k \notin [0, p]$ and $u_k = 0$ if $k \notin [0, q]$.

For example, when $p = 2$ and $q = 3$, we have the matrix

$$\begin{pmatrix} t_2 & 0 & 0 & u_3 & 0 \\ t_1 & t_2 & 0 & u_2 & u_3 \\ t_0 & t_1 & t_2 & u_1 & u_2 \\ 0 & t_0 & t_1 & u_0 & u_1 \\ 0 & 0 & t_0 & 0 & u_0 \end{pmatrix}.$$

DEFINITION 1. — *With the above notation the determinant of the matrix $M(f, g, p, q)$ is called the resultant of the pair (f, g) for the degrees p and q, or simply the resultant of f and g if $p = \deg f$ and $q = \deg g$.*

The resultant is denoted by $\operatorname{res}_{p,q}(f, g)$ or simply $\operatorname{res}(f, g)$ when $p = \deg f$, $q = \deg g$.

Examples. — 1) Given λ, μ in A, we have the formulae

$$\operatorname{res}_{p,0}(f, \lambda) = \lambda^p, \quad \operatorname{res}_{0,q}(\mu, g) = \mu^q$$
$$\operatorname{res}_{p,1}(f, \lambda) = \lambda^p t_p, \quad \operatorname{res}_{1,q}(\mu, g) = (-1)^q \mu^q u_q,$$

whose proof is immediate.

2) When $p = q = 1$, we have

$$\operatorname{res}_{1,1}(t_1 X + t_0, u_1 X + u_0) = t_1 u_0 - t_0 u_1.$$

Remarks. — 1) The matrix $M(g, f, q, p)$ is obtained from $M(f, g, p, q)$ by pq transpositions of columns, whence

$$\operatorname{res}_{q,p}(g, f) = (-1)^{pq} \operatorname{res}_{p,q}(f, g).$$

2) Let $\rho : A \to B$ be a ring homomorphism. Def. 1 implies directly the formula

$$\mathrm{res}_{p,q}({}^{\rho}f, {}^{\rho}g) = \rho(\mathrm{res}_{p,q}(f, g)).$$

3) Given λ, μ in A, we have

$$\mathrm{res}_{p,q}(\lambda f, \mu g) = \lambda^q \mu^p \mathrm{res}_{p,q}(f, g). \tag{28}$$

4) Suppose that $p + q \geqslant 1$. By III, p. 532, formula (28), the image of φ contains the constant polynomial $\mathrm{res}_{p,q}(f, g)$. Hence there exist a pair of polynomials (u, v), with $u \in S_q$, $v \in S_p$ such that

$$\mathrm{res}_{p,q}(f, g) = uf + vg,$$

whence

$$\mathrm{res}_{p,q}(f, g) \in A \cap (f, g).$$

This pair (u, v) is unique when $\mathrm{res}_{p,q}(f, g)$ is cancellable in A: for then φ is injective (III, p. 524, Prop. 3).

5) Suppose that $p \geqslant q$, and let $h \in A[X]$ be a polynomial of degree $\leqslant p - q$. Let us show that

$$\mathrm{res}_{p,q}(f, g) = \mathrm{res}_{p,q}(f + gh, g). \tag{29}$$

For if we write $\omega(u, v) = (u, uh + v)$ for $(u, v) \in S_q \times S_p$, then ω is an automorphism of the A-module $S_q \times S_p$ and we have

$$\omega^{-1}(u, v) = (u, -uh + v).$$

The matrix of ω with respect to the basis B_1 is lower triangular and its diagonal elements are equal to 1. On the other hand, $\varphi \circ \omega$ maps (u, v) to $u(f + gh) + vg$. Now formula (29) means that the matrices representing φ and $\varphi \circ \omega$ have the same determinant and this follows from the relation $\det \omega = 1$.

Assume that f is monic of degree p; put $E = A[X]/(f)$ and denote by x the canonical image of X in E. We know (IV, p. 11) that E is a free A-module with basis $(1, x, \ldots, x^{p-1})$. We can thus define the norm $N_{E/A}(u)$ of any element u of E (III, p. 543, Def. 2).

PROPOSITION 7. — *Assume that f is monic of degree p; with the above notations we have*[1]

$$\mathrm{res}_{p,q}(f, g) = N_{E/A}(g(x)). \tag{30}$$

[1] The resultant $\mathrm{res}_{p,q}(f, g)$ is thus independent of q when f is monic of degree p. We then denote it simply by $\mathrm{res}(f, g)$.

Let us define an A-linear mapping θ of $S_q \times S_p$ into S_{p+q} by $\theta(u, v) = uf + v$. Then θ maps the basis B_1 of $S_q \times S_p$ into the sequence

$$(fX^{q-1}, \dots, fX, f, X^{p-1}, \dots, X, 1)$$

of elements of S_{p+q}; hence the matrix M_θ of θ with respect to the bases B_1, B_2 is lower triangular and its diagonal elements are equal to 1, whence det $M_\theta = 1$.

It follows that θ is bijective and $\operatorname{res}_{p,q}(f, g)$ is equal to the determinant of the endomorphism $\varphi' = \varphi \circ \theta^{-1}$ of S_{p+q}. Explicitly, we have

$$\varphi'(uf + v) = uf + vg \tag{31}$$

for any pair (u, v) in $S_q \times S_p$. Now we have $A[X] = S_{p+q} + (f)$ and $fS_q = S_{p+q} \cap (f)$, hence the canonical injection of S_{p+q} into $A[X]$ defines by the passage to quotients an isomorphism γ of S_{p+q}/fS_q onto E. Let ψ be the multiplication by $g(x)$ in E. Formula (31) shows that φ' induces the identity on fS_q and $\gamma^{-1}\psi\gamma$ on S_{p+q}/fS_q. We thus have $\det \varphi' = \det \psi$, whence (30) follows, because $\det \varphi' = \operatorname{res}_{p,q}(f, g)$ and $\det \psi = N_{E/A}(g(x))$, by definition.

Corollary 1. — *Let $f \in A[X]$ be a monic polynomial; for every polynomial $g \in A[X]$ the following conditions are equivalent:*

(i) $\operatorname{res}(f, g)$ *is invertible in* A;

(ii) *there exist polynomials u, v in* $A[X]$ *such that* $uf + vg = 1$;

(iii) $g(x)$ *is invertible in the algebra* $A[X]/(f)$.

The equivalence of (i) and (iii) follows from Prop. 3 of III, p. 545 and Prop. 7; that of (ii) and (iii) follows trivially.

Corollary 2. — *Suppose that A is a field and let $f, g \in A[X]$, then the following conditions are equivalent when f, g are non-zero:*

(i) $\operatorname{res}(f, g) \neq 0$;

(ii) *the polynomials f and g are relatively prime in* $A[X]$;

* (iii) *for any extension L of A the polynomials f and g have no common root in* L. *

We can reduce at once to the case where f is monic (IV, p. 77, remark 3).

The equivalence of (i) and (ii) is just a translation of Cor. 1, by Def. 1 of IV, p. 12; the equivalence of (ii) and (iii) is just Cor. 7 of IV, p. 13.

Corollary 3. — *For every $\lambda \in A$ we have*

$$\operatorname{res}_{p,1}(f, \lambda - X) = f(\lambda), \quad \operatorname{res}_{1,q}(X - \lambda, g) = g(\lambda). \tag{32}$$

When $f(X) = X - \lambda$, the algebra E is equal to A and we have $x = \lambda$; the second formula (32) now follows from Prop. 7 (IV, p. 77). By remarks 1 and 3 (IV, p. 76, 77) we conclude

$$\operatorname{res}_{p,1}(f, \lambda - X) = (-1)^p \operatorname{res}_{1,p}(\lambda - X, f) = (-1)^{p+p} \operatorname{res}_{1,p}(X - \lambda, f) = f(\lambda).$$

Suppose now that f and g are monic. We denote by F the A-algebra $A[X, Y]/(f(X), g(Y))$ and denote by x (resp. y) the canonical image of X (resp. Y) in F.

PROPOSITION 8. — *Suppose that f and g are monic of degrees p and q respectively. With the above notation, the* A*-module* F *is free with basis* $(x^i y^j)_{0 \leqslant i < p, 0 \leqslant j < q}$ *and we have*

$$\operatorname{res}(f, g) = N_{F/A}(x - y) . \tag{33}$$

Put $E = A[X]/(f)$ and $E' = A[Y]/(g)$. By II, p. 253, Cor. 1, the homomorphism σ of $E \otimes E'$ into F derived from the canonical homomorphism $A[X] \otimes A[Y] \to A[X, Y]$ is bijective ; this proves the assertion about the basis of F. We now identify E with its image in F under σ. Then the E-algebra homomorphism of $E[Y]/(g(Y))$ into F which maps Y to y is an isomorphism.

By the transitivity of the norm (III, p. 546) we have

$$N_{F/A}(x - y) = N_{E/A}(N_{F/E}(x - y)) . \tag{34}$$

By Prop. 7 (IV, p. 77), $N_{F/E}(x - y)$ is the resultant of the polynomials $g(Y)$ and $x - Y$ in $E[Y]$, hence equal to $g(x)$ (IV, p. 78, Cor. 3). By formula (34) and Prop. 7 (IV, p. 77) we thus have

$$N_{F/A}(x - y) = N_{E/A}(g(x)) = \operatorname{res}(f, g) .$$

PROPOSITION 9. — *Let p_1 and q_1 be positive integers and f_1, g_1 polynomials in* $A[X]$ *such that* $\deg f_1 \leqslant p_1$, $\deg g_1 \leqslant q_1$; *then we have*

$$\operatorname{res}_{p,q+q_1}(f, gg_1) = \operatorname{res}_{p,q}(f, g) . \operatorname{res}_{p,q_1}(f, g_1) \tag{35}$$

$$\operatorname{res}_{p+p_1,q}(ff_1, g) = \operatorname{res}_{p,q}(f, g) . \operatorname{res}_{p_1,q}(f_1, g) . \tag{36}$$

We have $\operatorname{res}_{p,q}(f, g) = (-1)^{pq} \operatorname{res}_{q,p}(g, f)$ (IV, p. 76) ; so it is enough to prove (35). Likewise, Remark 3 (loc. cit.) shows that if formula (35) is established for a polynomial f, then it holds for all polynomials of the form λf, where $\lambda \in A$. Finally, when f is monic of degree p, formula (35) follows from Prop. 7 (IV, p. 77) by the formula $N_{E/A}(ab) = N_{E/A}(a) . N_{E/A}(b)$. In all we conclude that (35) holds when the coefficient t_p of X^p in f is *invertible*.

Lemma 5. — *Let t be an element of* A. *There exists a commutative ring* C *containing* A *as subring, a subring* B *of* C *containing* A, *an element τ of* B *invertible in* C *and a ring homomorphism* $\rho : B \to A$ *such that* $\rho(\tau) = t$ *and the restriction of* ρ *to* A *is equal to* Id_A.

It suffices to take for B the algebra $A^{(\mathbf{N})}$ of the monoid $\mathbf{N}$, that is, the polynomial algebra $A[\tau]$ in one indeterminate τ, for C the algebra $A^{(\mathbf{Z})}$ of the group $\mathbf{Z}$ and for ρ the homomorphism $P \mapsto P(t)$ of $A(\tau)$ into A.

With the notation of Lemma 5, where we have taken $t = t_p$, let us put

$$F = \tau X^p + t_{p-1} X^{p-1} + \cdots + t_1 X + t_0 \tag{37}$$

in $B[X]$. The coefficient of X^p in F is *invertible in* C; if we consider F, g, g_1 as polynomials of $C[X]$ we thus have

$$\operatorname{res}_{p,q+q_1}(F, gg_1) = \operatorname{res}_{p,q}(F, g) \cdot \operatorname{res}_{p,q_1}(F, g_1) \tag{38}$$

by what has been said. The resultants are unchanged if we consider F, g and g_1 as polynomials in $B[X]$. Since ${}^{\rho}F = f$, ${}^{\rho}g = g$, ${}^{\rho}g_1 = g_1$, formula (35) follows from (38) and Remark 2 (IV, p. 77).

COROLLARY 1. — (i) *Let* $\lambda, \alpha_1, \ldots, \alpha_p$ *be elements of* A *and suppose that* $f(X) = \lambda(X - \alpha_1) \ldots (X - \alpha_p)$. *We have*

$$\operatorname{res}_{p,q}(f, g) = \lambda^q g(\alpha_1) \ldots g(\alpha_p)\,. \tag{39}$$

(ii) *Let* $\mu, \beta_1, \ldots, \beta_q$ *be elements of* A *and suppose further that* $g(X) = \mu(X - \beta_1) \ldots (X - \beta_q)$. *Then we have*

$$\operatorname{res}_{p,q}(f, g) = \lambda^q \mu^p \prod_{\substack{1 \leqslant i \leqslant p \\ 1 \leqslant j \leqslant q}} (\alpha_i - \beta_j)\,. \tag{40}$$

Assertion (i) follows directly from the formulae (28), (32) and (36). Now Assertion (ii) follows from (i).

COROLLARY 2. — *For every integer* $r \geqslant 0$ *we have*

$$\operatorname{res}_{p,q+r}(f, g) = t_p^r \cdot \operatorname{res}_{p,q}(f, g)\,. \tag{41}$$

Bearing in mind Example 1 (IV, p. 76) we need only take $q_1 = r$, $g_1 = 1$ in (35).

Assume that f is monic, and let $\rho : A \to B$ be a ring homomorphism and $\xi_1, \ldots, \xi_p$ elements of B such that we have the decomposition

$${}^{\rho}f(X) = (X - \xi_1) \ldots (X - \xi_p)\,.$$

By Remark 2 (IV, p. 77) and Cor. 1 above we have

$$\rho(\operatorname{res}(f, g)) = g(\xi_1) \ldots g(\xi_p)\,.$$

This remark applies in particular to the universal decomposition of f (IV, p. 73) and since ρ is then injective, this provides a means of calculating $\operatorname{res}(f, g)$.

Example 3. — Let us prove the formula

$$\begin{aligned} \operatorname{res}_{2,2}(aX^2 + bX + c, a'X^2 + b'X + c') &= \\ &= (ac' - ca')^2 + (bc' - cb')(ba' - ab')\,. \end{aligned} \tag{42}$$

Arguing as for Prop. 9 (IV, p. 79) we see that it is enough to prove the formula when a is *invertible*. There exists then a decomposition of the form

$$aX^2 + bX + c = a(X - x)(X - y) \tag{43}$$

in $B[X]$, where B is an appropriate ring containing A as subring. By Cor. 1 above, the required resultant is equal to

$$R = a^2(a'x^2 + b'x + c')(a'y^2 + b'y + c') . \tag{44}$$

Now we have

$$ax + ay = -b , \quad axy = c$$

by (43), whence $(ax)^2 + (ay)^2 = b^2 - 2ac$.

By (44) we have

$$\begin{aligned} R &= a'^2(axy)^2 + ab'^2(axy) + a^2c'^2 + a'b'(axy)(ax + ay) + \\ &\qquad + a'c'((ax)^2 + (ay)^2) + ab'c'(ax + ay) \\ &= a'^2c^2 + ab'^2c + a^2c'^2 - a'b'cb + a'c'(b^2 - 2ac) - ab'c'b \\ &= (ac' - ca')^2 + (ab' - a'b)(b'c - c'b) , \end{aligned}$$

from which the desired result follows.

7. The discriminant

DEFINITION 2. — *Let f be a monic polynomial of* $A[X]$ *of degree m, and denote by* E *the* A*-algebra* $A[X]/(f)$ *and by x the canonical image of* X *in* E. *We define the discriminant of f, written* $\operatorname{dis}(f)$, *as the discriminant* $D_{E/A}(1, x, \dots, x^{m-1})$ *of the basis* $(1, x, \dots, x^{m-1})$ *of the* A*-algebra* E.

For every positive integer k we write $p_k = \operatorname{Tr}_{E/A}(x^k)$. By III, p. 549, Def. 2 amounts to the formula

$$\operatorname{dis}(f) = \det(p_{i+j})_{0 \leqslant i,j < m} . \tag{45}$$

Examples. — 1) If f is a monic polynomial of degree 0 or 1, we have $\operatorname{dis}(f) = 1$, by (45).

2) Let $f(X) = X^2 + \alpha X + \beta$ be a monic polynomial of degree 2. Newton's relations may be written in the form (IV, p. 75)

$$\begin{gathered} p_0 = 2 \\ p_1 + \alpha = 0 \\ p_2 + \alpha p_1 + 2\beta = 0 , \end{gathered}$$

whence $p_1 = -\alpha$, $p_2 = \alpha^2 - 2\beta$. It follows that we have

$$\operatorname{dis}(f) = \det\begin{pmatrix} 2 & -\alpha \\ -\alpha & \alpha^2 - 2\beta \end{pmatrix} = \alpha^2 - 4\beta .$$

Let B be a commutative ring, ρ a homomorphism from A to B and $\xi_1, \ldots, \xi_m$ elements of B such that

$$^{\rho}f(X) = (X - \xi_1) \ldots (X - \xi_m) .$$

Let us write M for the matrix $(\rho(p_{i+j}))_{0 \leqslant i,j < m}$ and D for the Van der Monde matrix $(\xi^j_{i+1})_{0 \leqslant i,j < m}$. By Example 4 of IV, p. 75 we have

$$\rho(p_k) = \xi_1^k + \cdots + \xi_m^k ,$$

whence $M = {}^tD . D$; by III, p. 532 we have $D = \prod_{i > j} (\xi_i - \xi_j)$ and $\det M = (\det D)^2$, that is

$$\rho(\text{dis}(f)) = \prod_{i<j} (\xi_i - \xi_j)^2 . \tag{46}$$

Moreover (on writing D for the derivation $\frac{d}{dX}$) we have

$$D(^{\rho}f)(\xi_i) = (\xi_i - \xi_1) \ldots (\xi_i - \xi_{i-1})(\xi_i - \xi_{i+1}) \ldots (\xi_i - \xi_m)$$

for $1 \leqslant i \leqslant m$, whence

$$\rho(\text{dis}(f)) = (-1)^{m(m-1)/2} \prod_{i \neq j} (\xi_i - \xi_j) = (-1)^{m(m-1)/2} \prod_{i=1}^{m} D(^{\rho}f)(\xi_i) .$$

By Cor. 1 of IV, p. 80 applied to the universal decomposition of f, we finally obtain

$$\text{res}(f, Df) = \text{res}(Df, f) = (-1)^{m(m-1)/2}\text{dis}(f) . \tag{47}$$

PROPOSITION 10. — *Let $m \geqslant 1$. There exists a unique polynomial $\Delta \in \mathbf{Z}[A_1, \ldots, A_m]$ with the following property: for any commutative ring A and monic polynomial $f = X^m + \sum_{i=1}^{m} a_i X^{m-i}$ in A[X] we have*

$$\text{dis}(f) = \Delta(a_1, \ldots, a_m) . \tag{48}$$

Moreover Δ is of degree $\leqslant 2m - 2$ and if we assign weight i to A_i then Δ is isobaric of weight $m(m-1)$.

a) Uniqueness of Δ: if Δ satisfies (48), we have in particular $\Delta = \text{dis}(F)$, where F is the polynomial $X^m + \sum_{i=1}^{m} A_i X^{m-i}$ with coefficients in $\mathbf{Z}[A_1, \ldots, A_m]$.

b) Existence of Δ: let $s_1, \ldots, s_m$ be the elementary symmetric polynomials in the indeterminates $X_1, \ldots, X_m$. There exists a polynomial $\Delta \in \mathbf{Z}[A_1, \ldots, A_m]$, *isobaric of weight* $m(m-1)$, such that

$$\Delta(-s_1, s_2, \ldots, (-1)^m s_m) = \prod_{i<j} (X_i - X_j)^2 ; \tag{49}$$

for the right-hand side is a symmetric polynomial P, homogeneous of degree $m(m-1)$ in $\mathbf{Z}[X_1, \dots, X_m]$ (IV, p. 62, Th. 1). Now formula (46) means that $\operatorname{dis}(f) = P\ (f)$, in the notation of IV, p. 74, and (48) follows directly from this.

c) Degree of Δ : the relation (47) and the definition of the resultant (IV, p. 76) imply the formula

$$(50) \qquad (-1)^{m(m-1)/2}\Delta = \det(a_{ij})_{0 \leqslant i,j \leqslant 2m-2}$$

with the following values of the a_{ij}

$$\begin{aligned} &a_{00} = 1\,, \quad a_{0,m-1} = m\,, && a_{0j} = 0 \quad \text{if} \quad j \neq 0\,, \quad j \neq m-1 \\ &a_{ij} = A_{i-j} && \text{for} \quad 1 \leqslant i \leqslant 2m-2\,, \quad 0 \leqslant j \leqslant m-2 \\ &a_{ij} = (j-i+1)\,A_{m+i-j-1} && \text{for} \quad 1 \leqslant i \leqslant 2m-2\,, \quad m-1 \leqslant j \leqslant 2m-2\,. \end{aligned}$$

In these formulae it is understood that $A_0 = 1$ and $A_i = 0$ for $i < 0$ or $i > m$. Now (50) shows at once that Δ is of degree $\leqslant 2m-2$, as we had to prove.

Prop. 10 allows us to extend the definition of the discriminant to non-monic polynomials. Let $m \geqslant 1$ be an integer, then there exists a unique homogeneous polynomial of degree $2m-2$, $\tilde{\Delta}$ say, in $\mathbf{Z}[A_0, A_1, \dots, A_m]$ such that

$$(51) \qquad \Delta(A_1, \dots, A_m) = \tilde{\Delta}(1, A_1, \dots, A_m)\,;$$

for, since Δ is of degree $\leqslant 2m-2$, the rational fraction

$$A_0^{2m-2}\Delta(A_1/A_0, \dots, A_m/A_0)$$

belongs to the subring $\mathbf{Z}[A_0, A_1, \dots, A_m]$ of $\mathbf{Q}(A_0, A_1, \dots, A_m)$. If A_i has weight i for $0 \leqslant i \leqslant m$, then $\tilde{\Delta}$ is isobaric of weight $m(m-1)$. If f is a polynomial of degree $\leqslant m$, say

$$f = a_0X^m + a_1X^{m-1} + \cdots + a_{m-1}X + a_m\,,$$

we shall put

$$(52) \qquad \operatorname{dis}_m(f) = \tilde{\Delta}(a_0, a_1, \dots, a_m)\,.$$

When $m = \deg f$, we shall simply write $\operatorname{dis}(f)$ for $\operatorname{dis}_m(f)$; if f is monic, $\operatorname{dis}(f)$ agrees with the discriminant defined in Def. 2, by (48), (51) and (52).

PROPOSITION 11. — *Let f in $A[X]$ be of degree $\leqslant m$.*

(i) *If $\rho : A \to B$ is a ring homomorphism, we have* $\operatorname{dis}_m(\rho_f) = \rho(\operatorname{dis}_m(f))$.

(ii) *Let $\lambda, \alpha_1, \dots, \alpha_m$ be elements of* A. *If $f = \lambda(X-\alpha_1)\dots(X-\alpha_m)$; then we have*

$$(53) \qquad \operatorname{dis}_m(f) = \lambda^{2m-2}\prod_{i<j}(\alpha_i - \alpha_j)^2\,.$$

(iii) *Let a_0 be the coefficient of X^m in f; then we have*

(54) $$\operatorname{res}_{m,m-1}(f, \mathrm{D}f) = \operatorname{res}_{m-1,m}(\mathrm{D}f, f) = (-1)^{m(m-1)/2} a_0 \operatorname{dis}_m(f).$$

Assertion (i) is clear.

Since $\tilde{\Delta}$ is homogeneous of degree $2m-2$, we have

(55) $$\operatorname{dis}_m(\lambda f) = \lambda^{2m-2}\operatorname{dis}_m(f)$$

for every polynomial $f \in A[X]$ of degree $\leqslant m$. Assertion (ii) now follows from formulae (46) and (55).

When f is monic of degree m, we have $a_0 = 1$ and Assertion (iii) is reduced to formula (47). Bearing in mind (55) and the relation

(56) $$\operatorname{res}_{m,n}(\lambda f, \mu g) = \lambda^n \mu^m \operatorname{res}_{m,n}(f, g),$$

(IV, p. 77), we can pass from there to the case where a_0 is invertible in A. Now the general case follows by Prop. 11, (i) and Lemma 5 (IV, p. 79).

Corollary 1. — *Let $g \in A[X]$ and let n be a positive integer such that $\deg g \leqslant n$. We have*

(57) $$\operatorname{dis}_{m+n}(fg) = \operatorname{dis}_m(f) \cdot \operatorname{dis}_n(g) \cdot \operatorname{res}_{m,n}(f,g)^2.$$

By reasoning as before we are reduced to the case where f and g are monic of degrees m and n respectively. Now put $B = E_f \otimes E_g$, where E_f (resp. E_g) is the universal decomposition algebra of f (resp. g) (IV, p. 73). Then A is a subring of B, and in B[X] we have the decompositions

$$f = \prod_{i=1}^{m} (X - \alpha_i), \quad g = \prod_{j=1}^{n} (X - \beta_j).$$

Hence we have

$$fg = \prod_{k=1}^{m+n} (X - \gamma_k),$$

with $\gamma_i = \alpha_i$ for $1 \leqslant i \leqslant m$ and $\gamma_{m+j} = \beta_j$ for $1 \leqslant j \leqslant n$. We have the obvious identity

$$\prod_{k<k'} (\gamma_k - \gamma_{k'}) = \prod_{i<i'} (\alpha_i - \alpha_{i'}) \cdot \prod_{j<j'} (\beta_j - \beta_{j'}) \cdot \prod_{i,j} (\alpha_i - \beta_j).$$

On squaring this relation we obtain (57), by (40) and (46).

Corollary 2. — *If a_0 is the coefficient of X^m in f, we have*

(58) $$\operatorname{dis}_{m+1}(f) = a_0^2 \operatorname{dis}_m(f).$$

This follows from Cor. 1 on taking $n = 1$, $g = 1$, by the formula $\mathrm{res}_{m,1}(f,1) = a_0$ (IV, p. 76, Example 1).

COROLLARY 3. — *Let* A *be a field, and let* f *be a non-constant polynomial in* A[X]. *For* f *and* Df *to be relatively prime, it is necessary and sufficient that* $\mathrm{dis}(f) \neq 0$.

This follows from Prop. 11, (iii) and Cor. 2 of IV, p. 78.

Remark. — A double application of Cor. 2 above shows that we have $\mathrm{dis}_m(f) = 0$ for every polynomial f of degree $\leqslant m - 2$.

Examples. — 3) Let $m = 2$. By Example 2 (IV, p. 81) we have $\Delta(A_1, A_2) = A_1^2 - 4A_2$, whence $\tilde{\Delta}(A_0, A_1, A_2) = A_1^2 - 4A_0A_2$. In other words, we have

$$\mathrm{dis}_2(a_0X^2 + a_1X + a_2) = a_1^2 - 4a_0a_2 .$$

4) Consider the polynomial

$$F = A_0X^3 + 3A_1X^2 + 3A_2X + A_3$$

with coefficients in $\mathbf{Q}[A_0, A_1, A_2, A_3]$. We have

$$DF = 3(A_0X^2 + 2A_1X + A_2) ,$$

$$F - 1/3X \, . \, DF = A_1X^2 + 2A_2X + A_3 .$$

By the formulae (54) (IV, p. 84) and (29) (IV, p. 77) we have

$$A_0 \, . \, \mathrm{dis}_3(F) = -\,\mathrm{res}_{2,3}(DF, F) = -\,\mathrm{res}_{2,3}(DF, F - 1/3X \, . \, DF) .$$

Applying Cor. 2 of IV, p. 80 we finally obtain

$$\mathrm{dis}_3(F) = -\,27\mathrm{res}_{2,2}(A_0X^2 + 2A_1X + A_2, A_1X^2 + 2A_2X + A_3) .$$

By Example 3 of IV, p. 80 we thus have

$$\tilde{\Delta}(A_0, 3A_1, 3A_2, A_3) = -\,27(A_0A_3 - A_1A_2)^2 - 108(A_1A_3 - A_2^2)(A_1^2 - A_0A_2) .$$

After some calculations we find, that if $f = a_0X^3 + a_1X^2 + a_2X + a_3$, we have

$$\mathrm{dis}_3(f) = a_1^2a_2^2 + 18a_0a_1a_2a_3 - 4a_1^3a_3 - 4a_0a_2^3 - 27a_0^2a_3^2 .$$

In particular, we have

$$\mathrm{dis}(X^3 + pX + q) = -\,(4p^3 + 27q^2) .$$

Exercises

§ 1

¶ 1) * When M is a subgroup of the additive group **R** of real numbers, generalize the Euclidean division of polynomials in one indeterminate to the group algebra of M.$_*$

2) Let f be a polynomial $\neq 0$ in $A[X]$ of degree n and leading coefficient α_0. Let M be the submodule of $A[X]$ (qua A-module) consisting of the polynomials in which the coefficient of the term of degree m (for arbitrary m in **N**) is divisible by α_0^{μ}, where $\mu = \max(m - n + 1, 0)$; show that for any polynomial g in M there exist two polynomials u, v in $A[X]$ such that $\deg v < n$ and $g = uf + v$.

¶ 3) *a*) In the algebra $A[X]$ of polynomials in one indeterminate over A, the mapping $(u, v) \mapsto u(v)$ is an internal law of composition ; show that this law is associative and distributive on the left with respect to addition and multiplication in $A[X]$.
b) If A is an integral domain show that the relation $u(v) = 0$ implies that $u = 0$ or v is constant, and that when $u \neq 0$ and $\deg v > 0$, the degree of $u(v)$ is equal to the product of the degrees of u and v.
c) From now on assume that A is a *field*. Let u and v be two polynomials of degree > 0 in $A[X]$ and f a polynomial of degree > 0 ; show that if q is the quotient and r the remainder in the Euclidean division of u by v, $q(f)$ and $r(f)$ are the quotient and remainder in the Euclidean division of $u(f)$ by $v(f)$.
d) For each polynomial f of $A[X]$ denote by $I(f)$ the set of all polynomials of the form $u(f)$, where u ranges over $A[X]$; this is a subring of $A[X]$. For $I(f) = I(g)$ it is necessary and sufficient that $g = \lambda f + \mu$, where $\lambda \neq 0$ and μ are in A.
e) Let f and g be two polynomials of degree > 0 in $A[X]$; show that $I(f) \cap I(g)$ is either equal to A or of the form $I(h)$, where h is a polynomial of degree > 0 (excluding the first possibility, consider in $I(f) \cap I(g)$ a polynomial h of least possible degree > 0).

¶ 4) *a*) Suppose that for every non-zero $n \in \mathbf{Z}$, the relation $n\xi = 0$ implies $\xi = 0$ in A. Show that the mapping $f \mapsto f(D_1, \ldots, D_n)$ is an injective homomorphism of the polynomial algebra $A[X_1, \ldots, X_n] = E$ into the algebra (over A) of endomorphisms of the A-module E (use induction on n).
b) Let m be an integer $\geqslant 1$ such that $m \,.\, 1 = 0$ in A. Show that we have $D_i^m = 0$ for every partial derivation D_i $(1 \leqslant i \leqslant n)$ in $A[X_1, \ldots, X_n]$.

5) Let $u \in A[(X_i)_{i \in I}]$ and r an integer $\geqslant 0$, and suppose that in A for each $n > 0$, the relation $n\xi = 0$ implies $\xi = 0$. If we have the relation

$$\sum_{i \in I} (D_i u)(\mathbf{X}) X_i = r u(\mathbf{X}),$$

then u is homogeneous of degree r.

¶ 6) Let K be a commutative field, I a set and L the free associative algebra on the set I over the ring K (III, p. 448, Def. 2) with the graduation defined in III, p. 458. We define the degree of an element of L and the notion of homogeneous element as in IV, p. 2.
a) Show that $\deg(uv) = \deg(u) + \deg(v)$ and $\deg(u+v) \leqslant \sup(\deg(u), \deg(v))$ for $u, v \in L$.
b) Given $u, v, q, q' \in L$ such that $\deg(u) \geqslant \deg(v)$ and $\deg(qu - q'v) < \deg(qu)$; show that there exists $q'' \in L$ such that $\deg(u - q''v) < \deg(u)$.
c) Given $u, v, q, q' \in L$ with $\deg u \geqslant \deg v > \deg(qu - q'v)$; show that there exists $q'' \in L$ such that $\deg(u - q''v) < \deg v$.
d) Given $u, v, u', v' \in L$, homogeneous and such that $uv = u'v'$ and $\deg v \geqslant \deg v' \geqslant 0$; there exists $t \in L$ such that $u' = ut$, $v = tv'$.
e) Given $u, v \in L$, homogeneous, such that $uv = vu$, there exists $w \in L$, $a, b \in K$ and m, $n \in K$ such that $u = aw^m$, $v = bw^n$.
f) Let $u, v \in L$ be such that there exist $a, b \in L$, non-zero, with $au = bv$; there exist m, $d \in L$ such that the left multiples (resp. right divisors) common to u and v are left multiples (resp. right divisors) of m (resp. d).

§ 2

1) Suppose A is an integral domain; let f be a polynomial of the ring $A[X_1, X_2, \ldots, X_n]$ of degree $\leqslant k_i$ in X_i (for $1 \leqslant i \leqslant n$). For each value of i ($1 \leqslant i \leqslant n$) let H_i be a set of $k_i + 1$ elements of A. Show that if $f(x_1, x_2, \ldots, x_n) = 0$ for each element $(x_i) \in \prod_{i=1}^{n} H_i$, then $f = 0$.

2) Assume that A is an infinite integral domain; let Φ be a set of polynomials $\neq 0$ in the ring $A[X_1, X_2, \ldots, X_n]$. Show that if the power of Φ is *strictly less* than that of A, then there exists a subset H of A^n equipotent with A such that for each $x = (x_i) \in H$ and for each $f \in \Phi$ we have $f(x) \neq 0$.

3) Suppose that A is an integral domain and let B be an infinite subset of A. Show that if the polynomial $f \in A[X]$ has degree > 0, then the image of B under the polynomial function $x \mapsto f(x)$ is equipotent with B.

4) Suppose that the additive group of A contains an infinite subgroup G whose non-zero elements are not divisors of zero in A. Show that the mapping $f \mapsto \tilde{f}$ of $A[X_1, \ldots, X_p]$ into the algebra of mappings of A^p into A is injective (observe that a polynomial of degree n in one indeterminate cannot have more than n distinct roots lying in G). This holds in particular when there is in A an element x_0, not a divisor of 0 and of infinite order in the additive group of A.

5) Let K be a field of infinitely many elements and let Q be the quaternion algebra over K, of type $(-1, 0, -1)$ (III, p. 445). Show that the polynomial $X^2 + 1$ has infinitely many roots in Q.

¶ 6) Let K be a finite field with q elements.
a) In the ring $K[X_1, X_2, \ldots, X_n]$ let $\mathfrak{a}$ be the ideal generated by the n polynomials

$X_i^q - X_i$ $(1 \leqslant i \leqslant n)$. Show that if $f \in \mathfrak{a}$, then $f(x_1, x_2, \ldots, x_n) = 0$ for all $(x_i) \in K^n$ (observe that the multiplicative group K^* is of order $q - 1$).
b) If f is any polynomial of $K[X_1, X_2, \ldots, X_n]$, show that there exists precisely one polynomial $\bar{f}$ such that $\deg_i \bar{f} \leqslant q - 1$ for $1 \leqslant i \leqslant n$ and such that $f \equiv \bar{f}$ (mod $\mathfrak{a}$). We thus have $\deg \bar{f} \leqslant \deg f$; if f is such that $f(x_1, x_2, \ldots, x_n) = 0$ for each $(x_i) \in K^n$, f belongs to the ideal $\mathfrak{a}$, which is thus the inverse image of 0 under the homomorphism $f \mapsto \bar{f}$ (use Exercise 1).
c) Let $f_1, f_2, \ldots, f_m$ be non-zero polynomials of $K[X_1, X_2, \ldots, X_n]$ such that $f_i(0, 0, \ldots, 0) = 0$ $(1 \leqslant i \leqslant m)$ and the sum of the total degrees of the f_i is $< n$. Show that there exists an element $(x_1, x_2, \ldots, x_n) \in K^n$ distinct from $(0, 0, \ldots, 0)$ such that

$$f_i(x_1, x_2, \ldots, x_n) = 0 \quad \text{for} \quad 1 \leqslant i \leqslant m .$$

(Observe that if this were not so, the polynomial $\prod_{i=1}^{m} (1 - f_i^{q-1})$ would be congruent mod $\mathfrak{a}$ to the polynomial $\prod_{i=1}^{n} (1 - X_i^{q-1})$, and use *b*)).

7) Let K be a finite field with q elements, and B the product ring K^I where I is an infinite set. Give an example of a non-zero polynomial f of $B[X]$ such that $f(x) = 0$ for all $x \in B$.

8) Let B be the additive group $\mathbf{Z} \oplus (\mathbf{Z}/2\mathbf{Z})$, with the commutative ring structure obtained by putting $(a, b) \cdot (a', b') = (aa', ab' + ba')$ for $(a, b) \in B$ and $(a', b') \in B$. Let $\varepsilon = (0, 1) \in B$, and $f(X) = \varepsilon X(X - 1) \in B[X]$. Then $f(\alpha) = 0$ for all $\alpha \in B$.

9) Let k be an integer > 0 such that $k \cdot 1 = 0$ in A. Let h be an integer > 0 and f the polynomial $(X - \alpha)^{k+h} + (X - \alpha)^k$. Then α is a root of order k of f and a root of order $\geqslant k + h - 1$ of Df.

* 10) Let f be a non-zero element of $\mathbf{Z}[X_1, \ldots, X_n]$. Let $N(q)$ be the number of zeros of f in the cell $|x_1| \leqslant q, \ldots, |x_n| \leqslant q$. Then $N(q)/q^n$ tends to 0 as $q \to +\infty$. *

11) Let P be the set of all prime numbers, P′ an infinite subset of P and $f \in \mathbf{Z}[(X_i)]$. For each $p \in P'$ let f_p be the element of $(\mathbf{Z}/p\mathbf{Z})[(X_i)]$ obtained by applying to the coefficients of f the canonical homomorphism of $\mathbf{Z}$ onto $\mathbf{Z}/p\mathbf{Z}$. Suppose that for each $p \in P'$ and each $x \in (\mathbf{Z}/p\mathbf{Z})^I$ we have $f_p(x) = 0$; then $f = 0$.

12) Let M be an A-module, and let $\alpha_0, \ldots, \alpha_n \in A$ be such that for $i \neq j$ the homothety with respect to $\alpha_i - \alpha_j$ in M is bijective. For any $m_0, \ldots, m_n \in M$ there exists one and only one $f \in M[X]$ such that $f(\alpha_i) = m_i$ for $0 \leqslant i \leqslant n$ and $\deg f \leqslant n$.

13) Let α_i $(1 \leqslant i \leqslant n)$ be n distinct elements of a commutative field K, β_i $(1 \leqslant i \leqslant n)$ and γ_i $(1 \leqslant i \leqslant n)$ $2n$ arbitrary elements of K. Show that there exists one and only one polynomial $f \in K[X]$ of degree $\leqslant 2n - 1$ such that $f(\alpha_i) = \beta_i$ and $f'(\alpha_i) = \gamma_i$ for $1 \leqslant i \leqslant n$ (Hermite interpolation formula) (to begin with consider the case where $2n - 1$ of the $2n$ elements β_i, γ_i are zero).

§ 3

1) Let E be an associative commutative and unital algebra over a commutative field K ; let $\mathbf{x} = (x_i)_{i \in I}$ be a family of pairwise permutable elements of E. Let U be the subring of $K(X_i)_{i \in I}$ consisting of all $f \in K(X_i)_{i \in I}$ such that $\mathbf{x}$ is substitutable in f. Show that if in E the set of non-invertible elements is an ideal, then the same is true of U. Show that the ideal of non-invertible elements in U is then maximal.

2) *a*) Let K be a commutative field and $u = a_m X^m + a_{m+1} X^{m+1} + \cdots + a_n X^n$ a polynomial of $K[X]$ such that $a_m \neq 0$, $a_n \neq 0$ $(0 \leqslant m \leqslant n)$. Show that if g is a rational fraction of degree $d \neq 0$ in $K(X)$, then $u(g)$ is non-zero and is of degree nd if $d > 0$, and of degree md if $d < 0$.
b) Deduce that a non-constant rational fraction g of $K(X)$ is substitutable in every rational fraction of $K(X)$ (observe that if g is of degree 0, there exists $\alpha \in K$ such that $g - \alpha$ has degree < 0).

3) A rational fraction $f \in K(X_1, X_2, \ldots, X_n)$ is said to be *homogeneous* if it is equal to the quotient of two homogeneous polynomials (the denominator being $\neq 0$). Show that for f to be homogeneous it is necessary and sufficient that

$$f(ZX_1, ZX_2, \ldots, ZX_n) = Z^d f(X_1, \ldots, X_n)$$

where d is the degree of f.

4) Show that the Lagrange interpolation formula (IV, p. 16, Prop. 6) may be written

$$f(X) = \omega(X) \sum_{i=1}^{n} \frac{\beta_i}{\omega'(\alpha_i)(X - \alpha_i)}$$

where $\omega(X) = (X - \alpha_1)(X - \alpha_2) \ldots (X - \alpha_n)$. Deduce that if g is a polynomial of $K[X]$ of degree $\leqslant n - 2$, then

$$\sum_{i=1}^{n} \frac{g(\alpha_i)}{\omega'(\alpha_i)} = 0$$

(consider the polynomial $f(X) = Xg(X)$).

* 5) Let K be a commutative field of characteristic 0. Show that if a rational fraction $u \in K(X)$ is such that $Du = 0$, then u is a constant.

6) Generalize Euler's identity to homogeneous rational fractions (Exercise 3) over a field of characteristic 0 (consider the rational fraction $\frac{1}{Z^m} f(ZX_1, ZX_2, \ldots, ZX_n)$ with respect to Z and use Exercise 5). *

7) Let K be a commutative field. Show that

$$[K(T) : K] = \operatorname{Sup}(\operatorname{Card}(K), \operatorname{Card}(\mathbf{N})).$$

(Use the linear independence of the $1/(T-a)$, $a \in K$, to show that

$$[K(T):K] \geqslant \operatorname{Card}(K)\,.)$$

§ 4

1) Let $u(X) = \sum_{n=0}^{\infty} \alpha_n X^n$ be a formal power series over a commutative field K.

a) For u to be a rational fraction of $K(X)$ it is necessary and sufficient that there should exist a finite sequence of elements $(\lambda_i)_{1 \leqslant i \leqslant q}$ of elements of K, not all zero, and an integer $d \geqslant 0$ such that for all $n \geqslant d$ we have

$$\lambda_1 \alpha_n + \lambda_2 \alpha_{n+1} + \cdots + \lambda_q \alpha_{n+q-1} = 0\,.$$

b) Put

$$H_n^{(k)} = \begin{vmatrix} \alpha_n & \alpha_{n+1} & \cdots & \alpha_{n+k-1} \\ \alpha_{n+1} & \alpha_{n+2} & \cdots & \alpha_{n+k} \\ \alpha_{n+2} & \alpha_{n+3} & \cdots & \alpha_{n+k+1} \\ & & & \\ \alpha_{n+k-1} & \alpha_{n+k} & \cdots & \alpha_{n+2k-2} \end{vmatrix}$$

(« Hankel determinant ») show that if $H_{d+j}^{(q+1)} = 0$ and $H_{d+j}^{(q)} \neq 0$ for every integer $j \geqslant 0$, then $u(X)$ is a rational fraction (use *a*)).

c) Prove the identity

$$H_n^{(k)} H_{n+2}^{(k)} - H_n^{(k+1)} H_{n+2}^{(k-1)} = (H_{n+1}^{(k)})^2$$

(cf. III, p. 639, Exercise 10). Deduce that if $H_{m+j}^{(k+1)} = 0$ for $0 \leqslant j \leqslant r-1$, then the r determinants $H_{m+j}^{(k)}$ where $1 \leqslant j \leqslant r$ are either all zero or all non-zero.

d) Deduce from *b*) and *c*) that for $u(X)$ to be a rational fraction it is necessary and sufficient that there exist two integers d and q such that $H_{d+j}^{(q+1)} = 0$ for all integers $j \geqslant 0$.

e) Let k be a subfield of K, and identify $K(X)$, $k(X)$ and $k[[X]]$ with subrings of $K((X))$. Prove that $k[[X]] \cap K(X) = k[[X]] \cap k(X)$.

2) Let $a_1, a_2, \ldots, a_p$ be integers > 0; denote by α_n the number of finite sequences $(x_i)_{1 \leqslant i \leqslant p}$ of integers $\geqslant 0$ satisfying the equation

$$a_1 x_1 + a_2 x_2 + \cdots + a_p x_p = n\,.$$

Show that the formal power series $\sum_{n=0}^{\infty} \alpha_n X^n$ (over **Q**) is the expansion of the rational fraction

$$\frac{1}{(1-X^{a_1})(1-X^{a_2})\ldots(1-X^{a_p})}\,.$$

3) Let F be a finite set of integers > 0, and let β_n be the number of finite sequences (x_i) of at most n terms, all of whose terms lie in F and such that $\sum_i x_i = n$. Show that the formal power series $\sum_{n=0}^{\infty} \beta_n X^n$ over $\mathbf{Q}$ is the expansion of the rational fraction

$$\frac{1}{1 - \sum_{a \in F} X^a}.$$

4) Let E be a vector space with an infinite basis over a field K of characteristic 2 ; let B be the exterior algebra $\bigwedge E$ on this space, which is a commutative ring. Give an example of a formal power series $u \in B[[X]]$ such that $u^2 = 0$ but such that there exists no element $\gamma \neq 0$ of B for which $\gamma u = 0$.

5) Let $E = A[X_1, \ldots, X_p]$, $F = A[[X_1, \ldots, X_p]]$.
a) If D is a **Z**-derivation of E into F, we have $\omega(Du) \geqslant \omega(u) - 1$ for every non-zero u in E.
b) If D′ is a **Z**-derivation of F, then $\omega(D'v) \geqslant \omega(v') - 1$ for every non-zero v in F.
c) If E is equipped with the topology induced by that of F, then D and D′ are continuous.
d) Every **Z**-derivation of E into F extends in a unique fashion to a **Z**-derivation of F.
e) Let $\mathscr{D}$ be the F-module of A-derivations of F. If $D \in \mathscr{D}$ and $DX_i = u_i$, then $D = \sum_{i=1}^{p} u_i D_i$.
f) $(D_1, \ldots, D_p)$ is a basis of the F-module $\mathscr{D}$.

6) Let $(u_\lambda)_{\lambda \in L}$ be a family of elements of $A[[X_i]]_{i \in I}$. If the family $(1 + u_\lambda)_{\lambda \in L}$ is multipliable, the family $(u_\lambda)_{\lambda \in L}$ is summable.

* 7) If A is reduced, then so is $A[[(X_i)_{i \in I}]]$. *

8) We denote by $(1 + X)^T$ the element $\exp(T \log(1 + X))$ of $\mathbf{Q}[[X, T]]$.
a) Show that $(1 + X)^T = \sum_{n \geqslant 0} \frac{T(T-1)\ldots(T-n+1)}{n!} X^n$. (The coefficient of X^n is a polynomial in T whose value is known for $T = n \in \mathbf{N}$.)
b) Show that $(1 + X)^T (1 + X)^{T'} = (1 + X)^{T+T'}$. Write down explicitly the identity between binomial coefficients so obtained. Show that $(1 + X + Y + XY)^T = (1 + X)^T (1 + Y)^T$.

§ 5

1) (Algebra of sequences of exponential type) Given two integers $p, q \in \mathbf{N}$, we denote by $((p, q))$ the binomial coefficient $\frac{(p+q)!}{p!\,q!}$. Let k be a commutative ring and B an associative commutative unital k-algebra. By a *sequence of exponential type* of elements of B we understand a sequence $a = (a_p)_{p \in \mathbf{N}}$ such that $a_0 = 1_B$ and $a_p a_q = ((p, q))\, a_{p+q}$ for any $p, q \in \mathbf{N}$. The set of all sequences of exponential type of B is written $\mathscr{E}(B)$.
a) Verify that there is on $\mathscr{E}(B)$ a unique k-algebra structure such that

$$(a + b)_p = \sum_{r+s=p} a_r b_s,$$

$$(ab)_p = p!\, a_p b_p,$$

$$(\lambda a)_p = \lambda^p a_p,$$

for any sequence $a, b \in \mathscr{E}(\mathrm{B})$ and $p \in \mathbf{N}$, $\lambda \in k$. The mapping $a \mapsto a_1$ is a homomorphism of the algebra $\mathscr{E}(\mathrm{B})$ into B.

b) Suppose that k contains a subring isomorphic to $\mathbf{Q}$. Show that for every $x \in \mathrm{B}$ the sequence $f(x)$ defined by $f(x)_p = \frac{1}{p!} x^p$ is of exponential type. Show that f is an isomorphism of B onto $\mathscr{E}(\mathrm{B})$.

c) Let B' be an associative commutative unital k-algebra and let h be a homomorphism of B into B'. For each $a \in \mathscr{E}(\mathrm{B})$ the sequence a' defined by $a'_p = h(a_p)$ for any $p \in \mathbf{N}$ belongs to $\mathscr{E}(\mathrm{B}')$. Show that the mapping $\mathscr{E}(h) : a \mapsto a'$ is a homomorphism of $\mathscr{E}(\mathrm{B})$ into $\mathscr{E}(\mathrm{B}')$. Verify that $\mathscr{E}(g \circ h) = \mathscr{E}(g) \circ \mathscr{E}(h)$ when g and h are composable k-algebra homomorphisms.

d) Let E be a k-module and **TS**(E) the algebra of symmetric tensors on E. Show that for each $x \in \mathrm{E}$, $(\gamma_p(x))$ is a sequence of exponential type in the algebra **TS**(E) (IV, p. 45).

2) (Functor Γ) Denote by k a commutative ring and by E a k-module. By the *gamma algebra* of E we understand the associative commutative unital algebra defined by the generating system $\mathbf{N} \times \mathrm{E}$ and by the relators (cf. III, p. 450)

$$\begin{gathered}(0, x) - 1 ,\\ (p, \lambda x) - \lambda^p (p, x) ,\\ (p, x + y) - \sum_{r+s=p} (r, x)(s, y) ,\\ (p, x)(q, x) - ((p, q))(p + q, x) ,\end{gathered}$$

where p, q range over $\mathbf{N}$, x, y range over E and λ over k. This algebra is denoted by $\Gamma(\mathrm{E})$. For each $p \in \mathbf{N}$ we denote by γ_p the mapping of E into $\Gamma(\mathrm{E})$ composed of the injection $x \mapsto (p, x)$ and the canonical homomorphism of the free commutative algebra on $\mathbf{N} \times \mathrm{E}$ into $\Gamma(\mathrm{E})$.

a) Verify that for each $x \in \mathrm{E}$, the sequence $\gamma_p(\mathrm{X})$ is a sequence of exponential type of elements of $\Gamma(\mathrm{E})$ (Exercise 1) and that the mapping $\gamma : x \mapsto (\gamma_p(x))_{p \in \mathbf{N}}$ is a homomorphism of the k-module E into the k-algebra $\mathscr{E}(\Gamma(\mathrm{E}))$.

b) (Universal property of $\Gamma(\mathrm{E})$). Let B be an associative commutative unital k-algebra and φ a homomorphism of the k-module E into $\mathscr{E}(\mathrm{B})$. Show that there exists a unique algebra homomorphism $h : \Gamma(\mathrm{E}) \to \mathrm{B}$ such that $\varphi = \mathscr{E}(h) \circ \gamma$.

c) Show that there exists a unique unital homomorphism of algebras $\varepsilon : \Gamma(\mathrm{E}) \to k$ such that $\varepsilon(\gamma_p(x)) = 0$ for all $p > 0$ and all $x \in \mathrm{E}$.

d) For each $p \in \mathbf{N}$ let $\Gamma_p(\mathrm{E})$ be the submodule of $\Gamma(\mathrm{E})$ generated by the elements

$$\gamma_{\nu_1}(x_1) \ldots \gamma_{\nu_s}(x_s)$$

where $s \in \mathbf{N}$, $\nu \in \mathbf{N}^s$ and $|\nu| = \sum_i \nu_i = p$. Verify that the submodules $\Gamma_p(\mathrm{E})$ form a graduation of the algebra $\Gamma(\mathrm{E})$. Show that γ_1 is an isomorphism of E onto $\Gamma_1(\mathrm{E})$ (to verify that γ_1 is injective, consider the algebra $\mathrm{B} = k \times \mathrm{E}$ with the product defined by the formula $(\lambda, x)(\mu, y) = (\lambda\mu, \lambda y + \mu x)$ and show that there exists a homomorphism of $\Gamma(\mathrm{E})$ into B which maps $\gamma_1(x)$ to $(0, x)$ for any $x \in \mathrm{E}$).

e) Let E and F be two k-modules and h a homomorphism of E into F. Show that there exists a unique homomorphism of graded algebras $\Gamma(h) : \Gamma(\mathrm{E}) \to \Gamma(\mathrm{F})$ such that $\gamma_p \circ h = \Gamma(h) \circ \gamma_p$ for all $p \in \mathbf{N}$. Verify that if g and h are homomorphisms of k-modules such that $g \circ h$ is defined, then $\Gamma(g \circ h) = \Gamma(g) \circ \Gamma(h)$.

f) Let $h : \mathrm{E} \to \mathrm{F}$ be a surjective homomorphism of k-modules. Show that $\Gamma(h)$ is surjective and that its kernel is the ideal of $\Gamma(\mathrm{E})$ generated by the elements $\gamma_p(x)$ with $p > 0$ and $x \in \mathrm{Ker}(h)$.
g) Let E and F be two k-modules. Show that there exists a unique homomorphism φ of the algebra $\Gamma(\mathrm{E} \times \mathrm{F})$ into the algebra $\Gamma(\mathrm{E}) \otimes_k \Gamma(\mathrm{F})$ such that

$$(\varphi \circ \gamma_p)(x, y) = \sum_{r+s=p} \gamma_r(x) \otimes \gamma_s(y)$$

for any $p \in \mathbf{N}$, $x \in \mathrm{E}$ and $y \in \mathrm{F}$. Show that this homomorphism is an isomorphism compatible with the graduations.

3) Let k be a commutative ring and E a k-module. Show that there exists a unique homomorphism Δ of the algebra $\Gamma(\mathrm{E})$ (Exercise 2) into the algebra $\Gamma(\mathrm{E}) \otimes_k \Gamma(\mathrm{E})$ such that

$$(\Delta \circ \gamma_p)(x) = \sum_{r+s=p} \gamma_r(x) \otimes \gamma_s(x)$$

for all $x \in \mathrm{E}$ and all $p \in \mathbf{N}$. Show that the coproduct Δ is coassociative, cocommutative (III, p. 580-581) and defines on $\Gamma(\mathrm{E})$ a graded bigebra structure (III, p. 585). Show that the homomorphism ε defined in Exercise 2, *c*) is a counit.

4) Let E_1 and E_2 be two modules over a commutative ring k and let h be a homomorphism of E_1 into E_2. Verify that $\Gamma(h)$ (Exercise 2, *e*)) is a homomorphism of graded bigebras (cf. Exercise 3).

5) Let $\Gamma(\mathbf{Z})$ be the gamma algebra of the $\mathbf{Z}$-module $\mathbf{Z}$ (Exercise 2). For each $p \in \mathbf{N}$ put $\mathrm{T}^{(p)} = \gamma_p(1)$.
a) Show that $(\mathrm{T}^{(p)})_{p \in \mathbf{N}}$ is a basis of the $\mathbf{Z}$-module $\Gamma(\mathbf{Z})$. Show that there exists a unique homomorphism θ of the algebra $\Gamma(\mathbf{Z})$ into the algebra $\Gamma(\mathbf{Z}) \otimes \Gamma(\mathbf{Z})$ such that

$$\theta(\mathrm{T}^{(p)}) = p!\, \mathrm{T}^{(p)} \otimes \mathrm{T}^{(p)}$$

for all $p \in \mathbf{N}$.
b) Let B be a commutative ring and H the set of homomorphisms of the ring $\Gamma(\mathbf{Z})$ into the ring B. Show that for every $f \in \mathrm{H}$ the sequence $f(\mathrm{T}^{(p)})$, $p \in \mathbf{N}$ is a sequence of exponential type of elements of B. Show that the mapping $\varphi : f \mapsto (f(\mathrm{T}^{(p)}))_{p \in \mathbf{N}}$ is a bijection of H onto $\mathscr{E}(\mathrm{B})$ (Exercise 1).
c) For $f, g \in \mathrm{H}$ denote by $f * g$ the homomorphism of $\Gamma(\mathbf{Z}) \otimes \Gamma(\mathbf{Z})$ into B defined by $(f * g)(u \otimes v) = f(u)\, g(v)$ for any $u, v \in \Gamma(\mathbf{Z})$. Verify that for any $f, g \in \mathrm{H}$ we have

$$\varphi(f) + \varphi(g) = \varphi((f * g) \circ \Delta),$$
$$\varphi(f)\, \varphi(g) = \varphi((f * g) \circ \theta).$$

6) Let n be an integer > 0 and let $\Gamma(\mathbf{Z}/n\mathbf{Z})$ be the gamma algebra of the $\mathbf{Z}$-module $\mathbf{Z}/n\mathbf{Z}$ (Exercise 2). Show that for every $p > 0$, $\Gamma_p(\mathbf{Z}/n\mathbf{Z})$ is isomorphic to the quotient of $\mathbf{Z}$ by the subgroup generated by the integers $n^k((k, p-k))$, where $1 \leqslant k \leqslant p$. Show that if p and n are distinct primes, then $\Gamma_p(\mathbf{Z}/n\mathbf{Z})$ is cyclic of order n. Show that if n is even, then $\Gamma_2(\mathbf{Z}/n\mathbf{Z})$ is cyclic of order $2n$.

7) (Extension of scalars) Let k be a commutative ring, L an associative commutative unital k-algebra and E a k-module.
a) Let $\Gamma(\mathrm{L} \otimes_k \mathrm{E})$ be the gamma algebra of the L-module $\mathrm{L} \otimes_k \mathrm{E}$ (Exercise 2). Show that there exists a unique homomorphism θ of $\mathrm{L} \otimes_k \Gamma(\mathrm{E})$ into $\Gamma(\mathrm{L} \otimes_k \mathrm{E})$ compatible with the algebra structure on L and such that

$$\theta(1 \otimes \gamma_p(x)) = \gamma_p(1 \otimes x)$$

for all $x \in \mathrm{E}$ and $p \in \mathbf{N}$.
b) Show that θ is an isomorphism of graded bigebras over L.

8) Let k be a commutative ring and E a k-module.
a) Show that there exists a unique homomorphism g of the algebra $\Gamma(\mathrm{E})$ (Exercise 2) into the algebra $\mathbf{TS}(\mathrm{E})$ of symmetric tensors over E (IV, p. 43) such that

$$(g \circ \gamma_p)(x) = x \otimes x \otimes \dots \otimes x \quad (p \text{ factors } x),$$

for any $x \in \mathrm{E}$, $p \in \mathbf{N}$. Show that if E is a free k-module with basis $(e_i)_{i \in \mathrm{I}}$, then g is an isomorphism of graded bigebras (cf. Exercise 3 and IV, p. 51) and the elements $\gamma_\alpha = \prod_i \gamma_{\alpha(i)}(e_i)$, where $\alpha \in \mathbf{N}^{(\mathrm{I})}$, form a basis of $\Gamma(\mathrm{E})$.

b) Let ξ be a mapping of $[1, n]$ into $\mathbf{N}$ and let $p = \sum_{i=1}^{n} \xi(i)$. Show that for every sequence $x_1, x_2, \dots, x_n$ of elements of E, the image of $\gamma_{\xi(1)}(x_1) \dots \gamma_{\xi(n)}(x_n)$ under g is $\sum_\varphi x_{\varphi(1)} \otimes \dots \otimes x_{\varphi(p)}$ where φ runs over the set of all mappings of $[1, p]$ into $[1, n]$ such that $\mathrm{Card}\, \varphi^{-1}(i) = \xi(i)$ for each $i \in [1, n]$ (cf. IV, p. 45, Prop. 3).
c) Let f be the mapping of $\otimes^p \mathrm{E}$ into $\Gamma_p(\mathrm{E})$ defined by $f(x_1 \otimes \dots \otimes x_p) = \gamma_1(x_1) \dots \gamma_1(x_p)$ for any $x_1, \dots, x_p \in \mathrm{E}$. Show that for each tensor $\mathbf{t} \in \otimes^p \mathrm{E}$, $(g \circ f)(\mathbf{t})$ is the symmetrization of $\mathbf{t}$. Show that for each $u \in \Gamma_p(\mathrm{E})$ we have $(f \circ g)(u) = p!\, u$.

¶ 9) *(Polynomial laws)* Denote by E and F two modules over the commutative ring k. By a *polynomial law of* E *in* F we understand, for each (associative commutative and unital) k-algebra L, a mapping φ_{L} of $\mathrm{L} \otimes_k \mathrm{E}$ into $\mathrm{L} \otimes_k \mathrm{F}$ satisfying the condition :

If L and R are two k-algebras and f is a unital homomorphism of L into R, then

$$(f \otimes 1) \circ \varphi_{\mathrm{L}} = \varphi_{\mathrm{R}} \circ (f \otimes 1).$$

a) Let φ be a polynomial law of E into F, let $\mathrm{L} = k[(\mathrm{T}_i)_{i \in \mathrm{I}}]$ be the polynomial algebra in a family of indeterminates T_i over k and let $(x_i)_{i \in \mathrm{I}}$ be a family of finite support of elements of E. Show that there exists a unique family of finite support $(y_\xi)_{\xi \in \mathbf{N}^{(\mathrm{I})}}$ of elements of F such that

$$\varphi_{\mathrm{L}}\left(\sum_i \mathrm{T}_i \otimes x_i\right) = \sum_{\xi \in \mathbf{N}^{(\mathrm{I})}} \mathrm{T}^\xi \otimes y_\xi .$$

Show that if R is an algebra over k, then for every family $(a_i)_{i \in \mathrm{I}}$ of elements of R we have

$$(1) \qquad \varphi_{\mathrm{R}}\left(\sum_i a_i \otimes x_i\right) = \sum_{\xi \in \mathbf{N}^{(\mathrm{I})}} a^\xi \otimes y_\xi .$$

b) A family $(\varphi^j)_{j \in J}$ of polynomial laws of E into F is called summable if for every algebra L over k and every $u \in L \otimes E$, the family $\varphi^j_L(u)$ has finite support. Show that if $(\varphi^j)_{j \in J}$ is a summable family of polynomial laws of E into F, then there exists a unique polynomial law φ of E into F such that for every k-algebra L and every $u \in L \otimes E$ we have $\sum_j \varphi^j_L(u) = \varphi_L(u)$. This polynomial law is called the sum of the laws φ^j and is denoted by $\sum_j \varphi^j$.

c) Let E, F, G be three k-modules, φ a polynomial law of E into F and ψ a polynomial law of F into G. Show that there exists a unique polynomial law η of E into G such that for every k-algebra L we have $\eta_L = \psi_L \circ \varphi_L$. This law η is called the composition of φ and ψ and is denoted by $\psi \circ \varphi$.

d) Let p be an integer $\geqslant 0$. A polynomial law φ of E into F is said to be *homogeneous of degree p* if for every algebra L over k we have $\varphi_L(au) = a^p \varphi_L(u)$ for any $a \in L$ and $u \in L \otimes E$. Show that if φ is homogeneous of degree p, then in formula (1) we have $y_\xi = 0$ for $|\xi| = \sum_i \xi(i) \neq p$. Determine all homogeneous polynomial laws of degrees 0 and 1.

e) With the same notation as in *c*), show that if φ and ψ are homogeneous laws of degrees p and q respectively, then $\psi \circ \varphi$ is a homogeneous law of degree pq.

f) Let p be an integer $\geqslant 0$. For every commutative k-algebra L we denote by ε_L the injection $a \mapsto aT$ of L into the polynomial algebra L[T] and by β^p_L the homomorphism of the L-module L[T] into L which associates with each element $\sum_k a_k T^k \in L[T]$ the coefficient a_p. Let φ be a polynomial law of E into F. Show that the mappings $\varphi^p_L = (\beta^p_L \otimes Id_F) \circ \varphi_{L[T]} \circ (\varepsilon_L \otimes Id_E)$ form a polynomial law of E into F which is homogeneous of degree p. This polynomial law is called the *homogeneous component of degree p* of the law φ. Show that the homogeneous components of φ form a summable family with sum φ.

10) Denote by k a commutative ring and by E a k-module. For each integer $p \geqslant 0$ and every k-algebra L we denote by $\gamma_{p,L}$ the mapping of $L \otimes E$ into $L \otimes \Gamma_p(E)$ composed of $\gamma_p : L \otimes E \to \Gamma_p(L \otimes E)$ and the canonical isomorphism of $\Gamma_p(L \otimes E)$ onto $L \otimes \Gamma_p(E)$ (cf. Exercises 2 and 7).

a) Verify that the mappings $\gamma_{p,L}$ form a polynomial law (Exercise 9) of E into $\Gamma_p(E)$, which is homogeneous of degree p ; this law will in what follows be denoted by γ_p. Show that for every k-module F and every $\sigma \in \mathrm{Hom}(\Gamma_p(E), F)$ the mappings $\varphi_L = (Id_L \otimes \sigma) \circ \gamma_{p,L}$ form a polynomial law of E into F which is homogeneous of degree p. This law is called the *law associated with* σ.

Let $(x_i)_{i \in I}$ be a family of elements of E and $(a_i)_{i \in I}$ a family with finite support of elements of L. Show that

$$\varphi_L\left(\sum_i a_i \otimes x_i\right) = \sum_{\xi \in \mathbf{N}^{(I)}, |\xi| = p} a^\xi \otimes \sigma(\gamma_\xi(x))$$

where $|\xi| = \sum_i \xi(i)$, $a^\xi = \prod_i a_i^{\xi(i)}$ and $\gamma_\xi(x) = \prod_i \gamma_{\xi(i)}(x_i)$.

b) Let φ be a polynomial law of E into a k-module F, homogeneous of degree p. Show that there exists a unique linear mapping σ of $\Gamma_p(E)$ into F such that φ is the law associated with

σ. (Suppose first that E is free on a basis $(x_i)_{i \in I}$; using Exercise 9, show that in this case there exists a unique mapping $\sigma \in \mathrm{Hom}(\Gamma_p(E), F)$ such that Relation (1) of Exercise 9 holds for every algebra L and every family $(a_i)_{i \in I}$ with finite support of elements of L. Treat the general case by introducing a free k-module E' and a surjective homomorphism $h : E' \to E$. The polynomial law φ, composed with the law defined by h is a polynomial law of E' into F which is associated to a linear mapping $\sigma' : \Gamma_p(E') \to F$; show that σ' is the composite of $\Gamma_p(h)$ and a mapping $\sigma : \Gamma_p(E) \to F$ such that φ is associated with σ (cf. Exercise 2, *f*)).)

11) We keep the notation of Exercise 10. A linear mapping of $\Gamma(E)$ into F will be called a *series over* E *with values in* F. The series over E with values in F form a k-module denoted by $S(E, F)$. For $n \in \mathbf{N}$ we say that the series $\sigma \in S(E, F)$ is of order $\geqslant n$ if $\mathrm{Ker}(\sigma) \supset \Gamma_p(E)$ for all $p < n$. In what follows we shall equip $S(E, F)$ with the topological group structure having as neighbourhood base of 0 the k-modules consisting of the series of order $\geqslant n$ $(n \in \mathbf{N})$.

Let φ be a polynomial law of E into F. Show that there exists a unique series $\sigma \in S(E, F)$ such that:
(i) for every k-algebra L and every $u \in L \otimes E$ the family $(\mathrm{Id}_L \otimes \sigma)\, \gamma_{p,L}(u)$ $(p \in \mathbf{N})$ has finite support;
(ii) $\varphi_L(u) = \sum_p (\mathrm{Id}_L \otimes \sigma) \circ \gamma_{p,L}(u)$.
(Apply the result of Exercise 10, *b*)).
This series σ is said to be *associated* to the law φ.

12) Let k be a commutative ring, E a k-module and B a k-algebra (not necessarily associative). Let Δ be the coproduct of $\Gamma(E)$ (cf. Exercise 3). For any series σ, $\sigma' \in S(E, B)$ we denote by $\sigma \,.\, \sigma'$ the series $m \circ (\sigma \otimes \sigma') \circ \Delta$, where m denotes the linear mapping of $B \otimes_k B$ into B defined by the product in B.
a) Show that the mapping $(\sigma, \sigma') \mapsto \sigma \,.\, \sigma'$ defines on $S(E, B)$ a k-algebra structure (cf. Exercise 11). Show that if B is associative (resp. commutative, unital) then $S(E, B)$ is associative (resp. commutative, unital) (cf. III, p. 578-584).
b) Assume that B is associative and unital. Show that the invertible elements of $S(E, B)$ are the series σ such that $\sigma(1)$ is invertible in B.
c) Suppose that E is a free k-module and $(e_i)_{i \in I}$ is a basis of E. For each $\alpha \in \mathbf{N}^{(I)}$ put $e^{(\alpha)} = \Pi \gamma_{\alpha(i)}(e_i)$ and denote by f^α the series over E with values in k defined by $f^\alpha(e^{(\beta)}) = \delta_{\alpha,\beta}$ (Kronecker index) for each $\beta \in \mathbf{N}^{(I)}$ (cf. Exercise 8). Show that $f^\alpha f^\beta = f^{\alpha+\beta}$ for any $\alpha, \beta \in \mathbf{N}^{(I)}$. Deduce that when I is finite, the algebra $S(E, k)$ is isomorphic to the algebra of formal power series $k[[(X_i)_{i \in I}]]$.

13) (Divided powers of a series.) The notation is as in Exercises 10 and 11. Let E and F be two k-modules and σ a series over E with values in F. Denote by $\boldsymbol{\sigma}$ the homogeneous polynomial law of degree 1 of $\Gamma(E)$ into F defined by σ. For any integers p, $m \in \mathbf{N}$ we denote by $\sigma_m^{(p)}$ the series associated with the polynomial law $\gamma_p \circ \boldsymbol{\sigma} \circ (\gamma_0 + \gamma_1 + \dots + \gamma_m)$ (cf. Exercise 10).
a) Show that as m tends to ∞, the sequence $\sigma_m^{(p)}$ converges in $S(E, \Gamma_p(F))$ to a series $\sigma^{(p)}$. This series $\sigma^{(p)}$ is called the *divided p-th power of the series* σ.

b) Show that if $(x_i)_{i \in I}$ is a family of elements of E and $\eta \in \mathbf{N}^{(I)}$, then

$$\sigma^{(p)}(\gamma_\eta(x)) = \sum_{\xi \in \mathscr{L}} \prod_{\alpha \in \mathbf{N}^{(I)}} \gamma_{\xi(\alpha)}(\sigma(\gamma_\alpha(x))),$$

where $\mathscr{L}$ is the set of all mappings with finite support of $\mathbf{N}^{(I)}$ into $\mathbf{N}$ satisfying the conditions $\sum_\alpha \xi(\alpha) = p$ and $\sum_\alpha \xi(\alpha)\,\alpha = \eta$ (Apply Ex. 10, *a*)).

c) Show that if $\sigma, \tau \in S(E, F)$, then for each $p \in \mathbf{N}$ we have

$$(\sigma + \tau)^{(p)} = \sum_{r+s=p} \sigma^{(r)} \sigma^{(s)}$$

(product calculated in the algebra $S(E, \Gamma(E))$, cf. Exercise 12).

d) Show that for every integer $p \in \mathbf{N}$ we have $\sigma^p = p!\, \sigma^{(p)}$, where σ^p is the p-th power of σ in the algebra $S(E, \Gamma(F))$.

e) Let Δ_E and Δ_F be the respective coproducts in $\Gamma(E)$ and $\Gamma(F)$. Show that for all $p \in \mathbf{N}$ we have

$$\Delta_F \circ \sigma^{(p)} = \sum_{r+s=p} (\sigma^{(r)} \otimes \sigma^{(s)}) \circ \Delta_E .$$

14) With the notation as in Ex. 13, suppose that $\sigma \in S(E, F)$ is of order $\geqslant 1$.

a) Show that for each $p \in \mathbf{N}$, $\sigma^{(p)}$ is of order $\geqslant p$. Deduce that the family $(\sigma^{(p)})_{p \in \mathbf{N}}$ is summable. Writing $A(\sigma) = \sum_p \sigma^{(p)}$, verify that for every $n \in \mathbf{N}$,

$$A(\sigma)(\Gamma_n(E)) \subset \Gamma_0(F) + \cdots + \Gamma_n(F).$$

b) Show that if $\operatorname{Ker}(\sigma) \supset \Gamma_m(E)$ for all $m \neq 1$, then $A(\sigma) = \Gamma(\sigma \circ \gamma_1)$.

c) Show that $A(\sigma)$ is a homomorphism of the cogebra $\Gamma(E)$ into the cogebra $\Gamma(F)$, compatible with the counits (cf. Ex. 13, *e*)).

d) Show that if $\sigma, \tau \in S(E, F)$ are two series of order $\geqslant 1$, then $A(\sigma + \tau) = A(\sigma)\,A(\tau)$, the product $A(\sigma)\,A(\tau)$ being defined by the product in $\Gamma(E)$ (cf. Ex. 13, *c*)).

15) With the notation as in Ex. 14, let E, F, G be three k-modules, $\sigma \in S(E, F)$, $\tau \in S(F, G)$, and suppose that σ is of order $\geqslant 1$. The series $\tau \circ A(\sigma) \in S(E, G)$ is called the *composition* of σ and τ; it is denoted by $\tau \circ \sigma$.

a) Let φ be a polynomial law of E into F (Ex. 9) and ψ a polynomial law of F into G, and denote by σ_φ and σ_ψ the series associated with φ and ψ respectively. Show that if the homogeneous component of degree 0 of φ is zero, then σ_φ is of order $\geqslant 1$ and $\sigma_\psi \circ \sigma_\varphi$ is the series associated with the polynomial law $\psi \circ \varphi$.

b) Show that if σ is of order $\geqslant p$ and τ is of order $\geqslant q$, then $\tau \circ \sigma$ is of order $\geqslant pq$.

c) Show that for every integer p, $(\tau \circ \sigma)^{(p)} = \tau^{(p)} \circ A(\sigma)$. Deduce that if τ is of order $\geqslant 1$ and if H is a k-module, then for any $\rho \in S(G, H)$ we have $(\rho \circ \tau) \circ \sigma = \rho \circ (\tau \circ \sigma)$.

d) Suppose that G is a k-algebra. Show that the mapping $\tau \mapsto \tau \circ A(\sigma)$ is a homomorphism of the algebra $S(F, G)$ into the algebra $S(E, G)$.

16) Let E and F be two modules over a commutative ring k and let p be an integer $\geqslant 0$.

a) Show that if $p \leqslant 2$, the mapping λ_p of $\mathrm{Hom}(\Gamma_p(\mathrm{E}), \mathrm{F})$ into F^{E} which associates with σ the mapping $\sigma \circ \gamma_p$ is an injective mapping.
b) Show that λ_1 has as image $\mathrm{Hom}_k(\mathrm{E}, \mathrm{F})$. Show that λ_2 has as image the set of all mappings g of E into F such that :

(i) $g(ax) = a^2 g(x)$ for any $a \in k$ and $x \in \mathrm{E}$;

(ii) the mapping $(x, y) \mapsto g(x + y) - g(x) - g(y)$ is a bilinear mapping of $\mathrm{E} \times \mathrm{E}$ into F.

c) Show that if E is a free k-module, then the image of λ_p is the module $\mathrm{Pol}_k^p(\mathrm{E}, \mathrm{F})$ of homogeneous polynomial mappings of degree p of E into F.

§ 6

1) Let r_k be the polynomial with integer coefficients such that

$$p_k = r_k(s_1, \ldots, s_k) .$$

a) Show that $r_k(\mathrm{Y}_1, \ldots, \mathrm{Y}_k)$ contains the terms Y_1^k and $(-1)^{k-1} k \mathrm{Y}_k$.
b) Show that

$$r_k(\mathrm{Y}_1, \ldots, \mathrm{Y}_k) = \sum_{p=0}^{k-1} (-1)^p \begin{vmatrix} \mathrm{Y}_{p+1} & \mathrm{Y}_{p+2} & \mathrm{Y}_{p+3} & \ldots \mathrm{Y}_{k-1} & \mathrm{Y}_k \\ 1 & \mathrm{Y}_1 & \mathrm{Y}_2 & \ldots \mathrm{Y}_{k-p-2} & \mathrm{Y}_{k-p-1} \\ 0 & 1 & \mathrm{Y}_1 & \ldots \mathrm{Y}_{k-p-3} & \mathrm{Y}_{k-p-2} \\ & & & & \\ 0 & 0 & 0 & \ldots \mathrm{Y}_1 & \mathrm{Y}_2 \\ 0 & 0 & 0 & \ldots 1 & \mathrm{Y}_1 \end{vmatrix} .$$

2) Show that $n!\, s_n \equiv s_1^n$ modulo the ideal of $\mathrm{A}[\mathrm{X}_1, \ldots, \mathrm{X}_n]$ generated by $\mathrm{X}_1^2, \mathrm{X}_2^2, \ldots, \mathrm{X}_n^2$.

3) Let f be a symmetric polynomial of $\mathrm{K}[\mathrm{X}_1, \ldots, \mathrm{X}_n]$ and φ the unique polynomial of $\mathrm{K}[\mathrm{Y}_1, \ldots, \mathrm{Y}_n]$ such that $f(\mathrm{X}_1, \ldots, \mathrm{X}_n) = \varphi(s_1, \ldots, s_n)$. Show that the total degree of φ is equal to the degree of f in any one of the X_i.

* 4) Let K be a field of characteristic 0. Show that if α_i $(1 \leqslant i \leqslant n)$ are n elements of K such that $\alpha_1^k + \alpha_2^k + \cdots + \alpha_n^k = 0$ for n consecutive values $h, h+1, \ldots, h+n-1$ of k, then $\alpha_1 = \alpha_2 = \cdots = \alpha_n = 0$. Show by an example that the property no longer holds if the n values of k are not consecutive or if K is not of characteristic 0. *

¶ 5) Let K be a commutative field and $f \in \mathrm{K}[\mathrm{X}_1, \ldots, \mathrm{X}_n, \mathrm{Y}_1, \ldots, \mathrm{Y}_n]$; for every permutation $\sigma \in \mathfrak{S}_n$ denote by σf the rational fraction

$$\sigma f(\mathrm{X}_1, \ldots, \mathrm{X}_n, \mathrm{Y}_1, \ldots, \mathrm{Y}_n) = f(\mathrm{X}_{\sigma(1)}, \ldots, \mathrm{X}_{\sigma(n)}, \mathrm{Y}_{\sigma(1)}, \ldots, \mathrm{Y}_{\sigma(n)}) .$$

Then f is said to be symmetric with respect to the pairs $(\mathrm{X}_i, \mathrm{Y}_i)$ if $\sigma f = f$ for all $\sigma \in \mathfrak{S}_n$.
a) For every element $\nu = (\lambda_1, \ldots, \lambda_n, \mu_1, \ldots, \mu_n)$ of $\mathbf{N}^{2n}$ and every permutation $\sigma \in \mathfrak{S}_n$ write

$$\sigma^{-1}(\nu) = (\lambda_{\sigma(1)}, \ldots, \lambda_{\sigma(n)}, \mu_{\sigma(1)}, \ldots, \mu_{\sigma(n)}) ,$$

so that $\mathfrak{S}_n$ in this way becomes a group of operators on $\mathbf{N}^{2n}$. Let γ be any orbit of $\mathfrak{S}_n$ in $\mathbf{N}^{2n}$, and denote by u_γ the symmetric polynomial

$$\sum X_1^{\lambda_1} X_2^{\lambda_2} \dots X_n^{\lambda_n} Y_1^{\mu_1} \dots Y_n^{\mu_n}$$

where $(\lambda_1, \dots, \lambda_n, \mu_1, \dots, \mu_n)$ ranges over γ. Show that the polynomials u_γ form a basis over K of the vector space of all polynomials that are symmetric in the (X_i, Y_i).

* *b*) Suppose that K has characteristic 0. Show that every polynomial u_γ is equal to a polynomial with rational coefficients in the symmetric functions

$$v_{\lambda\mu} = \sum_{i=1}^{n} X_i^\lambda Y_i^\mu$$

(argue by induction on the number of pairs (λ_i, μ_i) which are not $(0,0)$ in $\nu = (\lambda_1, \dots, \lambda_n, \mu_1, \dots, \mu_n)$ in considering the products $u_\gamma v_{\lambda\mu}$).

c) By *elementary* symmetric functions of the (X_i, Y_i) we shall understand the $n(n+3)/2$ polynomials $w_{hk} = u_\gamma$ corresponding to the orbits γ of elements $(\lambda_1, \dots, \lambda_n, \mu_1, \dots, \mu_n)$ other than $(0, \dots, 0)$ and such that h of the pairs (λ_i, μ_i) are equal to $(1, 0)$, k others are $(0, 1)$ and the remaining $n - h - k$ are $(0, 0)$. Show that if K is of characteristic 0, then every symmetric polynomial in the (X_i, Y_i) is equal to a polynomial in the elementary symmetric functions with coefficients in K (consider the sums $\sum_{i=1}^{n} (UX_i + VY_i)^k$, where U and V are two indeterminates, and use *b*)).

d) If K is a field of characteristic 3, and $n \geqslant 4$, show that $\sum_{i=1}^{n} X_i^2 Y_i^2$ is not equal to any polynomial in the symmetric functions w_{hk} with coefficients in K. *

6) Let n be an integer $\geqslant 1$, and let $X_1, \dots, X_n$ be indeterminates. For $1 \leqslant j \leqslant n$ let s_j be the elementary symmetric polynomial of degree j in the X_i and $u_j = (-1)^{j+1} s_j$. Put $S(T) = \prod_{i=1}^{n} (1 - X_i T) = 1 - \sum_{j=1}^{n} u_j T^j$ and for each $k \geqslant 1$ put $p_k = \sum_{i=1}^{n} X_i^k$; show that

$$p_k = \sum a_{\lambda_1 \lambda_2 \dots \lambda_n} u_1^{\lambda_1} u_2^{\lambda_2} \dots u_n^{\lambda_n}$$

the summation being extended to all systems $(\lambda_1, \dots, \lambda_n)$ of integers $\geqslant 0$ such that $\lambda_1 + 2\lambda_2 + \dots + n\lambda_n = k$, and the coefficient $a_{\lambda_1 \lambda_2 \dots \lambda_n}$ being equal to

$$\frac{k(\lambda_1 + \lambda_2 + \dots + \lambda_n - 1)!}{\lambda_1! \, \lambda_2! \dots \lambda_n!}$$

(expand the series $S'(T)/S(T)$ in a formal power series in T).

7) With the notation of Exercise 6 show that

$$\sum a_{\lambda_1 \lambda_2 \dots \lambda_n} = -1 + k \sum_{j=0}^{k/(n+1)} \frac{(-1)^j}{k - jn} \binom{k - jn}{j} 2^{k - j(n+1)} .$$

8) Put

$$s_k = \sum A_{m_1 \dots m_n} p_1^{m_1} \dots p_n^{m_n},$$

where the $A_{m_1 \dots m_n}$ are rational numbers and the sum is extended over all families $(m_1, \dots, m_n)$ such that $\sum i \cdot m_i = k$. Show that $\sum A_{m_1 \dots m_n} = 0$. for $k > 1$ and $\sum |A_{m_1 \dots m_n}| = 1$ for $k \geqslant 1$. Deduce that on writing $q_j = -p_j$ we have

$$(-1)^k s_k = \sum |A_{m_1 \dots m_n}| \, q_1^{m_1} \dots q_n^{m_n}.$$

9) *a*) Let K be a field and $X_1, \dots, X_n$ indeterminates. Show that for any exponents $\nu_1 < \nu_2 < \dots < \nu_n$ the determinant $\det(X_i^{\nu_j})$ is non-zero in $K(X_1, \dots, X_n)$.
b) Let s_k $(1 \leqslant k \leqslant n)$ be the elementary symmetric polynomials in the X_j and $r_1, \dots, r_n$ n elements of the field $K(X_1, \dots, X_n)$. We put (with a new indeterminate T)

$$R(T) = \frac{1 + r_1 T + \dots + r_n T^n}{1 + s_1 T + \dots + s_n T^n} = 1 + u_1 T + \dots + u_k T^k + \dots .$$

Show that if n of the coefficients u_k are zero, then all the u_k are zero. (Decompose R(T) into simple fractions.)
* *c*) Suppose that K is of characteristic zero. Let $\{\nu_1, \nu_2, \dots, \nu_n\}$ be a set of n elements $\geqslant 1$ of N, and let M be the complement of this set in N ; suppose also that M is *stable* under addition. Show that if $y_1, \dots, y_n$ are n elements of $K(X_1, \dots, X_n)$ such that

$$p_{\nu_j}(y_1, \dots, y_n) = p_{\nu_j}(X_1, \dots, X_n)$$

for $1 \leqslant j \leqslant n$, then $y_j = X_j$ for $1 \leqslant j \leqslant n$ except possibly for a permutation. (Consider the expansion of $\log R(T) = v_1 T + v_2 T^2 + \dots + v_k T^k + \dots$, observe that $v_{\nu_j} = 0$ for $1 \leqslant j \leqslant n$ and examine the relations between the v_k and the u_k.) *

10) Let $a \in A$. For each polynomial $h \in A[X]$ put $h_a(X) = h(X + a)$; show that res $(f_a, g_a) = \operatorname{res}(f, g)$, $\operatorname{dis}(f_a) = \operatorname{dis}(f)$.

11) In this exercise we assume that A is a field, $f(0) \neq 0$ and $\deg g < \deg f = p$. Let $\sum \alpha_n X^n$ be the expansion in a formal power series of the rational fraction $g(X)/f(X)$; for each integer $k \geqslant 0$ let $H_0^{(k)}$ be the Hankel determinant defined in Exercise 1 of IV, p. 90.
a) Show that $H_0^{(k)} = 0$ for $k > p$ and calculate $H_0^{(p)}$ as function of $\operatorname{res}(f, g)$.
b) Under what conditions do we have $H_0^{(p)} = H_0^{(p-1)} = \dots = H_0^{(p-r+1)} = 0$?

12) There exists a unique polynomial $P_{p,q}(T_0, \dots, T_p, U_0, \dots, U_q)$ with integer coefficients such that for every ring A and any polynomials $f = t_p X^p + \dots + t_0$, $g = u_q X^q + \dots + u_0$ of A[X] we have

$$\operatorname{res}_{p,q}(f, g) = P_{p,q}(t_0, \dots, t_p, u_0, \dots, u_q).$$

If we assign the weight i to T_i and U_i, then $P_{p,q}$ is isobaric of weight pq.

13) Let $n \geqslant 1$ be an integer, and order the elements of $\mathbf{N}^n$ lexicographically (notation $\alpha \preccurlyeq \beta$). For each $\alpha \in \mathbf{N}^n$ let T_α be the set of elements of $\mathbf{Z}[X_1, ..., X_n]$ of the form $X^\alpha + \sum_{\beta \prec \alpha} u_{\alpha\beta} X^\beta$ (with $u_{\alpha\beta} \in \mathbf{Z}$).

a) Show that the product of an element of T_α by an element of T_β lies in $T_{\alpha+\beta}$.

b) Let ε_k be the element $(\underbrace{1, 1, ..., 1}_{k}, \underbrace{0, ..., 0}_{n-k})$ of $\mathbf{N}^n$. Show that $s_k \in T_{\varepsilon_k}$. Deduce that $S(\alpha) = \prod_{k=1}^{n} s_k^{\alpha_k - \alpha_{k+1}}$ belongs to T_α, for $\alpha \in \mathbf{N}^n$ such that $\alpha_1 \geqslant \alpha_2 \geqslant \ldots \geqslant \alpha_n$. (Set $\alpha_{n+1} = 0$.)

c) With the notation of IV, p. 66, deduce that $c_{\alpha\alpha} = d_{\alpha\alpha} = 1$ for $\alpha \in D_k$ and $c_{\alpha\beta} = d_{\alpha\beta} = 0$ if $\alpha \prec \beta$.

14) Let L be a finite ordered set and A a commutative ring. In the ring M of square matrices over A with index set L, the set of matrices $m = (m_{uv})$ such that $m_{uv} = 0$ except when $u \leqslant v$ is a subring B.

a) An element $m = (m_{uv})$ of B is invertible in B if and only if m_{uu} is invertible in A for all $u \in L$.

b) Suppose henceforth that $A = \mathbf{Z}$. Let ζ be the element of B defined by $\zeta(u, v) = 1$ if $u \leqslant v$ and $\zeta(u, v) = 0$ otherwise. Show that ζ has an inverse μ (also written μ_L) in B (called the *Möbius function* of L). Let f and g be two mappings of L into $\mathbf{Z}$. Show that the relations « $f(x) = \sum_{y \geqslant x} g(y)$ for all $x \in L$ » and « $g(x) = \sum_{y \geqslant x} \mu(x, y)\, f(y)$ for all $x \in L$ » are equivalent.

c) Assume that L is the product of a finite family of ordered sets $L_1, ..., L_n$. Show that $\mu_L(x, y) = \prod_{i=1}^{n} \mu_{L_i}(x_i, y_i)$ for $x = (x_1, ..., x_n)$ and $y = (y_1, ..., y_n)$ such that $x \leqslant y$.

d) Let $m \geqslant 1$ be an integer and let D be the set of divisors of m, equipped with the order relation « x divides y ». Show that $\mu_D(x, y) = (-1)^s$ where s is the cardinal of the set of prime numbers p dividing y/x when y/x is not divisible by the square of an integer, and $\mu_D(x, y) = 0$ otherwise. (By applying *c*) reduce to the case where m is a power of a prime number.)

e) Suppose that the ordered set L is lattice-ordered. Let a, b, c be distinct elements of L such that $a \leqslant b \leqslant c$, and let Φ be the set of $x \in L$ such that $a \leqslant x$ and $\sup(x, b) = c$, and Ψ the corresponding set defined by $y \leqslant c$ and $\inf(y, b) = a$. Show that we have the relation $\sum_{x \in \Phi} \mu(a, x) = \sum_{y \in \Psi} \mu(y, c) = 0$.

f) Let T be a finite set and $\mathscr{R}(T)$ the set of equivalence relations on T, ordered by the relation « R is finer than S », written $R \leqslant S$. Show that $\mathscr{R}(T)$ is lattice-ordered. Let R, S in $\mathscr{R}$ be such that $R \leqslant S$; let $U_1, ..., U_k$ be the equivalence classes under S and for $1 \leqslant i \leqslant k$, let α_i be the number of equivalence classes under R contained in U_i. Show that $\mu_{\mathscr{R}(T)}(R, S) = \prod_{i=1}^{k} (-1)^{\alpha_i - 1} (\alpha_i - 1)!$ (if A_i is the set of equivalence classes under R contained in U_i, observe that the interval $[R, S]$ of $\mathscr{R}(T)$ is isomorphic to $\mathscr{R}(A_1) \times \ldots \times \mathscr{R}(A_k)$, apply *c*) and reduce to the case where R is the least element of $\mathscr{R}(T)$ and S the greatest ; treat this last case by using *e*)).

15) Let T be a finite set with k elements. We denote by D_k the set of elements $(\alpha_1, ..., \alpha_k)$ of $\mathbf{N}^k$ such that $\alpha_1 + \ldots + \alpha_k = k$ and $\alpha_1 \geqslant \alpha_2 \geqslant \ldots \geqslant \alpha_k$. With every equivalence

relation R in T we associate an element $\lambda(R)$ of D_k as follows : if $U_1, \ldots, U_l$ are the equivalence classes under R, numbered so that $\operatorname{Card}(U_1) \geqslant \ldots \geqslant \operatorname{Card}(U_l)$, put $\lambda(R) = (\alpha_1, \ldots, \alpha_k)$ with $\alpha_i = \operatorname{Card}(U_i)$ if $1 \leqslant i \leqslant l$ and $\alpha_i = 0$ if $l < i \leqslant k$. Let X_n $(n \in N)$ be indeterminates ; for each $R \in \mathscr{R}(T)$ put $r(R) = \sum_f \prod_{t \in T} X_{f(t)}$, where the summation is extended to the set of mappings $f : T \to N$ such that R is the equivalence relation associated with f. Put $s(R) = \sum_{S \geqslant R} r(S)$.

a) Prove that $r(R) = \prod_{i=1}^{k} r_i \, . \, M(\lambda(R))$ if r_i of the equivalence classes under R are of cardinal i for $1 \leqslant i \leqslant k$. (See formula (11) of IV, p. 65 for the definition of $M(\alpha)$.)

b) Prove the relation $S(R) = \prod_{i=1}^{k} p_i^{r_i}$ where the r_i are as in *a*) and $p_m = \sum_{n \in N} X_n^m$ for every integer $m \geqslant 1$.

c) Show that $r(R) = \sum_{S \geqslant R} \mu(R, S)\, s(S)$, where $\mu(R, S)$ is defined as in Exercise 14.

d) Hence deduce an expression of the s_k as a polynomial in $p_1, \ldots, p_k$ with rational coefficients, where $s_k = \sum_{i_1 < \ldots < i_k} X_{i_1} \ldots X_{i_k}$.

e) For $\alpha \in D_k$ put $S'(\alpha) = \prod_{i=1}^{k} s_{\alpha_i}$. Show that $S'(\alpha) = \sum_{\beta \in D_k} c'_{\alpha\beta} \, . \, M(\beta)$, where the integer $c'_{\alpha\beta}$ is defined as follows : suppose that r_i (resp. s_i) of the components of α (resp. β) are equal to i (for $1 \leqslant 1 \leqslant k$). Then $c'_{\alpha\beta}$ is the product $N r_1! \ldots r_k! s_1! \ldots s_k!/k!$, where N is the number of pairs of equivalence relations R, S on T such that $\lambda(R) = \alpha$, $\lambda(S) = \beta$ and such that the intersection of an equivalence class under R with an equivalence class under S has at most one element. Deduce the symmetry relation $c'_{\alpha\beta} = c'_{\beta\alpha}$.

f) Show that there is an involution $\alpha \mapsto \tilde{\alpha}$ in D_k, characterized by the fact that the relations $i \leqslant \alpha_j$ and $j \leqslant \tilde{\alpha}_i$ are equivalent, for i, j in the range from 1 to k. Show that $S'(\alpha) = S(\tilde{\alpha})$ and $c'_{\alpha\beta} = c_{\tilde{\alpha}\beta}$. Hence obtain the relation $c_{\alpha\beta} = c_{\tilde{\alpha}\tilde{\beta}}$ (with the notation of IV, p. 66).

TABLE

Conventions (cf. page IV, 66).

A symbol such as $p^{\alpha} . q^{\beta} \ldots$, where p, q, α, β, ... are integers $\geqslant 1$, denotes an element of D_k where $k = p\alpha + q\beta + \cdots$ having α components equal to p, β components equal to q, etc. Thus $3 . 1^2$ denotes the element of D_5 equal to $(3, 1, 1, 0, 0)$, while 1^4 denotes the element $(1, 1, 1, 1)$ of D_4.

On the right of the row of index α of the matrix C the expression of $S(\alpha)$ as monomial in $s_1, \ldots, s_k$ is given. For example, for $k = 3$ and $\alpha = 2 . 1$ we have

$$S(2 . 1) = s_1 s_2 = M(2 . 1) + 3M(1^3) .$$

We have used a corresponding convention for the columns of the matrix D. For example, for $k = 4$ and $\alpha = 2^2$ we have

$$M(2^2) = 2s_4 - 2s_1 s_3 + s_2^2 .$$

We recall that $M(2^2)$ is the sum of all monomials of the form $X_i^2 X_j^2$ for $1 \leqslant i < j \leqslant n$ in the ring $\mathbf{Z}[X_1, \ldots, X_n]$.

An empty box indicates a zero element of the matrices C and D.

$k = 2$

	2	1^2	
2	1	2	S_1^2
1^2		1	S_2

	2	1^2
2	1	−2
1^2		1
	S_1^2	S_2

$k = 3$

	3	2.1	1^3	
3	1	3	6	S_1^3
2.1		1	3	S_1S_2
1^3			1	S_3

	3	2.1	1^3
3	1	−3	3
2.1		1	−3
1^3			1
	S_1^3	S_1S_2	S_3

$k = 4$

	4	3.1	2^2	2.1^2	1^4	
4	1	4	6	12	24	S_1^4
3.1		1	2	5	12	$S_1^2S_2$
2^2			1	2	6	S_2^2
2.1^2				1	4	S_1S_3
1^4					1	S_4

Matrix C

	4	3.1	2^2	2.1^2	1^4
4	1	−4	2	4	−4
3.1		1	−2	−1	4
2^2			1	−2	2
2.1^2				1	−4
1^4					1
	S_1^4	$S_1^2S_2$	S_2^2	S_1S_3	S_4

Matrix D

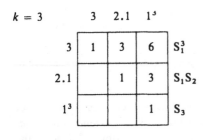

$k = 5$

	5	4.1	3.2	3.1^2	$2^2.1$	2.1^3	1^5	
5	1	5	10	20	30	60	120	S_1^5
4.1		1	3	7	12	27	60	$S_1^3S_2$
3.2			1	2	5	12	30	$S_1S_2^2$
3.1^2				1	2	7	20	$S_1^2S_3$
$2^2.1$					1	3	10	S_2S_3
2.1^3						1	5	S_1S_4
1^5							1	S_5

Matrix C

	5	4.1	3.2	3.1^2	$2^2.1$	2.1^3	1^5
5	1	−5	5	5	−5	−5	5
4.1		1	−3	−1	5	1	−5
3.2			1	−2	−1	5	−5
3.1^2				1	−2	−1	5
$2^2.1$					1	−3	5
2.1^3						1	−5
1^5							1
	S_1^5	$S_1^3S_2$	$S_1S_2^2$	$S_1^2S_3$	S_2S_3	S_1S_4	S_5

Matrix D

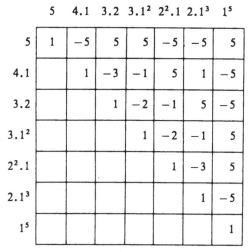

CHAPTER V

Commutative Fields

Except where the contrary is expressly stated, all the fields considered in this chapter are commutative ; all algebras are associative and unital and the algebra homomorphisms are unital, every subalgebra of an algebra contains the unit element of that algebra. Whenever a field K *is said to be contained in a ring* L *(in particular in a field) without further specification, it is understood that* K *is a subring of* L ; *we shall also say that* K *is a subfield of* L, *or also (if* L *is a field) that* L *is an extension field of* K.

§ 1. PRIME FIELDS. CHARACTERISTIC

1. Prime fields

The field of fractions of the ring $\mathbf{Z}$ of rational integers is called the field of rational numbers and is denoted by $\mathbf{Q}$ (I, p. 117). For every prime number p the quotient ring $\mathbf{Z}/(p)$ is a finite [1] field of p elements, denoted by $\mathbf{F}_p$ in the sequel. The field $\mathbf{Q}$ is infinite since it contains $\mathbf{Z}$, and is therefore not isomorphic to any field $\mathbf{F}_p$. If p and p' are distinct prime numbers, the fields $\mathbf{F}_p$ and $\mathbf{F}_{p'}$ have distinct cardinals and so are not isomorphic.

DEFINITION 1. — *A field is said to be prime if it is isomorphic either to* $\mathbf{Q}$ *or to one of the fields* $\mathbf{F}_p$.

Every subfield of $\mathbf{Q}$ contains the ring $\mathbf{Z}$ and hence the field of fractions $\mathbf{Q}$ of $\mathbf{Z}$; every subring of $\mathbf{F}_p$ is necessarily equal to $\mathbf{F}_p$. Therefore every subfield of a prime field is necessarily equal to it (cf. Cor. 2 of Th. 1 below). Let P be a prime field and A a ring ; if f and f' are two homomorphisms of P into A, then the set of $x \in \mathrm{P}$ such that $f(x) = f'(x)$ is a subfield of P, hence by what has been said we must have $f = f'$. In particular, the only endomorphism of a prime field is the identity mapping.

[1] By abuse of language one says that a ring or field is *finite* if its underlying set is finite.

THEOREM 1. — *Let* A *be a ring ; suppose that there exists a subfield of* A. *Then* A *has a unique subfield* P *which is a prime field. Moreover,* P *is contained in the centre of* A *and in every subfield of* A.

Let K be a subfield of A, C the centre of A, and put $K' = K \cap C$; then K′ is a subfield of A. Let f be the unique homomorphism of **Z** into A and $\mathfrak{p}$ its kernel. Every subring of A, in particular K′, contains $f(\mathbf{Z})$; hence the ideal $\mathfrak{p}$ is prime (I, p. 116-117). If $\mathfrak{p} = (0)$, the homomorphism f of **Z** into K′ is injective ; hence it extends (I, p. 116) to an isomorphism $\bar{f}$ of **Q** onto a subfield P of K′. If $\mathfrak{p} \neq (0)$, there is a strictly positive integer p such that $\mathfrak{p} = (p)$ (I, p. 111) ; if we had $p = ab$ with $a > 1$ and $b > 1$, this would mean $a \notin \mathfrak{p}$, $b \notin \mathfrak{p}$ and $ab \in \mathfrak{p}$ in contradiction with the fact that $\mathfrak{p}$ is prime. The number p is therefore prime and by passage to quotients f defines an isomorphism of $\mathbf{F}_p = \mathbf{Z}/\mathfrak{p}$ onto a subfield P of K′. In both cases P is a subfield of A contained in the centre C of A, and it is a prime field. Let L be a subfield of A ; then $P \cap L$ is a subfield of P, and since P is prime, we have $P \cap L = P$, whence $P \subset L$. If P′ is a subfield of A and is a prime field, then by what has been said, $P \subset P'$, whence $P = P'$ because P′ is a prime field.

COROLLARY 1. — *Let* K *be a field. There exists a unique subfield of* K *which is a prime field, and this is the least subfield of* K.

COROLLARY 2. — *For a field to be prime it is necessary and sufficient that it should contain no subfield other than itself.*

2. Characteristic of a ring and of a field

We shall define the characteristic of a ring A only when A has a subfield. When this is so, let f be the unique ring homomorphism of **Z** into A, and let n be the unique positive integer generating the ideal of **Z** which is the kernel of f (I, p. 111) ; then the integer n is called the *characteristic* of A.

Let A be a ring for which the characteristic is defined ; then A does not reduce to 0. By Th. 1 there exists a unique subfield P of A which is a prime field ; we shall call it the *prime subfield* of A. By the proof of Th. 1 there are the following two possibilities :

a) the characteristic of A is 0, P is isomorphic to **Q**,

b) the characteristic of A is a prime number p, P is isomorphic to $\mathbf{F}_p$.

If the characteristic of A is zero, there exists a unique ring homomorphism of **Q** into A ; its image is the prime subfield of A, contained in the centre of A. Therefore there exists a unique **Q**-algebra structure of A compatible with the ring structure. When the characteristic of A is a prime number p, we have the corresponding properties on replacing the field **Q** by the field $\mathbf{F}_p$.

PROPOSITION 1. — *Let* A *be a ring not reduced to* 0.

a) *For* A *to be of characteristic* 0 *it is necessary and sufficient that the mapping* $x \mapsto n \cdot x$ *of* A *into itself should be bijective, for every integer* $n \neq$

b) Let p be a prime number. For A to be of characteristic p it is necessary and sufficient that $p \cdot x = 0$ for all $x \in A$.

Let f be the unique homomorphism of **Z** into A ; we have $n \cdot x = f(n)x$ for every integer n and every x in A. For A to be of characteristic 0, it is necessary and sufficient that f should extend to a homomorphism of **Q** into A, that is, $f(n)$ should be invertible in A for every $n \neq 0$ (I, p. 113) ; this proves *a*). Similarly for A to be of characteristic p it is necessary and sufficient that f should annihilate $p\mathbf{Z}$, that is, $f(p) = 0$, or also that $p \cdot x = 0$ for all $x \in A$; this proves *b*).

Let us take for A a *not necessarily commutative* field. The centre of A is a (commutative) field ; therefore the characteristic and the prime subfield of A are defined.

Remarks. — 1) Let A and A′ be two rings not reduced to 0. Suppose that the characteristic of A is defined and that there is a homomorphism u of A into A′. The image under u of the prime subfield of A is a subfield P′ of A′, isomorphic to P, and hence prime. It follows that the characteristic of A′ is defined and is equal to that of A. If A and A′ are of characteristic 0 (resp. $p \neq 0$), the mapping u is a homomorphism of algebras over **Q** (resp. $\mathbf{F}_p$).

2) Remark 1 shows that if A is a ring of characteristic 0 (resp. $p \neq 0$), the same holds of any ring A′ containing A as subring, or of any quotient of A by a two-sided ideal $\mathfrak{a} \neq A$. In particular, if K is a field, every subfield of K and every extension field of K have the same characteristic as K.

3) Let A be an algebra not reduced to 0 over a field K. Since the mapping $\lambda \mapsto \lambda \cdot 1$ of K into A is a ring homomorphism, Remark 1 shows that the characteristic of A is defined and equal to that of K.

4) Since the field **Q** is infinite, every ring of characteristic 0 is infinite ; it follows that every finite field has non-zero characteristic.

5) Let A be a ring not reduced to 0, whose additive group is a torsion-free **Z**-module, and put $B = \mathbf{Q} \otimes_{\mathbf{Z}} A$. The mapping $x \mapsto 1 \otimes x$ of A into B is injective (II, p. 314), hence A is isomorphic to a subring of a ring of characteristic 0.

3. Commutative rings of characteristic *p*

In this No. and the following one p denotes a prime number.

THEOREM 2. — *Let A be a commutative ring of characteristic p. The mapping $a \mapsto a^p$ is an endomorphism of the ring A, that is, we have the relations*

$$(a + b)^p = a^p + b^p \tag{1}$$

$$(ab)^p = a^p b^p \tag{2}$$

for a, b in A.

Formula (2) follows from the commutativity of A. To prove (1) we use the binomial formula $(a+b)^p = a^p + b^p + \sum_{i=1}^{p-1} \binom{p}{i} \cdot a^i b^{p-i}$; since $p \cdot x = 0$ for all $x \in A$, it suffices to prove the following lemma :

Lemma 1. — *Let p be a prime number and i an integer in the range from* 1 *to* $p-1$, *then the binomial coefficient* $\binom{p}{i}$ *is an integer divisible by* p.

We argue by induction on i, the case $i = 1$ being immediate from the formula $\binom{p}{1} = p$. Suppose that $2 \leqslant i \leqslant p-1$ and that $\binom{p}{i-1}$ is divisible by p. Then the integer $i\binom{p}{i} = (p-i+1)\binom{p}{i-1}$ belongs to the prime ideal $p\mathbf{Z}$ of $\mathbf{Z}$; since $i \notin p\mathbf{Z}$, we have $\binom{p}{i} \in p\mathbf{Z}$ and the lemma follows.

Let A be a commutative ring of characteristic p and f an integer $\geqslant 0$. From Th. 2 we deduce by induction on f that the mapping $a \mapsto a^{p^f}$ is an endomorphism of the ring A. In particular we have the relation

$$(a_1 + \cdots + a_n)^{p^f} = a_1^{p^f} + \cdots + a_n^{p^f} \tag{3}$$

for any $a_1, \ldots, a_n$ in A. The mapping $a \mapsto a^p$ is sometimes called the *Frobenius endomorphism* of A. Taking $A = \mathbf{F}_p$ and $a_i = 1$, we obtain from (3) the relation :

$$n^{p^f} \equiv n \quad \text{mod. } p \quad (n \in \mathbf{Z}, f \in \mathbf{N}). \tag{4}$$

For each subset S of A we denote by S^{p^f} the set of elements of A of the form x^{p^f} with $x \in S$[1]. In particular, if K is a subring of A, the set K^{p^f} is a subring of A. If K is a subring of A and S a subset of A we denote by K[S] the subring of A generated by $K \cup S$; when A is a field, we denote by K(S) the field of fractions of K[S], that is, the subfield of A generated by $K \cup S$.

PROPOSITION 2. — *Let A be a commutative ring of characteristic p, K a subring of A, S a subset of A and f a positive integer.*

a) We have $K[S]^{p^f} = K^{p^f}[S^{p^f}]$, and if A is a field, $K(S)^{p^f} = K^{p^f}(S^{p^f})$.

b) If the K-module K[S] is generated by the family $(a_i)_{i \in I}$ of elements of A, then the K-module $K[S^{p^f}]$ is generated by the family $(a_i^{p^f})_{i \in I}$.

Since K[S] is the subring of A generated by $K \cup S$, its image $K[S]^{p^f}$ under the endomorphism $\pi : a \mapsto a^{p^f}$ of the ring A is the subring of A generated by the image $K^{p^f} \cup S^{p^f}$ of $K \cup S$ under π, whence $K[S]^{p^f} = K^{p^f}[S^{p^f}]$. The case of fields is treated similarly ; this proves *a*).

It is clear that the family $(a_i^{p^f})_{i \in I}$ generates the K^{p^f}-module $K[S]^{p^f}$. The K-module $K[S^{p^f}]$ is generated by products of the form $x_1^{p^f} \ldots x_n^{p^f} = (x_1 \ldots x_n)^{p^f}$ with $x_1, \ldots, x_n$ arbitrary in S, hence also by the set $K[S]^{p^f}$. Assertion *b*) follows directly from this.

[1] Of course the set S^{p^f} should not be confused with the set product of p^f sets equal to S, nor with the set of products of p^f elements belonging to S.

4. Perfect rings of characteristic p

DEFINITION 2. — *A ring* A *of characteristic* $p \neq 0$ *is said to be perfect if it is commutative and the mapping* $a \mapsto a^p$ *is bijective.*

If the ring A is perfect of characteristic p, the mapping $a \mapsto a^{p^f}$ is an automorphism of the ring A for every integer $f \geqslant 0$; the inverse automorphism is denoted by $a \mapsto a^{1/p^f}$ or $a \mapsto a^{p^{-f}}$ and the image of a subset S of A under this automorphism is written S^{1/p^f} or $S^{p^{-f}}$. It is clear that $(a^{p^e})^{p^f} = a^{p^{e+f}}$ for all $a \in A$ and any integers e and f (of whatever sign).

Let A be a commutative ring of characteristic p. For every integer $f \geqslant 0$ let us write $\mathfrak{n}_f$ for the kernel of the endomorphism $a \mapsto a^{p^f}$ of the ring A. Then $(\mathfrak{n}_f)_{f \geqslant 0}$ is an increasing sequence of ideals of A ; since every positive integer is majorized by a power of p, the ideal $\mathfrak{n} = \bigcup_{f \geqslant 0} \mathfrak{n}_f$ consists of all the nilpotent elements of A. In particular, if A is perfect, every nilpotent element of A is zero.

DEFINITION 3. — *Let* A *be a commutative ring of characteristic* $p \neq 0$. *By a perfect closure of* A *we understand a pair* $(\hat{A}, u)$ *where* $\hat{A}$ *is a perfect ring of characteristic* p *and* u *is a homomorphism of* A *into* $\hat{A}$ *satisfying the following universal property :*

(PC) *If* B *is a perfect ring of characteristic* p *and* v *is a homomorphism of* A *into* B, *then there exists a unique homomorphism* h *of* $\hat{A}$ *into* B *such that* $v = h \circ u$.

The universal property (PC) implies at once the *uniqueness* of the perfect closure, in the following sense : if $(\hat{A}, u)$ and $(\hat{A}', u')$ are two perfect closures of A, then there exists a unique isomorphism h of $\hat{A}$ onto $\hat{A}'$ such that $u' = h \circ u$ (cf. E, IV, p. 23). We shall now establish the *existence* of the perfect closure :

THEOREM 3. — *Let* A *be a commutative ring of characteristic* $p \neq 0$. *There exists a perfect closure* $(\hat{A}, u)$ *of* A. *Moreover, the kernel of* u *is the set of all nilpotent elements of* A *and for each* $x \in \hat{A}$ *there exists an integer* $n \geqslant 0$ *such that* $x^{p^n} \in u(A)$.

For each integer $n \geqslant 0$, put $A_n = A$; when $m \geqslant n$ we define a homomorphism $\pi_{m,n}$ of A_n into A_m by $\pi_{m,n}(a) = a^{p^{m-n}}$. We thus obtain a direct system of rings $(A_n, \pi_{m,n})$ (I, p. 120) ; let $\hat{A}$ be the direct limit of this system and u_n the canonical homomorphism of $A_n = A$ into $\hat{A}$; we also put $u = u_0$. By construction of the direct limit the kernel $\mathfrak{n}$ of u is the union of the kernels of the homomorphisms

$\pi_{n,0} : a \mapsto a^{p^n}$ of A into A, thus it consists of all the nilpotent elements of A. The ring $\hat{A}$ is commutative of characteristic p by Remark 1 of V, p. 3.

The ring $\hat{A}$ is the union of the ascending sequence $(u_n(A))_{n \geqslant 0}$ of subrings. We have $u_n(A)^{p^n} = A$; hence for each $x \in \hat{A}$ there exists an integer $n \geqslant 0$ such that $x^{p^n} \in u(A)$. We also have $u_n(A) = u_{n+1}(A)^p$, whence $\hat{A}^p = \hat{A}$. Let $x \in \hat{A}$ be such that $x^p = 0$; choose an integer $n \geqslant 1$ and an element $a \in A$ such that $x = u_n(a)$. Then we have $u_{n-1}(a) = u_n(a)^p = 0$; by definition of the direct limit there exists an integer $m \geqslant n$ such that $\pi_{m-1,n-1}(a) = 0$, that is, $a^{p^{m-n}} = 0$. We thus have $\pi_{m,n}(a) = 0$, whence $u_n(a) = 0$, that is, $x = 0$. Therefore the ring $\hat{A}$ is perfect of characteristic p.

Let v be a homomorphism of A into a perfect ring B of characteristic p. For every integer $n \geqslant 0$, the mapping $b \mapsto b^{p^n}$ is an automorphism of B and so there exists a homomorphism v_n of $A_n = A$ into B characterized by $v(a) = v_n(a)^{p^n}$. We then have $v_m \circ \pi_{m,n} = v_n$ for $m \geqslant n \geqslant 0$; by definition of the direct limit there exists a homomorphism h of $\hat{A}$ into B such that $v_n = h \circ u_n$ for all $n \geqslant 0$; in particular we have $v = v_0 = h \circ u_0 = h \circ u$. Finally let h' be a homomorphism of $\hat{A}$ into B such that $h' \circ u = v$. Let $x \in \hat{A}$; as we have seen, there exist an integer $n \geqslant 0$ and an element $a \in A$ such that $x^{p^n} = u(a)$. Then we have

$$h(x)^{p^n} = h(u(a)) = v(a) = h'(u(a)) = h'(x)^{p^n},$$

and since B is perfect, we find $h(x) = h'(x)$. Thus we have $h' = h$, and this completes the proof that $(\hat{A}, u)$ is a perfect closure of A.

PROPOSITION 3. — *Let* B *be a perfect ring of characteristic p and* A *a subring of* B. *Write* $A^{p^{-\infty}} = \bigcup_{f \geqslant 0} A^{p^{-f}}$ *and denote by j the canonical injection of* A *in* $A^{p^{-\infty}}$. *Then* $A^{p^{-\infty}}$ *is the smallest perfect subring of* B *containing* A *and* $(A^{p^{-\infty}}, j)$ *is a perfect closure of* A.

For each integer $f \in \mathbf{Z}$ denote by π_f the automorphism $b \mapsto b^{p^f}$ of B. The sequence of subrings $\pi_{-f}(A)$ of B (for $f \geqslant 0$) is ascending and its union $A^{p^{-\infty}}$ is thus a subring of B. We have $\pi_1(A^{p^{-\infty}}) = \bigcup_{f \geqslant 0} \pi_{-(f-1)}(A) = A^{p^{-\infty}}$, hence $A^{p^{-\infty}}$ is a perfect subring of B. Finally let B_0 be a perfect subring of B containing A; for every integer $f \geqslant 0$ we have $\pi_{-f}(A) \subset \pi_{-f}(B_0) = B_0$, whence $A^{p^{-\infty}} \subset B_0$.

If v is a homomorphism of A into a perfect ring B′ of characteristic p, then for every integer $f \geqslant 0$ we can define a homomorphism h_f of $\pi_{-f}(A)$ into B′ by $h_f(\pi_{-f}(a)) = v(a)^{p^{-f}}$ for all $a \in A$. We see at once that h_{f+1} agrees with h_f on $\pi_{-f}(A)$; thus there exists a homomorphism h of $A^{p^{-\infty}}$ into B′ which induces h_f on $\pi_{-f}(A)$ for all $f \geqslant 0$ and in particular, h extends $h_0 = v$. If h' is another extension of v to a homomorphism of $A^{p^{-\infty}}$ into B, the equality $h' = h$ may be established as in the proof of Th. 3.

5. Characteristic exponent of a field. Perfect fields

Let K be a field. By the *characteristic exponent* of K we understand the integer equal to 1 if K is of characteristic 0, and equal to the characteristic of K when this is non-zero.

PROPOSITION 4. — *Let K be a field of characteristic exponent q. For every integer $f \geqslant 0$ the mapping $x \mapsto x^{q^f}$ is an isomorphism of K on one of its subfields (denoted by K^{q^f}).*

This follows from Th. 2 when $q \neq 1$ and is trivial when $q = 1$.

Likewise one can extend Prop. 2 to the case where A is a field of characteristic exponent q, the case $q = 1$ being trivial.

DEFINITION 4. — *A field K of characteristic exponent q is said to be perfect if we have $K^q = K$. When $K^q \neq K$, K is called imperfect.*

By this definition a field is perfect if it is of characteristic 0, or if it is a perfect ring of characteristic $p \neq 0$ in the sense of Def. 2. If K is a field of characteristic $p \neq 0$ and $(\hat{K}, u)$ is a perfect closure of K, then $\hat{K}$ is a field by Prop. 3 (V, p. 6) and u is an isomorphism of K onto a subfield of $\hat{K}$. Frequently one identifies K with its image under u in $\hat{K}$, so that we have $\hat{K} = K^{p^{-\infty}}$ (Prop. 3).

Let K be a field of characteristic 0 ; the characteristic exponent of K is then 1. By convention the notation $x^{q^{-f}}$ and $S^{q^{-f}}$ is taken to mean x and S respectively (for an element x of K and a subset S of K). In particular we put $K^{q^{-\infty}} = K$ and we agree to take the perfect closure of K to be K.

PROPOSITION 5. — *If K is a field of characteristic 0, or is finite, * or algebraically closed $_*$, it is perfect. In particular every prime field is perfect.*

Suppose that K has characteristic $p \neq 0$. If K is finite, then the subfield K^p of K has the same cardinal as K, whence $K^p = K$. * If K is algebraically closed, the polynomial $X^p - a$ has a root x in K for each $a \in K$ (V, p. 20, Def. 1) whence $x^p = a$ and so $K^p = K$. $_*$ Finally a prime field is of characteristic 0 or finite.

Let K_0 be a field of characteristic $p \neq 0$ and $K = K_0(X)$ the field of rational fractions in an indeterminate X over K_0. Then K is *imperfect,* for there exists no element $u(X)/v(X)$ of K (u, v polynomials in $K_0[X]$) such that $(u(X)/v(X))^p = X$. This may be seen by writing this relation in the form $u(X)^p = Xv(X)^p$ and comparing the degrees of the two sides.

6. Characterization of polynomials with zero differential

PROPOSITION 6. — *Let K be a commutative ring, A the polynomial algebra $K[X_i]_{i \in I}$ and S the set of elements F of A such that $dF = 0$.*

a) If K *is a ring of characteristic* 0, *then* S = K.

b) If K *is a ring of characteristic* $p \neq 0$, *then* $S = K[X_i^p]_{i \in I}$; *if moreover* K *is perfect, then* $K = A^p$.

The mapping $F \mapsto dF$ of A into the module $\Omega_K(A)$ of K-differentials of A is K-linear and satisfies the relation

$$d(FF') = F \,.\, dF' + F' \,.\, dF$$

(III, p. 569). Therefore S is a subalgebra.

When K is of characteristic $p \neq 0$, put $T = K[X_i^p]_{i \in I}$; we thus have $T = A^p$ if K is perfect (V, p. 4, Prop. 2) ; moreover, we have $d(X_i^p) = pX_i^{p-1} \,.\, dX_i = 0$ for all $i \in I$, hence the subalgebra S of A contains T. If K is of characteristic 0, we put T = K, and still find that $T \subset S$. It remains to show that S is contained in T.

For every finite subset J of I let A_J be the subalgebra of A generated by the family $(X_j)_{j \in J}$. We have $A_\varnothing = K$ and $A = \bigcup_{J \in I} A_J$; so it suffices to prove the relation $S \cap A_J \subset T$, which we shall accomplish by induction on the cardinal of J. Thus let J be a finite subset of I such that $S \cap A_J \subset T$, let i be an element of $I - J$ and $J' = J \cup \{i\}$. Every element F of A_J may be written in just one way in the form

$$F = \sum_{n=0}^{\infty} F_n \,.\, X_i^n, \tag{5}$$

with $F_n \in A_J$ for all $n \geq 0$, and then

$$dF = \sum_{n=0}^{\infty} X_i^n \,.\, dF_n + \sum_{n=0}^{\infty} nX_i^{n-1}F_n \,.\, dX_i \,. \tag{6}$$

Suppose that F belongs to S ; the family $(dX_r)_{r \in I}$ is a basis of the A-module $\Omega_K(A)$ (III, p. 570) and dF_n is a linear combination of the differentials dX_j for $j \in J$ because $F_n \in A_J = K[X_j]_{j \in J}$. By (6) we then have $dF_n = 0$ and $nF_n = 0$ for each integer $n \geq 0$. By the induction hypothesis $F_n \in T$ for all n, since $dF_n = 0$.

a) If K is of characteristic 0, we have $nF_n = 0$ for all $n \geq 1$, whence $F_n = 0$ by Prop. 1 (V, p. 2) ; so we have $F = F_0$, whence $F \in T$.

b) If K is of characteristic $p \neq 0$, then A is an algebra over the field F_p and the relation $nF_n = 0$ implies $F_n = 0$ for every integer n not divisible by p. So we have $F = \sum_{m=0}^{\infty} F_{mp} X_i^{mp}$, whence $F \in T$.

Remark. — We still have S = K when the additive group of K is torsion-free ; this follows from the above proof or from Remark 5 of V, p. 3.

COROLLARY. — *Let* K *be a field and* F(X) *a polynomial with coefficients in* K, *whose derivative* F′(X) *is zero.*

a) If K *is of characteristic* 0, *then* $F \in K$.

b) If K *is of characteristic* $p \neq 0$, *there exists a polynomial* G(X) *such that* $F(X) = G(X^p)$.

For we have $dF = F' \,.\, dX = 0$.

§ 2. EXTENSIONS

1. The structure of an extension

DEFINITION 1. — *Let* K *be a field. By an extension of* K *we understand a* K*-algebra whose underlying ring is a field. By a subextension (or sub-*K*-extension) of the extension* E *we understand a sub-*K*-algebra of* E *which is a field.*

Let E be an extension of K. The mapping $u : \lambda \mapsto \lambda \,.\, 1$ of K into E is a ring homomorphism ; by I, p. 115, u induces an isomorphism of K onto a subfield $u(K)$ of E.

Conversely, if K, E are fields and u is a homomorphism of K into E, then u defines on E a structure of extension of K (III, p. 433). By abuse of language one sometimes says that (E, u) is *an extension of* K.

An extension is said to be *trivial* if $u(K) = E$, that is, if E is a vector space of dimension 1 over K.

Let L be an extension field of K. When we consider L as extension of K, we understand by this the extension (L, j) of K, where j is the canonical injection of K into L, or also L with the corresponding K-algebra structure. The subextensions of L are then the *intermediate* fields between K and L, that is, the subfields of L containing K. If L′ is another extension field of K, a K-homomorphism of L into L′ is then a homomorphism f of L into L′ such that $f(x) = x$ for all $x \in K$. We note that if f is any endomorphism of the field L, then the set of elements of L invariant under f is a subfield K′ of L, and that f is then a K′-endomorphism of L.

In particular let P be the prime subfield of a field L. We can consider L as an extension of P, and every endomorphism of L is then a P-endomorphism.

Let (E, u) be an extension of K ; since u defines an isomorphism of K onto a subfield K_1 of E, there is in general no difficulty in identifying K with K_1 by means of u. One case where we cannot allow such an identification is that where $K = E$ and where u is thus an endomorphism of K ; most frequently u will be an automorphism of K, or the mapping $u \mapsto u^p$, when the field K is of characteristic $\neq 0$.

It is clear that every extension of K is isomorphic to an extension (L, j), where L is a field containing K as subfield and j is the canonical injection of K into L.

2. Degree of an extension

Let A be an algebra over a field K. It is in particular a vector space over K ; the dimension of this vector space is called the *degree* of A over K and is written $[A : K]$ (II, p. 293). By definition $[A : K]$ is thus the cardinal of any basis of A over K. This definition applies in particular to the case of extensions of K.

An extension of degree 1 is trivial. An extension of degree 2 (resp. 3 etc.) is called *quadratic* (resp. *cubic* etc.). An extension of finite degree is sometimes called, by abuse of language, a *finite extension.*

THEOREM 1. — *Let* E *be an extension of* K *and* A *an algebra over* E. *Then we have* $[A : K] = [A : E] . [E : K]$. *In particular, if* F *is an extension of* E, *we have*

$$[F : K] = [F : E] . [E : K] . \tag{1}$$

The theorem is just a special case of II, p. 222, Prop. 25 ; more precisely, if $(a_\lambda)_{\lambda \in L}$ is a basis of A over E and $(b_\mu)_{\mu \in M}$ a basis of E over K, then the family $(a_\lambda b_\mu)_{(\lambda,\mu) \in L \times M}$ is a basis of A over K.

COROLLARY 1. — *Let* K, E, F *be three fields such that* $K \subset E \subset F$ *and* $[F : K]$ *is finite. Then the degrees* $[E : K]$ *and* $[F : E]$ *are divisors of* $[F : K]$.

If the degree $[F : K]$ is prime, there is thus no subextension of F other than K and F. But note that when $[F : K]$ is not prime, there is not necessarily a subextension of F other than K and F (cf. V, p. 146, Exercise 1).

COROLLARY 2. — *Let* K, E *and* F *be three fields such that* $K \subset E \subset F$. *Suppose that* $[F : K]$ *is finite, then the relation* $[E : K] = [F : K]$ *is equivalent to* E = F *and the relation* $[F : E] = [F : K]$ *is equivalent to* E = K.

For if L is an extension field of L′ then $[L : L'] = 1$ is equivalent to L′ = L.

PROPOSITION 1. — *Let* A *be an algebra of finite degree over a field* K. *If an element* $a \in A$ *is not a left (resp. right) divisor of zero in* A, *then it is invertible in* A.

For the vector space A over K is of finite dimension, by hypothesis, and the linear mapping $x \mapsto ax$ (resp. $x \mapsto xa$) of A into A injective ; it is therefore bijective (II, p. 298, Cor.) and hence (I, p. 16, Remark) a is invertible in A.

COROLLARY. — *Let* A *be a commutative algebra of finite degree over a field* K. *If* A *is an integral domain, then it is a field.*

3. Adjunction

Let E be an extension of a field K. Given a family $\mathbf{x} = (x_i)_{i \in I}$ of elements of E, we denote by $K(x_i)_{i \in I}$ (or $K(\mathbf{x})$ or also $K(x_1, \ldots, x_n)$ when I is the interval

$(1, n)$ of N) the least subextension of E containing the members of the family (x_i); we say that $K(x_i)_{i \in I}$ is obtained by *adjunction* to K of the elements of the family $(x_i)_{i \in I}$ and that the family $(x_i)_{i \in I}$ (or the set of its elements) is a *generating family of* $K(x_i)_{i \in I}$ *with respect to* K (or *over* K). The field $K(x_i)_{i \in I}$ depends only on the set A of elements of the family $(x_i)_{i \in I}$; we also denote it by $K(A)$. In particular we have $K(E) = E$ and $K(\varnothing) = K \cdot 1$. All that has been said applies in particular when E is a field containing K as subfield.

It should be observed that A is not in general a generating set of the *algebra* $K(A)$, in other words that $K(A) \neq K[A]$. * Nevertheless we shall see that $K(A) = K[A]$ when $K(A)$ is an algebraic extension of K (V, p. 18, Cor. 1). *

PROPOSITION 2. — *If* M *and* N *are any two subsets of an extension of a field* K, *then* $K(M \cup N) = K(M)(N) = K(N)(M)$.

For $K(M \cup N)$ contains $K(M)$ and N and hence $K(M)(N)$; since $K(M)(N)$ is a field containing $K \cup M \cup N$, it contains $K(M \cup N)$, whence the proposition.

We shall sometimes write $K(M, N)$ instead of $K(M \cup N)$.

Remark. — Let P be the prime subfield of a field E (V, p. 2); for every subset A of E, $P(A)$ is the least subfield of E containing A. In particular if K is a subfield of E, we have $P(K \cup A) = K(A)$. If K and K′ are two subfields of E, we thus have $P(K \cup K') = K(K') = K'(K)$; this field is the least subfield of E containing K or K′, or also the *upper bound* of K and K′ in the set of subfields of E, ordered by inclusion; we sometimes say that this field is the field *generated* by K and K′ in E.

PROPOSITION 3. — *Let* $\mathcal{F}$ *be a set of subfields of a field* E, *directed with respect to the relation* $\subset$. *The union* L *of the fields of* $\mathcal{F}$ *is a field.*

For if x and y are two elements of L, there exist two fields R, S of $\mathcal{F}$ such that $x \in R$, $y \in S$; let T be a field of $\mathcal{F}$ containing R and S; then $x \in T$, $y \in T$, hence $x + y$, xy and x^{-1} (if $x \neq 0$) belong to T, hence to L.

COROLLARY. — *Let* E *be an extension of a field* K *and* $A \subset E$. *The field* $K(A)$ *is the union of the fields* $K(F)$ *where* F *ranges over the set of all finite subsets of* A.

For the set of fields $K(F)$ is directed by the relation $\subset$, because $F \subset F'$ implies $K(F) \subset K(F')$. The union L of these fields is thus a field containing $K \cup A$ and contained in $K(A)$, and hence identical with $K(A)$.

DEFINITION 2. — *An extension* E *of a field* K *is said to be finitely generated if it has a finite generating family. It is called monogenous if there exists* x *in* E *such that* $E = K(x)$.

The Cor. of Prop. 3 shows that *every* extension E of a field K is a directed union of finitely generated extensions contained in E. It is clear that every extension E of K of *finite degree* is also *finitely generated* because a basis of E (considered as vector space over K) is also a generating family of E over K; we shall see later that the converse does not hold.

4. Composite extensions

Let E and F be two extensions of a field K. By a *composite extension* of E and F we understand any triple (L, u, v), where L is an extension of K, u is a K-homomorphism of E into L and v is a K-homomorphism of F into L, and where the field L is generated by $u(\mathrm{E}) \cup v(\mathrm{F})$ (cf. Fig. 1).

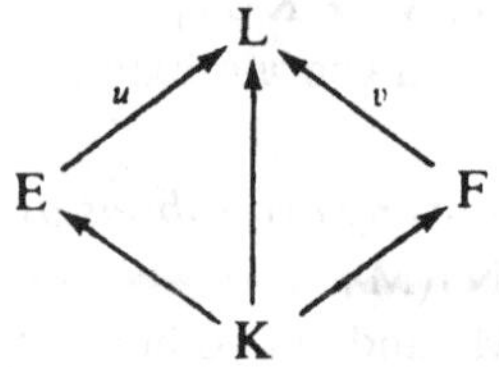

FIG. 1.

In agreement with the general definitions (E, IV, p. 6) an *isomorphism* of a composite extension (L, u, v) of E and F onto a composite extension (L′, u', v') of E and F is a K-isomorphism φ of L onto L′ such that $u' = \varphi \circ u$ and $v' = \varphi \circ v$.

Let (L, u, v) be a composite extension of E and F. The K-linear mapping w of $\mathrm{E} \otimes_{\mathrm{K}} \mathrm{F}$ into L which sends $x \otimes y$ to $u(x)\,v(y)$ is a K-algebra homomorphism ; in this No. we shall denote it by $u * v$. Its image is the subring of L generated by $u(\mathrm{E}) \cup v(\mathrm{F})$.

PROPOSITION 4. — *Let* E, F *be two extensions of* K.

a) Let (L, u, v) *be a composite extension of* E *and* F ; *then the kernel* $\mathfrak{p}$ *of the homomorphism* $u * v$ *of* $\mathrm{E} \otimes_{\mathrm{K}} \mathrm{F}$ *into* L *is a prime ideal.*

b) Let $\mathfrak{p}$ *be a prime ideal of* $\mathrm{E} \otimes_{\mathrm{K}} \mathrm{F}$; *then there exists a composite extension* (L, u, v) *of* E *and* F *such that* $\mathfrak{p}$ *is the kernel of* $u * v$, *and any two such composite extensions are isomorphic.*

Assertion *a*) follows from the fact that the kernel of a homomorphism of a ring into a field is a prime ideal (I, p. 116-117).

Let $\mathfrak{p}$ be a prime ideal of $\mathrm{E} \otimes_{\mathrm{K}} \mathrm{F}$, A the quotient ring $(\mathrm{E} \otimes_{\mathrm{K}} \mathrm{F})/\mathfrak{p}$ and L the field of fractions of A. For $x \in \mathrm{E}$ (resp. $y \in \mathrm{F}$) we denote by $u(x)$ (resp. $v(y)$) the residue class mod $\mathfrak{p}$ of $x \otimes 1$ (resp. $1 \otimes y$). Then u (resp. v) is a K-homomorphism of E (resp. F) into L and $u(\mathrm{E}) \cup v(\mathrm{F})$ generates A as a ring, hence L as a field. Therefore (L, u, v) is a composite extension of E and F ; we see at once that $u * v$ is the canonical homomorphism of $\mathrm{E} \otimes_{\mathrm{K}} \mathrm{F}$ into L, and its kernel is thus equal to $\mathfrak{p}$.

Let (L′, u', v') be a composite extension of E and F such that the kernel of $u' * v'$ is equal to $\mathfrak{p}$. Since $u * v$ and $u' * v'$ have the same kernel, there exists an isomorphism ψ of A onto the image A′ of $u' * v'$, characterized by $u' * v' = \psi \circ (u * v)$. But A′ is the subring of L′ generated by $u'(\mathrm{E}) \cup v'(\mathrm{F})$, hence L′ is the field of fractions of A′. Therefore ψ extends to an homomorphism φ of L onto L′ and it is clear that φ is an isomorphism of (L, u, v) onto (L', u', v').

Remark. — If $\mathfrak{p}$ and $\mathfrak{p}'$ are two distinct prime ideals of $E \otimes_K F$, the corresponding *composite extensions* of E and F (constructed by the procedure of the above proof) are not isomorphic. However, they may nevertheless be isomorphic as *extensions* of K (V, p. 146, Ex. 2).

COROLLARY. — *There exist composite extensions of* E *and* F.

For since the commutative ring $E \otimes_K F$ is not reduced to 0, it has prime ideals : Krull's theorem (I, p. 104) proves the existence of maximal ideals and every maximal ideal is prime.

We can make this corollary more precise as follows. Let (E, u) and (F, v) be two extensions of K ; choose a maximal ideal $\mathfrak{m}$ of the commutative ring $E \otimes_K F$ and put $L = (E \otimes_K F)/\mathfrak{m}$; then L is an extension of K. For $x \in E$ write $u'(x)$ for the residue class of $x \otimes 1$ mod $\mathfrak{m}$ and similarly put $v'(y)$ for the residue class of $1 \otimes y$ mod $\mathfrak{m}$ for all $y \in F$. We then have a commutative diagram of field homomorphisms

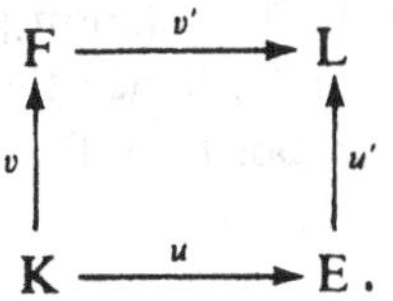

By replacing (L, u') by an isomorphic extension of E we may suppose that L contains E as subfield and that u' is the canonical injection of E in L. By changing notation we thus obtain the following scholium :

SCHOLIUM. — *Let* K *and* E *be two fields and u a homomorphism of* K *into* E. *If* K′ *is a field containing* K *as subfield, there exists a field* E′ *containing* E *as subfield and a homomorphism u′ of* K′ *into* E′ *extending u′.*

5. Linearly disjoint extensions

Throughout this No. Ω *denotes an extension of the field* K.

Let A and B be two sub-K-algebras of Ω. There exists an algebra homomorphism $\varphi : A \otimes_K B \to \Omega$ which maps $x \otimes y$ to xy. The image of φ is a subring C of Ω generated by $A \cup B$. Moreover by II, p. 256, if (b_μ) is a basis of B over K and (a_λ) a basis of A over K, then C coincides with the set of linear combinations $\sum_\mu \alpha_\mu b_\mu$ where $\alpha_\mu \in A$, the set of all $\sum_\lambda \beta_\lambda a_\lambda$ where $\beta_\lambda \in B$ and also the set of all $\sum_{\lambda,\mu} \gamma_{\lambda\mu} a_\lambda b_\mu$, where $\gamma_{\lambda\mu} \in K$.

We shall say that A and B are *linearly disjoint over* K, if φ is an *isomorphism* of $A \otimes_K B$ onto C. We then have $A \cap B = K$; every free subset of B (resp. A) with respect to K is then free with respect to A (resp. B) ; conversely, for A and B to be linearly disjoint over K, it is sufficient that there should exist *one* basis of B over K (for example) which is free with respect to A (II, p. 256 and III, p. 469).

Consider particularly the case where A and B are *subextensions* of Ω.

PROPOSITION 5. — *Let* E *and* F *be two extensions of* K *contained in* Ω.

a) If F *has finite degree over* K, *then the subring of* Ω *generated by* $E \cup F$ *is a field, coinciding with* $E(F)$ *and the degree of* $E(F)$ *over* E *is finite ; we have* $[E(F):E] \leqslant [F:K]$, *with equality if and only if* E *and* F *are linearly disjoint over* K. *In that case* $E(F)$ *is E-isomorphic to* $E \otimes_K F$.

b) If further E *is of finite degree over* K, *then* $E(F) = K(E \cup F)$ *is of finite degree over* K. *We have* $[K(E \cup F):K] \leqslant [E:K][F:K]$ *with equality if and only if* E *and* F *are linearly disjoint over* K.

For let C be the subring of Ω generated by $E \cup F$; if $(b_j)_{1 \leqslant j \leqslant n}$ is a basis of F over K, then C is the vector sub-E-space of Ω generated by the b_j hence C is an algebra of *finite* rank $\leqslant n$ over E ; since the ring C is contained in a field, it is an integral domain, and hence a *field,* by the Cor. to Prop. 1 (V, p. 10), whence $C = E(F)$ and $[E(F):E] \leqslant [F:K]$. The relation $[E(F):E] = [F:K]$ means that the b_j are linearly independent over E, hence that E and F are linearly disjoint over K ; this proves part *a)* of the proposition. Part *b)* now follows at once because $[E(F):K] = [E(F):E][E:K]$.

Let E and F be extensions of K contained in Ω ; if E and F are of infinite degree over K, the subring $C = K[E \cup F]$ is not necessarily a field [1] ; however the *field of fractions* of C then coincides with $K(E \cup F)$. More generally let A be a subring of E such that E is the field of fractions of A, and let B be a subring of F such that F is the field of fractions of B ; then if C is the subring of Ω generated by $A \cup B$, $K(E \cup F)$ coincides with the *field of fractions* of C, because the latter field is the least subfield of Ω containing C and it contains E and F. Moreover :

PROPOSITION 6. — *Let* E *and* F *be two extensions of* K *contained in* Ω, *and* A *and* B *two subalgebras of* Ω *over* K *such that* E *is the field of fractions of* A *and* F *the field of fractions of* B. *Then for* E *and* F *to be linearly disjoint over* K *it is necessary and sufficient that* A *and* B *be linearly disjoint over* K.

The condition is clearly necessary. Conversely, if A and B are linearly disjoint over K, A and F are so too, because if a family of elements of Ω is free with respect to B, it is free with respect to the field of fractions F of B (II, p. 315, Cor. 1 and p. 316, Cor. 3) ; now the same argument shows that E and F are linearly disjoint over K.

[1] It suffices to consider for example the case where Ω is the field $K(X, Y)$ of rational fractions in two indeterminates X and Y and $E = K(X)$, $F = K(Y)$.

PROPOSITION 7. — *Let* E *and* F *be two extensions of* K *contained in* Ω ; *if* E *and* F *are linearly disjoint over* K, *then every subextension of* E *and every subextension of* F *are linearly disjoint over* K. *Conversely, if for every pair of finitely generated subextensions* E′, F′ *of* E *and* F *respectively,* E′ *and* F′ *are linearly disjoint over* K, *then* E *and* F *are linearly disjoint over* K.

For the condition for E and F to be linearly disjoint over K may be expressed thus : if (a_α) is a free family in E and (b_β) a free family in F, then the relation $\sum_{\alpha,\beta} \lambda_{\alpha\beta} a_\alpha b_\beta = 0$, where $\lambda_{\alpha\beta} \in K$, implies $\lambda_{\alpha\beta} = 0$ for each pair of suffixes. But this condition is satisfied for each pair of free families if it holds for each pair of *finite* free families.

Thus we may say, speaking intuitively, that linear disjunction is a property « of finite character ».

PROPOSITION 8. — *Let* E, F, G *be three extensions of a field* K *contained in* Ω, *such that* $F \subset G$. *For* E *and* G *to be linearly disjoint over* K *it is necessary and sufficient that* E *and* F *should be linearly disjoint over* K *and that* E(F) *and* G *be linearly disjoint over* F.

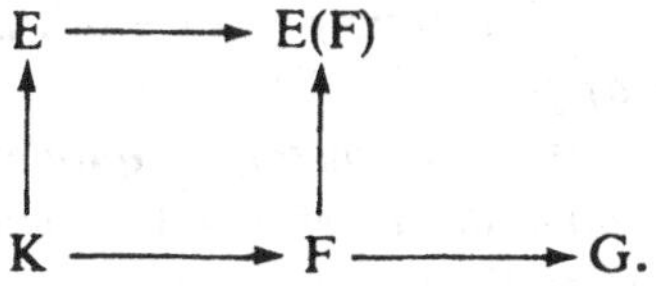

FIG. 2.

The condition is *necessary :* suppose that E and G are linearly disjoint over K. The same then holds of E and F (Prop. 7) ; on the other hand, if B is a basis of E over K, this is also a basis of the algebra F[E] over F ; since B is by hypothesis free over G, F[E] and G are linearly disjoint over F, and the same is true of E(F) = F(E) and G, by Prop. 6.

The condition is *sufficient :* with the same notation it implies that B is free over F, hence it is a basis of F[E] over F ; since F[E] and G are by hypothesis linearly disjoint over F, B is free over G, and this shows that E and G are linearly disjoint over K.

§ 3. ALGEBRAIC EXTENSIONS

1. Algebraic elements of an algebra

Let A be an algebra over a field K and x an element of A. Two cases are possible :

a) The family of monomials $(x^n)_{n \in \mathbf{N}}$ *is free over* K. Then we say that x is *transcendental* over K. There is an isomorphism of the polynomial algebra K[X] onto the subalgebra K[x] of A generated by x, and the latter is of infinite degree over K.

b) There exists an integer $n \geqslant 1$ *such that the monomials* $1, x, \ldots, x^{n-1}, x^n$ *are linearly dependent*; it comes to the same to say that there exists a polynomial $f \neq 0$ in K[X] such that $f(x) = 0$. Then we say that x is *algebraic* over K. The least integer $n \geqslant 1$ satisfying the above property is called the *degree* of x over K. If the degree of x over K is n, then the monomials $1, x, \ldots, x^{n-1}$ are linearly independent over K and there exist elements $a_0, a_1, \ldots, a_{n-1}$ of K such that

$$x^n = a_0 + a_1 x + \cdots + a_{n-1} x^{n-1}.$$

The polynomial $f(X) = X^n - \sum_{k=0}^{n-1} a_k X^k$ is the unique monic polynomial of degree n in K[X] such that $f(x) = 0$; it is called the *minimal polynomial* of x over K.

THEOREM 1. — *Let* A *be an algebra over a field* K, x *an element of* A *algebraic over* K, n *the degree and* f *the minimal polynomial of* x *over* K.

a) For a polynomial $g \in$ K[X] *to satisfy* $g(x) = 0$ *it is necessary and sufficient that* g *should be a multiple of* f.

b) The mapping $g \mapsto g(x)$ *defines by passage to quotients an isomorphism of the quotient algebra* K[X]/(f) *onto the algebra* K[x], *and the elements* $1, x, \ldots, x^{n-1}$ *form a basis of* K[x] *over* K. *In particular,* $[K[x] : K] = n$.

c) Assume that A *is an integral domain. Then* K[x] *is a field and* f *the unique monic irreducible polynomial in* K[X] *such that* $f(x) = 0$.

d) For x *to be invertible in* A *it is necessary and sufficient that* $f(0) \neq 0$; *then we have* $x^{-1} \in$ K[x].

There exists a unique algebra homomorphism $\varphi : K[X] \to A$ such that $\varphi(X) = x$; we have $\varphi(P) = P(x)$ for every $P \in K[X]$ and the image of φ equals K[x]. Let $\mathfrak{a}$ be the kernel of φ; by construction, the minimal polynomial f of x over K belongs to $\mathfrak{a}$ and it is the monic polynomial of least degree in $\mathfrak{a}$. Therefore (IV, p. 11, Prop. 11) we have $\mathfrak{a} = (f)$, whence *a*). Assertion *b*) follows at once from *a*) and the Cor. of Prop. 10 of IV, p. 11.

Suppose that A is an integral domain. The algebra K[x] is then an integral domain and of finite degree over K, hence it is a field (V, p. 10, Cor.). The ideal (f) of K[x] is thus maximal, that means that f is irreducible in K[X] (IV, p. 13). Finally, let g be a monic irreducible polynomial in K[X] such that $g(x) = 0$; by *a*) it is a multiple of f, hence $g = f$, and this proves *c*).

It remains to prove *d*). There exists a polynomial $g \in$ K[X] of degree $n - 1$ and an element a of K such that $f(X) = Xg(X) + a$, whence $f(0) = a$. If $a = 0$, we have $xg(x) = f(x) = 0$, and $g(x) \neq 0$, so then x is not invertible in A. If on the other hand $a \neq 0$, then we have $x \,.\, [-a^{-1} g(x)] = 1$, so x is then invertible in A and $x^{-1} = -a^{-1} g(x)$.

COROLLARY 1. — *Let* A *be an algebra over a field* K. *For an element* x *of* A *to be algebraic over* K *it is necessary and sufficient that the subalgebra* $K[x]$ *of* A *generated by* x *should be of finite degree over* K. *In particular, if* A *is of finite degree over* K, *every element of* A *is algebraic over* K.

COROLLARY 2. — *Let* E *be an extension of* K, A *an algebra over* E *and* x *an element of* A *algebraic over* K. *Then* x *is algebraic over* E, *the minimal polynomial of* x *over* E *divides the minimal polynomial of* x *over* K *and the degree of* x *over* E *is at most equal to the degree of* x *over* K.

For let f be the minimal polynomial of x over K; we have $f(x) = 0$ and $f \in E[X]$, hence x is algebraic over E and f is a multiple of the minimal polynomial of x over E (Th. 1, *a*)).

Remark. — Let E be an extension of a field K and x an element of E which is root of a monic *irreducible* polynomial $f \in K[X]$. Th. 1, *c*) then shows that f is the minimal polynomial of x over K.

Examples. — * 1) In the field of complex numbers **C**, the number i is algebraic of degree 2 over the prime field **Q**; for if $f(X) = X^2 + 1$, then $f(i) = 0$, and $x^2 + 1 \neq 0$ for all $x \in \mathbf{Q}$, hence $i \notin \mathbf{Q}$. The field $\mathbf{Q}(i)$ is thus an extension of degree 2 of **Q**; it consists of all numbers $a + bi$, where a, b are rational. Likewise i is algebraic of degree 2 over the field **R** of real numbers, and **C** is an extension of degree 2 of **R**. *

2) Let K be a field and F the field $K(X)$ of rational functions in one indeterminate over K. Let E be the subfield $K(X^3)$ of F; we have $F = E(X)$ and X is algebraic over E, since it is a root of the polynomial $Y^3 - X^3$ of the ring $E[Y]$; this polynomial is irreducible in $E[Y]$, for in the contrary case it would have at least one factor of the first degree, and there would then exist two non-zero polynomials $u(X)$, $v(X)$ of $K[X]$ such that $(u(X^3))^3 = X^3(v(X^3))^3$ which is absurd, for if m and n are the degrees of u and v, this would imply $9m = 9n + 3$, or $3m = 3n + 1$. The field F is thus of degree 3 over E, and every element of F may be written in just one way as a linear combination $f(X^3) + Xg(X^3) + X^2 h(X^3)$, where f, g, h are three rational fractions of $K(X)$.

* 3) In the field **R** of real numbers it may be shown [1] that the number π is transcendental over the prime field **Q**. *

2. Algebraic extensions

DEFINITION 1. — *An extension* E *of a field* K *is said to be algebraic* (*over* K) *if every element of* E *is algebraic over* K. *An extension* E *of* K *which is not algebraic is called transcendental* (*over* K).

PROPOSITION 1. — *For an extension* E *of* K *to be algebraic it is necessary and sufficient that every sub-*K*-algebra* A *of* E *should be a field.*

[1] Cf. for example D. Hilbert, *Gesammelte Abhandlungen*, vol. 1, p. 1 (Berlin (Springer), 1932).

The condition is *necessary* : if E is algebraic over K and $x \neq 0$ is an element of a sub-K-algebra A of E, then $x^{-1} \in K[x] \subset A$ by V, p. 16, Th. 1, *d*). Therefore A is a field.

The condition is *sufficient* : if it is satisfied and x is an element $\neq 0$ in E, then the ring $K[x]$ is a field, hence $x^{-1} \in K[x]$; in other words, there exists a polynomial $g \in K[X]$ such that $x^{-1} = g(x)$, or also $xg(x) - 1 = 0$; this shows x to be algebraic over K, hence E is an algebraic extension of K.

PROPOSITION 2. — *If an extension* E *of* K *is of finite degree n, then it is algebraic and the degree over* K *of each element of* E *divides n.*

For, given $x \in E$, $[K(x) : K]$ is finite and divides n (V, p. 10, Cor. 1) and hence x is algebraic over K (V, p. 17, Cor. 1).

* There exist algebraic extensions of infinite degree, for example the algebraic closure of a finite field (V, p. 24, Remark 4). *

THEOREM 2. — *Let* E *be a finitely generated extension of* K, *generated by the elements* $a_1, \ldots, a_m$ *which are algebraic over* K ; *then* E *is an extension of finite degree of* K. *If the degree of* a_i *over* $K(a_1, a_2, \ldots, a_{i-1})$ *is* n_i *(for* $1 \leqslant i \leqslant m$*), then the degree of* E *over* K *is* $n_1 n_2 \ldots n_m$ *and the elements* $a_1^{\nu_1} a_2^{\nu_2} \ldots a_m^{\nu_m}$ $(0 \leqslant \nu_i \leqslant n_i - 1)$ *form a basis of* E *over* K.

The elements $a_i^{\nu_i}$ $(0 \leqslant \nu_i \leqslant n_i - 1)$ form a basis of $K(a_1, a_2, \ldots, a_i)$ over $K(a_1, a_2, \ldots, a_{i-1})$ by Th. 1, *b*) of V, p. 16 ; the theorem therefore follows by induction on m from Prop. 25 of II, p. 222.

COROLLARY 1. — *Let* E *be an extension of* K *and* A *a subset of* E *consisting of elements that are algebraic over* K. *Then* K(A) *is algebraic over* K *and we have* $K[A] = K(A)$.

For each $x \in K(A)$ belongs to a field K(F), where F is a finite subset of A (V, p. 11, Cor.) ; now K(F) is algebraic over K and equal to K[F] by Th. 2, hence x is algebraic over K and $K(A) = K[A]$.

COROLLARY 2. — *Let* L *be an extension of* K *and* E, F *subextensions of* L. *If* F *is algebraic over* K, *the subring* K[E, F] *of* L *generated by* $E \cup F$ *is a field coinciding with* E(F) *and algebraic over* E.

For each element of F, being algebraic over K, is also algebraic over E (V, p. 17, Cor. 2) hence E(F) is an algebraic extension of E, and we have $E(F) = E[F]$ by Cor. 1.

Remarks. — 1) With the notation of Th. 2, $E = K[a_1, a_2, \ldots, a_m]$ and hence E is isomorphic to a quotient $K[X_1, X_2, \ldots, X_m]/\mathfrak{a}$; since E is a field, $\mathfrak{a}$ is a maximal ideal in $K[X_1, \ldots, X_m]$.

2) Let E be an algebraic extension of K, of infinite degree. By Th. 2 there exists an infinite sequence $(a_n)_{n \geqslant 1}$ of elements of E such that $a_n \notin K(a_1, a_2, \ldots, a_{n-1})$; Th. 2 shows further that the degree of $K(a_1, a_2, \ldots, a_n)$ over K takes arbitrarily large

values. In other words, if E is an algebraic extension of K such that the degrees $[F : K]$ of the subextensions F of E of finite degree over K are *bounded*, then E is an extension of finite degree of K.

3. Transitivity of algebraic extensions. Fields that are relatively algebraically closed in an extension field

PROPOSITION 3. — *Let* E *and* F *be two extension fields of a field* K *such that* $K \subset E \subset F$. *For* F *to be algebraic over* K *it is necessary and sufficient that* E *should be algebraic over* K *and* F *algebraic over* E.

The condition is necessary, by V, p. 17, Cor. 2. Let us show that it is sufficient. Let x be any element of F ; it is algebraic over E ; let $g \in E[X]$ be its minimal polynomial over E. If A is the (finite) set of coefficients of g, then $g \in K(A)[X]$, hence x is algebraic over $K(A)$ and $K(A \cup \{x\}) = K(A)(x)$ is of finite degree over $K(A)$. Now $A \subset E$ and E is algebraic over K, hence $K(A)$ is of finite degree over K, by Th. 2. Therefore (V, p. 10, Th. 1), $K(A \cup \{x\})$ is of finite degree over K, which shows x to be algebraic over K (V, p. 18, Prop. 2).

DEFINITION 2. — *A subfield* K *of a field* E *is said to be relatively algebraically closed in* E *if every element of* E *which is algebraic over* K, *belongs to* K.

It comes to the same to say that K is the only algebraic extension of K contained in E. Every field K is relatively algebraically closed in itself. In § 4 we shall study the fields which are relatively algebraically closed in each extension field.

PROPOSITION 4. — *Let* E *be an extension of a field* K ; *the set* L *of elements of* E *which are algebraic over* K *forms a subextension of* E *which is relatively algebraically closed in* E.

For the field $K(L)$ is algebraic over K (Cor. 1 of Th. 2), hence $K(L) \subset L$; it follows that $K(L) = L$ and L is a field. On the other hand, if $x \in E$ is algebraic over L, it is so also over K (Prop. 3) and hence belongs to L.

The extension L of K consisting of all the elements of E that are algebraic over K is called the *relative algebraic closure of* K *in* E ; it is *the largest algebraic extension of* K contained in E.

§ 4. ALGEBRAICALLY CLOSED EXTENSIONS

1. Algebraically closed fields

PROPOSITION 1. — *Let* K *be a field ; then the following properties are equivalent :*

(AC) *Every non-constant polynomial of* K[X] *splits in* K[X] *into a product of polynomials of degree* 1 (distinct or not).

(AC') *Every non-constant polynomial of* K[X] *has at least one root in* K.

(AC") *Every irreducible polynomial in* K[X] *is of degree* 1.

(AC‴) *Every algebraic extension of* K *is of degree* 1 *(in other words,* K *is relatively algebraically closed in every extension field of* K*).*

Let us first prove the equivalence of the properties (AC), (AC′) and (AC″). Clearly (AC) implies (AC″). Since every non-constant polynomial of K[X] is divisible by an irreducible polynomial (IV, p. 13, Prop. 13) and every polynomial of degree 1 in K[X] clearly admits a root in K, we see that (AC″) implies (AC′). The condition (AC′) implies by induction on n, that every polynomial of degree n in K[X] is a product of n polynomials of degree 1 (IV, p. 14, Prop. 1), hence (AC′) implies (AC).

It remains to see that (AC″) and (AC‴) are equivalent. If (AC″) holds, every element of an extension field L of K which is algebraic over K is of degree 1 (V, p. 16, Th. 1), hence belongs to K, which establishes (AC‴). Conversely, let f be an irreducible polynomial of degree $n \geqslant 1$ in K[X]; the quotient algebra $K[X]/(f)$ is of degree n over K and is a field, hence an algebraic extension of degree n of K (V, p. 18, Prop. 2). Now it is clear that (AC‴) implies (AC″).

DEFINITION 1. — *A field* K *is said to be algebraically closed if it possesses the (equivalent) properties* (AC), (AC′), (AC″), (AC‴).

* *Example* 1. — The field **C** of complex numbers is algebraically closed (Gen. Top., VIII, p. 100). *

A field K which is *relatively algebraically closed in an extension field* E of K is not necessarily algebraically closed (in effect every field is relatively algebraically closed in itself, and there exist fields that are not algebraically closed, for example **Q** or $\mathbf{F}_p$ * or **R** *). However :

PROPOSITION 2. — *Let* Ω *be an algebraically closed field and* K *a subfield of* Ω*. Then the relative algebraic closure* $\bar{K}$ *of* K *in* Ω *is an algebraically closed field.*

Let f be a non-constant polynomial in $\bar{K}[X] \subset \Omega[X]$. Since Ω is algebraically closed, the polynomial f has at least one root in Ω, and since this root is algebraic over $\bar{K}$, it belongs to $\bar{K}$ (V, p. 19, Prop. 4). Therefore $\bar{K}$ satisfies (AC′).

* *Example* 2. — By Prop. 2 the set of all complex numbers that are algebraic over **Q** (often called briefly *algebraic numbers*) is an algebraically closed field. *

PROPOSITION 3. — *Every algebraically closed field is infinite.*

Let K be a finite field and put $f(X) = 1 + \prod_{a \in K} (X - a)$. The polynomial $f \in K[X]$ is non-constant and $f(a) = 1$ for each $a \in K$. So the field K does not satisfy (AC′) and hence is not algebraically closed.

THEOREM 1 (Steinitz). — *Let* K *be a field,* E *an algebraic extension of* K *and* Ω *an algebraically closed extension of* K *; then there exists a* K*-homomorphism of* E *into* Ω.

By V, p. 13, Scholium, there exists an extension field Ω′ of Ω and a K-homomorphism u of E into Ω′. Let $x \in E$; since x is algebraic over K,

$u(x)$ is algebraic over $u(\mathrm{K})$ and *a fortiori* over Ω (V, p. 17, Cor. 2) ; since Ω is algebraically closed, we thus have $u(x) \in \Omega$. It follows that u maps E into Ω.

2. Splitting extensions

DEFINITION 2. — *Let* K *be a field and* $(f_i)_{i \in \mathrm{I}}$ *a family of non-constant polynomials in* K[X]. *By a splitting extension of* $(f_i)_{i \in \mathrm{I}}$ *we understand any extension* E *of* K *with the following properties :*

a) For each $i \in \mathrm{I}$, *the polynomial* f_i *splits in* E[X] *into a product of polynomials of degree* 1.

b) For each $i \in \mathrm{I}$ *let* R_i *be the set of roots of* f_i *in* E, *then* $\mathrm{E} = \mathrm{K}\left(\bigcup_{i \in \mathrm{I}} \mathrm{R}_i\right)$.

Sometimes the term « splitting field » is used instead of « splitting extension ».

Remarks. — 1) For each $i \in \mathrm{I}$ let c_i be a non-zero element of K and let $f'_i = c_i f_i$. Then it is clear that every splitting extension for the family $(f_i)_{i \in \mathrm{I}}$ is also a splitting extension for the family $(f'_i)_{i \in \mathrm{I}}$ and conversely. In particular, in studying splitting extensions we may limit ourselves to the case of monic polynomials.

2) Suppose that I is finite and put $f = \prod_{i \in \mathrm{I}} f_i$. Using the uniqueness of the decomposition of a polynomial into irreducible factors in E[X] (IV, p. 13, Prop. 13), we can easily show that a splitting extension for the polynomial f is a splitting extension for the family $(f_i)_{i \in \mathrm{I}}$, and conversely. In other words, the case of a finite family may be reduced to the case of a single polynomial.

3) Let $f \in \mathrm{K}[\mathrm{X}]$ be a polynomial of degree $\geqslant 1$ and let E be a splitting extension of f. If $x_1, \ldots, x_n$ are the roots of f in E, we thus have $\mathrm{E} = \mathrm{K}(x_1, \ldots, x_n)$ and $[\mathrm{E} : \mathrm{K}]$ is finite (V, p. 18, Th. 2) ; but it may happen that E is distinct from the subfields $\mathrm{K}(x_1), \ldots, \mathrm{K}(x_n)$ generated by a single root ; this can happen even when f is irreducible [1]. We note however that when f is irreducible, the fields $\mathrm{K}(x_i)$ all have the same degree n over K, and whenever E is equal to one of them, we have $[\mathrm{E} : \mathrm{K}] = n$ and hence $\mathrm{E} = \mathrm{K}(x_1) = \cdots = \mathrm{K}(x_n)$.

PROPOSITION 4. — *Let* K *be a field and* $(f_i)_{i \in \mathrm{I}}$ *a family of non-constant polynomials in* K[X] ; *then there exists a splitting extension for the family* $(f_i)_{i \in \mathrm{I}}$.

We may take the polynomials f_i to be monic (Remark 1). Let $i \in \mathrm{I}$ and let the degree of f_i be d_i. By IV, p. 73, Prop. 5 there exists a commutative algebra A_i over K, not reduced to 0, and elements $\xi_{i,1}, \ldots, \xi_{i,d_i}$ of A_i such that :

a) the algebra A_i is generated by $(\xi_{i,1}, \ldots, \xi_{i,d_i})$;

[1] Take for example $\mathrm{K} = \mathbf{Q}$ and $f = \mathrm{X}^3 - 2$.

b) we have $f_i(X) = \prod_{k=1}^{d_i} (X - \xi_{i,k})$ in $A_i[X]$.

Let A be the tensor product of the family of algebras $(A_i)_{i \in I}$ and let φ_i be the canonical homomorphism of A_i into A (III, p. 470). The algebra A is then commutative and not reduced to 0 ; by Krull's theorem (I, p. 104), there exists thus a maximal ideal $\mathfrak{a}$ in A and $E = A/\mathfrak{a}$ is an extension of the field K.

Denote by ψ the canonical homomorphism of A into E and put $x_{i,k} = \psi(\varphi_i(\xi_{i,k}))$ for $i \in I$ and $1 \leqslant k \leqslant d_i$. Since the algebra A is generated by $\bigcup_{i \in I} \varphi_i(A_i)$, the extension E is generated by the family $(x_{i,k})$. Further, we have $f_i(X) = \prod_{k=1}^{d_i} (X - x_{i,k})$ in E[X]. Therefore E is a splitting extension of the family $(f_i)_{i \in I}$.

PROPOSITION 5. — *Let* K *be a field,* $(f_i)_{i \in I}$ *a family of non-constant polynomials in* K[X], E *an extension of* K, *and* F, F′ *subextensions of* E *which are each a splitting extension of* $(f_i)_{i \in I}$. *Then* F = F′.

Let R_i be the set of roots of f_i in E and $R = \bigcup_{i \in I} R_i$. Since f_i is a product of polynomials of the first degree lying in F[X], we have $R_i \subset F$. By Def. 2, we have $F = K(R)$; in the same way we find that $F' = K(R)$.

COROLLARY. — *Let* K *be a field,* $(f_i)_{i \in I}$ *a family of non-constant polynomials in* K[X] *and* F, F′ *splitting extensions of* $(f_i)_{i \in I}$. *Then there exists a* K*-isomorphism of* F *onto* F′.

This follows from Prop. 5 and V, p. 13, Cor. of Prop. 4.

3. Algebraic closure of a field

DEFINITION 3. — *Let* K *be a field. By an algebraic closure of* K *we understand any extension of* K *which is algebraic and algebraically closed.*

Examples. — * 1) The field **C** of complex numbers is an algebraic closure of the field **R** of real numbers (Gen. Top., VIII, p. 100) *

2) Let K be a field and Ω an algebraically closed extension of K. If $\bar{K}$ is the relative algebraic closure of K in Ω, then by V, p. 20, Prop. 2, $\bar{K}$ is an algebraic closure of K. * In particular the field of all algebraic numbers (V, p. 20, Ex. 2) is an algebraic closure of the field **Q** of rational numbers. *

PROPOSITION 6. — *Let* Ω *be an extension of a field* K. *For* Ω *to be an algebraic closure of* K *it is necessary and sufficient that it should be algebraic and that each non-constant polynomial in* K[X] *should split in* Ω[X] *into a product of factors of degree* 1.

The condition is necessary by (AC). Conversely, suppose that Ω is algebraic over K and every non-constant polynomial over K[X] is a product in Ω[X] of factors of degree 1. Let Ω' be an algebraic extension of Ω and let $x \in \Omega'$. Since x is algebraic over Ω and Ω is algebraic over K, x is algebraic over K (V, p. 19, Prop. 3). Let f be the minimal polynomial of x over K. By hypothesis the polynomial $f \in$ K[X] splits in Ω[X] into a product of factors of degree 1, whence $x \in \Omega$. Thus we have $\Omega' = \Omega$ and Ω is algebraically closed because it satisfies (AC‴).

Remark 1. — If Ω is algebraic over K and if every non-constant polynomial of K[X] has a root in Ω then Ω is an algebraic closure of K (V, p. 156, Ex. 20).

PROPOSITION 7. — *Let Ω be an algebraic extension of a field* K.

a) If Ω is algebraically closed, then every algebraic extension of K *is isomorphic to a subextension of Ω.*

b) Conversely suppose that every algebraic extension of finite degree of K *is isomorphic to a subextension of Ω ; then Ω is algebraically closed.*

Assertion *a*) follows from Th. 1 (V, p. 20). Now assume the hypotheses of *b*) and consider a non-constant polynomial $f \in$ K[X]. Let E be a splitting field of f (V, p. 21, Prop. 4) ; since E is algebraic of finite degree over K (V, p. 18, Th. 2), we may suppose that E is a subextension of Ω. Then the polynomial f is a product of polynomials of degree 1 in Ω[X] and Prop. 6 shows that Ω is algebraically closed.

We can now prove the existence and uniqueness (up to isomorphism) of the algebraic closure of a field.

THEOREM 2 (Steinitz). — *Let* K *be a field ; then there exists an algebraic closure of* K. *If Ω and Ω' are two algebraic closures of* K, *there exists a* K-*isomorphism of Ω onto Ω'.*

By Prop. 6 an algebraic closure of K is nothing other than a splitting extension for the set of all non-constant polynomials in K[X]. Th. 2 therefore follows from V, p. 21, Prop. 4 and V, p. 22, Cor.

COROLLARY. — *Let* K *and* K' *be two fields, Ω an algebraic closure of* K *and Ω' an algebraic closure of* K'. *For every isomorphism u of* K *onto* K' *there exists an isomorphism v of Ω onto Ω' extending u.*

It is enough to apply Th. 2 to the algebraic closures Ω and (Ω', u) of K.

Remarks. — 2) In the notation of the preceding Corollary there exist in general K-automorphisms of Ω distinct from the identity. Hence there is in general no uniqueness about the isomorphism v of Ω onto Ω' extending the isomorphisms u of K onto K'. For similar reasons there is in general more than one isomorphism of a splitting extension E onto a splitting extension E' for the same family $(f_i)_{i \in I}$ of polynomials. We recall that by contrast, for the perfect closure we have uniqueness (V, p. 5).

3) Let K be a field and Ω an algebraic closure of K. Then the following construction may be given for a splitting extension for a family $(f_i)_{i \in I}$ of non-constant polynomials in K[X]: let R_i be the set of roots of f_i in Ω and let $R = \bigcup_{i \in I} R_i$. Then K(R) is the unique subextension of Ω which is a splitting extension for $(f_i)_{i \in I}$ (V, p. 22, Prop. 5).

4) Let K be a finite field and Ω an algebraic closure of K. Then Ω is infinite (V, p. 20, Prop. 3) ; since every extension of finite degree of K is a finite field, Ω is an algebraic extension of *infinite* degree of K.

§ 5. *p*-RADICAL EXTENSIONS

Throughout this paragraph the letter p denotes an integer which is either 1 *or a prime number. All the fields considered are of characteristic exponent p. All the results stated in this paragraph are trivial when* $p = 1$.

1. *p*-radical elements

DEFINITION 1. — *Let* K *be a field and* E *an extension of* K. *An element x of* E *is said to be p-radical over* K *if there exists an integer* $m \geq 0$ *such that* $x^{p^m} \in K$; *the least of these integers is called the height of x (over* K).

PROPOSITION 1. — *Let* E *be an extension of a field* K *and x a p-radical element of height e over* K ; *put* $a = x^{p^e}$. *Then* $a \in K$ *and the minimal polynomial of x over* K *is* $X^{p^e} - a$. *Further we have* $[K(x) : K] = p^e$.

It suffices to prove that the polynomial $X^{p^e} - a$ is irreducible in K[X]. By the definition of height of x we have $a \notin K^p$, so that the proposition follows from the lemma :

Lemma 1. — *Let* K *be a field and a an element of* K *such that* $a \notin K^p$. *Then for every integer* $e \geq 0$, *the polynomial* $f(X) = X^{p^e} - a$ *is irreducible in* K[X].

Let Ω be an algebraic closure of K and let b be the element $a^{p^{-e}}$ of Ω ; we denote by g the minimal polynomial of b over K. We have $f(X) = (X - b)^{p^e}$ and hence every irreducible polynomial in K[X] which divides f admits b as root, and hence is equal to g. There exists thus (IV, p. 13, Prop. 13) an integer $q \geq 1$ such that $f = g^q$; since q divides the degree p^e of f, there exists an integer e' such that $0 \leq e' \leq e$ and $q = p^{e'}$. If c is the constant term of g, we have $-a = c^q$; since we assumed that a does not lie in K^p, we must have $q = 1$, that is, $f = g$, so the lemma follows.

2. p-radical extensions

DEFINITION 2. — *Let E be an extension of a field K. We shall say that E is p-radical*[1] *(over K) if every element of E is p-radical over K. When this is so, E is said to be of finite height when the set of heights of elements of E is bounded above and we call the height of E the maximum of the heights of its elements.*

We note that every p-radical extension of a perfect field (in particular of a field of characteristic 0) is trivial.

Let K be a field. The p-radical extensions of height 0 of K are the trivial extensions. Every p-radical extension of K is algebraic. If E is a p-radical extension of K and F a p-radical extension of E, then F is a p-radical extension of K: for, given any $x \in F$, there exists an integer $m \geqslant 0$ such that $x^{p^m} \in E$ and an integer $n \geqslant 0$ such that $(x^{p^m})^{p^n} \in K$, that is, $x^{p^{m+n}} \in K$.

PROPOSITION 2. — *Let E be an extension of a field K. For every integer $n \geqslant 0$ let E_n be the set of elements of E which are p-radical of height $\leqslant n$ over K, and let E_∞ be the set of all elements of E that are p-radical over K. Then $(E_n)_{n \geqslant 0}$ is an ascending sequence of subextensions of E whose union is E_∞, and E_∞ is the largest p-radical extension of K contained in E.*

For each integer $n \geqslant 0$ the set E_n consists of all elements x of E such that $x^{p^n} \in K$; since the mapping $x \mapsto x^{p^n}$ is an endomorphism of the field E, we deduce that E_n is a subextension of E. The sequence $(E_n)_{n \geqslant 0}$ is ascending, with union E_∞ and hence E_∞ is a subextension of E (V, p. 11, Prop. 3). It is clear that E_∞ is a p-radical extension of K and that E_∞ contains every subextension of E which is p-radical over K.

COROLLARY. — *If an extension E of a field K is generated by a set of p-radical elements over K, then it is p-radical over K.*

For E_∞ is a subextension of E and by hypothesis $E = K(E_\infty)$ whence $E = E_\infty$; so E is a p-radical extension of K.

Under the conditions of Proposition 2, E_∞ is said to be the *relative p-radical closure* of K in E.

We shall apply Prop. 2 particularly to the case where E is an algebraically closed extension of K; then E is perfect and we have $E_n = K^{p^{-n}}$ for all $n \geqslant 0$. In this case we denote by $K^{p^{-\infty}}$ the set of elements of E that are p-radical over K; it is the subfield of E which is the union of the ascending sequence $(K^{p^{-n}})_{n \geqslant 0}$ of subfields of E. By Prop. 2, $K^{p^{-\infty}}$ is the largest subextension of E which is p-radical over K; by Prop. 3 of V, p. 6, $K^{p^{-\infty}}$ is a perfect closure of K and it is also the smallest perfect subfield of E containing K. When K is perfect we clearly have $K = K^{p^{-n}} = K^{p^{-\infty}}$ for all n. If K is imperfect, we have $K \neq K^p$, whence

[1] A p-radical extension is also called a purely inseparable extension, or a radical extension.

$$K^{p^{-n}} \neq (K^p)^{p^{-n}} = K^{p^{-(n-1)}}$$

for $n \geqslant 1$; the subfields $K^{p^{-n}}$ of E are then pairwise distinct and $K^{p^{-\infty}}$ is an algebraic extension of *infinite* degree of K.

PROPOSITION 3. — *Let* K *be a field,* E *an extension field of* K *which is p-radical over* K *and u a homomorphism of* K *into a perfect field* F. *Then there exists a unique homomorphism v of* E *into* F *extending u.*

Let E_n be the set of elements of E which are p-radical of height $\leqslant n$ over K. By Prop. 2 the field E is the union of the ascending sequence $(E_n)_{n \geqslant 0}$ of subfields. Let v be a homomorphism of E into F extending u ; for any $x \in E_n$ we have $x^{p^n} \in K$, whence $v(x)^{p^n} = v(x^{p^n}) = u(x^{p^n})$; we thus have $v(x) = u(x^{p^n})^{p^{-n}}$ for all $n \geqslant 0$ and all $x \in E_n$, whence the uniqueness of the extension of u to E.

Let n be a positive integer ; for any $x \in E_n$ we have $x^{p^n} \in K$ and since F is perfect, we can define an element $v_n(x)$ of F by $v_n(x) = u(x^{p^n})^{p^{-n}}$. It is clear that v_n is a homomorphism of E_n into F, that $v_0 = u$ and that v_{n+1} induces v_n on E_n. Hence there exists a homomorphism v of E into F inducing v_n on E_n for all $n \geqslant 0$ and, in particular, coinciding with $v_0 = u$ on $E_0 = K$.

COROLLARY. — *For an extension* E *of a field* K *to be a perfect closure of* K *it is necessary and sufficient that it should be a p-radical extension of* K *and that the field* E *should be perfect.*

The Corollary is trivial when $p = 1$; suppose then that $p \neq 1$. The stated conditions are necessary by V, p. 5, Th. 3 ; they are sufficient by Prop. 3.

PROPOSITION 4. — *Let* E *be a p-radical extension of finite degree of a field* K. *Then* $[E : K]$ *is a power of the characteristic exponent p of* K.

Since E is a p-radical extension of finite degree of K, there are elements $a_1, \ldots, a_m$ of E, p-radical over K, such that $E = K(a_1, \ldots, a_m)$. Let i be in the range 1 to m ; since a_i is *a fortiori* p-radical over $K(a_1, \ldots, a_{i-1})$, the degree

$$n_i = [K(a_1, \ldots, a_i) : K(a_1, \ldots, a_{i-1})]$$

is a power of p (V, p. 24, Prop. 1). We have $[E : K] = n_1 \ldots n_m$ by V, p. 18, Th. 2, whence the result.

§ 6. ETALE ALGEBRAS

Throughout this paragraph K *denotes a field.*

1. Linear independence of homomorphisms

Let L be an extension of K and V a vector space over K. In this paragraph we shall denote by $\mathrm{Hom}_K(V, L)$ the set of all K-*linear mappings of* V *into* L, equipped with the vector space structure on L such that :

$$(f+g)(x) = f(x) + g(x), \quad (\alpha f)(x) = \alpha f(x) \tag{1}$$

for $x \in V$, $\alpha \in L$ and f, g in $\mathrm{Hom}_K(V, L)$. Let $V_{(L)} = L \otimes_K V$ be the vector space on L derived from V by extension of scalars, and $(V_{(L)})^*$ its dual. By II, p. 277 we have a canonical isomorphism $u \mapsto \tilde{u}$ of vector L-spaces from $(V_{(L)})^*$ onto $\mathrm{Hom}_K(V, L)$ such that $\tilde{u}(x) = u(1 \otimes x)$ for $x \in V$ and u in $(V_{(L)})^*$. If V is of *finite* dimension n over K, the vector space $(V_{(L)})$ over L is of dimension n, as well as its dual $(V_{(L)})^* = V^*_{(L)}$, whence the formula

$$[\mathrm{Hom}_K(V, L) : L] = [V : K]. \tag{2}$$

THEOREM 1. — *Let* L *be an extension of a field* K *and* A *an algebra over* K ; *let* $\mathscr{H}$ *be the set of all* K*-algebra homomorphisms of* A *into* L. *Then* $\mathscr{H}$ *is a free subset of the vector space* $\mathrm{Hom}_K(A, L)$ *over* L.

Let us show by induction on the integer $n \geqslant 0$ that every sequence $(u_1, \ldots, u_n)$ of distinct elements of $\mathscr{H}$ is free. The case $n = 0$ being trivial, we may henceforth suppose that $n \geqslant 1$; let $\alpha_1, \ldots, \alpha_n$ be elements of L such that $\sum_{i=1}^{n} \alpha_i u_i = 0$. For x, y in A we have

$$\sum_{i=1}^{n-1} \alpha_i [u_i(x) - u_n(x)] \,.\, u_i(y) = \sum_{i=1}^{n} \alpha_i u_i(xy) - u_n(x) \sum_{i=1}^{n} \alpha_i u_i(y) = 0,$$

whence $\sum_{i=1}^{n-1} \alpha_i [u_i(x) - u_n(x)] \,.\, u_i = 0$. By the induction hypothesis, the elements $u_1, \ldots, u_{n-1}$ of $\mathscr{H}$ are linearly independent, whence $\alpha_i [u_i(x) - u_n(x)] = 0$ for $1 \leqslant i \leqslant n-1$ and for all x in A. Since the u_i are distinct, this implies that $\alpha_i = 0$ for $i \neq n$, hence $\alpha_n u_n = 0$ and so $\alpha_n = \alpha_n u_n(1) = 0$ (on denoting by 1 the unit element of A). We have thus shown that $\alpha_1, \ldots, \alpha_{n-1}, \alpha_n$ are zero, and this proves the theorem.

COROLLARY 1. — *Let* Γ *be a monoid,* L *a field and* X *a set of homomorphisms of* Γ *into the multiplicative monoid of* L. *Then* X *is a free subset of the vector* L*-space* L^Γ *of mappings of* Γ *into* L.

Let A be the algebra of the monoid Γ with coefficients in L and $(e_\gamma)_{\gamma \in \Gamma}$ the canonical basis of A over L (III, p. 446). For every L-linear mapping u of A into L let us write $\tilde{u}(\gamma) = u(e_\gamma)$ (for $\gamma \in \Gamma$) ; then the mapping $u \mapsto \tilde{u}$ is an isomorphism of vector L-spaces of $\mathrm{Hom}_L(A, L)$ onto L^Γ which maps onto X the set of L-algebra homomorphisms of A into L. Now it suffices to apply Th. 1 with K = L.

COROLLARY 2 (Dedekind's theorem). — *Let* E *and* L *be two extensions of* K. *The set of* K*-homomorphisms of* E *into* L *is free over* L. *If* E *is of finite degree over* K, *the number of* K*-homomorphisms of* E *into* L *is at most equal to* [E : K].

The last assertion follows from the first, taking account of Formula (2).

2. Algebraic independence of homomorphisms

THEOREM 2. — *Let* K *be an* infinite *field,* L *an extension of* K *and* A *an algebra over* K. *Let* $u_1, \dots, u_n$ *be distinct* K*-algebra homomorphisms of* A *into* L *and* f *a polynomial in* $L[X_1, \dots, X_n]$. *If we have* $f(u_1(x), \dots, u_n(x)) = 0$ *for all* $x \in A$, *then* $f = 0$.

Let B be the set of elements of L^n of the form $(u_1(x), \dots, u_n(x))$ with $x \in A$. By Th. 1, there is no sequence $(\alpha_1, \dots, \alpha_n)$ of elements not all zero in L such that $\sum_{i=1}^{n} \alpha_i u_i(x) = 0$ for all $x \in A$; therefore (II, p. 301, Th. 7) B generates the vector space L^n over L. So there exist elements $a_1, \dots, a_n$ of A such that the matrix $(u_i(a_j))_{1 \leqslant i,j \leqslant n}$ is invertible.

Let us define the polynomial $g \in L[Y_1, \dots, Y_n]$ by

$$g(Y_1, \dots, Y_n) = f\left(\sum_{j=1}^{n} u_1(a_j) Y_j, \dots, \sum_{j=1}^{n} u_n(a_j) Y_j\right). \tag{3}$$

Let $y_1, \dots, y_n$ be in K ; writing $x = \sum_{i=1}^{n} y_i a_i$, we have

$$g(y_1, \dots, y_n) = f(u_1(x), \dots, u_n(x)), \quad \text{whence} \quad g(y_1, \dots, y_n) = 0$$

by the hypothesis on f. Since the field K is infinite, we have $g = 0$ (IV, p. 18, Cor. 2) ; now the matrix $(u_i(a_j))$ has an inverse (b_{ij}) and we have

$$f(X_1, \dots, X_n) = g\left(\sum_{j=1}^{n} b_{1j} X_j, \dots, \sum_{j=1}^{n} b_{nj} X_j\right), \tag{4}$$

whence $f = 0$.

Th. 2 has no analogue for finite fields. Thus let K be a finite field with q elements, $A = L = K$ and $f(X) = X^q - X$. We have $x^q = x$ for all $x \in K$ (V, p. 93, Prop. 2) ; therefore if u is the identity automorphism of K, we have $f(u(x)) = 0$ for all $x \in K$, even though f is not zero.

3. Diagonalizable algebras and etale algebras

DEFINITION 1. — *Let* A *be an algebra over* K ; *then* A *is said to be diagonalizable if there exists an integer* $n \geqslant 0$ *such that* A *is isomorphic to the product algebra* K^n. *We say that* A *is diagonalized by an extension* L *of* K *if the algebra* $A_{(L)}$ *over* L *derived from* A *by extension of scalars is diagonalizable. We shall say that* A *is etale if there exists an extension of* K *which diagonalizes* A.

We recall that the product algebra K^n is the vector space K^n equipped with the product defined by

(5) $$(x_1, \dots, x_n) \cdot (y_1, \dots, y_n) = (x_1 y_1, \dots, x_n y_n) .$$

If $\varepsilon_1, \dots, \varepsilon_n$ is the canonical basis of K^n, we have

(6) $$\varepsilon_i^2 = \varepsilon_i , \quad \varepsilon_i \varepsilon_j = 0 \text{ if } i \neq j$$

and $1 = \varepsilon_1 + \dots + \varepsilon_n$.

Every etale algebra over K is *commutative* and of *finite degree* over K.

PROPOSITION 1. — *Let* A *be an algebra of finite degree n over the field* K ; *then the following conditions are equivalent :*

a) The algebra A *is diagonalizable.*

b) There is a basis $(e_1, \dots, e_n)$ *of* A *such that* $e_i^2 = e_i$ *and* $e_i e_j = 0$ *for* $i \neq j$.

c) The K-*algebra homomorphisms of* A *into* K *generate the dual of the vector* K-*space* A.

d) Every A-*module is a direct sum of submodules which are of dimension* 1 *over* K.

The equivalence of *a*) and *b*) follows from Formula (6) ; on the other hand the n projections $K^n \to K$ are algebra homomorphisms, hence *a*) implies *c*). If *c*) holds, the algebra homomorphisms of A into K form a basis of the dual of A (V, p. 27, Th. 1), we denote them by $u_1, \dots, u_n$; then $a \mapsto (u_i(a))$ is an isomorphism of A onto the algebra K^n, whence *a*). We have thus established the equivalence of the conditions *a*), *b*) and *c*).

Suppose that *b*) holds and let M be an A-module ; then the homotheties $(e_i)_M$ of ratio e_i are projectors of M, and M is a direct sum of the $e_i M$, which are sub-A-modules. We may thus suppose that there exists an index i such that $(e_j)_M = 0$ for $j \neq i$. Hence every vector subspace of M is a sub-A-module, whence *d*).

Conversely suppose that *d*) holds and consider the A-module A_s. There exists then a basis (f_i) of the vector K-space A such that $Af_i = Kf_i$ for $i = 1, \dots, n$. After replacing each f_i by a suitable scalar multiple, if necessary, we may suppose that $1 = f_i + \dots + f_n$. If $i \neq j$, then $f_i f_j$ belongs to $Af_i \cap Af_j = Kf_i \cap Kf_j$ hence it is zero. Now $f_i = f_i f_1 + \dots + f_i f_n = f_i^2$, whence *b*).

COROLLARY. — *Let* L *be an extension of* K *and* $\mathscr{H}$ *the set of algebra homomorphisms of* A *into* L. *We have* Card $\mathscr{H} \leqslant [A : K]$, *with equality if and only if* A *is diagonalized by* L. *If* A *is diagonalized by* L, *then* $\mathscr{H}$ *is a basis of the vector* L-*space* $\mathrm{Hom}_K(A, L)$.

The vector space $\mathrm{Hom}_K(A, L)$ over L has dimension $[A : K]$, by Formula (2), and $\mathscr{H}$ is a free subset of $\mathrm{Hom}_K(A, L)$ by Th. 1 (V, p. 27). We thus have Card $\mathscr{H} \leqslant [A : K]$ with equality if and only if $\mathscr{H}$ is a basis of $\mathrm{Hom}_K(A, L)$. There exists an isomorphism of vector L-spaces, say $\pi : \mathrm{Hom}_K(A, L) \to A^*_{(L)}$, characterized by $u(x) = (\pi u)(1 \otimes x)$ for $x \in A$, and π maps $\mathscr{H}$ onto the set $\mathscr{H}_L$ of L-algebra homomorphisms of $A_{(L)}$ into L. Finally the equivalence of *a*) and *c*) in Prop. 1 shows that the algebra $A_{(L)}$ over L is diagonalizable if and only if $\mathscr{H}_L$ generates the vector space $A^*_{(L)}$ over L. This completes the proof of the Corollary.

PROPOSITION 2. — *Let* A *be an algebra over* K *and* Ω *an algebraically closed extension of* K. *The following assertions are equivalent :*

a) *The algebra* A *is etale.*

b) *There exists an extension of finite degree which diagonalizes* A.

c) *The extension* Ω *of* K *diagonalizes* A.

Suppose that A is etale. Let n be the degree of A over K, let L be an extension of K which diagonalizes A and let $\mathscr{H}$ be the set of algebra homomorphisms of A into L. By the Cor. to Prop. 1 we have Card $\mathscr{H} = n$. On the other hand, for each $u \in \mathscr{H}$, we have $[u(A) : K] \leqslant n$. By V, p. 18, Th. 2, the subextension L′ of L generated by the images of elements of $\mathscr{H}$ is of finite degree over K. Since there exist n distinct homomorphisms of A into L′, the extension L′ diagonalizes A, by the Cor. 1 of Prop. 1. This shows that *a*) implies *b*).

Since every extension of finite degree of K is isomorphic to a subextension of Ω (V, p. 20, Th. 1), *b*) implies *c*). Finally *c*) clearly implies *a*).

4. Subalgebras of an etale algebra

PROPOSITION 3. — *Let* A *be an etale algebra over* K. *There exist only a finite number of subalgebras and ideals of* A. *Moreover, every extension of* K *which diagonalizes* A *also diagonalizes every subalgebra and every quotient algebra of* A ; *in particular these algebras are etale.*

It suffices to show that an algebra K^n has only a finite number of subalgebras and ideals, and that the subalgebras and quotient algebras of K^n are diagonalizable. We denote by $(\varepsilon_1, \ldots, \varepsilon_n)$ the canonical basis of K^n.

Let A be a subalgebra of K^n and let $v_1, \ldots, v_n$ be the restrictions to A of the n projections $K^n \to K$. Since the intersection of the kernels of the v_i is clearly 0, the v_i generate the vector K-space dual to A (II, p. 302, Cor. 1) ; hence the K-algebra A is diagonalizable (V, p. 29, Prop. 1).

For every subset I of $\{1, 2, \ldots, n\}$ put $\varepsilon_I = \sum_{i \in I} \varepsilon_i$. It is clear that the elements ε_I are idempotents of K^n ; we have $\varepsilon_I = 0$ if and only if I is empty, and $\varepsilon_I \varepsilon_J = \varepsilon_{I \cap J}$. By what has been said, every subalgebra A of K^n is diagonalizable ; by condition *b*) of Prop. 1 every subalgebra A of K^n therefore admits a basis $(\varepsilon_{I_1}, \ldots, \varepsilon_{I_p})$, where $(I_1, \ldots, I_p)$ is a partition of $\{1, 2, \ldots, n\}$, and there are only a finite number of such subalgebras.

For every subset I of $\{1, 2, \dots, n\}$ let $\mathfrak{a}_I$ be the vector subspace of K^n having as basis the idempotents ε_i for $i \in I$; then it is clear that $\mathfrak{a}_I$ is an ideal of K^n; moreover if $J = \{1, 2, \dots, n\} - I$, then the residue classes $\bar{\varepsilon}_j$ of ε_j mod $\mathfrak{a}_I$ for $j \in J$ form a basis of $K^n/\mathfrak{a}_I$. We have $\bar{\varepsilon}_j^2 = \bar{\varepsilon}_j$ and $\bar{\varepsilon}_j \bar{\varepsilon}_k = 0$ if $j \neq k$, hence the algebra $K^n/\mathfrak{a}_I$ is diagonalizable, by Prop. 1 of V, p. 29.

It remains to show that every ideal of K^n is of the form $\mathfrak{a}_I$. Let I be the set of integers i such that $1 \leqslant i \leqslant n$ and $\varepsilon_i \in \mathfrak{a}$, then $\mathfrak{a}_I \subset \mathfrak{a}$. Let $x = x_1\varepsilon_1 + \dots + x_n\varepsilon_n$ be an element of $\mathfrak{a}$ (with $x_1, \dots, x_n$ in K) and let i be in $\{1, 2, \dots, n\} - I$. We have $x_i\varepsilon_i = x\varepsilon_i \in \mathfrak{a}$, and $\varepsilon_i \notin \mathfrak{a}$, whence $x_i = 0$. Thus $x = \sum_{i \in I} x_i\varepsilon_i$ and this shows that $x \in \mathfrak{a}_I$. We have now shown $\mathfrak{a} \subset \mathfrak{a}_I$, whence finally $\mathfrak{a} = \mathfrak{a}_I$.

COROLLARY. — *Let* $A_1, \dots, A_m$ *be algebras over* K *and* $A = A_1 \times \dots \times A_m$. *For* A *to be etale it is necessary and sufficient that* $A_1, \dots, A_m$ *are etale.*

Suppose that A is etale; each of the algebras $A_1, \dots, A_m$ is isomorphic to a quotient of A, and hence is etale by Prop. 3. Conversely, every extension of K which diagonalizes $A_1, \dots, A_m$ clearly diagonalizes A.

5. Separable degree of a commutative algebra

Let A be a commutative algebra of finite degree n over K. For any extension L of K, the number $h(L)$ of algebra homomorphisms of A into L is finite and is bounded above by n (V, p. 29, Cor.).

Lemma 1. — *Let* Ω *be an algebraic closure of* K; *then we have* $h(L) \leqslant h(\Omega)$ *for every extension* L *of* K, *with equality when* L *is algebraically closed.*

Let L′ be the algebraic closure of K in L. For every homomorphism u of A into L we have $[u(A) : K] \leqslant n$, hence $u(A) \subset L'$ by V, p. 18, Prop. 2; we thus have $h(L') = h(L)$. Since the extension L′ of K is isomorphic to a subextension of Ω (V, p. 20, Th. 1), we have $h(L') \leqslant h(\Omega)$. If L is algebraically closed, then L′ is an algebraic closure of K; the extensions L′ and Ω of K are then isomorphic (V, p. 23, Th. 2) and so $h(L') = h(\Omega)$; the Lemma follows directly from this.

By Lemma 1, the number $h(L)$ has the same value for all algebraically closed extensions L of K; this number will be denoted by $[A : K]_s$ and called the *separable degree* of A.

Let A and B be two commutative algebras of finite degree over K. We shall establish the formula

$$[A \otimes_K B : K]_s = [A : K]_s . [B : K]_s . \tag{7}$$

Let L be an algebraically closed extension of K and denote by $\mathscr{H}(A)$ the set of algebra homomorphisms of A into L, and similarly define $\mathscr{H}(B)$ and $\mathscr{H}(A \otimes_K B)$. By definition we have Card $\mathscr{H}(A) = [A : K]_s$ and corresponding formulae for $[B:K]_s$ and $[A \otimes_K B : K]_s$. Moreover (III, p. 465, Formula (6)), the formula $(u * v)(a \otimes b) = u(a)\,v(b)$ defines a bijection $(u, v) \mapsto u * v$ of $\mathscr{H}(A) \times \mathscr{H}(B)$ onto $\mathscr{H}(A \otimes_K B)$, whence Formula (7) follows.

Let K' be an extension of K ; we shall prove the formula

$$[A_{(K')} : K']_s = [A : K]_s \,. \tag{8}$$

For take L to be an algebraic closure of K'. The formula $\tilde{u}(x) = u(1 \otimes x)$ (for $x \in A$) defines a bijection $u \mapsto \tilde{u}$ between the set of K'-homomorphisms of $A_{(K')}$ into L and the set of K-homomorphisms of A into L, whence (8).

Finally, suppose that K' is an extension of finite degree of K ; if A' is a commutative K'-algebra of finite degree, then it is also a commutative K-algebra of finite degree and we have $[A' : K] = [A' : K'] \,.\, [K' : K]$ (V. p. 10, Th. 1). We shall prove the formula

$$[A' : K]_s = [A' : K']_s \,.\, [K' : K]_s \,. \tag{9}$$

For let S (resp. T) be the set of K-homomorphisms of K' (resp. A') into an algebraic closure L of K ; for every $\sigma \in S$ we denote by T_σ the set of elements f of T such that $f(\alpha \,.\, 1) = \sigma(\alpha)$ for all $\alpha \in K'$. Then the family $(T_\sigma)_{\sigma \in S}$ is a partition of T and we have Card S $= [K' : K]_s$; now for each $\sigma \in S$ the set T_σ consists of K'-homomorphisms of A' into the algebraically closed extension (L, σ) of K', hence Card $T_\sigma = [A' : K']_s$, and so we have proved (9).

PROPOSITION 4. — *Let* A *be a commutative algebra of finite degree over* K ; *then* $[A : K]_s \leqslant [A : K]$ *with equality if and only if* A *is etale.*

Let Ω be an algebraic closure of K and $\mathscr{H}$ the set of algebra homomorphisms of A into Ω. We have Card $\mathscr{H} = [A : K]_s$, and A is etale if and only if A is diagonalized by the extension Ω of K (V, p. 30, Prop. 2). Thus Prop. 4 follows from the Cor. of V, p. 29.

COROLLARY 1. — *Let* A, B *be two commutative algebras over* K, *of finite non-zero degrees. Then for the algebra* $C = A \otimes_K B$ *to be etale it is necessary and sufficient for* A *and* B *to be etale.*

We have $[C : K] = [A : K] \,.\, [B : K]$ and the corresponding Formula (7) for the separable degrees. Further we have $[A : K]_s \leqslant [A : K]$ and corresponding formulae for B and C. It follows that $[C : K] = [C : K]_s$ if and only if we have both $[A : K] = [A : K]_s$ and $[B : K] = [B : K]_s$; now it is enough to apply Prop. 4.

COROLLARY 2. — *Let* K' *be an extension of* K.

a) For a K*-algebra* A *to be etale, it is necessary and sufficient that the* K'*-algebra* $A_{(K')}$ *should be etale.*

b) Let A' *be an algebra over* K', *not reduced to* 0. *For* A' *to be etale over* K *it is necessary and sufficient that* A' *should be etale over* K' *and* K' *etale over* K.

We argue as for Cor. 1, applying this time (8) for *a*) and (9) for *b*).

6. Differential characterization of etale algebras

THEOREM 3. — *Let* A *be a commutative algebra of finite degree over* K. *For* A *to be etale it is necessary and sufficient that the module* $\Omega_K(A)$ *of* K*-differentials of* A *should be reduced to* 0.

A) Let L be an algebraic closure of K (V, p. 23, Th. 2). For A to be etale it is necessary and sufficient that the algebra $A_{(L)}$ over L should be diagonalizable (V, p. 30, Prop. 2). Further, the A-module $\Omega_L(A_{(L)})$ is isomorphic to $\Omega_K(A) \otimes_A A_{(L)}$ (III, p. 572, Prop. 20), hence to $\Omega_K(A) \otimes_K L$, by the associativity of the tensor product ; therefore $\Omega_K(A) = 0$ is equivalent to $\Omega_L(A_{(L)}) = 0$. To prove Th. 3 it is therefore enough to consider the case where K is algebraically closed and to show that the algebra A is diagonalizable if and only if $\Omega_K(A) = 0$.

B) Suppose that A is diagonalizable ; then (V, p. 29, Prop. 1), the vector space A is generated by the idempotents of A. The assertion $\Omega_K(A) = 0$ is thus a consequence of the following lemma :

Lemma 2. — *Let* A *be a commutative algebra over* K *and* e *an idempotent of* A ; *then we have* $de = 0$ *in* $\Omega_K(A)$.

From the relation $e = e^2$ we deduce $de = 2e \,.\, de$; on multiplying by e we obtain $e \,.\, de = 2e \,.\, de$, hence $e \,.\, de = 0$, and so finally, $de = 2e \,.\, de = 0$.

C) We first prove two lemmas :

Lemma 3. — *Let* A *be a commutative algebra of finite degree over the algebraically closed field* K, *such that* $\Omega_K(A) = 0$. *Then we have* $\mathfrak{m} = \mathfrak{m}^2$ *for every maximal ideal* $\mathfrak{m}$ *of* A.

The algebra $A/\mathfrak{m}$ is an extension of finite degree of the algebraically closed field K, whence $[A/\mathfrak{m} : K] = 1$. Hence for each $a \in A$ there exists a unique scalar λ such that $a - \lambda \,.\, 1 \in \mathfrak{m}$; write $D(a)$ for the residue class of $a - \lambda \,.\, 1$ mod $\mathfrak{m}^2$. It is clear that D is a K-derivation of A into the A-module $\mathfrak{m}/\mathfrak{m}^2$. The universal property of $\Omega_K(A)$ (III, p. 569) and the hypothesis $\Omega_K(A) = 0$ now imply that $D = 0$, whence $\mathfrak{m}/\mathfrak{m}^2 = 0$ and so $\mathfrak{m} = \mathfrak{m}^2$.

Lemma 4. — *Let* A *be a commutative ring and let* $\mathfrak{a}$ *be a finitely generated ideal of* A *such that* $\mathfrak{a} = \mathfrak{a}^2$. *Then there exists an idempotent* e *in* A *such that* $\mathfrak{a} = Ae$.

Let $(a_1, \ldots, a_r)$ be a generating system of the ideal $\mathfrak{a}$; since $\mathfrak{a} = \mathfrak{a}^2$, there exist elements x_{ij} in $\mathfrak{a}$ such that $a_i = \sum_{j=1}^{r} x_{ij} a_j$ for $1 \leqslant i \leqslant r$. Write M for the square matrix of order r whose elements are $\delta_{ij} - x_{ij}$ and let D be its determinant. There exists (III, p. 532, Formula (26)) a square matrix N of order r with elements in A such that $N \,.\, M = D \,.\, I_r$, whence immediately $Da_j = 0$ for $1 \leqslant j \leqslant r$ and so finally $D\mathfrak{a} = 0$. Now the matrix M is congruent to I_r mod $\mathfrak{a}$, hence $D \equiv 1$ mod $\mathfrak{a}$. Put $e = 1 - D$; then $e \in \mathfrak{a}$ and $ex = x$ for all $x \in \mathfrak{a}$. It follows that e is an idempotent and $\mathfrak{a}$ is equal to Ae.

With these lemmas established, let us show by induction on the degree of A that A is diagonalizable if K is algebraically closed and $\Omega_K(A) = 0$. Let $\mathfrak{m}$ be a maximal ideal of A (I, p. 104). By Lemmas 3 and 4, there exists an idempotent e such that $\mathfrak{m} = Ae$; we have seen that $A/\mathfrak{m}$ is of degree 1 over K. Hence A is a direct sum of the ideals $\mathfrak{a} = (1 - e)A$ and $\mathfrak{m}$ and we have $[\mathfrak{a} : K] = 1$, hence A is isomorphic to $K \times A/\mathfrak{a}$. Since $\Omega_K(A/\mathfrak{a})$ is isomorphic to a quotient of $\Omega_K(A)$ (III, p. 573, Prop. 22), it is zero and the induction hypothesis shows that $A/\mathfrak{a}$ is diagonalizable. This then shows A to be diagonalizable.

7. Reduced algebras and etale algebras

DEFINITION 2. — *Let* A *be a commutative ring ; then* A *is said to be reduced if every nilpotent element* (I, p. 98) *of* A *is zero.*

If A is a field, or an integral domain, or a product of reduced rings, it is a reduced ring. For a commutative ring A to be reduced it is necessary and sufficient that $a^2 \neq 0$ for every $a \neq 0$ in A : for from this we obtain by induction on n, $a^{2^n} \neq 0$, hence $a^n \neq 0$ for any $a \neq 0$ in A.

An algebra is said to be *reduced* if its underlying ring is reduced.

PROPOSITION 5. — *Let* A *be a commutative algebra of finite degree over* K. *For* A *to be reduced it is necessary and sufficient that there exist extensions* $L_1, \dots, L_n$ *of finite degree over* K *such that* A *is* K*-isomorphic to* $L_1 \times \cdots \times L_n$.

The stated condition is clearly sufficient.

Conversely, assume that A is reduced ; arguing by induction on the degree of A we see that it suffices to prove that if A is not a field, there exist two non-zero algebras A_1 and A_2 such that A is isomorphic to $A_1 \times A_2$, or also that there exists in A an idempotent different from 0 and 1.

Suppose from now on that A is reduced and is not a field. Among the ideals of A different from 0 and A let $\mathfrak{a}$ be an ideal whose dimension as vector K-space is minimal. For any $x \neq 0$ in $\mathfrak{a}$ we have $x^2 \neq 0$, because A is reduced, whence $\mathfrak{a}^2 \neq \{0\}$. We have $\mathfrak{a}^2 \subset \mathfrak{a}$ and by the minimality of $\mathfrak{a}$ we find that $\mathfrak{a}^2 = \mathfrak{a}$. By Lemma 4 there exists an idempotent e such that $\mathfrak{a} = Ae$, and we have $e \neq 0$, $e \neq 1$ because $\mathfrak{a}$ is distinct from 0 and A.

THEOREM 4. — *Let* A *be a commutative algebra of finite degree over* K. *Then the following assertions are equivalent :*

a) The algebra A *is etale.*

b) For every extension L *of* K, *the ring* $L \otimes_K A$ *is reduced.*

c) There exists a perfect extension field P *of* K *such that the ring* $P \otimes_K A$ *is reduced.*

* *d) There exist separable algebraic extensions* $L_1, \dots, L_n$ *of* K *such that* A *is isomorphic to* $L_1 \times \cdots \times L_n$. *

In particular, every etale algebra is reduced.

A) Let us first prove the equivalence of *a*), *b*) and *c*).

Suppose that A is etale and let L be an extension of K. Let Ω be an algebraically closed extension field of L (V, p. 23, Th. 2). Then $L \otimes_K A$ is isomorphic to a subring of $\Omega \otimes_K A$ and the latter is isomorphic to a ring Ω^n by Prop. 2 (V, p. 30). Therefore the ring $L \otimes_K A$ is reduced.

We have thus shown that *a*) implies *b*), and *c*) is a particular case of *b*). Assume now that *c*) holds. For the K-algebra A to be etale it is necessary and sufficient that the P-algebra $A_{(P)}$ should be etale (V, p. 32, Cor. 2). The algebra A is therefore etale by the following lemma :

Lemma 5. — *Let* B *be a reduced algebra of finite degree over a perfect field* P ; *then* B *is etale.*

By Prop. 5 there exist extensions $L_1, \dots, L_n$ of P such that B is isomorphic to the algebra $L_1 \times \cdots \times L_n$. Since a finite product of etale algebras is etale (V, p. 31, Cor.) it is enough to examine the case where B is an extension of P. By Th. 3 (V, p. 33) it is enough to show that $dx = 0$ in $\Omega_P(B)$ for all $x \in B$.

Let $x \in B$; since B is of finite degree over K, x is algebraic over K (V, p. 18, Prop. 2). Let f be the minimal polynomial of x and let f' be the derivative of f. The polynomial f is non-constant. Suppose that $f' = 0$; by V, p. 9, Cor. the field P is of characteristic $p \neq 0$ and we have $f \in P[X^p]$; since P is perfect, we have $P[X^p] = P[X]^p$, but the irreducible polynomial f cannot lie in $P[X]^p$.

We thus have $f' \neq 0$ and since the degree of f' is strictly less than that of f, we have $f'(x) \neq 0$. Now from $f(x) = 0$ we deduce $f'(x) \, . \, dx = 0$ in $\Omega_P(B)$, whence $dx = 0$, as we had to show.

* *B*) Suppose that A is etale ; by the equivalence of *a*) and *b*) the algebra A is reduced, so there exist extensions $L_1, \dots, L_n$ of K such that A is isomorphic to $L_1 \times \cdots \times L_n$ (Prop. 5). Since A is etale, each of the extensions L_i is an etale algebra (V, p. 31, Cor.), and hence by definition is a separable algebraic extension of K.

The implication *d*) $\Rightarrow$ *a*) follows from V, p. 31, Cor. *

COROLLARY. — *Suppose that* K *has characteristic* $p \neq 0$. *For* A *to be etale it is necessary and sufficient that* $A = K[A^p]$. *For every basis* $(a_i)_{i \in I}$ *of* A *over* K, *the family* $(a_i^p)_{i \in I}$ *is then a basis of* A *over* K.

Let us choose an algebraic closure Ω of K. Given two K-homomorphisms u and v of A into Ω, if u and v have the same restriction to $K[A^p]$, we have

$$u(x)^p = u(x^p) = v(x^p) = v(x)^p \,, \quad \text{whence} \quad u(x) = v(x)$$

for all $x \in A$. We thus have the inequality $[A : K]_s \leqslant [K[A^p] : K]_s$; if A is etale, we thus have

$$[A : K] = [A : K]_s \leqslant [K[A^p] : K]_s \leqslant [K[A^p] : K] \,,$$

whence $A = K[A^p]$.

Conversely suppose that we have $A = K[A^p]$; let $(a_i)_{i \in I}$ be a basis of A over K. By V, p. 4, Prop. 2, *b*), the family $(a_i^p)_{i \in I}$ generates the vector K-space $K[A^p]$ and since $A = K[A^p]$ is of finite dimension equal to the cardinal of I, the family $(a_i^p)_{i \in I}$ is a basis of A over K. Let u be an element of $\Omega \otimes_K A$ such that $u^2 = 0$, whence $u^p = 0$; since $(a_i)_{i \in I}$ is a basis of A over K, there exists a family $(\lambda_i)_{i \in I}$ of elements of Ω such that $u = \sum_{i \in I} \lambda_i \otimes a_i$, whence $u^p = \sum_{i \in I} \lambda_i^p \otimes a_i^p$. Since $(a_i^p)_{i \in I}$ is a basis of A over K and $u^p = 0$, we must have $\lambda_i^p = 0$, whence $\lambda_i = 0$ for all $i \in I$, and so $u = 0$. This shows the ring $\Omega \otimes_K A$ to be reduced ; since Ω is perfect, the algebra A over K is etale by Th. 4.

For another characterization of etale algebras see V, p. 48, Prop. 1.

§ 7. SEPARABLE ALGEBRAIC EXTENSIONS

Throughout this paragraph K *denotes a field.*

1. Separable algebraic extensions

DEFINITION 1. — *Let* E *be an algebraic extension of* K ; *then* E *is said to be separable* (*over* K) *if every subextension* F *of* E *of finite degree over* K *is an etale algebra over* K (V, p. 28, Def. 1).

Let E be an extension of finite degree of K. Since every subalgebra of an etale algebra is etale (V, p. 30, Prop. 3), it comes to the same to suppose that E is a separable extension of K, or that E is an etale algebra over K.

PROPOSITION 1. — *Let* E *be an algebraic extension of* K. *If* E *is separable, then every subextension* E′ *of* E *is separable. Conversely, if every subextension of finite degree of* E *is separable then* E *is separable.*

This follows immediately from Definition 1.

PROPOSITION 2. — *For a field* K *to be perfect it is necessary and sufficient that every algebraic extension of* K *should be separable.*

Suppose first that K is perfect. Since a field is a reduced ring, it follows from Lemma 5 (V, p. 35) that every extension of finite degree of K is an etale algebra over K ; therfore every algebraic extension of K is separable.

Suppose now that K is an imperfect field of characteristic $p \neq 0$. Let Ω be an algebraic closure of K. Since K is imperfect, there exists $b \in K$ not belonging to K^p ; put $a = b^{1/p.}$ Then the extension $K(a)$ of K is p-radical of finite degree. By V, p. 26, Prop. 3, there exists a single K-homomorphism of $K(a)$ into Ω, and since $[K(a) : K] > 1$, the algebra $K(a)$ is not etale over K (V, p. 32, Prop. 4). In other words, the extension $K(a)$ of finite degree of K is not separable.

COROLLARY. — *Every algebraic extension of a field of characteristic* 0, *or of a finite field, is separable.*

This follows from V, p. 7, Prop. 5.

2. Separable polynomials

PROPOSITION 3. — *Let f be a non-zero polynomial in* $K[X]$ *and let* Ω *be an algebraically closed extension of* K. *The following conditions are equivalent :*

a) The polynomial f is relatively prime to its derivative f' in $K[X]$.

b) Either $\deg(f) = 0$, *or* $\deg(f) > 0$ *and* $\operatorname{dis}(f) \neq 0$ (IV, p. 83).

c) There exists an extension L *of* K *such that f splits in* $L[X]$ *into a product of distinct polynomials of degree* $\leqslant 1$.

d) The roots of f in Ω *are simple.*

e) The K-*algebra* $K[X]/(f)$ *is etale* (V, p. 28, Def. 1).

a) $\Rightarrow$ *d)* : Under the hypothesis *a)* there exist two polynomials g and h in $K[X]$ such that $fg + f'h = 1$ (IV, p. 12). Let a be a root of f in Ω ; we have

$$f'(a)\,h(a) = f(a)\,g(a) + f'(a)\,h(a) = 1\ ,$$

whence $f'(a) \neq 0$; therefore a is a simple root of f in Ω (IV, p. 17, Prop. 7).

d) $\Rightarrow$ *c)* : If *d)* holds, f splits in $\Omega[X]$ into a product of distinct factors of degree $\leqslant 1$.

c) $\Rightarrow$ *b)* : Under the hypothesis *c)* there exist an element $\lambda \neq 0$ in L and *distinct* elements $\alpha_1, \dots, \alpha_n$ of L such that $f(X) = \lambda(X - \alpha_1) \dots (X - \alpha_n)$. If $\deg(f) > 0$, we have (IV, p. 83, Prop. 11),

$$\operatorname{dis}(f) = \lambda^{2n-2} \prod_{i<j} (\alpha_i - \alpha_j)^2 \neq 0\ .$$

b) $\Rightarrow$ *a)* : Let c be the leading coefficient and D the discriminant of f; the resultant of f' and f is equal to $\pm\, c\mathrm{D}$ (IV, p. 84, Formula (54)), hence is non-zero ; therefore (IV, p. 78, Cor. 2) the polynomials f and f' are relatively prime in $K[X]$.

a) $\Rightarrow$ *e)* : Let A be the K-algebra $K[X]/(f)$ and x the image of X in A ; by III, p. 573, Prop. 22, the A-module $\Omega_K(A)$ is generated by the elements dx, subject to the single relation $f'(x)\,dx = 0$. By V, p. 33, Th. 3, the K-algebra A is thus etale if and only if $f'(x)$ is an invertible element in A, which means that f and f' are relatively prime in $K[X]$.

DEFINITION 2. — *A polynomial* $f \in K[X]$ *is said to be separable if it is non-zero and satisfies the equivalent conditions a), b), c), d) and e) of Prop.* 3.

Remarks. — 1) Let L be an extension of K and f a non-constant polynomial in $K[X]$. By *e)* of Prop. 3 and V, p. 32, Cor. 2 it comes to the same to suppose that f is separable, whether considered as element of $K[X]$ or of $L[X]$. On the other hand, it may well happen that f is irreducible in $K[X]$ but not in $L[X]$.

2) Let $f \in K[X]$; we know (IV, p. 13, Prop. 13) that there exist irreducible polynomials $f_1, \dots, f_m$ in K[X] such that $f = f_1 \dots f_m$. Let Ω be an algebraic closure of K ; since an irreducible polynomial $g \in K[X]$ is the minimal polynomial over K of each of its roots in Ω, two distinct irreducible polynomials in K[X] have no common root in Ω. Condition *d*) of Prop. 3 now shows that f is separable if and only if the polynomials $f_1, \dots, f_m$ are separable and pairwise distinct.

PROPOSITION 4. — *Let f be an irreducible polynomial in* K[X]. *Then the following conditions are equivalent :*

a) *f is separable.*

b) *There exists an extension* L *of* K *in which f has a simple root.*

c) *The derivative f' of f is not zero.*

d) *The field* K *is of characteristic* 0, *or it is of characteristic* $p \neq 0$ *and* $f \notin K[X^p]$.

We note first that an irreducible polynomial in K[X] is not constant. It is clear that *a*) implies *b*) (take an algebraic closure of K for L). If x is a simple root of f in an extension L of K, we have $f'(x) \neq 0$ (IV, p. 17, Prop. 7), so *b*) implies *c*), and the equivalence of *c*) and *d*) follows from V, p. 9, Cor.

Suppose finally that $f' \neq 0$; let x be a root of f in an algebraically closed extension Ω of K. Since f is the minimal polynomial of x over K and $\deg f' < \deg f$, we have $f'(x) \neq 0$, and so x is a simple root of f (IV, p. 17, Prop. 7). Therefore f is separable and we have shown that *c*) implies *a*).

COROLLARY 1. — *For a field* K *to be perfect it is necessary and sufficient that every irreducible polynomial of* K[X] *should be separable.*

If the field K is of characteristic 0, K is perfect and every irreducible polynomial of K[X] is separable, by *d*) above. Suppose then that K has characteristic $p \neq 0$.

Suppose first that K is perfect. We have $K[X^p] = K[X]^p$, hence there exists no irreducible polynomial of K[X] belonging to $K[X^p]$. By Prop. 4, every irreducible polynomial of K[X] is then separable.

Suppose next that K is imperfect, whence $K \neq K^p$. Let a be an element of K not belonging to K^p ; the polynomial $X^p - a$ is irreducible in K[X] (V, p. 24, Lemma 1), and it belongs to $K[X^p]$, and so is not separable.

COROLLARY 2. — *Let* $f \in K[X]$ *be a non-zero polynomial. For f to be separable it is necessary and sufficient that there should exist an extension* L *of* K *which is a perfect field and such that f has no repeated factor in* L[X].

Let Ω be an algebraic closure of K ; if f is separable, f has no repeated factors in $\Omega[X]$ (Prop. 3, *d*)). Conversely, if L is a perfect extension of K such that f has no repeated factor in L[X], then f is separable in L[X] (Cor. 1 and Remark 2), hence in K[X] (Remark 1).

3. Separable algebraic elements

DEFINITION 3. — *Let* E *be an extension of* K. *An element* x *of* E *which is algebraic over* K *is said to be separable over* K *if the algebraic extension* $K(x)$ *of* K *is separable.*

PROPOSITION 5. — *Let* E *be an extension of* K, x *an element of* E *algebraic over* K *and* f *the minimal polynomial of* x *over* K. *Then the following conditions are equivalent :*

a) x *is separable over* K ;

b) the polynomial f *is separable ;*

c) x *is a simple root of* f.

The equivalence of *a*) and *b*) follows from Prop. 3, that of *b*) and *c*) from Prop. 3 and 4 (cf. V, p. 37 and 38).

COROLLARY 1. — *If an element* x *of* E *is a simple root of a polynomial* g *of* $K[X]$, *it is separable over* K.

For the minimal polynomial f of x over K divides g in $K[X]$ (V, p. 16, Th. 1), so x is a simple root of f.

COROLLARY 2. — *If an element* x *of* E *is algebraic and separable over* K, *it is algebraic and separable over every extension* K' *of* K *contained in* E.

Let f be the minimal polynomial of x over K. Then x is a simple root of f by Prop. 5, and since f belongs to $K'[X]$, the element x of E is separable over K' by Cor. 1.

COROLLARY 3. — *Suppose that* K *has characteristic* $p \neq 0$. *For an element* x *of* E *to belong to* K *it is necessary and sufficient for it to be both separable algebraic and* p*-radical over* K.

The stated condition is clearly necessary. Conversely suppose that x is separable algebraic over K and p-radical of height e over K. Since x is separable over K, the minimal polynomial f of x over K does not belong to $K[X^p]$ (Prop. 4 and 5) ; since x is p-radical of height e over K, we have $f(X) = X^{p^e} - x^{p^e}$ (V, p. 24, Prop. 1) ; we conclude that $e = 0$, so $x \in K$.

PROPOSITION 6. — *Let* E *be an extension of* K.

a) If E *is algebraic and separable over* K, *every element of* E *is algebraic and separable over* K.

b) Conversely, let A *be a set of elements of* E, *algebraic and separable over* K *and such that* $E = K(A)$; *then* E *is algebraic and separable over* K.

If E is algebraic and separable over K, the same is true of the extension $K(x)$ of K for every $x \in E$, whence *a*).

Under the hypothesis *b*), the extension E is algebraic over K (V, p. 18, Cor. 1).

Let F be a subextension of E of finite degree over K. By V, p. 11, Cor., there exist elements $x_1, \ldots, x_m$ of A such that $F \subset K(x_1, \ldots, x_m)$ and we have

$$K(x_1, \ldots, x_m) = K[x_1, \ldots, x_m] \quad (\text{V, p. 18, Cor. 1}) .$$

By the hypothesis on A, the algebras $K[x_1], \ldots, K[x_m]$ are etale over K ; hence the same is true of $K[x_1] \otimes \cdots \otimes K[x_m]$ (V, p. 32, Cor. 1). Now F is isomorphic to a subalgebra of a quotient algebra of $K[x_1] \otimes \cdots \otimes K[x_m]$, and so is etale (V, p. 30, Prop. 3).

COROLLARY. — *For an algebraic extension* E *to be separable over* K *it is necessary and sufficient that every element of* E *should be a simple root of its minimal polynomial over* K. *It is enough to apply Prop.* 5 *and* 6.

4. The theorem of the primitive element

Let E be an extension of K ; an element x of E is said to be *primitive* if $E = K[x]$. For the extension E to possess a primitive element it is necessary for $[E : K]$ to be finite.

THEOREM 1. — *Let* E *be an extension of* K. *Then the following conditions are equivalent :*

a) E *possesses a primitive element ;*

b) *there exist only a finite number of subextensions of* E.

These conditions are satisfied when E *is a separable extension of finite degree.*

Suppose first that E possesses a primitive element x, and let f be the minimal polynomial of x over K. For each monic polynomial $g \in E[X]$ dividing f in $E[X]$ denote by E_g the subextension of E generated by the coefficients of g. Since the possible polynomials g are finite in number (if f splits in $E[X]$ into a product of r monic irreducible polynomials, the number is bounded by 2^r), the subextensions E_g are finite in number. To prove *b*) it is therefore enough to show that every subextension L of E is one of the E_g. Now if L is a subextension of E, then we have $L[x] = E$; if g is the minimal polynomial of x over L, we have $[E : L] = \deg(g)$. Besides, g is a divisor of f in $L[X]$, hence in $E[X]$; so we have $E_g \subset L$ and $E = E_g[x]$. Since $g(x) = 0$, we have $[E : E_g] \leqslant \deg(g)$, therefore $[E : E_g] \leqslant [E : L]$ and so $L = E_g$ as we wished to show.

Next we observe that Condition *b*) implies that the extension E is of finite degree : by Remark 2 of V, p. 18 it suffices to prove that it is algebraic ; now if z is an element of E transcendental over K then the subextensions $K(z^n)$, $n \in \mathbf{N}$ are pairwise distinct.

To show that *b*) $\Rightarrow$ *a*) we now distinguish two cases :

A) If the field K is *finite*, the field E is a vector space of finite dimension over K and hence is a finite set. Therefore [1] (V, p. 78, Lemma 1) there exists an element x of E generating the multiplicative group of E, and we have $E = K[x]$.

B) Suppose now that the field K is *infinite.* If *b*) holds, the extension E is of finite degree, so *b*) can also be expressed by saying that E possesses only a finite number of subalgebras. This being so, the implication *b*) $\Rightarrow$ *a*) is a consequence of the following more general proposition (for which the hypothesis that the field K be infinite is indispensable, cf. V, p. 153, Ex. 5 of § 7):

PROPOSITION 7. — *Suppose that* K *is infinite; let* A *be a commutative* K*-algebra possessing only a finite number of sub-algebras (for example an etale* K*-algebra,* V, p. 30, Prop. 3) *and let* V *be a vector subspace of* A *generating* A. *Then there exists* $x \in V$ *such that* $A = K[x]$.

Let $A_1, \ldots, A_n$ be the subalgebras of A distinct from A. If $x \notin A_1 \cup \ldots \cup A_n$, then the sub-algebra $K[x]$ cannot equal any of the A_i and so must coincide with A. Further, since V generates A, it is not contained in any of the subspaces A_i. Prop. 7 is therefore a consequence of the following lemma:

Lemma 1. — *Let* A *be a vector* K*-space,* V, $A_1, \ldots, A_n$ *subspaces of* A. *If* $\mathrm{Card}(K) \geqslant n$ *and if* V *is not contained in any of the* A_i, *then* V *is not contained in* $A_1 \cup \ldots \cup A_n$.

Arguing by induction on n, we need only prove that if $V \not\subset A_n$ and $V \subset A_1 \cup \ldots \cup A_n$, then $V \subset A_1 \cup \ldots \cup A_{n-1}$. Let $x \in V$, $x \notin A_n$, and let y be arbitrary in V. If $y \in Kx$, we have $y \in A_1 \cup \ldots \cup A_{n-1}$; if not, then the elements x and $y + \lambda x$, $\lambda \in K$ are strictly greater than n in number and belong to $A_1 \cup \ldots \cup A_n$, so two of them belong to the same A_i. Thus there exists i, $1 \leqslant i \leqslant n$ such that either $x \in A_i$ and $y + \lambda x \in A_i$ for some $\lambda \in K$, or $y + \lambda x \in A_i$ and $y + \mu x \in A_i$ for two distinct scalars $\lambda, \mu \in K$. In both cases we conclude that $x \in A_i$ and $y \in A_i$; but this implies that $i \neq n$, hence $y \in A_1 \cup \ldots \cup A_{n-1}$ as we wished to show.

This completes the proof of the equivalence of *a*) and *b*) in Th. 1. Finally if the extension E is separable and of finite degree, condition *b*) holds, by V, p. 30, Prop. 3.

5. Stability properties of separable algebraic extensions

PROPOSITION 8. — *Let* E *be an extension of* K *and* $(E_i)_{i \in I}$ *a family of subextensions of* E *such that* $E = K\left(\bigcup_{i \in I} E_i\right)$. *If each extension* E_i *is algebraic and separable over* K, *the same is true of* E.

This follows at once from Prop. 6 (V, p. 39).

[1] The reader may convince himself that Th. 1 is used nowhere before the proof of Lemma 1 of V, p. 78.

PROPOSITION 9. — *Let* F *be an algebraic extension of* K *and* E *a subextension of* F. *For* F *to be separable over* K *it is necessary and sufficient that* F *should be separable over* E *and* E *separable over* K.

Suppose first that F is separable over K ; then E is separable over K by Prop. 1 (V, p. 36). Moreover, every element of F is separable over K (V, p. 39, Prop. 6) hence over E (V, p. 39, Cor. 2) and so F is separable over E (V, p. 39, Prop. 6).

Conversely, suppose that F is separable over E and E separable over K. Denote by x an element of F and by $f \in E[X]$ the minimal polynomial of x over E. Since E is algebraic over K, Th. 2 (V, p. 18) shows that there exists a subextension E′ of E of finite degree over K such that $f \in E'[X]$; then f is at the same time the minimal polynomial of x over E and over E′, and since x is separable over E (V, p. 39, Prop. 6) it is so also over E′ (V, p. 39, Prop. 5). Write $F' = E'(x)$; then F′ is separable and of finite degree over E′, and since E is separable over K, E′ is separable and of finite degree over K (V, p. 36, Prop. 1). Hence F′ is separable and of finite degree over K, by V, p. 32, Cor. 2. Therefore (V, p. 39, Prop. 6) x is separable over K. We have now shown that every element of F is separable over K, hence F is separable over K (V, p. 39, Prop. 6).

PROPOSITION 10. — *Let* E *and* K′ *be two subextensions of the same extension of* K *and let* $E' = K'(E)$. *Suppose that* E *is algebraic over* K, *hence* E′ *algebraic over* K′ (V, p. 18, Cor. 2).

a) *If* E *is separable over* K, *then* E′ *is separable over* K′.

b) *Conversely if* E′ *is separable over* K′ *and* E *and* K′ *are linearly disjoint over* K, *then* E *is separable over* K.

Assertion *a*) follows directly from Prop. 6 (V, p. 39).

Under the hypothesis *b*), let F be a subextension of E of finite degree over K. Then F and K′ are linearly disjoint over K, hence the K′-algebra $F_{(K')} = K' \otimes_K F$ is isomorphic to $K'(F)$. Since $K'(F)$ is a subextension of E′ of finite degree over K′ and E′ is algebraic and separable over K′, the K′-algebra $K'(F)$ is etale. In other words, the K′-algebra $F_{(K')}$ is etale, and now Cor. 2 of Prop. 4 (V, p. 32) shows that F is etale over K. Thus we have shown E to be separable over K.

6. A separability criterion

PROPOSITION 11. — *Suppose that* K *is of characteristic exponent* p, *and let* E *be an algebraic extension of* K, *generated by a set* S. *If* E *is separable over* K, *then* $E = K(S^{p^n})$ *for all integers* $n \geqslant 0$; *conversely if* E *is of finite degree over* K *and* $E = K(S^p)$, *then* E *is separable over* K.

The case $p = 1$ is trivial by the Cor. of V, p. 37. Suppose from now on that $p \neq 1$.

By hypothesis E is algebraic over K and we have $E = K(S)$, hence

$$K(S^p) = K(E^p) = K[E^p] \quad \text{by V, p.18, Cor. 1 .}$$

If E is of finite degree over K, it is a separable extension of K if and only if it is an etale algebra over K ; the Cor. of V, p. 35 shows that this happens if and only if $E = K[E^p]$.

Suppose now that E is separable and of infinite degree over K. Then $K[E^p]$ is the union of the subrings $K[E'^p]$ where E′ ranges over the set of subextensions of E of finite degree over K ; but such an extension E′ is separable over K (V, p. 36, Prop. 1), whence $E' = K[E'^p] \subset K[E^p]$ by what has been said ; finally we have $E = K[E^p]$. By induction on $n \geqslant 0$, the relation $E = K[E^p]$ implies that $E = K[E^{p^n}]$.

Corollary 1. — *Every algebraic extension of a perfect field is a perfect field.*

Let K be a perfect field of characteristic exponent p, and let E be an algebraic extension of K. Then E is separable over K (V, p. 36, Prop. 2) whence $E = K(E^p)$ by Prop. 11 ; but we have $K = K^p \subset E^p$, hence $E = K(E^p) = E^p$, and so E is perfect.

Corollary 2 (Mac Lane). — *Let* $\bar{K}$ *be an algebraic closure of* K *and* $K^{p^{-\infty}}$ *the perfect closure of* K *in* $\bar{K}$. *For a subextension* E *of* $\bar{K}$ *to be separable over* K *it is necessary and sufficient that it should be linearly disjoint from* $K^{p^{-\infty}}$ *over* K.

We can immediately reduce to the case where $[E:K]$ is finite. Let $(x_i)_{i \in I}$ be a basis of E over K. For E to be linearly disjoint from $K^{p^{-\infty}}$ it is necessary and sufficient that it should be so from $K^{p^{-n}}$ for all $n \geqslant 0$, and this just means that the relation $\sum_{i \in I} x_i a_i^{p^{-n}} = 0$ implies $a_i = 0$ for all $i \in I$, for any family $(a_i)_{i \in I}$ of elements of K. This in turn means that the family $(x_i^{p^n})_{i \in I}$ is free over K, or also that it is a basis of the vector space E over K. Put differently, E is linearly disjoint from $K^{p^{-n}}$ if and only if $E = K(E^{p^n})$. Now it suffices to apply Prop. 11.

Remarks. — 1) When E is algebraic and of infinite degree over K, the condition $E = K(E^p)$ does not always ensure that E is separable over K. For example, if K is imperfect and E is a perfect closure of K, then we have $E = K(E^p)$ but E is not a separable extension of K (V, p. 39, Cor. 3).

2) Let E a separable algebraic extension of a field K of characteristic exponent p. Then we have $E^p \cap K = K^p$ (Cor. 2) ; therefore if E is perfect, the same is true of K.

7. The relative separable algebraic closure

Proposition 12. — *Let* E *be an extension of* K *and* E_s *the set of elements of* E *which are algebraic and separable over* K. *Then* E_s *is the largest subextension of* E *which is algebraic and separable over* K.

By Prop. 6, *a*) (V, p. 39) every subextension of E which is algebraic and separable over K is contained in E_s. By Prop. 6, *b*) (*loc. cit.*) the extension $K(E_s)$ of K is algebraic and separable, whence $K(E_s) \subset E_s$ and so $K(E_s) = E_s$.

With the notation of the preceding proposition, E_s is called the *relative separable* (algebraic) *closure of* K *in* E. When K is perfect, E_s is the relative algebraic closure of K in E (V, p. 36, Prop. 2).

PROPOSITION 13. — *Let* E *be an algebraic extension of* K *and let* E_s *be the relative separable algebraic closure of* K *in* E.

a) E *is a p-radical extension of* E_s.

b) If F *is a subextension of* E *such that* E *is p-radical over* F, *then* $F \supset E_s$.

c) E_s *is the unique subextension of* E *which is separable over* K *and over which* E *is p-radical.*

It suffices to prove *a*) in the case where K is of characteristic $p \neq 0$. Let x be an element of E and f its minimal polynomial over K. There exists an integer $m \geqslant 0$ such that f belongs to $K[X^{p^m}]$ but not to $K[X^{p^{m+1}}]$; in other words, we have $f(X) = g(X^{p^m})$ with $g \in K[X]$, $g \notin K[X^p]$. Since f is irreducible, the same is true of g, hence g is the minimal polynomial of x^{p^m} over K. By V, p. 38, Prop. 4 and p. 39, Prop. 5 we thus have $x^{p^m} \in E_s$, so E is p-radical over E_s.

Assume now the hypothesis *b*) and let $x \in E_s$. Since x is separable over K, it is so also over F (V, p. 39, Cor. 2), but since E is p-radical over F, x is also p-radical over F, hence $x \in F$ (V, p. 39, Cor. 3).

Finally *c*) follows from *a*) and *b*) and Prop. 12.

COROLLARY 1. — *Let* E *and* K′ *be two extensions of* K *contained in the same extension of* K. *Suppose that* E *is algebraic over* K *and denote by* E_s *the relative separable algebraic closure of* K *in* E. *Then* $K'(E_s)$ *is the relative separable algebraic closure of* K′ *in* K′(E).

For $K'(E_s)$ is a separable algebraic extension of K′ by Prop. 10 (V, p. 42); since E is p-radical over E_s, the extension K′(E) of $K'(E_s)$ is p-radical (V, p. 25, Cor.). Now it suffices to apply Prop. 13.

COROLLARY 2. — *If* E *has finite degree over* K, *then* $E_s = \bigcap_{n \geqslant 0} K(E^{p^n})$.

For each integer $n \geqslant 0$ denote by F_n the subextension $K(E^{p^n})$ of E. The sequence $(F_n)_{n \geqslant 0}$ of vector subspaces of E is descending and E is of finite dimension over K. Hence there exists an integer $m \geqslant 0$ such that $F_m = F_n$ for all $n \geqslant m$. We thus have $K(F_m^p) = F_{m+1} = F_m$, hence F_m is a separable extension of K (V, p. 42, Prop. 11); it is clear that E is p-radical over F_m and so Prop. 13 implies $E_s = F_m = \bigcap_{n \geqslant 0} F_n$.

Remark. — Let E be an algebraic extension of K and E_r the relative p-radical closure of E in K (V, p. 25). Then E_r is the largest subextension of E which is p-radical over K (V, p. 25, Prop. 2). However, E is in general not separable over E_r (V, p. 152, Ex. 2); for the case of quasi-Galois extensions see V, p. 76.

8. The separable closure of a field

DEFINITION 4. — *A field* K *is said to be separably closed if every separable algebraic extension of* K *is trivial.*

An algebraically closed field is separably closed. Conversely, if a *perfect* field K is separably closed, it is algebraically closed, because every algebraic extension of K is separable (V, p. 3 , Prop. 2).

DEFINITION 5. — *Let* K *be a field. By a separable algebraic closure, or (by abuse of language) separable closure of* K *we understand any extension* E *of* K *which is algebraic and separable over* K, *and such that the field* E *is separably closed.*

When K is perfect, there is complete identity between the notions of separable closure and algebraic closure of K (V, p. 36, Prop. 2 and p. 43, Cor. 1).

PROPOSITION 14. — *Let* Ω *be an algebraically closed extension of* K.

a) The relative separable algebraic closure Ω_s *of* K *in* Ω *is a separable closure of* K.

b) If E *and* E′ *are two separable closures of* K, *there exists a* K*-isomorphism of* E *onto* E′.

Let F be a separable algebraic extension of Ω_s ; since Ω is algebraically closed, there exists an Ω_s-homomorphism u of F into Ω (V, p. 20, Th. 1). By Prop. 9 (V, p. 42) $u(\mathrm{F})$ is separable over K, hence $u(\mathrm{F}) = \Omega_s$. Therefore F is a trivial extension of Ω_s and so Ω_s is separably closed, whence *a*).

Let E be a separable closure of K. Since E is an algebraic extension of K, there exists a K-homomorphism v of E into Ω (V, p. 20, Th. 1). Hence $v(\mathrm{E})$ is separable algebraic over K, whence $v(\mathrm{E}) \subset \Omega_s$. By V, p. 42, Prop. 9, Ω_s is separable over $v(\mathrm{E})$ and since the field $v(\mathrm{E})$ is separably closed, we have $v(\mathrm{E}) = \Omega_s$. It follows that v is a K-isomorphism of E onto Ω_s. Now *b*) is an immediate consequence.

COROLLARY. — *Let* E *be a separably closed extension of* K *and* F *a separable algebraic extension of* K ; *then there exists a* K*-homomorphism of* F *into* E.

Let Ω be an algebraic closure of E ; we have $\Omega_s \subset \mathrm{E}$ and it suffices to treat the case where $\mathrm{E} = \Omega_s$. Since F is an algebraic extension of K, there exists a K-homomorphism u of F into Ω (V, p. 20, Th. 1). Since the field $u(\mathrm{F})$ is separable over K, we have $u(\mathrm{F}) \subset \Omega_s$ and u defines a K-homomorphism of F into $\Omega_s = \mathrm{E}$.

Remarks. — 1) Let E and E′ be two separable closures of K. If K is not separably closed, there exist several K-isomorphisms of E onto E′. * For E is then a non-trivial Galois extension of K, and so there exist K-automorphisms of E distinct from the identity (V, p. 56, Th. 1). *

* 2) Let E be an algebraic and separable extension of K. If every algebraic and separable extension of K is isomorphic to a subextension of E, then E is a separable

closure of K. For if E′ is a separable closure of K, then each of the extensions E and E′ is isomorphic to a subextension of the other ; hence E and E′ are isomorphic extensions of K (V, p. 52, Prop. 1, *a*)). $_*$

9. Separable and inseparable degrees of an extension of finite degree

Let E be an extension of finite degree of K and Ω an algebraic closure of K. Recall (V, p. 31) that by the *separable degree* of E over K, written $[E:K]_s$, we understand the number of K-homomorphisms of E into Ω.

PROPOSITION 15. — *Let* E_s *be the relative separable closure of* K *in* E ; *then* $[E:K]_s = [E_s:K]$.

The field Ω is perfect and E is p-radical over E_s, by V, p. 44, Prop. 13 ; therefore Prop. 3 (V, p. 26) shows that every K-homomorphism of E_s into Ω extends in a unique fashion to a K-homomorphism of E into Ω ; we thus have $[E:K]_s = [E_s:K]_s$. Since E_s is a separable extension of finite degree of K, it is an etale algebra over K ; so we have $[E_s:K]_s = [E_s:K]$ by V, p. 32, Prop. 4, and the result follows.

With the preceding notation, the degree of E over E_s is called the *inseparable degree* of E over K and is denoted by $[E:K]_i$. We thus have

$$(1) \qquad [E:K] = [E:K]_s \,.\, [E:K]_i$$

by Prop. 15.

When K is of characteristic 0, then $E = E_s$, and so $[E:K]_s = [E:K]$ and $[E:K]_i = 1$. If K is of characteristic $p \neq 0$, the number $[E:K]_i$ is a power of p because E is p-radical over E_s (V, p. 44, Prop. 13 and p. 26, Prop. 4). It should be noted that $[E:K]_i$ is not necessarily equal to the highest power of p dividing $[E:K]$, nor equal to the degree $[E_r:K]$ of the relative p-radical closure of E in K (V, p. 152, Ex. 3 and 2).

PROPOSITION 16. — *Let* Ω *be an extension of* K *and* E, F *two subextensions of* Ω, *of finite degree over* K.

a) *If* $E \subset F$, *then* $[F:K]_s = [F:E]_s \,.\, [E:K]_s$ *and* $[F:K]_i = [F:E]_i \,.\, [E:K]_i$.

b) *Let* K′ *be a subextension of* Ω ; *then we have*

$$[K'(E):K']_s \leqslant [E:K]_s \quad \textit{and} \quad [K'(E):K']_i \leqslant [E:K]_i \,,$$

and equality holds if K′ *is linearly disjoint from* E *over* K.

c) *We have* $[K(E \cup F):K]_s \leqslant [E:K]_s \,.\, [F:K]_s$ *and* $[K(E \cup F):K]_i \leqslant [E:K]_i \,.\, [F:K]_i$, *and equality holds if* E *and* F *are linearly disjoint over* K.

The assertion about the separable degrees in *a*) follows from (9) (V, p. 32). Since $[F:K] = [F:E] \,.\, [E:K]$, the assertion about the inseparable degrees follows from this and (1).

By Cor. 1 of Prop. 13 (V, p. 44) and Prop. 15, we have

(2) $[K'(E):K']_s = [K'(E_s):K']$, $[K'(E):K']_i = [F'(E):K'(E_s)]$;

when K′ is linearly disjoint from E over K, then E_s is linearly disjoint from K′ over K and E is linearly disjoint from $K'(E_s)$ over E_s (V, p. 15, Prop. 8). Assertion *b*) now follows from Prop. 5 (V, p. 14).

By *a*) we have $[K(E \cup F):K]_s = [F(E):F]_s \cdot [F:K]_s$; by *b*) we have $[F(E):F]_s \leqslant [E:K]_s$ with equality if E and F are linearly disjoint over K. Hence we obtain the inequality $[K(E \cup F):K]_s \leqslant [E:K]_s \cdot [F:K]_s$ with equality if E and F are linearly disjoint over K. The assertion of *c*) about the inseparable degrees is proved in a similar fashion.

§ 8. NORMS AND TRACES

Throughout this paragraph K *denotes a field.*

1. Recall

Let A be an algebra of finite degree n over K. For each $x \in A$ we denote by L_x the linear mapping $a \mapsto xa$ of A into itself. We recall (III, p. 543) that the trace of L_x is called the *trace* of x relative to A and is written $\mathrm{Tr}_{A/K}(x)$; likewise the determinant of L_x is called the *norm* of x relative to A and is denoted by $N_{A/K}(x)$. The *discriminant* of a sequence $(x_1, \dots, x_n)$ of n elements of A is by definition the determinant $D_{A/K}(x_1, \dots, x_n)$ of the matrix $(\mathrm{Tr}_{A/K}(x_i x_j))_{1 \leqslant i,j \leqslant n}$ (III, p. 549).

Let K′ be an extension of K and let $A' = A_{(K')}$ be the K′-algebra derived from A by extension of scalars. We have the formulae

(1) $\mathrm{Tr}_{A'/K'}(1 \otimes x) = \mathrm{Tr}_{A/K}(x) \cdot 1$, $N_{A'/K'}(1 \otimes x) = N_{A/K}(x) \cdot 1$

for all $x \in A$ (III, p. 544). For every sequence $(x_1, \dots, x_n)$ of elements of A we have

(2) $$D_{A'/K'}(1 \otimes x_1, \dots, 1 \otimes x_n) = D_{A/K}(x_1, \dots, x_n) \cdot 1,$$

as follows from the first formula (1).

2. Norms and traces in etale algebras

Let A be an etale algebra of (finite) degree n over K. By definition there exist then an extension L of K and distinct homomorphisms $u_1, \dots, u_n$ of A into L with the following properties.

a) every homomorphism of A into L is equal to one of the u_i (V, p. 29, Cor.) ;
b) there exists an L-algebra isomorphism $u : A_{(L)} \to L^n$ such that

$$u(1 \otimes x) = (u_1(x), \ldots, u_n(x)) \quad \text{for all} \quad x \in A .$$

Moreover, every algebraically closed extension L of K has the above properties (V, p. 30, Prop. 2).

We fix L, $u_1, \ldots, u_n$ in what follows. Let $x \in A$; we shall prove the formulae

$$\text{(3)} \qquad \mathrm{Tr}_{A/K}(x) \,.\, 1 = \sum_{i=1}^{n} u_i(x) \,, \quad \mathrm{N}_{A/K}(x) \,.\, 1 = \prod_{i=1}^{n} u_i(x) \,.$$

Let v be multiplication by $1 \otimes x$ in $A_{(L)}$; with respect to the basis of $A_{(L)}$ which is the image under u^{-1} of the canonical basis of L^n, the matrix of the linear mapping v is diagonal, with diagonal elements $u_1(x), \ldots, u_n(x)$. We conclude that

$$\mathrm{Tr}_{A_{(L)}/L}(1 \otimes x) \,.\, 1 = \sum_{i=1}^{n} u_i(x) \,, \quad \text{whence} \quad \mathrm{Tr}_{A/K}(x) \,.\, 1 = \sum_{i=1}^{n} u_i(x) \text{ by (1) ;}$$

the case of norms is treated similarly.

Further let $(x_1, \ldots, x_n)$ be a sequence of elements of A, let U be the matrix

$$(u_i(x_j))_{1 \leqslant i,j \leqslant n}$$

and let $(t_{ij}) = {}^t\mathrm{U} \,.\, \mathrm{U}$. By the first formula (3) we have

$$\mathrm{Tr}_{A/K}(x_i x_j) \,.\, 1 = \sum_{k=1}^{n} u_k(x_i x_j) = \sum_{k=1}^{n} u_k(x_i)\, u_k(x_j) = t_{ij} \,;$$

passing to determinants, we obtain

$$\text{(4)} \qquad \mathrm{D}_{A/K}(x_1, \ldots, x_n) \,.\, 1 = [\det u_i(x_j)]^2 \,.$$

PROPOSITION 1. — *Let* A *be a commutative algebra of finite degree over* K. *Then the following conditions are equivalent :*

a) *The algebra* A *is etale.*

b) *There exists a basis of* A *whose discriminant is non-zero.*

c) *For each* $x \neq 0$ *in* A *there exists* y *in* A *such that* $\mathrm{Tr}_{A/K}(xy) \neq 0$.

Further, when these conditions are satisfied, the discriminant of any basis of A *is non-zero.*

We shall show that when A is assumed etale, the discriminant of A with respect to any basis $(x_1, \ldots, x_n)$ of A over K is non-zero ; this will in particular establish the implication *a*) ⇒ *b*). By (4), with the above notation, it suffices to show that the matrix U is invertible, or equivalently, that the system of linear equations

$$\text{(5)} \qquad \sum_{i=1}^{n} \lambda_i u_i(x_j) = 0 \quad (\text{for } 1 \leqslant j \leqslant n)$$

has only the solution $\lambda_1 = \cdots = \lambda_n = 0$ in L. Now the relation (5) implies $\sum_{i=1}^{n} \lambda_i u_i(x) = 0$ for all $x \in \mathrm{A}$, whence $\lambda_i = 0$ for $1 \leqslant i \leqslant n$, by the theorem on the linear independence of homomorphisms (V, p. 27, Th. 1).

The equivalence of *b*) and *c*) is a consequence of the following general lemma :

Lemma 1. — *Let* V *be a vector space of finite dimension over* K *and* B *a bilinear form on* V × V. *Let* $(v_1, \ldots, v_n)$ *be a basis of* V *over* K *and* $\Delta = \det B(v_i, v_j)$. *Then* $\Delta \neq 0$ *if and only if, for each* $x \neq 0$ *in* V *there exists* y *in* V *such that* $\mathrm{B}(x, y) \neq 0$.

We have $\Delta \neq 0$ if and only if the system of linear equations

$$\sum_{i=1}^{n} \lambda_i \mathrm{B}(v_i, v_j) = 0 \quad (1 \leqslant j \leqslant n)$$

has only the solution $\lambda_1 = \cdots = \lambda_n = 0$ in K. If we put $x = \sum_{i=1}^{n} \lambda_i v_i$, the above system is equivalent to $\mathrm{B}(x, v_j) = 0$ for $1 \leqslant j \leqslant n$, or also, since $(v_1, \ldots, v_n)$ is a basis of V over K, to $\mathrm{B}(x, y) = 0$ for all $y \in \mathrm{V}$, whence the lemma.

Let us show that condition *c*) implies that A is reduced. Let x be a nilpotent element of A ; for any $y \in \mathrm{A}$ the element xy is nilpotent, and so the endomorphism L_{xy} of the vector space A is nilpotent. Now the following lemma implies that $\mathrm{Tr}(xy) = 0$ for all $y \in \mathrm{A}$, whence $x = 0$ under hypothesis *c*).

Lemma 2. — *Let* V *be a vector space of finite dimension over* K *and* u *a nilpotent endomorphism of* V, *then* $\mathrm{Tr}(u) = 0$.

For each integer $n \geqslant 0$ let V_n be the image of u^n. Since u is nilpotent, there exists an integer $r \geqslant 0$ such that $\mathrm{V}_0 = \mathrm{V}$, $\mathrm{V}_r = 0$ and $\mathrm{V}_i \neq \mathrm{V}_{i+1}$ for $0 \leqslant i \leqslant r - 1$. Let d_i be the dimension of V_{i-1} (for $1 \leqslant i \leqslant r$). There exists a basis $(x_1, \ldots, x_d)$ of V such that the vectors x_j with $d - d_i < j \leqslant d$ form a basis of V_{i-1} (for $1 \leqslant i \leqslant r$). We have $u(\mathrm{V}_{i-1}) \subset \mathrm{V}_i$ and so the diagonal elements of the matrix of u for the basis $(x_1, \ldots, x_n)$ are zero. Thus we have $\mathrm{Tr}(u) = 0$ and the lemma follows.

Finally let us show that *b*) implies *a*). Let $(x_1, \ldots, x_n)$ be a basis of A over K such that $\mathrm{D}_{\mathrm{A}/\mathrm{K}}(x_1, \ldots, x_n) \neq 0$. Let K′ be an extension of K, A′ the K′-algebra derived from A by extension of scalars and $x'_i = 1 \otimes x_i$ for $1 \leqslant i \leqslant n$. By Formula (2) (V, p. 47) we have $\mathrm{D}_{\mathrm{A}'/\mathrm{K}'}(x'_1, \ldots, x'_n) \neq 0$. Applying the preceding result to A′ we see that A′ is reduced, hence the algebra A is etale (V, p. 34, Th. 4).

COROLLARY. — *Let* E *be an extension of finite degree of* K. *For* E *to be separable it is necessary and sufficient that there exist* a *in* E *such that* $\mathrm{Tr}_{\mathrm{E}/\mathrm{K}}(a) \neq 0$.

The condition is necessary by Prop. 1. Conversely, assume that there exists $a \in \mathrm{E}$ such that $\mathrm{Tr}_{\mathrm{E}/\mathrm{K}}(a) \neq 0$. Given $x \neq 0$ in E, if we put $y = ax^{-1}$, then $\mathrm{Tr}_{\mathrm{E}/\mathrm{K}}(xy) \neq 0$. Now Prop. 1 shows that E is an etale algebra over K, hence a separable extension of K.

3. Norms and traces in extensions of finite degree

The transitivity formulae in algebras (III, p. 548) imply the following proposition in the case of extensions of finite degree.

PROPOSITION 2. — *Let* F *be an extension of finite degree of* K *and* E *a subextension of* F. *Then for every* $x \in F$ *we have*

$$\mathrm{Tr}_{F/K}(x) = \mathrm{Tr}_{E/K}(\mathrm{Tr}_{F/E}(x)) \tag{6}$$

$$\mathrm{N}_{F/K}(x) = \mathrm{N}_{E/K}(\mathrm{N}_{F/E}(x))\,. \tag{7}$$

COROLLARY. — *Put* $m = [F : E]$; *then for every* $x \in E$ *we have*

$$\mathrm{Tr}_{F/K}(x) = m\,.\,\mathrm{Tr}_{E/K}(x) \tag{8}$$

$$\mathrm{N}_{F/K}(x) = \mathrm{N}_{E/K}(x)^m\,. \tag{9}$$

PROPOSITION 3. — *Let* E *be an extension of finite degree n of* K *and x an element of* E, *of degree d over* K. *Write* $f(X) = X^d + \sum_{i=1}^{d} a_i X^{d-i}$ *for the minimal polynomial of* X *over* K. *Then we have*

$$\mathrm{Tr}_{E/K}(x) = -\frac{n}{d}a_1 \tag{10}$$

$$\mathrm{N}_{E/K}(x) = ((-1)^d a_d)^{n/d} = (-1)^n a_d^{n/d}\,. \tag{11}$$

Prop. 3 follows directly from the Cor. of Prop. 2 and the lemma :

Lemma 3. — *Let* R *be a commutative ring,* $f(X) = X^d + \sum_{i=1}^{d} a_i X^{d-i}$ *a monic polynomial of* R[X], A *the* R*-algebra* R[X]/(*f*) *and x the residue class of* X *in* A. *Then* $\mathrm{Tr}_{A/R}(x) = -a_1$ *and* $\mathrm{N}_{A/R}(x) = (-1)^d a_d$.

By the Cor. (IV, p. 11) the sequence $(1, x, \ldots, x^{d-1})$ is a basis of A ; further we have

$$x\,.\,1 = x\,,\quad x\,.\,x = x^2, \ldots, x\,.\,x^{d-2} = x^{d-1}\,,$$
$$x\,.\,x^{d-1} = -a_d\,.\,1 - a_{d-1}\,.\,x - \cdots - a_1\,.\,x^{d-1}\,.$$

The matrix which expresses the multiplication by x with respect to the basis $(1, x, \ldots, x^{d-1})$ of A is thus of the following form (we have taken $d = 5$ to fix the ideas) :

$$\begin{pmatrix} 0 & 0 & 0 & 0 & -a_5 \\ 1 & 0 & 0 & 0 & -a_4 \\ 0 & 1 & 0 & 0 & -a_3 \\ 0 & 0 & 1 & 0 & -a_2 \\ 0 & 0 & 0 & 1 & -a_1 \end{pmatrix}.$$

The trace of this matrix is clearly $-a_1$; the determinant may be calculated by expanding by the first row, and we then find

$$(-1)^{d-1}(-a_d) = (-1)^d a_d .$$

For the rest of this No. we denote by E an extension of finite degree of K and by x an element of E. We shall indicate how to calculate the norm and trace of x in various cases.

a) Case of a separable extension : suppose that E is separable of degree n over K, denote by Ω an algebraic closure of K and by $\sigma_1, \ldots, \sigma_n$ the n distinct K-homomorphisms of E into Ω. By Formula (3) (V, p. 48) we have in Ω

$$\mathrm{Tr}_{E/K}(x) = \sum_{i=1}^{n} \sigma_i(x), \quad N_{E/K}(x) = \prod_{i=1}^{n} \sigma_i(x). \tag{12}$$

b) Case of a p-radical extension : suppose that K is of characteristic $p > 0$ and that the extension E is p-radical ; there exists an integer $e \geqslant 0$ such that $[E:K] = p^e$ (V, p. 26, Prop. 4). If f is the height of x over K, the minimal polynomial of x over K is $X^{p^f} - x^{p^f}$ (V, p. 24, Prop. 1). By Prop. 3 we have $N_{E/K}(x) = (x^{p^f})^{p^e/p^f}$, whence

$$N_{E/K}(x) = x^{p^e} = x^{[E:K]}. \tag{13}$$

For the trace we find $\mathrm{Tr}_{E/K}(x) = -p^{e-f}a$, where a is the coefficient of X^{p^f-1} in the polynomial $X^{p^f} - x^{p^f}$; in other words, we have

$$\mathrm{Tr}_{E/K}(x) = p^e . x = [E:K]\,x = \begin{cases} x & \text{if } [E:K] = 1 \\ 0 & \text{if } [E:K] > 1 . \end{cases} \tag{14}$$

c) General case : we can summarize the calculation of norm and trace in the following proposition :

PROPOSITION 4. — *Let p be the characteristic exponent of* K *and* E *an extension of finite degree of* K. *Let* $\sigma_1, \ldots, \sigma_n$ *be the distinct* K*-homomorphisms of* E *into an algebraic closure* Ω *of* K, *and let* $p^e = [E:K]_i$. *For each* $x \in E$ *we have in* Ω

$$\mathrm{Tr}_{E/K}(x) = p^e . \sum_{i=1}^{n} \sigma_i(x), \quad N_{E/K}(x) = \left(\prod_{i=1}^{n} \sigma_i(x)\right)^{p^e}. \tag{15}$$

Let E_s be the relative separable closure of K in E ; then E_s is a separable extension of degree n of K and $\sigma_1, \ldots, \sigma_n$ induce distinct K-homomorphisms of E_s into Ω ; further, E is a p-radical extension of E_s of degree p^e (V, p. 44, Prop. 13 and p. 46). So Prop. 4 follows from the Formulae (6), (7), (13), (14) and (12).

§ 9. CONJUGATE ELEMENTS AND QUASI-GALOIS EXTENSIONS

Throughout this paragraph K *denotes a field and* Ω *an algebraic closure of* K.

1. Extension of isomorphisms

PROPOSITION 1. — *Let* E *be an extension of* K *contained in* Ω *and* u *a* K-*homomorphism of* E *into* Ω.

a) If u *maps* E *into* E, u *induces a* K-*automorphism of* E.

b) There exists a K-*automorphism* v *of* Ω *extending* u.

Suppose that $u(\mathrm{E}) \subset \mathrm{E}$; to prove *a)* it is enough to show that $u(\mathrm{E}) = \mathrm{E}$. Let x be an element of E, f the minimal polynomial of x over K and Φ the set of roots of f *in* E. The set Φ is finite, the mapping u of E into E is injective and we have $u(\Phi) \subset \Phi$; therefore $u(\Phi) = \Phi$, whence $x \in u(\Phi) \subset u(\mathrm{E})$; this shows that $\mathrm{E} = u(\mathrm{E})$.

It is clear that Ω is an algebraic closure of E and of $u(\mathrm{E})$; therefore (V, p. 23, Cor.), the isomorphism u of E onto $u(\mathrm{E})$ extends to an isomorphism v of Ω onto Ω.

2. Conjugate extensions. Conjugate elements

DEFINITION 1. — *Let* E *and* F *be two extensions of* K *contained in* Ω. *We shall say that* E *and* F *are conjugate (in* Ω*) if there exists a* K-*automorphism* u *of* Ω *such that* $u(\mathrm{E}) = \mathrm{F}$. *Two elements* x *and* y *of* Ω *are said to be conjugate over* K *if there exists a* K-*automorphism* u *of* Ω *such that* $u(x) = y$.

Let E and F be two extensions of K contained in Ω. By Prop. 1, E and F are conjugate over K if and only if they are isomorphic extensions of K. This is so in particular when there exist two subsets A and B of Ω such that $\mathrm{E} = \mathrm{K}(\mathrm{A})$ and $\mathrm{F} = \mathrm{K}(\mathrm{B})$ and a K-automorphism u of Ω such that $u(\mathrm{A}) = \mathrm{B}$.

The relation « x and y are conjugate over K » is an equivalence relation in Ω ; the equivalence classes are called the *conjugacy classes* in Ω ; they are the orbits in Ω of the group of all K-automorphisms of Ω.

PROPOSITION 2. — *Let* x *and* y *be two elements of* Ω. *Then the following conditions are equivalent :*

a) x *and* y *are conjugate over* K.

b) There exists a K-*isomorphism* v *of* $\mathrm{K}(x)$ *onto* $\mathrm{K}(y)$ *such that* $v(x) = y$.

c) x *and* y *have the same minimal polynomial over* K.

Suppose first that x and y are conjugate over K ; let u be a K-automorphism of Ω such that $u(x) = y$ and let f be the minimal polynomial of x over K. We have

$$f(y) = f(u(x)) = u(f(x)) = 0 ,$$

and f is a monic irreducible polynomial in $K[X]$; therefore (V, p. 16, Th. 1, c)), f is the minimal polynomial of y over K. Thus a) implies c).

Suppose now that x and y have the same minimal polynomial f over K. There exists a K-isomorphism of the field $K[X]/(f)$ onto $K(x)$ (resp. onto $K(y)$) mapping the residue class of X modulo (f) to x (resp. y) (V, p. 16, Th. 1, b)) ; hence there exists a K-isomorphism v of $K(x)$ onto $K(y)$ such that $v(x) = y$. Therefore c) implies b).

Finally, under the hypothesis b), Prop. 1 implies the existence of a K-automorphism u of Ω extending v ; we then have $u(x) = y$, hence x and y are conjugate over K, so b) implies a).

COROLLARY 1. — *Let x be an element of Ω of degree n over* K, *and let f be the minimal polynomial of x over* K. *The conjugates of x over* K *are the roots of f in Ω and their number is at most equal to n.*

COROLLARY 2. — *Let x be an element of Ω of degree n over* K. *For x to be separable over* K *it is necessary and sufficient that x should have n conjugates in Ω ; when this is so, all the conjugates of x over* K *are separable over* K.

Let f be the minimal polynomial of x over K ; its roots are the conjugates of x over K, and each of these roots admits f as minimal polynomial over K. Now x is separable over K if and only if the polynomial f is separable (V, p. 39, Prop. 5), that is, if f has n distinct roots in Ω. Cor. 2 follows from this.

COROLLARY 3. — *Let* G *be the group of* K*-automorphisms of Ω. The set of invariants of* G *in Ω is the relative p-radical closure of* K *in Ω* (V, p. 25).

In other words, an element x of Ω is p-radical over K if and only if it has no conjugate other than itself.

Let p be the characteristic exponent and let $x \in \Omega$. By V, p. 44, Prop. 13, there exists an integer $m \geqslant 0$ such that $y = x^{p^m}$ is algebraic and separable over K. Now x is invariant under G if and only if y is invariant under G ; by Cor. 2 this is the case if and only if y is of degree 1 over K, which amounts to saying that $y \in K$. Now the corollary is an immediate consequence.

3. Quasi-Galois extensions

DEFINITION 2. — *Let* E *be an extension of* K. *Then* E *is said to be quasi-Galois or normal (over* K*), if it is algebraic and if every irreducible polynomial of* $K[X]$ *which has at least one root in* E, *splits into a product of polynomials of degree* 1 *(distinct or not) in* $E[X]$.

If E is an algebraic closure of K, it is a quasi-Galois extension of K ; for the condition (AC) of Prop. 1 (V, p. 19) asserts that every non-constant polynomial in $E[X]$ is a product of polynomials of degree 1.

PROPOSITION 3. — *Let* E *be an extension of* K *contained in* Ω. *Then the following conditions are equivalent :*

a) E *is a quasi-Galois extension of* K.

b) For each $x \in E$ *the conjugates of* x *over* K *in* Ω *all belong to* E.

c) Every K*-automorphism of* Ω *maps the field* E *into itself.*

d) Every K*-homomorphism of* E *into* Ω *maps* E *into itself.*

e) E *is the splitting extension in* Ω *of a family* $(f_i)_{i \in I}$ *of non-constant polynomials in* K[X] (V, p. 24, Remark 3).

The equivalence of *c*) and *d*) stems from the fact that every K-homomorphism of E into Ω is induced by a K-automorphism of Ω (V, p. 52, Prop. 1).

By definition a quasi-Galois extension is the splitting field of the family of minimal polynomials (over K) of its elements, so *a*) implies *e*). Under the hypothesis *e*) let u be an automorphism of Ω ; for each $i \in I$, u permutes the set R_i of roots of f_i and since $E = K\left(\bigcup_{i \in I} R_i\right)$, we have $u(E) = E$; thus *e*) implies *c*).

The definition of conjugate elements shows that *c*) implies *b*). Finally assume that *b*) holds ; let f be a monic irreducible polynomial in K[X] having at least one root x in E ; since Ω is algebraically closed, there exist elements a_k of Ω $(1 \leq k \leq n)$ such that $f(x) = \prod_{k=1}^{n} (X - a_k)$ and since the a_k are the conjugates of x over K (V, p. 53, Cor. 1), they belong to E by hypothesis. So we have shown that *b*) implies *a*).

COROLLARY 1. — *For an extension* E *of* K *contained in* Ω *to be quasi-Galois it is necessary and sufficient that it should be identical to all its conjugates over* K.

This follows from Prop. 1, *a*) (V, p. 52) and the equivalence of the conditions *a*) and *c*) of Prop. 3.

COROLLARY 2. — *Let* E *and* F *be two algebraic extensions of* K *such that* $E \subset F$. *If* F *is quasi-Galois over* K, *then it is quasi-Galois over* E.

We may assume that $F \subset \Omega$. Let u be an E-automorphism of Ω. Since u is a K-automorphism of Ω and F is quasi-Galois over K, we have $u(F) = F$, therefore F is quasi-Galois over E.

COROLLARY 3. — *Let* N *be a quasi-Galois extension of* K *contained in* Ω, *and* E *a subextension of* N. *Let* u *be a* K*-homomorphism of* E *into* Ω ; *then* $u(E) \subset N$ *and there exists a* K*-automorphism* v *of* N *which induces* u *on* E.

Let w be a K-automorphism of Ω extending u (V, p. 52, Prop. 1) ; since N is quasi-Galois over K, we have $w(N) = N$, whence $w(E) \subset N$ and so w induces a K-automorphism v of N.

COROLLARY 4. — *Let* E′ *be an extension of* K *and* E, K′ *two subextensions of* E′. *Suppose that* E *is quasi-Galois over* K *and* $E' = K'(E)$; *then* E′ *is quasi-Galois over* K′.

Let $(f_i)_{i \in I}$ be a family of non-constant polynomials in K[X] whose splitting field over K is E. Then it is clear that E′ is the splitting field of the family $(f_i)_{i \in I}$ over K′, so it is quasi-Galois over K′.

Remarks. — 1) Let F be an extension of K and E a subextension of F, and suppose that E is quasi-Galois over K. Let us show that every K-automorphism u of F leaves E invariant. For let $x \in E$ and f the minimal polynomial of x over K. Since E is quasi-Galois over K, there exist $a_1, \ldots, a_n \in E$ such that $f(X) = \Pi(X - a_i)$; we have $f(u(x)) = u(f(x)) = 0$, hence $u(x)$ is one of the a_i and so belongs to E. We have proved $u(E) \subset E$, and now $u(E) = E$ follows by V, p. 52, Prop. 1.

2) Suppose that E is a quasi-Galois extension of K and F a quasi-Galois extension of E ; it is not always the case that F is quasi-Galois over K (V, p. 153, Ex. 1). The reason is the following : let u be a K-automorphism of Ω ; we have $u(E) = E$, but if f is the minimal polynomial over E of an element x of F, it is not necessarily invariant under u ; therefore $u(x)$ is not necessarily conjugate to x over E, and so need not necessarily belong to F. In that case F and $u(F)$ are two *distinct* quasi-Galois extensions of E which are K-*isomorphic* but not E-*isomorphic.*

3) Let E be an algebraic extension of K and x, y two elements of E. If there exists a K-automorphism of E which maps x to y then x and y have the same minimal polynomial over K and so are conjugate over K by Prop. 2 (V, p. 52). The converse holds if E is quasi-Galois, because every K-automorphism of Ω induces a K-automorphism of E. The hypothesis that E be quasi-Galois is not superfluous as the following example shows : * let $K = \mathbf{Q}$ and Ω the field of all algebraic numbers ; put $E = \Omega \cap \mathbf{R}$. It may be shown (V, p. 153, Ex. 2) that every automorphism of E is the identity mapping and that $\sqrt{2}$ and $-\sqrt{2}$ are elements of E conjugate over $\mathbf{Q}$. *

4. The quasi-Galois extension generated by a set

PROPOSITION 4. — *Let $(N_i)_{i \in I}$ be a family of quasi-Galois extensions of* K *contained in* Ω. *Let* $N = \bigcap_{i \in I} N_i$ *and* $M = K\left(\bigcup_{i \in I} N_i\right)$; *then the extensions* N *and* M *of* K *are quasi-Galois.*

Let u be a K-automorphism of Ω. We have $u(N_i) = N_i$ for all $i \in I$, hence we obtain the equalities $u(N) = N$ and $u(M) = M$; now the proposition follows from Cor. 1 (V, p. 54).

Let A be a set of elements of Ω, and let B be the set of elements of Ω which are conjugates of elements of A ; expressed differently, if G is the group of K-automorphisms of Ω, we have $B = \bigcup_{u \in G} u(A)$. For each $u \in G$ we have $u(B) = B$ whence $u(K(B)) = K(B)$. Therefore K(B) is a quasi-Galois extension of K containing A and it is immediate that every quasi-Galois extension N of K containing A contains B, hence K(B). We shall say that K(B) is the *quasi-Galois extension generated by* A. This definition applies in particular when A is an extension of K.

The next proposition follows directly from the preceding remarks.

PROPOSITION 5. — *Let* E *be an extension of* K *contained in* Ω *and* N *the quasi-Galois extension generated by* E. *If* A *is a subset of* Ω *such that* $E = K(A)$ *then* $N = K(B)$, *where* B *is the set of elements of* Ω *which are conjugate to an element of* A.

COROLLARY 1. — *If* E *is an extension of finite degree of* K, *then the quasi-Galois extension* N *of* K *generated by* E *is of finite degree over* K.

For we have $E = K(A)$ where A is finite, hence the set B of conjugates of elements of A is finite, whence the corollary follows by Th. 2 (V, p. 18).

COROLLARY 2. — *Every quasi-Galois extension* N *of* K *is the union of quasi-Galois subextensions of* N *of finite degree over* K.

Let $x \in N$ and let N_x be the quasi-Galois extension of K generated by $\{x\}$. Since $K(x)$ is of finite degree over K, the same is true of N_x (Cor. 1) and we have $x \in N_x$.

§ 10. GALOIS EXTENSIONS

Throughout this paragraph K *denotes a field.*

1. Definition of Galois extensions

THEOREM 1. — *Let* N *be an algebraic extension of* K *and* Γ *the group of* K-*automorphisms of* N. *Then the following assertions are equivalent* :

a) Every element of N *invariant under* Γ *belongs to the image of* K *in* N.

b) N *is a separable quasi-Galois extension of* K.

c) For each $x \in N$ *the minimal polynomial of* x *over* K *splits in* N[X] *into a product of distinct polynomials of degree* 1.

The equivalence of *b*) and *c*) follows from the Cor. of Prop. 6 (V, p. 40) and the definition of quasi-Galois extension (V, p. 53, Def. 2). Let us identify K with its canonical image in N.

a) $\Rightarrow$ *c*) : Suppose that K is the field of invariants of Γ. Let $x \in N$, with minimal polynomial f over K and let A be the set of all roots of f in N. Put

$$g(X) = \prod_{y \in A} (X - y).$$

Every automorphism $\sigma \in \Gamma$ induces a permutation of A, and hence leaves invariant the coefficients of the polynomial $g \in N[X]$. We thus have $g \in K[X]$ and since $g(x) = 0$, the polynomial g is a multiple of f in K[X] (V, p. 16, Th. 1). Further, f and g are monic and g divides f (IV, p. 16, Prop. 5) ; thus we have $f = g$, which means that the minimal polynomial f of x over K is a product in N[X] of distinct polynomials of degree 1.

c) $\Rightarrow$ *a)* : let x be an element of N not belonging to K. Denote by Ω an algebraic closure of K containing N as subextension (V, p. 23, Th. 2). Let f be the minimal polynomial of x over K, which is of degree $\geqslant 2$ by hypothesis and let A be the set of roots of $f(X)$ *in* N. If condition *c)* holds, we have $f(X) = \prod_{y \in A} (X - y)$ and so (V, p. 53, Cor. 1) A is the set of conjugates of x in Ω. Since f has degree $\geqslant 2$, there exists in A an element $y \neq x$, hence a K-automorphism u of Ω such that $u(x) = y$. Now under the hypothesis *c)*, the extension N of K is quasi-Galois, hence $u(N) = N$ (V, p. 54, Cor. 1) ; it follows that u induces a K-automorphism σ of N such that $\sigma(x) = y \neq x$, so K is the field of invariants of Γ.

DEFINITION 1. — *An extension* N *of* K *is said to be Galois if it is algebraic and satisfies the equivalent conditions a), b), c) of Th.* 1.

Let N be a field, Γ a group of automorphisms of N and N_0 the field of invariants of Γ. When N is *algebraic* over N_0 it is a Galois extension of N_0. This need not always be so : for example, suppose that K is infinite and take N to be the field of rational fractions $K(X)$; for each $a \in K$ let σ_a be the automorphism of $K(X)$ which maps $f(X)$ to $f(X + a)$. The set of all σ_a is a group of automorphisms of $K(X)$ whose field of invariants is easily seen to be K ; however $K(X)$ is not algebraic over K.

Let Ω be an algebraic closure of K, let A denote a set of elements of Ω separable over K and B the set of conjugates over K of elements of A. Then B consists of elements that are algebraic and separable over K. Therefore (V, p. 39, Prop. 6 and p. 56, Prop. 5) the field $K(B)$ is a separable quasi-Galois extension of K ; in other words, the quasi-Galois extension generated by A (V, p. 55) is a Galois extension of K ; we shall also say that it is *the Galois extension of* K *generated by the subset* A of Ω.

In particular, the splitting field in Ω of a family of separable polynomials over K, a separable closure of K, are Galois extensions of K.

PROPOSITION 1. — *Let* N *be an extension of* K *and* $(N_i)_{i \in I}$ *a non-empty family of subextensions of* N. *We put* $E = \bigcap_{i \in I} N_i$ *and* $F = K\left(\bigcup_{i \in I} N_i\right)$. *If all the extensions* N_i *are Galois over* K *then the same is true of* E *and* F.

In the first place E is algebraic and separable over K (V, p. 36, Prop. 1) and the same holds for F (V, p. 41, Prop. 8). Moreover, E and F are quasi-Galois over K by Prop. 4 (V, p. 55).

PROPOSITION 2. — *Let* N *be a Galois extension of* K *and* E *a subextension of* N, *of finite degree over* K. *There exists a subextension* F *of* N *containing* E, *Galois and of finite degree over* K.

Since N is quasi-Galois over K, Cor. 1 of V, p. 56 proves the existence of a quasi-Galois subextension F of N containing E and of finite degree over K. Since N is separable over K, the same holds for F (V, p. 36, Prop. 1), hence F is Galois over K.

Prop. 2 implies the following result : let Ω be an algebraic closure of K and $E_1, \ldots, E_n$ separable algebraic extensions of finite degree over K, contained in Ω. Then there exists a Galois extension N of K, of finite degree, contained in Ω and containing $E_1, \ldots, E_n$.

2. The Galois group

DEFINITION 2. — *Let* N *be a Galois extension of the field* K. *The group of all* K-*automorphisms of* N *will be called the Galois group of* N *over* K *and denoted by* Gal(N/K).

Let N be a *finite* Galois extension of K. Then N is a finite separable and quasi-Galois extension of K. Therefore (V, p. 32, Prop. 4 and V, p. 54, Prop. 3), the order of Gal(N/K) is equal to [N : K]. We shall prove later that if N is a Galois extension of K such that Gal(N/K) is finite, then N is of finite degree over K (V, p. 66, Th. 3).

Let Ω be an algebraically closed extension of K and let A be the set of roots in Ω of a separable polynomial $f \in K[X]$. Then the field $N = K(A)$ is a Galois extension of K. It is clear that every K-automorphism of N leaves A stable, and since A generates N over K, the mapping $\sigma \mapsto \sigma | A$ is an isomorphism of Gal(N/K) onto a subgroup Γ of the symmetric group $\mathfrak{S}_A$ of the set A, which will be called the *Galois group* of the polynomial f. From Remark 3 (V, p. 55) it follows that if x and y belong to A, the following properties are equivalent :

a) x and y are conjugate over K,

b) x and y belong to the same orbit under Γ,

c) x and y are roots of the same irreducible factor of f.

In particular f is irreducible if and only if A is non-empty and Γ operates transitively on A.

Examples. — 1) Suppose that the characteristic of K is different from 2 and let N be a quadratic extension of K. If $x \in N - K$, then we have $N = K(x)$ and the minimal polynomial of x over K is of the form $f(X) = X^2 - aX + b$, with $a, b \in K$. We thus have $f(X) = (X - x)(X - y)$ where $y = a - x$, so y is conjugate to x ; since $f(X)$ is separable, the extension N is Galois. The group Gal(N/K) has two elements which induce the two permutations of the set $\{x, y\}$.

* 2) Let $f = X^3 + X^2 - 2X - 1 \in \mathbf{Q}[X]$. The polynomial f is irreducible, for if not, then it would have a root $x \in \mathbf{Q}$; on writing $x = a/b$ with $a, b \in \mathbf{Z}$, a and b relatively prime, we should find $a(a^2 + ab - 2b^2) = b^3$ and $a^3 = b(b^2 + 2ab - a^2)$; but this implies that a divides b and b divides a, hence $x = \pm 1$, which is impossible. Put $\xi = e^{2\pi i/7} \in \mathbf{C}$; then the polynomial f has the roots $\alpha = \xi + \xi^{-1}$, $\beta = \xi^2 + \xi^{-2}$, $\gamma = \xi^3 + \xi^{-3}$. We have $\beta = \alpha^2 - 2$ and $\gamma = \alpha^3 - 3\alpha$, hence the extension $\mathbf{Q}(\alpha)$ is Galois over $\mathbf{Q}$. The Galois group of $\mathbf{Q}(\alpha)$ over $\mathbf{Q}$ is cyclic of order 3 and is generated by an element σ such that $\sigma(\alpha) = \beta$, $\sigma(\beta) = \gamma$, $\sigma(\gamma) = \alpha$. *

3) Suppose that $K = \mathbf{Q}$ and take $f = X^3 - 2$. Using the decomposition of integers into products of prime factors (I, p. 51), we easily see that 2 is not the cube of an element of $\mathbf{Q}$. Hence the polynomial f is irreducible, for if not, it would have a root in $\mathbf{Q}$. Let $A = \{x_1, x_2, x_3\}$ be the set of roots of f in Ω and Γ the Galois group of f. It acts transitively on A ; its order is therefore divisible by three. On the other hand, the quotient $j = \frac{x_2}{x_1}$ is different from 1 and we have $j^3 = 1$. Hence j satisfies the relation $j^2 + j + 1 = 0$; now the polynomial $T^2 + T + 1 = \left(T + \frac{1}{2}\right)^2 + \frac{3}{4}$ has no roots in $\mathbf{Q}$, which shows (Example 1) that $[\mathbf{Q}(j) : \mathbf{Q}] = 2$. Thus $[N : \mathbf{Q}]$ is divisible by 2, and it follows that the order of Γ is divisible by 6. Since Γ is contained in the group $\mathfrak{S}_A$ of order 6, we have $\Gamma = \mathfrak{S}_A$.

4) Suppose that K is of characteristic $p \neq 0$ and let K(T) be the field of rational fractions and $U = T^p - T$. Put $E = K[U]$ and $F = K[T]$, then the polynomial $f(X) = X^p - X - U$ of $E[X]$ has the roots $T, T+1, \dots, T+p-1$ in F. Let σ be the K-automorphism of F such that $\sigma(T) = T + 1$. We have $\sigma^i(T) = T + i$ and $\sigma(U) = U$. The group $G = \{1, \sigma, \dots, \sigma^{p-1}\}$ is cyclic of order p, and its field of invariants contains E ; since $[F : E] \leqslant p$, Dedekind's theorem (V, p. 27, Cor. 2) implies that E is the field of invariants of G and $[F : E] = p$. The polynomial f is thus irreducible in $E[X]$; the extension F of E is Galois, its Galois group G is cyclic of order p, and the group Γ is the group of cyclic permutations of T, $T+1, \dots, T+p-1$.

For a generalization of this example see V, p. 93, Example 2.

5) Let $F = K(X_1, \dots, X_n)$ be the field of rational fractions in n indeterminates $X_1, \dots, X_n$ with coefficients in K. Put

$$s_k = \sum_{1 \leqslant i_1 < \dots < i_k \leqslant n} X_{i_1} \dots X_{i_k}$$

for $1 \leqslant k \leqslant n$ and $E = K(s_1, \dots, s_n)$; we denote by $f(T)$ the polynomial

$$T^n - s_1 T^{n-1} + \dots + (-1)^n s_n .$$

We have $f(T) = \prod_{i=1}^{n} (T - X_i)$, so that F is a splitting field of the separable polynomial $f(T) \in E[T]$. Further, for every permutation $\sigma \in \mathfrak{S}_n$ there exists a unique K-automorphism h_σ of F such that $h_\sigma(X_i) = X_{\sigma(i)}$ for $1 \leqslant i \leqslant n$; we have $h_\sigma(s_k) = s_k$ for $1 \leqslant k \leqslant n$, hence h_σ is an E-automorphism of F. In other words, F is a Galois extension of E and the restriction to the set of roots $\{X_1, \dots, X_n\}$ of $f(T)$ defines an isomorphism of Gal(F/E) onto the group $\mathfrak{S}_n$. In particular, E consists of rational fractions f such that

$$f(X_{\sigma(1)}, \dots, X_{\sigma(n)}) = f(X_1, \dots, X_n)$$

for every $\sigma \in \mathfrak{S}_n$ (cf. IV, p. 67, Cor.).

6) Suppose that f is monic of degree >0 and K is of characteristic $\neq 2$. Define on A a total ordering, denoted by $\leqslant$, and put $\delta(f) = \prod_{\alpha<\beta} (\beta - \alpha)$, $(\alpha, \beta) \in A \times A$, and for each $\sigma \in \mathfrak{S}_A$ put $\delta_\sigma(f) = \prod_{\alpha<\beta} (\sigma(\beta) - \sigma(\alpha))$. We have $\delta_\sigma(f) = \varepsilon(\sigma)\,\delta(f)$, where $\varepsilon(\sigma)$ is the signature of σ (I, p. 64) and $\delta(f) \neq 0$. For every $\tau \in \mathrm{Gal}(N/K)$ we have $\tau(\delta(f)) = \delta_{\tau|A}(f)$. Hence Γ is contained in the alternating group $\mathfrak{A}_A$ if and only if $\delta(f) \in K$. Moreover $\delta(f)^2 = \prod_{\alpha<\beta} (\beta - \alpha)^2 = d(f)$ is the discriminant of the polynomial f (IV, p. 81). Hence $\Gamma \subset \mathfrak{A}_A$ if and only if $d(f)$ is the square of an element of K. Thus in Example 2 we have $d(f) = 49 = 7^2$ and in Example 3, $d(f) = -108$ (IV, p. 85).

Let N be a Galois extension of K and let L be a subextension of N which is Galois over K. Every K-automorphism σ of N induces a K-automorphism σ_L of L (V, p. 55, Remark 1). Therefore the mapping $\sigma \mapsto \sigma_L$ is a homomorphism of Gal(N/K) into Gal(L/K), called the *restriction homomorphism.*

PROPOSITION 3. — *The restriction homomorphism of* Gal(N/K) *into* Gal(L/K) *is surjective.*

More generally, consider two subextensions L and L′ of N, and a K-isomorphism u of L onto L′. Choose an algebraic closure Ω of K containing N as subextension (V, p. 23, Th. 2). There exists a K-automorphism v of Ω which coincides with u on L (V, p. 52, Prop. 1), and since N is a quasi-Galois extension of K, v induces a K-automorphism σ of N (V, p. 55, Remark 1). In other words, the element σ of Gal(N/K) coincides with u on L.

3. Topology of the Galois group

Let N be a Galois extension of K and Γ the Galois group of N over K. We equip N with the discrete topology, the set N^N of all mappings of N into itself with the product topology of the discrete topology of the factors (« topology of simple convergence in N ») and the group Γ with the topology induced from N^N.

Let Λ be the set of all subextensions of N of finite degree over K. For $\sigma \in \Gamma$ and $E \in \Lambda$ we shall write $U_E(\sigma)$ for the set of elements τ of Γ which have the same restriction as σ to E. If $E = K(x_1, \ldots, x_n)$, the set $U_E(\sigma)$ consists of the elements $\tau \in \Gamma$ such that $\tau(x_1) = \sigma(x_1), \ldots, \tau(x_n) = \sigma(x_n)$. It follows that the family $(U_E(\sigma))_{E \in \Lambda}$ is a base of the filter of neighbourhoods of σ in Γ.

When N is of finite degree over K, we have $N \in \Lambda$ and $U_N(\sigma) = \{\sigma\}$, hence the topology of Gal(N/K) is discrete ; we recall (V, p. 58), that the group Gal(N/K) is finite in this case.

This description of the topology of Gal(N/K) shows that the *restriction homomorphism of* Gal(N/K) *onto* Gal(L/K) *is continuous* for every subextension L of N which is Galois over N.

Let A be a subset of Γ. To say that A is *open* means that for each $\sigma \in A$ there exists E in Λ such that the set $U_E(\sigma)$ is contained in A. The *closure* $\bar{A}$ of A consists of all $\sigma \in \Gamma$ such that for any $E \in \Lambda$ there exists $\tau \in A$ having the same restriction to E as σ ; the field of invariants of $\bar{A}$ is the same as that of A.

Let ε be the neutral element of Γ and let Λ' be the set of subextensions of N which are Galois and of finite degree over K. By Prop. 2 (V, p. 57) the set Λ' is cofinal in Λ and the family $(U_E(\varepsilon))_{E \in \Lambda'}$ is thus a base of the filter of neighbourhoods of ε in Γ. Further, for $E \in \Lambda'$ the set $U_E(E)$ is the kernel of the restriction homomorphism of $\mathrm{Gal}(N/K) = \Gamma$ into $\mathrm{Gal}(E/K)$. Since $\mathrm{Gal}(E/K)$ is finite, it follows that $U_E(\varepsilon)$ is an open and closed subgroup, normal and of finite index in Γ.

Clearly we have $U_E(\sigma) = \sigma U_E(\varepsilon) = U_E(\varepsilon)\sigma$ for $\sigma \in \Gamma$ and $E \in \Lambda'$. Since $U_E(\varepsilon)$ is a normal subgroup of Γ for all $E \in \Lambda'$ and the family $(U_E(\varepsilon))_{E \in \Lambda'}$ is a base of neighbourhoods at ε, the topology of Γ is compatible with the group structure (Gen. Top., III, p. 223). In other words, the mapping $(\sigma, \tau) \mapsto \sigma\tau^{-1}$ of $\Gamma \times \Gamma$ into Γ is continuous.

PROPOSITION 4. — *Let* N *be a Galois extension of* K. *Then the Galois group* $\Gamma = \mathrm{Gal}(N/K)$ *is compact and totally disconnected.*

Every element σ of Γ has a base of neighbourhoods consisting of the open and closed sets $U_E(\sigma)$, hence Γ is totally disconnected (Gen. Top., I, p. 111). We have $\{\sigma\} = \bigcap_{E \in \Lambda} U_E(\sigma)$, hence Γ is separated. For each $x \in N$ the set of conjugates $\sigma(x)$ of x, as σ ranges over Γ, is *finite* because x is algebraic over K (V, p. Cor. 1) ; hence all the projections of Γ on the factor spaces of N^N are finite sets, and this shows Γ to be relatively compact in N^N (Gen. Top., I, p. 88). It remains to show that Γ is *closed* in N^N. Now if u is in the closure of Γ in N^N, then for each pair of points (x, y) of N there exists $\sigma \in \Gamma$ with $u(x) = \sigma(x)$, $u(y) = \sigma(y)$, $u(x+y) = \sigma(x+y)$, $u(xy) = \sigma(xy)$, whence $u(x+y) = u(x) + u(y)$ and $u(xy) = u(x)u(y)$. By the same reasoning we have $u(x) = x$ for all $x \in K$, hence u is a K-homomorphism of N into N ; since N is algebraic over K, u is a K-automorphism of N (V, p. 52, Prop. 1), hence $u \in \Gamma$.

Let N be a Galois extension of K and $(N_i)_{i \in I}$ an increasing directed family of subextensions of N. Suppose that N_i is Galois over K for all $i \in I$ and that $N = \bigcup_{i \in I} N_i$. For each $i \in I$ denote by Γ_i the Galois group of N_i over K ; for $i \leqslant j$ in I we have $N_i \subset N_j$ and the restriction homomorphism φ_{ij} of Γ_j into Γ_i is defined. It is continuous and so the family (Γ_i, φ_{ij}) is an inverse system of topological groups. Further, for each $i \in I$ denote by λ_i the restriction homomorphism of $\mathrm{Gal}(N/K)$ into $\mathrm{Gal}(N_i/K) = \Gamma_i$; it is continuous and we have $\lambda_i = \varphi_{ij} \circ \lambda_j$ for $i \leqslant j$, hence the family $(\lambda_i)_{i \in I}$ defines a continuous homomorphism λ of $\mathrm{Gal}(N/K)$ into $\varprojlim \Gamma_i$.

PROPOSITION 5. — *The homomorphism* λ *of* $\mathrm{Gal}(N/K)$ *into* $\varprojlim \mathrm{Gal}(N_i/K)$ *is an isomorphism of topological groups.*

Since $\mathrm{Gal}(N/K)$ is compact, λ continuous and the group $\varprojlim \mathrm{Gal}(N_i/K)$ separated, it suffices to show that λ is bijective (Gen. Top., I, p. 87, Cor. 2). Let $u = (u_i)_{i \in I}$ be an element of $\varprojlim \mathrm{Gal}(N_i/K)$; for each $i \in I$, u_i is a K-automorphism of N_i and u_i is the restriction of u_j to N_i for $i \leqslant j$. Since $N = \bigcup_{i \in I} N_i$ there exists a unique element σ of $\mathrm{Gal}(N/K)$ which coincides with u_i on N_i for all $i \in I$. Thus σ is the unique element of $\mathrm{Gal}(N/K)$ such that $\lambda(\sigma) = u$, hence λ is bijective.

This applies in particular when we take for the family (N_i) the family of all finite Galois subextensions of N ; then each group $\mathrm{Gal}(N_i/K)$ is discrete and finite. The topological group $\mathrm{Gal}(N/K)$ is thus isomorphic to a directed inverse limit of finite groups, equipped with the discrete topology ; this is sometimes called a *profinite* topological group.

4. Galois descent

In this No. we denote by N *a field,* Γ *a group of automorphisms of* N, ε *the neutral element of* Γ *and* K *the field of invariants of* Γ.

Let V be a vector space over N. We recall (II, p. 317) that a K-*structure* on V is a vector sub-K-space V_0 of V such that the K-linear mapping $\varphi : N \otimes_K V_0 \to V$ which maps $\lambda \otimes x$ to λx is bijective. Let V_0 be such a K-structure ; for each $\sigma \in \Gamma$ we put $u_\sigma = \varphi \circ (\sigma \otimes \mathrm{Id}_{V_0}) \circ \varphi^{-1}$; then we have $u_\sigma\left(\sum_{i \in I} \lambda_i e_i\right) = \sum_{i \in I} \sigma(\lambda_i)\, e_i$ for every family of elements λ_i of N and e_i of V_0 whence we obtain the relations

$$u_\sigma(x+y) = u_\sigma(x) + u_\sigma(y) \tag{1}$$

$$u_\sigma(\lambda x) = \sigma(\lambda)\, u_\sigma(x) \tag{2}$$

$$u_\sigma \circ u_\tau = u_{\sigma\tau} \tag{3}$$

$$u_\varepsilon = \mathrm{Id}_V \tag{4}$$

for σ, τ in Γ, x, y in V and λ in N.

PROPOSITION 6. — *a) Let* V *be a vector space over* N *with a* K*-structure. For a vector* $x \in V$ *to be rational over* K *it is necessary and sufficient that* $u_\sigma(x) = x$ *for all* $\sigma \in \Gamma$. *For a vector sub-*N*-space* W *of* V *to be rational over* K *it is necessary and sufficient that* $u_\sigma(W) \subset W$ *for all* $\sigma \in \Gamma$.

b) Let V_1 *and* V_2 *be two vector spaces over* N, *each with a* K*-structure. For a linear mapping* f *of* V_1 *into* V_2 *to be rational over* K *it is necessary and sufficient that* $f(u_\sigma(x)) = u_\sigma(f(x))$ *for all* $\sigma \in \Gamma$ *and all* $x \in V_1$.

It is clear that K is the set of $x \in N$ such that $\sigma(xy) = x\sigma(y)$ for all $\sigma \in \Gamma$ and all $y \in N$. The proposition therefore follows from Th. 1 (II, p. 324).

COROLLARY. — *Let* V_0 *be a vector space over* K *and let* W *be a vector sub-N-space of* $N \otimes_K V_0$. *Suppose that* W *is stable under the mappings* $\sigma \otimes \mathrm{Id}_{V_0}$ *for all* $\sigma \in \Gamma$. *Let* W_0 *be the set of* $x \in V_0$ *such that* $1 \otimes x \in W$; *then* W_0 *is the unique vector sub-K-space of* V_0 *such that* $W = N \otimes_K W_0$.

It suffices to remark that the set of elements of the form $1 \otimes x$ $(x \in V_0)$ is a K-structure on $N \otimes_K V_0$ for which we have $u_\sigma = \sigma \otimes \mathrm{Id}_{V_0}$ for $\sigma \in \Gamma$.

PROPOSITION 7. — *Let* V *be a vector space over* N, $(u_\sigma)_{\sigma \in \Gamma}$ *a family of mappings of* V *into itself satisfying* (1) *to* (4) *and* V_0 *the set of* $x \in V$ *such that* $u_\sigma(x) = x$ *for all* $\sigma \in \Gamma$.

a) V_0 *is a vector sub-K-space of* V *and the K-linear mapping* φ *of* $N \otimes_K V_0$ *into* V *which maps* $\lambda \otimes x$ *to* λx *is injective.*

b) If Γ *is finite, then* φ *is bijective and* V_0 *is a* K*-structure on* V.

It is clear that V_0 is a vector sub-K-space of V.

The formula $u_\sigma \circ \varphi = \varphi \circ (\sigma \otimes \mathrm{Id}_{V_0})$ shows that the kernel W of φ is stable under the mappings $\sigma \otimes \mathrm{Id}_{V_0}$; by the Cor. to Prop. 6 there exists therefore a subspace W_0 of V_0 such that $W = N \otimes_K W_0$. If x belongs to W_0 we then have $x = \varphi(1 \otimes x) = 0$, hence $W_0 = 0$ and so $W = 0$. This proves *a*).

Suppose now that Γ is finite ; we have to show that φ is surjective, or equivalently that V_0 generates the vector N-space V. Thus let f be an N-linear form on V whose restriction to V_0 is zero. Let $x \in V$; for every $\lambda \in N$ the element $y_\lambda = \sum_{\sigma \in \Gamma} u_\sigma(\lambda x)$ of V clearly belongs to V_0, whence $f(y_\lambda) = 0$, that is, $\sum_{\sigma \in \Gamma} f(u_\sigma(x))\, \sigma(\lambda) = 0$. By Dedekind's theorem (V, p. 27, Cor. 2) we thus have $f(u_\sigma(x)) = 0$ for each $\sigma \in \Gamma$; in particular, taking $\sigma = \varepsilon$ we find $f(x) = 0$, which means that $f = 0$. This proves *b*).

Let M be a vector space over N ; for each $\sigma \in \Gamma$ let M^σ be the vector space over N with the same underlying additive group as M, with the external law $(\lambda, x) \mapsto \sigma(\lambda)\, x$. Write $V = \prod_{\sigma \in \Gamma} M^\sigma$; the underlying additive group of V is that of all mappings of Γ into M, with the external law defined by

$$(\lambda \,.\, h)(\sigma) = \sigma(\lambda)\, h(\sigma) \quad (\lambda \in N, h \in V, \sigma \in \Gamma)\,. \tag{5}$$

(The product $\sigma(\lambda)\, h(\sigma)$ is calculated in the vector space M.) Further, we define on $N \otimes_K M$ a vector space structure over N by the formula

$$\lambda\Big(\sum_i \mu_i \otimes x_i\Big) = \sum_i \lambda\mu_i \otimes x_i\,.$$

Finally we denote by ψ the K-linear mapping of $N \otimes_K M$ into V characterized by the relation

$$\psi(\lambda \otimes x)(\sigma) = \sigma(\lambda) \,.\, x \tag{6}$$

for $\lambda \in N$, $x \in M$ and $\sigma \in \Gamma$. It is clear that ψ is N-linear.

PROPOSITION 8. — *The* N-*linear mapping* ψ *of* $N \otimes_K M$ *into* $V = \prod_{\sigma \in \Gamma} M^\sigma$ *is injective, and it is bijective if* Γ *is finite.*

For every $\sigma \in \Gamma$ we define a mapping u_σ of V into V by

$$(u_\sigma h)(\tau) = h(\tau\sigma) \tag{7}$$

for $h \in V$ and $\tau \in \Gamma$. The verification of (1)-(4) is immediate. Denote by V_0 the set of $h \in V$ such that $u_\sigma(h) = h$ for all $\sigma \in \Gamma$. For each $x \in M$ let $\theta(x)$ be the constant mapping of Γ into M with value x ; then θ is a K-isomorphism of M onto V_0. If we define the homomorphism $\varphi : N \otimes_K V_0 \to V$ as above, we have $\psi = \varphi \circ (\mathrm{Id}_N \otimes \theta)$ and now Prop. 8 follows from Prop. 7.

COROLLARY. — *Let* ψ *be the* K-*linear mapping of* $N \otimes_K N$ *into the product vector space* N^Γ *such that* $\psi(x \otimes y)\ (\sigma) = \sigma(x)y$ *for* x, y *in* N *and* $\sigma \in \Gamma$. *Then* ψ *is injective and it is bijective when* Γ *is finite.*

This is the particular case M = N of Prop. 8.

Remarks. — 1) Let F be an extension of K and N a subextension of F, and let Γ be a *finite* group of automorphisms of N, of which K is the field of invariants. Prop. 8 implies the existence of an isomorphism of K-algebras $\theta : N \otimes_K F \to F^\Gamma$ characterized by $\theta(x \otimes y)\ (\sigma) = \sigma(x)y$ for $x \in N$, $y \in F$ and $\sigma \in \Gamma$.

2) The notation K, N and Γ has the same meaning as before. For each integer $n \geqslant 1$, let A_n be the tensor product of n K-algebras identical with N ; let B_n be the set of mappings of Γ^{n-1} into N. By induction on n we can deduce from the Cor. of Prop. 8 the existence of an isomorphism $\varphi_n : A_n \to B_n$ mapping $x_1 \otimes \ldots \otimes x_n$ to the function $(\sigma_1, \ldots, \sigma_{n-1}) \mapsto \sigma_1(x_1) \ldots \sigma_{n-1}(x_{n-1})\, x_n$.

5. Galois cohomology

Let N be a field, Γ a *finite* group of automorphisms of N and K the field of invariants of Γ. For every integer $n \geqslant 1$ we denote by $\mathbf{GL}(n, N)$ the group of square matrices of order n with coefficients in N and non-zero determinant (II, p. 349). We let the group Γ operate on the group $\mathbf{GL}(n, N)$ by the rule $\sigma(A) = (\sigma(a_{ij}))$ for $A = (a_{ij})$.

PROPOSITION 9. — *Let* $(U_\sigma)_{\sigma \in \Gamma}$ *be a family of elements of* $\mathbf{GL}(n, N)$. *For* A *to exist in* $\mathbf{GL}(n, N)$ *such that* $U_\sigma = A^{-1} . \sigma(A)$ *for all* $\sigma \in \Gamma$ *it is necessary and sufficient that* $U_{\sigma\tau} = U_\sigma . \sigma(U_\tau)$ *for* σ, τ *in* Γ.

The condition is *necessary :* if $U_\sigma = A^{-1} . \sigma(A)$, then we have

$$U_\sigma . \sigma(U_\tau) = A^{-1} . \sigma(A)\,\sigma(A^{-1} . \tau(A)) = A^{-1}\sigma\tau(A) = U_{\sigma\tau} .$$

The condition is *sufficient* : we identify the elements of N^n with matrices of n rows and one column with coefficients in N. We let the groups Γ act on N^n by

$$\sigma(x) = (\sigma(x_i))_{1 \leq i \leq n} \quad \text{for} \quad x = (x_i)_{1 \leq i \leq n} .$$

For each $\sigma \in \Gamma$ we denote by u_σ the mapping $x \mapsto U_\sigma . \sigma(x)$ of N^n into itself. The verification of Formulae (1) to (3) of V, p. 62 is immediate. Moreover we have $u_\varepsilon \circ u_\varepsilon = u_\varepsilon$ and since u_ε is bijective, we have $u_\varepsilon = \mathrm{Id}_{N^n}$. Let V_0 be the set of vectors $x \in N^n$ such that $u_\sigma(x) = x$ for all $\sigma \in \Gamma$. By Prop. 7 (V, p. 63), V_0 is a K-structure on N^n ; in particular there exist in V_0 vectors $b_1, \ldots, b_n$ forming a basis of N^n over N. The matrix B with columns $b_1, \ldots, b_n$ is therefore invertible and the relation $u_\sigma(b_i) = b_i$ for $1 \leq i \leq n$ is equivalent to $U_\sigma . \sigma(B) = B$. Writing $A = B^{-1}$, we obtain $U_\sigma = A^{-1}\sigma(A)$ for all $\sigma \in \Gamma$.

COROLLARY 1. — *Let $(c_\sigma)_{\sigma \in \Gamma}$ be a family of non-zero elements of* N. *For $a \neq 0$ to exist in* N *such that $c_\sigma = \sigma(a) . a^{-1}$ for all $\sigma \in \Gamma$ it is necessary and sufficient that $c_{\sigma\tau} = c_\sigma . \sigma(c_\tau)$ for σ, τ in Γ.*

COROLLARY 2. — *Let $(c_\sigma)_{\sigma \in \Gamma}$ be a family of elements of* N. *For b to exist in* N *such that $a_\sigma = \sigma(b) - b$ for all $\sigma \in \Gamma$ it is nesssary and sufficient that $a_{\sigma\tau} = a_\sigma + \sigma(a_\tau)$ for σ, τ in Γ.*

We have $\sigma\tau(b) - b = [\sigma(b) - b] + \sigma[\tau(b) - b]$ for all b in N and σ, τ in Γ, whence the necessity.

Conversely suppose that $a_{\sigma\tau} = a_\sigma + \sigma(a_\tau)$ for any σ and τ in Γ. Put $U_\sigma = \begin{pmatrix} 1 & a_\sigma \\ 0 & 1 \end{pmatrix}$ for $\sigma \in \Gamma$; then we have $U_{\sigma\tau} = U_\sigma . \sigma(U_\tau)$ for σ, τ in Γ ; by Prop. 9, there exists thus a matrix $A = \begin{pmatrix} x & y \\ z & t \end{pmatrix}$ with non-zero determinant such that $\sigma(A) = AU_\sigma$ for all $\sigma \in \Gamma$; writing down the relation $\sigma(A) = AU_\sigma$ we find

$$\begin{pmatrix} \sigma(x) & \sigma(y) \\ \sigma(z) & \sigma(t) \end{pmatrix} = \begin{pmatrix} x & xa_\sigma + y \\ z & za_\sigma + t \end{pmatrix} \quad (\sigma \in \Gamma) .$$

In particular, x and z belong to K and we have

$$\sigma(y) = xa_\sigma + y , \quad \sigma(t) = za_\sigma + t \quad (\sigma \in \Gamma) .$$

If $x \neq 0$ we have $a_\sigma = \sigma(b) - b$ with $b = x^{-1}y$; if $z \neq 0$, we have the same relation with $b = z^{-1}t$. Now x and z cannot both be zero because

$$xt - yz = \det A \neq 0 .$$

6. Artin's theorem

THEOREM 2 (Artin). — *Let* N *be a field,* Γ *a group of automorphisms of* N *and* K *the field of invariants of* Γ. *Let* V *be a vector sub-K-space of* N *of finite dimension over* K. *Then every* K*-linear mapping u of* V *into* N *is a linear combination with coefficients in* N *of the restrictions to* V *of elements of* Γ.

Let u be a K-linear mapping of V into N and let $V_{(N)} = N \otimes_K V$ be the vector N-space derived from V by extension of scalars ; denote by $\tilde{u}$ the N-linear form on $V_{(N)}$ such that $\tilde{u}(x \otimes y) = x \cdot u(y)$ for $x \in N$ and $y \in V$. For each $\sigma \in \Gamma$ there exists an N-linear form h_σ on $V_{(N)}$ such that $h_\sigma(x \otimes y) = x\sigma(y)$ for $x \in N$ and $y \in V$. The canonical mapping of $V_{(N)} = N \otimes_K V$ into $N \otimes_K N$ is injective. Now the Cor. of Prop. 8 (V, p. 64) show that the intersection of the kernels of the linear forms h_σ on $V_{(N)}$ is reduced to 0. Therefore (II, p. 302, Cor. 1) there exist $\sigma_1, \ldots, \sigma_n$ in Γ and $a_1, \ldots, a_n$ in N such that $\tilde{u} = \sum_{i=1}^{n} a_i h_{\sigma_i}$ whence $u(x) = \sum_{i=1}^{n} a_i \sigma_i(x)$ for all $x \in V$.

Let us equip the set N^N of all mappings of N into N with the product topology of the discrete topologies of the factors. Th. 2 means that the set of linear combinations with coefficients in N of elements of Γ is dense in the set of K-linear mappings of N into itself.

THEOREM 3. — *Let* N *be a field,* Γ *a finite group of automorphisms of* N *and* K *the field of invariants of* Γ. *Let n be the cardinal of* Γ.

a) *We have* $[N : K] = n$ *and* N *is a Galois extension of* K *with Galois group* Γ.

b) *Let* $\sigma_1, \ldots, \sigma_n$ *be the elements of* Γ *and* $(x_1, \ldots, x_n)$ *a basis of* N *over* K, *then* $\det(\sigma_i(x_j)) \neq 0$.

c) *Let u be a* K*-linear mapping of* N *into* N. *There exists a unique family* $(a_\sigma)_{\sigma \in \Gamma}$ *of elements of* N *such that* $u(x) = \sum_{\sigma \in \Gamma} a_\sigma \sigma(x)$ *for all* $x \in N$.

We equip the ring $N \otimes_K N$ with the N-algebra structure whose external law is given by $\lambda(x \otimes y) = x \otimes \lambda y$ for λ, x, y in N. Then the dimension of the vector N-space $N \otimes_K N$ is $[N : K]$. The dimension of the product vector N-space N^Γ is equal to n. The mapping ψ defined in the Cor. of Prop. 8 (V, p. 64) is an N-isomorphism of $N \otimes_K N$ onto N^Γ, whence $[N : K] = n$. Let Δ be the group of K-automorphisms of N. We have $\Gamma \subset \Delta$, hence K is the field of invariants of Δ, and N is a Galois extension of K. Further, the order of Δ is at most equal to $[N : K]$ by Dedekind's theorem (V, p. 27, Cor. 2) and since the order of Γ equals $[N : K]$, we have $\Gamma = \Delta$. Hence Γ is the Galois group of N over K, and this proves *a*).

With the notation of *b*) put $f_i = \psi(x_i \otimes 1)$; we have $f_i(\sigma) = \sigma(x_i)$ for $1 \leqslant i \leqslant n$ and $\sigma \in \Gamma$. Since ψ is an isomorphism of vector N-spaces, the sequence $(f_1, \ldots, f_n)$ is a basis of N^Γ over N, whence $\det(f_j(\sigma_i)) \neq 0$, that is,

$$\det(\sigma_i(x_j)) \neq 0 .$$

This proves *b*).

Finally, *c*) follows from Th. 2 (V, p. 65) which proves the *existence* of a family $(a_\sigma)_{\sigma \in \Gamma}$ such that $u(x) = \sum_{\sigma \in \Gamma} a_\sigma \sigma(x)$ (for all $x \in \mathrm{N}$) and from Dedekind's theorem (V, p. 27, Cor. 2) which proves the *uniqueness* of $(a_\sigma)_{\sigma \in \Gamma}$.

7. The fundamental theorem of Galois theory

THEOREM 4. — *Let* N *be a Galois extension of* K *and* Γ *its Galois group. Let* $\mathcal{K}$ *be the set of subextensions of* N *and* $\mathcal{G}$ *the set of closed subgroups of* Γ. *For every subgroup* $\Delta \in \mathcal{G}$ *we denote by* $k(\Delta)$ *the field of invariants of* Δ *and for every subfield* $\mathrm{E} \in \mathcal{K}$ *we denote by* $g(\mathrm{E})$ *the group of* E*-automorphisms of* N. *Then* $\Delta \mapsto k(\Delta)$ *is a bijection of* $\mathcal{G}$ *onto* $\mathcal{K}$, *and* $\mathrm{E} \mapsto g(\mathrm{E})$ *is the inverse bijection.*

A) The relation $\mathrm{E} = k(g(\mathrm{E}))$ (for $\mathrm{E} \in \mathcal{K}$) is a consequence of the following more precise lemma :

Lemma 1. — *Let* E *be a subextension of* N. *Then* N *is a Galois extension of* E *and* Gal(N/E) *is a closed subgroup of* Gal(N/K) *with the induced topology.*

Let $x \in \mathrm{N}$; the minimal polynomial f of x over E divides in E[X] the minimal polynomial g of x over K (V, p. 17, Cor. 2). Since N is Galois over K, the polynomial g is a product in N[X] of distinct factors of degree 1 ; hence the same is true of f and so N is Galois over E.

Let Γ be the Galois group of N over K and Δ that of N over E. By definition Δ is the subgroup of Γ consisting of all σ such that $\sigma(x) = x$ for all $x \in \mathrm{E}$. Now for each $x \in \mathrm{E}$ the mapping $\sigma \mapsto \sigma(x)$ of Γ into the discrete space N is continuous, hence D is closed in Γ. Let $\sigma \in \Gamma$; for $x_1, \ldots, x_n$ in N let $\mathrm{U}(x_1, \ldots, x_n)$ be the set of all $\tau \in \Gamma$ such that $\tau(x_i) = \sigma(x_i)$ for $1 \leqslant i \leqslant n$; put

$$\mathrm{V}(x_1, \ldots, x_n) = \mathrm{U}(x_1, \ldots, x_n) \cap \Delta .$$

Then the family of sets $\mathrm{U}(x_1, \ldots, x_n)$ (resp. $\mathrm{V}(x_1, \ldots, x_n)$) is a base of neighbourhoods of σ in Γ (resp. Δ). Hence the topology on Δ is that induced by Γ.

B) The relation $\Delta = g(k(\Delta))$ (for $\Delta \in g$) is a consequence of the following more precise lemma :

Lemma 2. — *Let* Δ *be a subgroup of* Γ. *Let* E *be the field of invariants of* Δ ; *then the Galois group of* N *over* E *is the closure of* Δ *in* Γ.

The Galois group of N over E is closed in Γ (Lemma 1) and contains Δ, hence it contains the closure $\bar{\Delta}$ of Δ. Let σ be an E-automorphism of N and let $x_1, \ldots, x_n$ be in N. Since N is Galois over E (Lemma 1) there exist (V, p. 57, Prop. 2) a subextension N_0 of N, Galois of finite degree over E and containing $x_1, \ldots, x_n$. Let Δ_0 be the image of the subgroup Δ of Gal(N/E) under the restriction homomorphism of Gal(N/E) into Gal(N_0/E). Since $[\mathrm{N}_0 : \mathrm{E}]$ is finite, Dedekind's theorem (V, p. 27, Cor. 2) shows that Gal(N_0/E) is finite. Hence Δ_0 is finite, and since E is the field of invariants of Δ_0, we have $\Delta_0 =$

$\mathrm{Gal}(N_0/E)$ (V, p. 66, Th. 3). In particular, Δ_0 contains the restriction of σ to N_0. Therefore there exists $\tau \in \Delta$ such that σ and τ have the same restriction to N_0, whence $\sigma(x_1) = \tau(x_1), \ldots, \sigma(x_n) = \tau(x_n)$. It follows that σ is a limit point of Δ in Γ, and hence $\mathrm{Gal}(N/E) \subset \bar{\Delta}$.

COROLLARY 1. — *Let* E *and* E′ *be two subfields of* N *containing* K ; *then* $E \subset E'$ *if and only if* $g(E) \supset g(E')$. *If* Δ *and* Δ' *are two closed subgroups of* Γ, *then* $\Delta \subset \Delta'$ *if and only if* $k(\Delta) \supset k(\Delta')$.

For the two inverse bijections $E \mapsto g(E)$ and $\Delta \mapsto k(\Delta)$ are inclusion-reversing.

COROLLARY 2. — *Let* $(E_i)_{i \in I}$ *be a family of subfields of* N *containing* K ; *put* $L = \bigcap_{i \in I} E_i$ *and* $M = K\left(\bigcup_{i \in I} E_i\right)$. *Then* $g(L)$ *is the smallest closed subgroup of* Γ *containing* $\bigcup_{i \in I} g(E_i)$ *and we have* $g(M) = \bigcap_{i \in I} g(E_i)$.

The first assertion follows from Cor. 1 and the second is immediate.

COROLLARY 3. — *For* $i = 1, 2$ *let* E_i *be a subfield of* N *containing* K *and let* $\Delta_i = g(E_i)$. *For any* $\sigma \in \Gamma$ *the relations* $\sigma(E_1) = E_2$ *and* $\sigma\Delta_1\sigma^{-1} = \Delta_2$ *are equivalent.*

For we have $\tau \in g(\sigma(E_1))$ if and only if $\tau\sigma(x) = \sigma(x)$, that is, $\sigma^{-1}\tau\sigma(x) = x$, for all $x \in E_1$; this amounts to saying that $\sigma^{-1}\tau\sigma \in \Delta_1$, whence $g(\sigma(E_1)) = \sigma\Delta_1\sigma^{-1}$.

COROLLARY 4. — *Let* E *be a subfield of* N *containing* K *and let* $\Delta = g(E)$. *For* E *to be Galois over* K *it is necessary and sufficient that* Δ *should be a normal subgroup of* Γ. *When this is so, the restriction homomorphism of* Γ *into* $\mathrm{Gal}(E/K)$ *defines by passage to quotients a topological group isomorphism of* Γ/Δ *onto* $\mathrm{Gal}(E/K)$.

Since N is separable over K, the same is true of E (V, p. 36, Prop. 1). Therefore E is Galois over K if and only if it is quasi-Galois over K ; this also means that $\sigma(E) = E$ for every K-automorphism σ of N (V, p. 52, Prop. 1 and p. 54, Prop. 3). By Cor. 3 this is equivalent to $\sigma\Delta\sigma^{-1} = \Delta$ for all $\sigma \in \Gamma$.

The restriction homomorphism $\varphi : \mathrm{Gal}(N/K) \to \mathrm{Gal}(E/K)$ is continuous and surjective (V, p. 60, Prop. 3) and its kernel is clearly equal to $\Delta = \mathrm{Gal}(N/E)$. Since Γ is compact, the homomorphism of Γ/Δ onto $\mathrm{Gal}(E/K)$ derived from φ by passage to quotients is an isomorphism of topological groups (Gen. Top., I, p. 87, Cor. 2).

COROLLARY 5. — *Let* E *be a subfield of* N *containing* K. *For* E *to have finite degree over* K *it is necessary and sufficient that* $g(E)$ *should be open in* Γ. *When this is so, the index* $(\Gamma : g(E))$ *is finite and equal to* $[E : K]$.

For $g(E)$ to be open it is necessary and sufficient that there should exist a subextension F of N, of finite degree over K, such that in the notation of V, p. 60,

$g(E)$ contains $U_F(Id_N) = g(F)$. The relation $g(E) \supset g(F)$ is equivalent to $E \subset F$ by Cor. 1 (V, p. 68), whence the first assertion of Cor. 5.

Suppose that $[E : K]$ is finite. Let Ω be an algebraic closure of K containing N as subextension (V, p. 23, Th. 2) and let $\mathscr{H}$ be the set of K-homomorphisms of E into Ω. Every element of $\mathscr{H}$ is induced by a K-automorphism of Ω (V, p. 52, Prop. 1), and since N is quasi-Galois over K, the mapping $\sigma \mapsto \sigma | E$ of Γ into $\mathscr{H}$ is surjective. For σ and σ' in Γ to have the same restriction to E it is necessary and sufficient that $\sigma^{-1}\sigma' \in g(E)$, whence Card $\mathscr{H} = (\Gamma : g(E))$. Finally since E is an etale algebra over K, we have Card $\mathscr{H} = [E : K]$ (V, p. 32, Prop. 4), so in conclusion we have $(\Gamma : g(E)) = [E : K]$.

COROLLARY 6. — *For $i = 1, 2$ let E_i be a subextension of N and Γ_i the Galois group of N over E_i. The following conditions are equivalent :*

a) The group Γ is the direct product of the subgroups Γ_1 and Γ_2.

b) The extensions E_1 and E_2 are Galois over K, we have $E_1 \cap E_2 = K$ and

$$K(E_1 \cup E_2) = N .$$

For Γ to be the direct product of the subgroups Γ_1 and Γ_2 it is necessary and sufficient that the following conditions hold (I, p. 48, Prop. 15) :

(i) the subgroups Γ_1 and Γ_2 are normal in Γ ;

(ii) $\Gamma_1 \cap \Gamma_2 = \{\varepsilon\}$, where ε is the neutral element of Γ ;

(iii) $\Gamma = \Gamma_1 . \Gamma_2$.

Now (i) means that E_1 and E_2 are Galois over K (Cor. 4). By Cor. 2, condition (ii) is equivalent to $N = K(E_1 \cup E_2)$, and finally if (i) and (ii) hold, $\Gamma_1\Gamma_2$ is the least subgroup of Γ containing $\Gamma_1 \cup \Gamma_2$; it is closed because Γ_1 and Γ_2 are compact and the mapping $(\sigma, \tau) \mapsto \sigma\tau$ of $\Gamma_1 \times \Gamma_2$ into Γ is continuous (Gen. Top., I, p. 63, Cor. 1). Cor. 2 now shows (iii) to be equivalent to $E_1 \cap E_2 = K$, and this proves the equivalence of *a*) and *b*).

Remark. — With the notation of Cor. 6 suppose that conditions *a*) and *b*) hold. The restriction homomorphisms $\varphi_i : \Gamma \to \mathrm{Gal}(E_i/K)$ for $i = 1, 2$ induce topological group isomorphisms

$$\Psi_1 : \Gamma_2 \to \mathrm{Gal}(E_1/K) , \quad \Psi_2 : \Gamma_1 \to \mathrm{Gal}(E_2/K) .$$

By *a*) we see that the mapping $\sigma \mapsto (\varphi_1(\sigma), \varphi_2(\sigma))$ is a topological group isomorphism of $\mathrm{Gal}(N/K)$ onto $\mathrm{Gal}(E_1/K) \times \mathrm{Gal}(E_2/K)$.

8. Change of base field

Let N be a Galois extension of K and Γ the Galois group of N over K ; further let N' be a Galois extension of K' with Galois group Γ'. We shall identify K (resp. K') with its image in N (resp. N'). Let *u* be a homomorphism of K into K' and *v* a homomorphism of N into N' whose restriction to K equals *u* (*cf.*

Fig. 1). Let $\sigma \in \Gamma'$; since $u(\mathrm{K}) \subset \mathrm{K}'$, σ is a $u(\mathrm{K})$-automorphism of N′ ; moreover $v(\mathrm{N})$ is a Galois extension of $u(\mathrm{K})$, hence σ induces a $u(\mathrm{K})$-automorphism of $v(\mathrm{N})$ (V, p. 55, Remark 1). In other words, for every $\sigma \in \Gamma'$ there exists a unique element $v^*(\sigma)$ of Γ such that

$$v \circ v^*(\sigma) = \sigma \circ v . \tag{8}$$

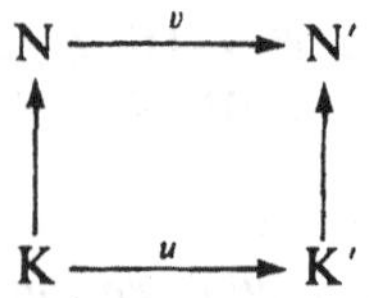

FIG. 1.

The mapping v^* *is a homomorphism of* Gal(N′/K′) *into* Gal(N/K). For every $x \in \mathrm{N}$, the mapping $\sigma \mapsto v^*(\sigma)\ (x) = v^{-1}(\sigma(v(x)))$ of Γ' into the discrete space N is continuous, so v^* is *continuous.*

Three particular cases are of importance :

a) If F is a Galois extension of K and E a subextension of F, we know (V, p. 67, Lemma 1) that F is a Galois extension of E. Let us apply what has been said to the case where N = F, K′ = E, N′ = F and $v = \mathrm{Id}_{\mathrm{F}}$. Then v^* is merely the canonical injection

$$j : \mathrm{Gal}(\mathrm{F}/\mathrm{E}) \to \mathrm{Gal}(\mathrm{F}/\mathrm{K}) .$$

This is sometimes called the *inflation homomorphism.*

b) Suppose that in addition E is Galois over K. Let us apply what has been said to the case where N = E, K′ = K, N′ = F and v is the canonical injection of E into F. Then v^* is just the *restriction homomorphism*

$$\pi : \mathrm{Gal}(\mathrm{F}/\mathrm{K}) \to \mathrm{Gal}(\mathrm{E}/\mathrm{K}) .$$

We know (V, p. 60, Prop. 3) that π is surjective, with kernel Gal(F/E) and that by taking quotients it defines a topological group isomorphism of Gal(F/K)/Gal(F/E) onto Gal(E/K) (V, p. 68, Cor. 4).

c) Suppose that $v^{-1}(\mathrm{K}') = \mathrm{K}$ and $\mathrm{N}' = \mathrm{K}'(v(\mathrm{N}))$; let us show that the homomorphism

$$v^* : \mathrm{Gal}(\mathrm{N}'/\mathrm{K}') \to \mathrm{Gal}(\mathrm{N}/\mathrm{K}) ,$$

is a topological group isomorphism, sometimes called *translation.* For the group Gal(N′/K′) is compact, the group Gal(N/K) is separated and v^* is continuous ; so it is enough (Gen. Top., I, p. 87, Cor. 2) to prove that v^* is bijective. Now every element σ of the kernel of v^* is an automorphism of N′ which induces the identity on K′ and on $v(\mathrm{N})$, hence $\sigma = \varepsilon$ because $\mathrm{N}' = \mathrm{K}'(v(\mathrm{N}))$; it follows that v^* is injective. Further, the image of v^* is a closed subgroup Δ of Gal(N/K)

(Gen. Top., I, p. 81, *ibid.*) and the field of invariants of Δ is equal to $v^{-1}(K') = K$; therefore we have $\Delta = \mathrm{Gal}(N/K)$ (V, p. 67, Th. 4) and so v^* is surjective.

The general case may be reduced to the preceding ones by composition. To begin with we note that $K'(v(N))$ is the field of invariants in N' of the kernel Δ of v^*; since Δ is a normal subgroup of $\mathrm{Gal}(N'/K')$, the extension $K'(v(N))$ of K' is Galois (V, p. 68, Cor. 4). Thus v^* is composed of the homomorphisms

$$\mathrm{Gal}(N'/K') \to \mathrm{Gal}(K'(v(N))/K') \xrightarrow{\psi} \mathrm{Gal}(N/v^{-1}(K')) \xrightarrow{j} \mathrm{Gal}(N/K);$$

in this sequence π is the restriction homomorphism associated with the triple $K' \subset K'(v(N)) \subset N'$, ψ is the translation isomorphism associated with the central square of the diagram (Fig. 2) and j is the inflation homomorphism associated with the triple $K \subset v^{-1}(K') \subset N$.

The next theorem gives more detailed information about the structure of translation isomorphisms.

$$\begin{array}{ccccc} N & \to & K'(v(N)) & \to & N' \\ & & \uparrow & & \uparrow \\ K & \to & v^{-1}(K') & \to & K' \end{array}$$

FIG. 2.

THEOREM 5. — *Let* N' *be an extension of* K *generated by two subextensions* K' *and* N. *Suppose that* N *is Galois over* K, *with Galois group* Γ *and that* $K' \cap N = K$. *Then the extension* N' *of* K' *is Galois and the canonical homomorphism* φ *of* $K' \otimes_K N$ *into* N' *is an isomorphism. Let* $\sigma \in \mathrm{Gal}(N/K)$ *and let* σ' *be the element of* $\mathrm{Gal}(N'/K')$ *which corresponds to it under the translation isomorphism; then we have* $\sigma' \circ \varphi = \varphi \circ (\mathrm{Id}_{K'} \otimes \sigma)$.

We have $N' = K'(N)$ and N is algebraic and separable over K; hence (V, p. 42, Prop. 10), the extension N' of K' is algebraic and separable. By Cor. 4 of V, p. 54 the extension N' of K' is quasi-Galois. Therefore the extension N' of K' is Galois. By *c)* above the mapping $\sigma \mapsto \sigma | N$ is an homomorphism λ of $\mathrm{Gal}(N'/K')$ onto $\mathrm{Gal}(N/K)$.

We have $N = K'[N]$ because N is algebraic over K (V, p. 18, Cor. 1), hence φ is surjective. If σ belongs to $\mathrm{Gal}(N/K)$, we have

$$\lambda^{-1}(\sigma) \circ \varphi = \varphi \circ (\mathrm{Id}_{K'} \otimes \sigma). \tag{9}$$

Therefore the kernel of φ is stable under the mappings $\mathrm{Id}_{K'} \otimes \sigma$, hence of the form $K' \otimes_K N_0$ with $N_0 \subset N$ (V, p. 63, Cor.). For x in N_0 we have $x = \varphi(1 \otimes x) = 0$, hence $N_0 = 0$ and so φ is injective.

COROLLARY 1. — *Let* E' *be a subfield of* N' *containing* K'. *There exists a unique subfield* E *of* N *containing* K *and such that* $E' = K'(E)$. *We have* $E = E' \cap N$.

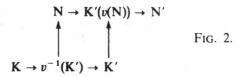

Put $E = E' \cap N$, then $E' \supset K'(E)$. Now put $\Gamma = \mathrm{Gal}(N/K)$ and $\Delta = \mathrm{Gal}(N/E)$, and define Γ' and Δ' similarly. The mapping $\lambda : \sigma \mapsto \sigma \mid N$ is an isomorphism of Γ' onto Γ and also of Δ' onto Δ ; in other words, Δ' consists of those $\sigma \in \Gamma'$ for which $\lambda(\sigma)$ belongs to Δ. If $\sigma \in \Gamma'$ leaves the elements of $K'(E)$ fixed, we have $\lambda(\sigma) \in \Delta$, whence $\sigma \in \Delta'$ and σ leaves the elements of E' fixed ; by Cor. 1 of V, p. 68 we thus have $K'(E) \supset E'$.

We have proved the equality $E' = K'(E)$, whence $\varphi^{-1}(E') = K' \otimes_K E$. If F is a subfield of N containing K and such that $E' = K'(F)$, we have likewise $\varphi^{-1}(E') = K' \otimes_K F$, whence $F = E$.

COROLLARY 2. — *Let* N *be a Galois extension of* K. *Suppose that the Galois group* Γ *of* N *over* K *is the direct product of two closed subgroups* Γ_1 *and* Γ_2 *and denote by* E_i *the field of invariants of* Γ_i *for* $i = 1, 2$. *Then the canonical homomorphism of* $E_1 \otimes_K E_2$ *into* N *is an isomorphism.*

We have $E_1 \cap E_2 = K$ and $N = K(E_1 \cup E_2)$ by Cor. 6 (V, p. 69), and so it is enough to apply Theorem 5.

Remark. — Let K and K′ be two fields and u a homomorphism of K into K′. Let K_s (resp. K'_s) be a separable closure (V, p. 45, Prop. 14) of K (resp. K′) and Π (resp. Π') the Galois group of K_s over K (resp. K'_s over K′). Since K_s is a separable algebraic extension of K and the extension (K'_s, u) of K is separably closed, there exists (V, p. 45, Cor.) a homomorphism v of K_s into K'_s extending u. From v we obtain a continuous homomorphism v^* of Π' into Π. Let v_1 be another extension of u ; since K_s is a quasi-Galois extension of K, there exists an element σ_0 of Π such that $v_1 = v \circ \sigma_0$. We conclude that $v_1^*(\tau) = \sigma_0^{-1} v^*(\tau) \sigma_0$ for all $\tau \in \Pi$.

9. The normal basis theorem

Let N be a Galois extension of K, with Galois group Γ. We identify Γ with the canonical basis of the group algebra $K^{(\Gamma)}$ (III, p. 446) ; then N may be considered as a left $K^{(\Gamma)}$-module (III, p. 447, Example), so that

$$u.x = \sum_{\sigma \in \Gamma} a_\sigma \sigma(x) \quad \text{for} \quad x \in N \quad \text{and} \quad u = \sum_{\sigma \in \Gamma} a_\sigma \sigma \quad \text{in} \quad K^{(\Gamma)} .$$

If N is of finite degree over K, the group Γ is finite by Dedekind's theorem (V, p. 27, Cor. 2) and we can define the element $t = \sum_{\sigma \in \Gamma} \sigma$ in $K^{(\Gamma)}$; then we have

$$\mathrm{Tr}_{N/K}(x) = \sum_{\sigma \in \Gamma} \sigma(x) ,$$

that is, $\mathrm{Tr}_{N/K}(x) = t.x$ for all $x \in N$.

Let us define an action on the right by Γ on N by $x^\sigma = \sigma^{-1}(x)$. In a similar way we can consider the multiplicative group N^* as a right $\mathbf{Z}^{(\Gamma)}$-module, the external law being written $(x, u) \mapsto x^u$. For example, the notation $x^{2\sigma+3\tau+\pi}$, where σ, τ, π are elements of Γ, indicates the product $(x^\sigma)^2 \, . \, (x^\tau)^3 \, . \, x^\pi$. If N is of finite degree over K and $t = \sum_{\sigma \in \Gamma} \sigma$ as above, then we have $N_{N/K}(x) = \prod_{\sigma \in \Gamma} x^\sigma$, that is, $N_{N/K}(x) = x^t$ for all $x \in N^*$.

Suppose henceforth that N is of finite degree over K. Given $x \in N$, for $\{x\}$ to be a basis of the $K^{(\Gamma)}$-module N it is necessary and sufficient that the family $(\sigma(x))_{\sigma \in \Gamma}$ should be a basis of N over K. Such a basis is called a *normal basis of* N *over* K.

THEOREM 6. — *Let* N *be a Galois extension of finite degree over* K *and let* Γ *be the Galois group of* N *over* K. *Then there exists a normal basis of* N *over* K ; *in other words, the* $K^{(\Gamma)}$*-module* N *is free of rank* 1.

We shall give two proofs of this statement. The first uses the following lemma which will be proved in Chapter VIII (§ 2, No. 5).

* *Lemma* 3. — *Let* A *be a* K*-algebra,* M_1 *and* M_2 *two* A*-modules of finite rank over* K *and suppose that there exists an extension* L *of* K *such that the modules* $L \otimes_K M_1$ *and* $L \otimes_K M_2$ *over the ring* $L \otimes_K A$ *are isomorphic. Then the* A*-modules* M_1 *and* M_2 *are isomorphic.*

We shall apply Lemma 3 in the case where $A = K^{(\Gamma)}$, $M_1 = N$, $M_2 = A_s$ and $L = N$. By the Cor. of V, p. 64 there exists a K-isomorphism φ of $N \otimes_K N$ onto $N \otimes_K K^{(\Gamma)}$ which maps $x \otimes y$ to $\sum x\sigma^{-1}(y) \otimes \sigma$. It is clear that φ is an isomorphism of $N \otimes_K K^{(\Gamma)}$-modules and so the theorem follows from Lemma 3. *

For the second proof we shall use the following proposition :

PROPOSITION 10. — *Let* $x \in N$, *then for* $\{x\}$ *to form a basis of the* $K^{(\Gamma)}$*-module* N *it is necessary and sufficient that* $\det(\sigma\tau(x))_{\sigma,\tau \in \Gamma} \neq 0$.

Since $K^{(\Gamma)}$ and N have the same dimension over K, to say that $\{x\}$ is a basis of N over $K^{(\Gamma)}$ means that the mapping $a \mapsto ax$ of $K^{(\Gamma)}$ into N is injective. This in turn means that the mapping $b \mapsto b(1 \otimes x)$ of $N \otimes_K K^{(\Gamma)}$ into $N \otimes_K N$ is injective (II, p. 306, Prop. 14). Now there is an $N \otimes_K K^{(\Gamma)}$-module isomorphism of $N \otimes_K N$ onto $N \otimes_K K^{(\Gamma)}$ which maps $1 \otimes x$ to $\sum_\sigma \sigma^{-1}(x) \otimes \sigma$. It follows that $\{x\}$ is a basis of N over $K^{(\Gamma)}$ if and only if, for every non-zero family of elements $(n_\tau)_{\tau \in T}$ of N we have $\left(\sum n_\tau \otimes \tau\right)\left(\sum \sigma^{-1}(x) \otimes \sigma\right) \neq 0$. But this last relation means that there exists $\sigma \in \Gamma$ such that $\sum_\tau n_\tau \sigma^{-1}\tau(x) \neq 0$, whence the proposition.

A) Suppose that K *is infinite* ; the mapping $x \mapsto \det(\sigma\tau(x))$ of N into N is a polynomial mapping over K (IV, p. 54). By extension of scalars from K to N we get a corresponding mapping for the vector N-space $N \otimes_K N$ and we have just seen that the latter is not identically zero (because $N \otimes_K N$ is free of rank 1 over $N \otimes_K K^{(\Gamma)}$). So there exists $x \in N$ such that $\det(\sigma\tau(x)) \neq 0$ (IV, p. 18, Th. 2) ; more generally, from the same reference we have :

PROPOSITION 11. — *Suppose that* K *is infinite and let* $P : N \to K$ *be a polynomial mapping which is non-zero on* K. *There exists* $x \in N$ *such that* $P(x) \neq 0$ *and* $\{x\}$ *is a basis of* N *over* $K^{(\Gamma)}$.

B) Suppose that K *is finite.* By Prop. 4 (V, p. 95)[1] every extension of finite degree over K has a cyclic Galois group. We shall therefore more generally consider the case where the group Γ is cyclic of order n ; we denote by γ a generator of Γ.

The following lemma is a particular case of more general results proved in Chapter VII. The ring A is either the ring **Z** of rational integers or the ring K[X] of polynomials over the field K.

Lemma 4. — *Let* M *be a torsion* A-*module generated by a finite number* $x_1, \dots, x_h$ *of elements ; then there exists an element* x *of* M *whose annihilator* (II, p. 219) *is equal to the annihilator of* M.

In both cases A is an integral domain and every ideal of A is principal. When $A = \mathbf{Z}$ (resp. $A = K[X]$), we denote by $\mathscr{P}$ the set of prime numbers (resp. the set of irreducible monic polynomials in K[X]). For every element $a \neq 0$ of A there exists then an invertible element u of A and a family $(v_p(a))_{p \in \mathscr{P}}$, with finite support, of positive integers such that $a = u \prod_{p \in \mathscr{P}} p^{v_p(a)}$ and u and the integers $v_p(a)$ are uniquely determined (I, p. 51 and IV, p. 13, Prop. 13).

Let $\mathfrak{a}_i$ be the annihilator of x_i (for $1 \leqslant i \leqslant h$) and $\mathfrak{a}$ the annihilator of M ; let $a_1, \dots, a_h, a$ be non-zero elements of A such that $\mathfrak{a}_i = Aa_i$ and $\mathfrak{a} = Aa$; since $\mathfrak{a} = \mathfrak{a}_1 \cap \dots \cap \mathfrak{a}_h$, it follows from what has been said that

$$(10) \qquad v_p(a) = \sup_{1 \leqslant i \leqslant h} v_p(a_i) \quad \text{for all } p \in \mathscr{P} .$$

Let us write a in the form $up_1^{n(1)} \dots p_r^{n(r)}$, with $p_1, \dots, p_r$ distinct in $\mathscr{P}$, $n(1) > 0, \dots, n(r) > 0$ and u an invertible element of A. Let $j = 1, \dots, r$; by (10) there exists an integer $c(j)$ such that $1 \leqslant c(j) \leqslant h$ and $v_{p_j}(a_{c(j)}) = n(j)$; there exists b_j in A with $a_{c(j)} = p_j^{n(j)} b_j$ and the element $y_j = b_j x_{c(j)}$ has as annihilator the ideal $Ap_j^{n(j)}$.

Let us show that the annihilator $\mathfrak{b}$ of $y = y_1 + \cdots + y_r$ is equal to the annihilator $\mathfrak{a}$ of M. In any case we have $\mathfrak{a} \subset \mathfrak{b}$, so $\mathfrak{b}$ is of the form $Ap_1^{m(1)} \dots p_r^{m(r)}$ with $0 \leqslant m(j) \leqslant n(j)$ for $1 \leqslant j \leqslant r$. If we had $\mathfrak{a} \neq \mathfrak{b}$, there would exist an integer j such that $1 \leqslant j \leqslant r$ and $m(j) < n(j)$, and hence $d_j = a/p_j$ would annihilate y. Now we have $d_j y_k = 0$ for $k \neq j$, whence we obtain $d_j y_j = 0$; but the annihilator of y_j is $Ap_j^{m(j)}$ and d_j is not a multiple of $p_j^{m(j)}$. So the hypothesis $\mathfrak{a} \neq \mathfrak{b}$ is absurd.

[1] The reader can easily convice himself that the normal basis theorem is not used anywhere before the proof of that proposition.

We shall apply Lemma 4 to the case where A is the polynomial ring K[X] and M the abelian group N with the external law defined by $a \, . \, x = \sum_{k=0}^{\infty} c_k \gamma^k(x)$ for $a = \sum_{k=0}^{\infty} c_k X^k$ in K[X] and $x \in N$. Let $\mathfrak{a}$ be the annihilator of M, then $\gamma^n = 1$, hence the polynomial $X^n - 1$ lies in $\mathfrak{a}$. Let $F \in \mathfrak{a}$; by IV, p. 11, Cor. there exist elements $c_0, c_1, \ldots, c_{n-1}$ of K and $G \in K[X]$ such that

$$F(X) = c_0 + c_1 X + \cdots + c_{n-1} X^{n-1} + (X^n - 1) G(X) . \tag{11}$$

Thus we have $c_0 + c_1\gamma + \cdots + c_{n-1}\gamma^{n-1} = 0$ in $\mathrm{Hom}_K(N, N)$ and since the automorphisms $1, \gamma, \gamma^2, \ldots, \gamma^{n-1}$ of N are distinct, Dedekind's theorem (V, p. 27, Cor. 2) implies that $c_0 = c_1 = \cdots = c_{n-1} = 0$. Finally, we have

$$F(X) = (X^n - 1) G(X), \quad \text{that is,} \quad \mathfrak{a} = (X^n - 1) K[X] .$$

By Lemma 4 there exists an element x of N whose annihilator in K[X] is equal to $(X^n - 1) K[X]$. Since the monomials $1, X, \ldots, X^{n-1}$ form a basis of a vector subspace of K[X] supplementary to $(X^n - 1) K[X]$ (IV, p. 11, Cor.), the elements $x, \gamma(x), \ldots, \gamma^{n-1}(x)$ of N are linearly independent over K. Since $[N : K] = n$ (V, p. 66, Th. 3), the sequence $(x, \gamma(x), \ldots, \gamma^{n-1}(x))$ is therefore a (normal) basis of N over K.

10. Finite Γ-sets and etale algebras

Let K_s be a separable closure of K (V, p. 45, Prop. 14) and Γ the Galois group of K_s over K. By a *Γ-set* we understand a set X with an action $(\sigma, x) \mapsto \sigma x$ of the group Γ such that the stabilizer of each point of X is an *open* subgroup of Γ. It comes to the same to say that the mapping $(\sigma, x) \mapsto \sigma x$ of $\Gamma \times X$ into X is *continuous* when X is equipped with the discrete topology.

Let X be a *finite* Γ-set. We define an action of Γ on the K-algebra K_s^X of mappings of X into K_s by the formula

$$u_\sigma f(x) = \sigma(f(\sigma^{-1}x)) \tag{12}$$

for $\sigma \in \Gamma$, $f \in K_s^X$ and $x \in X$. Let $\Theta(X)$ be the set of invariants of Γ in K_s^X ; this is the sub-K-algebra of K_s^X consisting of mappings $f : X \to K_s$ such that $f(\sigma x) = \sigma(f(x))$ for $\sigma \in \Gamma$ and $x \in X$.

Lemma 5. — *Let* X *be a finite* Γ*-set and let* $x_1, \ldots, x_n$ *be points of* X *such that the orbits* $\Gamma x_1, \ldots, \Gamma x_n$ *form a partition of* X. *For* $1 \leqslant i \leqslant n$ *let* Δ_i *be the stabilizer of* x_i *in* Γ *and let* L_i *be the field of invariants of* Δ_i. *Then* $L_1, \ldots, L_n$ *are separable extensions of finite degree of* K, *and the mapping* $f \mapsto (f(x_1), \ldots, f(x_n))$ *is a* K-*algebra isomorphism of* $\Theta(X)$ *onto* $L_1 \times \cdots \times L_n$.

By hypothesis the subgroups $\Delta_1, \ldots, \Delta_n$ of Γ are open and Cor. 5 of V, p. 68, shows that the subextensions $L_1, \ldots, L_n$ of K_s are of finite degree over K. Clearly they are separable ; now the last assertion of Lemma 5 is immediate.

From Lemma 5 and Th. 4 (V, p. 34f.) we obtain immediately the following result.

PROPOSITION 12. — *For every finite* Γ*-set* X *the algebra* $\Theta(X)$ *is etale over* K, *of degree equal to the cardinal of* X. *Moreover, every etale algebra over* K *is isomorphic to an algebra of the form* $\Theta(X)$.

Remarks. — a) It is easy to show that for every K-algebra homomorphism φ of $\Theta(X)$ into K_s there exists a unique element x of X such that $\varphi(f) = f(x)$ for all $f \in \Theta(X)$.

2) Let X and Y be two finite Γ-sets. Let $\mathfrak{F}_\Gamma(X, Y)$ be the set of mappings u of X into Y such that $u(\sigma x) = \sigma u(x)$ for all $\sigma \in \Gamma$ and all $x \in X$. For $u \in \mathfrak{F}_\Gamma(X, Y)$ we define a K-algebra homomorphism $u^* : \Theta(Y) \to \Theta(X)$ by $u^*(f) = f \circ u$. For every homomorphism Ψ of $\Theta(Y)$ into $\Theta(X)$ there exists a unique element u of $\mathfrak{F}_\Gamma(X, Y)$ such that $\Psi = u^*$.

11. The structure of quasi-Galois extensions

PROPOSITION 13. — *Let* N *be a quasi-Galois extension of* K. *We denote by* N_r *the field of invariants of the group of all* K*-automorphisms of* N *and by* N_s *the relative separable algebraic closure of* K *in* N (V, p. 44). *Then :*

a) N_r *is the relative p-radical closure of* K *in* N (V, p. 25).

b) N_s *is a Galois extension of* K *and every* K*-automorphism of* N_s *extends in a unique fashion to an* N_r*-automorphism of* N.

c) The fields N_r *and* N_s *are linearly disjoint over* K *and we have* $N = K[N_r \cup N_s]$; *in other words, the canonical homomorphism of* $N_r \otimes_K N_s$ *into* N *is an isomorphism.*

Let Ω be an algebraic closure of K containing N as subextension (V, p. 23, Th. 2). Every K-automorphism of Ω induces an automorphism of N because N is quasi-Galois. Therefore every element of N_r is invariant under the group of K-automorphisms of Ω, hence p-radical over K (V, p. 53, Cor. 3). Conversely every element of N which is p-radical over K is clearly invariant under every K-automorphism of N and hence belongs to N_r. This proves *a*).

Every K-automorphism of Ω maps N into N, hence N_s into N_s, so N_s is a quasi-Galois extension of K (V, p. 54, Prop. 3). It follows that N_s is a Galois extension of K. Every element of $N_r \cap N_s$ is separable algebraic and p-radical over K, hence belongs to K (V, p.   Cor. 3) ; we thus have $K_r \cap K_s = K$. Now N is p-radical over N_s (V, p. 44, Prop. 13) and separable algebraic over N_r (V, p. 56, Th. 1) hence both p-radical and separable over $K(N_r \cup N_s)$. Hence we have $N = K(N_r \cup N_s)$ (V, p. 39, Cor. 3) and the assertions *b*), *c*) follow from Th. 5 (V, p. 71).

COROLLARY. — *Let p the characteristic exponent of* K, $\bar{K}$ *an algebraic closure of* K, K_s *the relative separable closure of* K *in* $\bar{K}$ *and* $K^{p^{-\infty}}$ *the perfect closure of* K. *Then the canonical homomorphism of* $K^{p^{-\infty}} \otimes K_s$ *into* $\bar{K}$ *is an isomorphism.*

Remark. — Let R (resp. S) be a p-radical (resp. separable algebraic) extension of K. Then the algebra $R \otimes_K S$ is a field : for R (resp. S) is isomorphic to a subextension of $K^{p^{-\infty}}$ (resp. K_s) and it suffices to apply the above Cor. and Prop. 1 of V, p. 17.

§ 11. ABELIAN EXTENSIONS

Throughout this paragraph K *denotes a field.*

1. Abelian extensions and the abelian closure

DEFINITION 1. — *An extension* E *of* K *is said to be abelian if it is Galois and its Galois group is commutative.*

Since every subgroup of a commutative group is normal, Cor. 4 of V, p. 68 shows that every subextension of an abelian extension is abelian.

PROPOSITION 1. — *Let* E *be a Galois extension of* K *and* Γ *its Galois group. Let* Δ *be the derived group of* Γ (I, p. 10, Def. 4) *and* F *the field of invariants of* Δ. *For a subextension* L *to* E *to be abelian it is necessary and sufficient that it should be contained in* F.

Firstly we note that F is also the field of invariants of the closure $\bar{\Delta}$ of Δ in Γ, and that $\bar{\Delta}$ is a closed normal subgroup of Γ. By V, p. 68, Cor. 4, F is thus a Galois extension of K. Further, the Galois group of F over K is isomorphic to $\Gamma/\bar{\Delta}$ and hence is commutative. Hence every subextension of F is abelian. Conversely let L be an abelian extension of K contained in E, and let Π be the Galois group of E over L. Since L is Galois, Π is a normal subgroup of Γ and the Galois group of L over K is isomorphic to Γ/Π (V, p. 68, Cor. 4). Therefore Γ/Π is commutative and Π contains Δ, whence $L \subset F$.

COROLLARY. — *Let* E *be an extension of* K *and let* $(E_i)_{i \in I}$ *be a family of subextensions of* E *such that* $E = K\left(\bigcup_{i \in I} E_i\right)$. *Suppose that each extension* E_i *is abelian, then the same is true of* E.

In the first place E is a Galois extension of K (V, p. 57, Prop. 1). If the field F is defined as in Prop. 1, then we have $E_i \subset F$ for all $i \in I$, whence $E = F$.

An extension E of K is said to be an *abelian closure* of K if it is an abelian extension of K and every abelian extension of K is isomorphic to a subextension of E. Prop. 1 implies the *existence* of an abelian closure of K: for let K_s be a separable closure of K, with Galois group Γ and let $(\overline{\Gamma, \Gamma})$ be the closure of the derived group of Γ; denote by K_{ab} the field of invariants of $(\overline{\Gamma, \Gamma})$; since every separable algebraic extension of K is isomorphic to a subextension of K_s (V, p. 45, Cor.). Prop. 1 shows that K_{ab} is an abelian closure of K. The Galois group of K_{ab} over K is canonically isomorphic to $\Gamma/(\overline{\Gamma, \Gamma})$. Let us now show the *uniqueness* of abelian closures: let E, E′ be two abelian closures of K; by definition there exist K-homomorphisms $u : E \to E'$ and $v : E' \to E$ and Prop. 1 (V, p. 52) implies that $v(u(E)) = E$ and $u(v(E')) = E'$, hence u is a K-isomorphism of E onto E'. Any other K-isomorphism of E onto E' is of the form $u_1 = \sigma_0 \circ u$ with $\sigma_0 \in \mathrm{Gal}(E'/K)$; since $\mathrm{Gal}(E'/K)$ is commutative, the isomorphism $\sigma \mapsto u \circ \sigma \circ u^{-1}$ of $\mathrm{Gal}(E/K)$ onto $\mathrm{Gal}(E'/K)$ is independent of u ; it is called the *canonical isomorphism of* Gal(E/K) *onto* Gal(E'/K).

2. Roots of unity

DEFINITION 2. — *An element ζ of* K *is said to be a root of unity if there exists an integer $n > 0$ such that $\zeta^n = 1$; for every integer $n > 0$ such that $\zeta^n = 1$, ζ is called an n-th root of unity.*

It amounts to the same to say that the roots of unity are the elements of *finite order* of the multiplicative group K^* of non-zero elements of K (I, p. 51). The roots of unity form a subgroup $\mu_\infty(K)$ of K^*, the n-th roots form a subgroup $\mu_n(K)$ of $\mu_\infty(K)$. We have $\mu_\infty(K) = \bigcup_{n \geqslant 1} \mu_n(K)$ and $\mu_n(K) \subset \mu_m(K)$ if n divides m. For every root of unity ζ there exists a least integer $n \geqslant 1$ such that ζ belongs to $\mu_n(K)$, namely the order of ζ in the group K^*.

The group $\mu_n(K)$, being the set of roots of the polynomial $X^n - 1$, is of finite order $\leqslant n$. Let p be the characteristic of K. When $p = 0$, or when $p \neq 0$ and n is not divisible by p, the derivative nX^{n-1} of $X^n - 1$ is relatively prime to $X^n - 1$ and so the polynomial $X^n - 1$ is then *separable* ; if moreover K is algebraically closed, $\mu_n(K)$ is thus a group of n elements.

Suppose that K is of non-zero characteristic p and let $r \geqslant 0$ be an integer ; since the mapping $x \mapsto x^{p^r}$ of K into K is injective, we have $\mu_{np^r}(K) = \mu_n(K)$ for every integer $n \geqslant 1$.

We remark that a field may contain no n-th root of unity other than 1 : this is the case for example with the prime fields **Q** and $\mathbf{F}_2$ for any odd integer n.

THEOREM 1. — *Let p be the characteristic exponent of* K *and let $n > 0$ be an integer. Then the group $\mu_n(K)$ of n-th roots of unity in* K *is cyclic and its order divides n ; when* K *is algebraically closed and n is prime to p, the group $\mu_n(K)$ is cyclic of order n.*

It is enough to prove the first assertion of the theorem which is a consequence of the following more precise lemma :

Lemma 1. — *Let* G *be a finite subgroup of* K^* *of order m ; then* G *is cyclic and we have* $G = \mu_m(K)$.

Consider G as **Z**-module ; we have $mx = 0$ for all $x \in G$, hence the annihilator of G is an ideal of the form $r\mathbf{Z}$, where the integer $r \geqslant 1$ divides m. We thus have $G \subset \mu_r(K)$. By Lemma 4 (V, p. 74) applied with $A = \mathbf{Z}$ and $M = G$ there exists an element x of G of order r ; let G' be the cyclic subgroup of G generated by x. We have $G' \subset \mu_r(K)$, Card $(G') = r$ and Card $\mu_r(K) \leqslant r$; therefore we have

$G' = \mu_r(K) \supset G$ and G is cyclic of order r, hence equal to $\mu_r(K)$. Since G has order m, we have $m = r$ and the lemma follows.

PROPOSITION 2. — *Let* K *be algebraically closed and let* p *be its characteristic exponent. There exists an isomorphism of* $\mu_\infty(K)$ *onto the group* $\mathbf{Q}/\mathbf{Z}[1/p]$.

We have denoted by $\mathbf{Z}[1/p]$ the subring of $\mathbf{Q}$ generated by $1/p$, that is, the set of rational numbers of the form a/p^n with $a \in \mathbf{Z}$ and $n \geqslant 1$; thus we have $\mathbf{Z}[1/p] = \mathbf{Z}$ if K is of characteristic 0.

Let $(\nu_n)_{n \geqslant 1}$ be the strictly increasing sequence consisting of all the integers which are not divisible by p, if $p \neq 1$; put $\lambda_n = \nu_1 \nu_2 \ldots \nu_n$ and denote by H_n the group of λ_n-th roots of unity; we have $H_{n+1} \supset H_n$ and $\mu_\infty(K) = \bigcup_n H_n$. Since H_n is cyclic of order λ_n (Th. 1), there exists a sequence $(\alpha_n)_{n \geqslant 1}$ of roots of unity such that α_n generates H_n and $\alpha_n = \alpha_{n+1}^{\nu_{n+1}}$.

Further let β_n be the residue class mod $\mathbf{Z}[1/p]$ of $1/\lambda_n$ and let H'_n be the cyclic subgroup of $\mathbf{Q}/\mathbf{Z}[1/p]$ generated by β_n. It is clear that $\beta_n = \nu_{n+1}\beta_{n+1}$ and H'_n is of order λ_n because λ_n is not divisible by p if $p \neq 1$. Thus there exists for all $n \geqslant 1$ an isomorphism $\varphi_n : H_n \to H'_n$ such that $\varphi_n(\alpha_n) = \beta_n$ and the relations $\alpha_n = \alpha_{n+1}^{\nu_{n+1}}$, $\beta_n = \nu_{n+1}\beta_{n+1}$ show that φ_{n+1} extends φ_n. Finally there exists a unique isomorphism φ of $\mu_\infty(K)$ onto $\mathbf{Q}/\mathbf{Z}[1/p]$ extending the isomorphisms φ_n, that is, $\varphi(\alpha_n) = \beta_n$ for all $n \geqslant 1$.

Remarks. — 1) When K is an algebraically closed field of characteristic 0, the group $\mu_\infty(K)$ is thus isomorphic (not canonically) to $\mathbf{Q}/\mathbf{Z}$. * When K is the field $\mathbf{C}$ of complex numbers, we can write down an explicit such isomorphism; in effect the mapping $x \mapsto e^{2\pi i x}$ is a homomorphism of $\mathbf{Q}$ into $\mathbf{C}^*$ with kernel $\mathbf{Z}$ and image $\mu_\infty(\mathbf{C})$; thus by passage to quotients it defines an isomorphism of $\mathbf{Q}/\mathbf{Z}$ onto $\mu_\infty(\mathbf{C})$. *

2) The following result may be shown (cf. V, p. 165, Ex. 21): let G and H be two commutative groups all of whose elements are of finite order. Suppose that for every integer $n \geqslant 1$, the equation $nx = 0$ has the same number of solutions, assumed *finite* in G and in H. Then the groups G and H are isomorphic. This provides a new proof of Prop. 2.

3) For every prime number l put $\mu_{l^\infty}(K) = \bigcup_{n \geqslant 0} \mu_{l^n}(K)$. When l is the characteristic p of K, we have $\mu_{p^\infty}(K) = \{1\}$. From I, p. 80, Th. 4, we deduce that $\mu_\infty(K)$ is a direct sum of the subgroups $\mu_{l^\infty}(K)$, where l runs over the set of all prime numbers distinct from p. For a given prime number l only two cases are possible: either $\mu_{l^\infty}(K)$ is finite and then $\mu_{l^\infty}(K)$ is isomorphic to $\mathbf{Z}/l^n\mathbf{Z}$ for a suitable n (Th. 1), or $\mu_{l^\infty}(K)$ is infinite and then $\mu_{l^\infty}(K)$ is isomorphic to $\mathbf{Z}[l^{-1}]/\mathbf{Z}$ (cf. Remark 2).

3. Primitive roots of unity

Let $n \geqslant 1$ be an integer. By the *Euler indicator* of n, written $\varphi(n)$, we understand the number of invertible elements of the ring $\mathbf{Z}/n\mathbf{Z}$ of integers modulo

n. By the next proposition $\varphi(n)$ is also the number of integers k prime to n and such that $0 \leqslant k < n$.

PROPOSITION 3. — *Let k and $n \geqslant 1$ be two integers. Then the following assertions are equivalent :*

a) the residue class of k mod n is invertible in the ring $\mathbf{Z}/n\mathbf{Z}$;

b) the residue class of k mod n generates the cyclic group $\mathbf{Z}/n\mathbf{Z}$;

c) the integers k and n are relatively prime (I, p. 112).

Each of the conditions *a*) and *b*) means that there exists an integer x such that $kx \equiv 1 \pmod{n}$, that is, there exist two integers x and y such that $kx + ny = 1$. This latter condition just means that k and n are relatively prime.

COROLLARY 1. — *Let G be a cyclic group of order n and let d be a divisor of n. Then the number of elements of order d in G is equal to $\varphi(d)$. In particular, $\varphi(n)$ is the number of generators of G.*

Since G is isomorphic to $\mathbf{Z}/n\mathbf{Z}$, the number of generators of G is equal to $\varphi(n)$ by Prop. 3. Let g be a generator of G ; then the elements h of G such that $h^d = 1$ constitute the subgroup H of G generated by $g^{n/d}$; this group is cyclic of order d and the elements of order d of G are the generators of H, hence their number is $\varphi(d)$.

COROLLARY 2.— *For every integer $n \geqslant 1$, we have*

$$\sum_{d \mid n} \varphi(d) = n , \tag{1}$$

where the integer d runs over the set of divisors > 0 of n [1].

With the notation of Cor. 1, every element of G has a finite order which is a divisor d of n and for a fixed d there are $\varphi(d)$ such elements.

The calculation of $\varphi(n)$ is based on the two formulae :

$$\varphi(mn) = \varphi(m)\,\varphi(n) \quad \text{if } m \text{ and } n \text{ are relatively prime} , \tag{2}$$

$$\varphi(p^a) = p^{a-1}(p-1) \quad (p \text{ prime}, a \geqslant 1) . \tag{3}$$

The first follows at once from the fact that the rings $\mathbf{Z}/mn\mathbf{Z}$ and $(\mathbf{Z}/m\mathbf{Z}) \times (\mathbf{Z}/n\mathbf{Z})$ are isomorphic (I, p. 112), and that $(\mathrm{A} \times \mathrm{B})^* = \mathrm{A}^* \times \mathrm{B}^*$ for two rings A and B. To prove (3) we note that the positive divisors of p^a are $1, p, p^2, \ldots, p^a$; hence the integer k has no common divisor with p^a other than 1 if and only if it is not divisible by p ; since there are p^{a-1} multiples of p between 0 and $p^a - 1$, we find indeed (3).

The formulae (2) and (3) show at once that

$$\varphi(n) = n \prod_p (1 - 1/p) , \tag{4}$$

where p runs over the set of prime divisors of n.

[1] The relation $d \mid n$ between integers > 0 means « d divides n » (VI, p. 5).

An n-th root of unity is said to be *primitive* if it is of order n ; if there exists such a root ζ, the group $\mu_n(K)$ is of order n and is generated by ζ. * For example, the primitive n-th roots of unity in $\mathbf{C}$ are the numbers $e^{2\pi ik/n}$ with $0 \leqslant k < n$ and k prime to n. $_*$ Cor. 1 of Prop. 3 now implies the following result.

PROPOSITION 4. — *Let $n \geqslant 1$ be an integer ; suppose that there exist n n-th roots of unity in* K (this is true for example if K is separably closed and $n \cdot 1_K \neq 0$). *Then the number of primitive n-th roots of unity in* K *is equal to* $\varphi(n)$.

4. Cyclotomic extensions

Let p be the characteristic exponent of K and let $n \geqslant 1$ be an integer prime to p ; by a *cyclotomic extension of level n* over K we understand any splitting extension E of the polynomial $X^n - 1$ over K (V, p. 21). Since this polynomial is separable, E is a *Galois* extension of K, of finite degree (V, p. 57). There exists a primitive n-th root of unity in E ; if ζ is such a root, then every n-th root of unity is a power of ζ, hence $E = K(\zeta)$.

For the rest of this No. we shall choose a separable closure K_s of K. For every $n \geqslant 1$ prime to p the group $\mu_n(K_s)$ is cyclic of order n and the field

$$R_n(K) = K(\mu_n(K_s))$$

is a cyclotomic extension of level n of K. We can consider $\mu_n(K_s)$ as a free module of rank 1 over the ring $\mathbf{Z}/n\mathbf{Z}$, and every element σ of $\mathrm{Gal}(K_s/K)$ induces an automorphism of $\mu_n(K_s)$; there exists therefore a homomorphism χ_n : $\mathrm{Gal}(K_s/K) \to (\mathbf{Z}/n\mathbf{Z})^*$ characterized by the formula $u(\zeta) = \zeta^j$ for every n-th root of unity ζ in K_s, every u in $\mathrm{Gal}(K_s/K)$ and every integer j in the residue class $\chi_n(u)$ mod n. Since $R_n(K) = K(\mu_n(K_s))$, the kernel of χ_n is the subgroup $\mathrm{Gal}(K_s/R_n(K))$ of $\mathrm{Gal}(K_s/K)$; hence we have $\chi_n = \varphi_n \circ \psi_n$ where ψ_n is the restriction homomorphism of $\mathrm{Gal}(K_s/K)$ over $\mathrm{Gal}(R_n(K)/K)$ and φ_n is an *injective* homomorphism of $\mathrm{Gal}(R_n(K)/K)$ in $(\mathbf{Z}/n\mathbf{Z})^*$. In particular we have the following result :

PROPOSITION 5. — *For every integer $n \geqslant 1$ prime to p the extension* $R_n(K)$ *of* K *is abelian of finite degree, its Galois group is isomorphic to a subgroup of* $(\mathbf{Z}/n\mathbf{Z})^*$ *and its degree divides the order* $\varphi(n)$ *of* $(\mathbf{Z}/n\mathbf{Z})^*$.

Let $\bar{\mathbf{Q}}$ be an algebraic closure of the field $\mathbf{Q}$ of rational numbers, and let n be an integer. Then the *cyclotomic polynomial* Φ_n *of level n* is defined by

$$\Phi_n(X) = \prod_{\zeta \in S_n} (X - \zeta), \tag{5}$$

where S_n is the set of primitive n-th roots of unity in $\bar{\mathbf{Q}}$. The polynomial Φ_n is of degree $\varphi(n)$ (Prop. 4). It is clear that $\Phi_n(X)$ is invariant under all automorphisms of $\bar{\mathbf{Q}}$, and so belongs to $\mathbf{Q}[X]$. Since every element ζ of S_n is a root of the polynomial $X^n - 1$, the polynomial $\Phi_n(X)$ divides $X^n - 1$, and the following lemma shows that $\Phi_n(X)$ is a monic polynomial with *integer* coefficients.

Lemma 2. — *Let f, g and h be monic polynomials in* $\mathbf{Q}[X]$ *such that* $f = gh$. *If f has integer coefficients, the same is true of g and h.*

Let a (resp. b) be the least of the integers $\alpha \geqslant 1$ (resp. $\beta \geqslant 1$) such that αg (resp. βh) has integer coefficients; we put $g' = ag$ and $h' = bh$ and show by *reductio ad absurdum* that $a = b = 1$. If this were not so, there would be a prime divisor p of ab. If $u \in \mathbf{Z}[X]$, let us write $\bar{u}$ for the polynomial with coefficients in the field $\mathbf{F}_p$ obtained by reduction mod p of the coefficients of u. We have $g'h' = abf$, whence $\bar{g}'\bar{h}' = 0$; since the ring $\mathbf{F}_p[X]$ is an integral domain (IV, p. 9, Prop. 8) we thus have $\bar{g}' = 0$ or $\bar{h}' = 0$. In other words, p divides all the coefficients of g' or all those if h' and this contradicts the hypothesis.

We have the relation

$$X^n - 1 = \prod_{d \mid n} \Phi_d(X) . \tag{6}$$

For we have $X^n - 1 = \prod_{\zeta \in \mu_n(\mathbf{Q})} (X - \zeta)$ and the sets S_d for d dividing n form a partition of $\mu_n(\mathbf{Q})$.

The formula (6) determines $\Phi_n(X)$ when we know $\Phi_d(X)$ for all the divisors $d < n$ of n; since $\Phi_1(X) = X - 1$, we thus have a recursive procedure for calculating Φ_n. For example for prime p we have

$$X^p - 1 = (X - 1)\, \Phi_p(X) ,$$

whence

$$\Phi_p(X) = X^{p-1} + X^{p-2} + \cdots + X + 1 , \tag{7}$$

and

$$\Phi_{p^{r+1}}(X) = \Phi_p(X^{p^r}) \quad \text{for} \quad r \geqslant 0 .$$

Let us list the values of the polynomials $\Phi_n(X)$ for $1 \leqslant n \leqslant 12$:

$$\Phi_1(X) = X - 1$$

$$\Phi_2(X) = X + 1$$

$$\Phi_3(X) = X^2 + X + 1$$

$$\Phi_4(X) = X^2 + 1$$

$$\Phi_5(X) = X^4 + X^3 + X^2 + X + 1$$

$$\Phi_6(X) = X^2 - X + 1$$

$$\Phi_7(X) = X^6 + X^5 + X^4 + X^3 + X^2 + X + 1$$

$$\Phi_8(X) = X^4 + 1$$

$$\Phi_9(X) = X^6 + X^3 + 1$$

$$\Phi_{10}(X) = X^4 - X^3 + X^2 - X + 1$$

$$\Phi_{11}(X) = X^{10} + X^9 + X^8 + X^7 + X^6 + X^5 + X^4 + X^3 + X^2 + X + 1$$

$$\Phi_{12}(X) = X^4 - X^2 + 1$$

The values of Φ_1, Φ_2, Φ_3, Φ_4, Φ_5, Φ_7, Φ_8, Φ_9 and Φ_{11} follow directly from (7) ; now we have $\Phi_1\Phi_2\Phi_3\Phi_6 = X^6 - 1$ and $\Phi_1\Phi_2\Phi_3\Phi_4\Phi_6\Phi_{12} = X^{12} - 1$, whence $\Phi_4\Phi_{12} = \dfrac{X^{12}-1}{X^6-1} = X^6 + 1$ and finally $\Phi_{12} = \dfrac{X^6+1}{X^2+1} = X^4 - X^2 + 1$. The cases $n = 6$ and $n = 10$ may be treated similarly.

Remark. — * For every integer $n > 0$ a function $\mu(n)$ is defined as follows : if n is divisible by the square of a prime number, we put $\mu(n) = 0$, otherwise $\mu(n) = (-1)^h$ if n is the product of h distinct prime numbers (« Möbius function »). It may be shown that

$$\Phi_n(X) = \prod_{d \mid n} (X^{n/d} - 1)^{\mu(d)}, \tag{8}$$

or more explicitly

$$\Phi_n(X) = \prod_{p_1 < \dots < p_h} (X^{n/p_1 \dots p_h} - 1)^{(-1)^h} \tag{9}$$

where $(p_1, \dots, p_h)$ runs over the set of all strictly increasing sequences of prime divisors of n (cf. *Lie*, II, p. 207, Ex. 1). *

5. Irreducibility of cyclotomic polynomials

Let p be the characteristic exponent of K and let n be an integer prime to p. We denote by $\Phi_n \in K[X]$ the image of the polynomial with integer coefficients Φ_n under the unique homomorphism of $\mathbf{Z}[X]$ into $K[X]$ which maps X to X.

Lemma 3. — *The roots of Φ_n in K_s are the primitive n-th roots of unity.*

Denote by S_n the set of roots of Φ_n in K_s. By Formula (6), the set $\mu_n(K_s)$ is the union of the S_d for d dividing n. Every primitive n-th root of unity thus belongs to S_n and the lemma now follows from Prop. 4 (V, p. 81).

PROPOSITION 6. — *Let p be the characteristic exponent of* K *and let $n \geqslant 1$ be an integer prime to p. For the polynomial $\Phi_n(X)$ to be irreducible in* K[X] *it is necessary and sufficient that the homomorphism* $\chi_n : \mathrm{Gal}(K_s/K) \to (\mathbf{Z}/n\mathbf{Z})^*$ *should be surjective.*

By Lemma 3 we have $R_n(K) = K(\zeta)$ for each root ζ of $\Phi_n(X)$ and hence $\Phi_n(X)$ is irreducible in K[X] if and only if the degree $\varphi(n)$ of $\Phi_n(X)$ is equal to $[R_n(K) : K]$. Further, the Galois group of $R_n(K)$ over K is of order $[R_n(K) : K]$ and it is isomorphic to the subgroup of $(\mathbf{Z}/n\mathbf{Z})^*$ which is the image of χ_n. Now Prop. 6 follows from the fact that $(\mathbf{Z}/n\mathbf{Z})^*$ is of order $\varphi(n)$.

THEOREM 2 (Gauss). — *Let $\bar{\mathbf{Q}}$ be an algebraic closure of* **Q** *and let $n \geqslant 1$ be an integer.*

a) The cyclotomic polynomial $\Phi_n(X)$ is irreducible in **Q**[X].

b) The degree of $R_n(\mathbf{Q})$ *over* **Q** *is* $\varphi(n)$.

c) The homomorphism χ_n of $\mathrm{Gal}(\bar{\mathbf{Q}}/\mathbf{Q})$ *into* $(\mathbf{Z}/n\mathbf{Z})^*$ *is surjective and defines by passage to quotients an isomorphism of* $\mathrm{Gal}(R_n(\mathbf{Q})/\mathbf{Q})$ *onto* $(\mathbf{Z}/n\mathbf{Z})^*$.

Taking account of Prop. 6, we need only prove *c*). Every integer r prime to n is a product of prime numbers $p_1, \dots, p_s$ not dividing n ; so it is enough to show that for every prime number p not dividing n, the mapping $x \mapsto x^p$ of $\mu_n(\mathbf{Q})$ into itself extends to an automorphism of $R_n(\mathbf{Q})$. It suffices to prove that if ζ is a primitive n-th root of unity, the minimal polynomial f of ζ over **Q** is equal to the minimal polynomial g of ζ^p over **Q**.

We shall assume that $f \neq g$ and argue by contradiction. The polynomials f and g are monic irreducible in **Q**[X] and divide $X^n - 1$, so there exists $u \in \mathbf{Q}[X]$ such that $X^n - 1 = fgu$ (IV, p. 13, Prop. 13). Lemma 2 (V, p. 82) shows that f, g and u have integer coefficients. Let us denote by $\bar{v}$ the polynomial with coefficients in $\mathbf{F}_p$ obtained from a polynomial $v \in \mathbf{Z}[X]$ by reduction mod p. We thus have $X^n - 1 = \bar{f}\bar{g}\bar{u}$ in $\mathbf{F}_p[X]$.

Further we have $g(\zeta^p) = 0$ and so $g(X^p)$ is a multiple of $f(X)$ in **Q**[X]. By Lemma 2 there exists $h \in \mathbf{Z}[X]$ such that $g(X^p) = f(X) \cdot h(X)$. Now we have $v(X^p) = v(X)^p$ for every polynomial $v \in \mathbf{F}_p[X]$. By reduction mod p we thus obtain $\bar{g}^p = \bar{f}\bar{h}$. If v is an irreducible polynomial in $\mathbf{F}_p[X]$ dividing $\bar{f}$, it must then divide $\bar{g}$. Since $\bar{f}\bar{g}$ divides $X^n - 1$, we conclude that v^2 divides $X^n - 1$ in $\mathbf{F}_p[X]$. This is absurd because the polynomial $X^n - 1$ is separable in $\mathbf{F}_p[X]$.

It may be shown that for every abelian extension E of finite degree over **Q** there exists an integer $n \geqslant 1$ such that E is isomorphic to a subextension of $R_n(\mathbf{Q})$. * In other words, the field $\mathbf{Q}(\mu_\infty(\mathbf{C}))$ is an abelian closure of **Q**. * (« Kronecker-Weber theorem »).

6. Cyclic extensions

DEFINITION 3. — *An extension* E *of* K *is said to be cyclic if it is Galois and its Galois group is cyclic.*

Examples. — 1) Every Galois extension of prime degree is cyclic, because every finite group G of prime order p is cyclic (for every element $x \neq 1$ of G is of order p, hence generates G).

2) Let $F(X) = X^2 + aX + b$ be an irreducible polynomial in K[X]. The only case where F(X) is not separable is that where K is of characteristic 2 and $a = 0$. We shall leave this case aside ; let E be an extension of K generated by a root x of F(X). We have $[E:K] = 2$ and $F(X) = (X - x)(X + a + x)$, hence E is a Galois extension of K. The Galois group of E over K is of order 2, hence cyclic.

3) Let F be a field and σ an automorphism of finite order. The field E of invariants of σ is also the field of invariants of the cyclic group of order n generated by σ, and hence (V, p. 66, Th. 3) F is a cyclic extension of degree n of E.

We know (I, p. 50) that every *subgroup* and every *quotient group* of a cyclic group is cyclic. Therefore (V, p. 68, Cor. 4), if E is a cyclic extension of a field K, of degree n, then every subextension F of E is *cyclic over* K, and E is *cyclic over* F. For every divisor d of n there exists a unique subfield F of degree d over K contained in E : for in a cyclic group of order n there exists a unique subgroup of index d.

THEOREM 3 (Hilbert). — *Let* E *be a cyclic extension of* K *and let* σ *be a generator of the Galois group* Γ *of* E *over* K.

a) For an element $x \in E$ *to satisfy* $N_{E/K}(x) = 1$ *it is necessary and sufficient that* $y \in E^*$ *exist such that* $x = y/\sigma(y)$; *every* $y_1 \in E^*$ *with* $x = y_1/\sigma(y_1)$ *is then of the form* λy *where* $\lambda \in K^*$.

b) For an element $x \in E$ *to satisfy* $Tr_{E/K}(x) = 0$ *it is necessary and sufficient that* $z \in E$ *exist with* $x = z - \sigma(z)$; *then every* $z_1 \in E$ *with* $x = z_1 - \sigma(z_1)$ *is of the form* $z + \mu$, *where* $\mu \in K$.

Let us first prove a lemma.

Lemma 4. — *Let* Γ *be a cyclic group of order* n, σ *a generator of* Γ *and* M *a commutative group on which* Γ *operates according to the rule* $\gamma . (m + m') = \gamma . m + \gamma . m'$ *for all* $\gamma \in \Gamma$ *and* $m, m' \in M$. *Let* Z *be the set of all mappings of* Γ *into* M *satisfying the relation*

$$f(\tau\tau') = f(\tau) + \tau . f(\tau') \quad \text{for} \quad \tau, \tau' \text{ in } \Gamma . \tag{10}$$

Put $u(f) = f(\sigma)$ *for* $f \in Z$ *and* $t(m) = \sum_{\tau \in \Gamma} \tau . m$ *for* $m \in M$. *Then the sequence*

$$0 \to Z \xrightarrow{u} M \xrightarrow{t} M$$

is exact.

Let $f \in Z$; taking $\tau = \tau' = 1$ in (10) we find that $f(1) = 0$. Further, by induction on $m \geqslant 0$ we deduce the relation

$$(11) \qquad f(\sigma^m) = f(\sigma) + \sigma \,.\, f(\sigma) + \cdots + \sigma^{m-2} \,.\, f(\sigma) + \sigma^{m-1} \,.\, f(\sigma) \,.$$

We have $\sigma^n = 1$, whence $f(\sigma^n) = 0$; the preceding relation with $m = n$ is equivalent to the equality $t(u(f)) = 0$, whence $\operatorname{Im} u \subset \operatorname{Ker} t$. Moreover, from (11) it follows that $\operatorname{Ker} u = 0$.

Let $m \in M$ be such that $t(m) = 0$, that is, $m + \sigma \,.\, m + \cdots + \sigma^{n-1} \,.\, m = 0$. Let us define the mapping f of Γ into M by

$$(12) \qquad f(\sigma^j) = m + \sigma \,.\, m + \cdots + \sigma^{j-2} \,.\, m + \sigma^{j-1} \,.\, m$$

for $0 \leqslant j \leqslant n-1$. It may be left to the reader to establish the relation (10). We clearly have $m = f(\sigma)$, whence $\operatorname{Im} u \supset \operatorname{Ker} t$.

With the lemma now proved, let us take $y \in E^*$ and $x = y/\sigma(y)$; we have $N_{E/K}(\sigma(y)) = N_{E/K}(y)$, whence $N_{E/K}(x) = 1$. Conversely, let x in E^* be such that $N_{E/K}(x) = 1$; by Lemma 4, applied to $M = E^*$, there exists a family of elements $(c_\tau)_{\tau \in \Gamma}$ of E^* satisfying the relation $c_{\tau\tau'} = c_\tau \,.\, \tau(c_{\tau'})$ for τ, τ' in Γ and $c_\sigma = x$. By Cor. 1 of Prop. 9 (V, p. 65) there exists $y \in E^*$ with $c_\tau = y/\tau(y)$ for all $\tau \in \Gamma$, whence in particular $x = c_\sigma = y/\sigma(y)$. If $y_1 \in E^*$ satisfies $x = y_1/\sigma(y_1)$, then we have

$$\sigma(y_1 y^{-1}) = y_1 y^{-1} \,,$$

hence $y_1 y^{-1}$ belongs to K^* because σ generates the Galois group of E over K. This proves *a*).

The assertion *b*) follows in similar fashion from Cor. 2 of Prop. 9 (V, p. 65).

7. Duality of $\mathbf{Z}/n\mathbf{Z}$-modules

In this No. we denote by n an integer > 0 and by T a cyclic group of order n. A group G is said to be *annihilated by n* if $g^n = 1$ for all $g \in G$; if moreover G is commutative, the group structure of G is underlying a unique $\mathbf{Z}/n\mathbf{Z}$-module structure.

For every group G we denote by $\operatorname{Hom}(G, T)$ the group of homomorphisms of G into T; it is a commutative group annihilated by n.

PROPOSITION 7. — *Let* G *be a commutative group annihilated by n and* H *a subgroup of* G. *Then the restriction homomorphism* $\operatorname{Hom}(G, T) \to \operatorname{Hom}(H, T)$ *is surjective.*

For let $f : H \to T$ be a homomorphism; we shall prove that there exists a homomorphism of G into T extending f. Suppose first that G is cyclic, generated by an element g of order r dividing n; denote by t a generator of T. There exists a divisor s of r such that H is generated by g^s (I, p. 50, Prop. 19), and for every $x \in \mathbf{Z}$ we have $f(g^{sx}) = t^{ax}$, where a is an integer such that n divides ar/s. Then $a/s = (ar/ns)(n/r)$ is an integer and the homomorphism $g^x \mapsto t^{(a/s)x}$, $x \in \mathbf{Z}$, of G into T extends f. In the general case consider the set of pairs (H', f') where H' is a subgroup of G containing H and f' is a homomorphism of H' into T extending f, and let us order this set by the relation $(\mathrm{H}', f') \leqslant (\mathrm{H}'', f'')$ if $\mathrm{H}' \subset \mathrm{H}''$ and the restriction of f'' to H' is f'. By *Set Theory*, III, p. 154, Def. 3 and Th. 2, this set has a maximal element (H_1, f_1) and it suffices to prove that $\mathrm{H}_1 = \mathrm{G}$; if this is not the case, there exists $g \in \mathrm{G}$, $g \notin \mathrm{H}_1$ and it is enough to prove that f_1 may be extended to a homomorphism T of the subgroup of G generated by H_1 and g. Now if C denotes the cyclic group generated by g, the restriction of f_1 to $\mathrm{C} \cap \mathrm{H}_1$ extends to a homomorphism f_2 of C into T and the homomorphism $xy \mapsto f_1(x)\, f_2(y)$, $x \in \mathrm{H}_1$, $y \in \mathrm{C}$, of $\mathrm{H}_1\mathrm{C}$ into T is the desired mapping.

COROLLARY 1. — *If* G *is a commutative group annihilated by n, and* $\mathrm{G} \neq \{1\}$, *then* $\mathrm{Hom}(\mathrm{G}, \mathrm{T}) \neq \{1\}$.

For it suffices to note that if H is a cyclic subgroup of G distinct from $\{1\}$, then $\mathrm{Hom}(\mathrm{H}, \mathrm{T}) \neq \{1\}$, and to apply Prop. 7.

COROLLARY 2. — *If* G *is a finite commutative group annihilated by n, then the groups* G *and* $\mathrm{Hom}(\mathrm{G}, \mathrm{T})$ *have the same order.*

If G is cyclic of order r, with generator g, then the mapping $f \mapsto f(g)$ is a bijection of $\mathrm{Hom}(\mathrm{G}, \mathrm{T})$ onto the set of elements t of T such that $t^r = 1$, hence the assertion follows in this case. On the other hand, if H is a cyclic subgroup of G, we have $\mathrm{Card}(\mathrm{G}) = \mathrm{Card}(\mathrm{H}) \,.\, \mathrm{Card}(\mathrm{G}/\mathrm{H})$; further, we have an exact sequence

$$\{1\} \to \mathrm{Hom}(\mathrm{G}/\mathrm{H}, \mathrm{T}) \to \mathrm{Hom}(\mathrm{G}, \mathrm{T}) \to \mathrm{Hom}(\mathrm{H}, \mathrm{T}) \to \{1\}$$

(II, p. 227, Th. 1 and Prop. 7 above), hence

$$\mathrm{Card}(\mathrm{Hom}(\mathrm{G}, \mathrm{T})) = \mathrm{Card}(\mathrm{Hom}(\mathrm{H}, \mathrm{T})) \,.\, \mathrm{Card}(\mathrm{Hom}(\mathrm{G}/\mathrm{H}, \mathrm{T})) \,.$$

Since $\mathrm{Card}(\mathrm{Hom}(\mathrm{H}, \mathrm{T})) = \mathrm{Card}(\mathrm{H})$, it comes to the same to prove the corollary for G or for G/H. Now the result follows by induction on Card(G).

Now let G and H be two groups and $f : \mathrm{G} \times \mathrm{H} \to \mathrm{T}$ a *bimultiplicative* mapping, that is, such that for any $g, g' \in \mathrm{G}$, $h, h' \in \mathrm{H}$ we have

$$f(gg', h) = f(g, h)\, f(g', h)\,, \quad f(g, hh') = f(g, h)\, f(g, h')\,.$$

We define group homomorphisms

$$s_f : \mathrm{G} \to \mathrm{Hom}(\mathrm{H}, \mathrm{T})\,, \quad d_f : \mathrm{H} \to \mathrm{Hom}(\mathrm{G}, \mathrm{T})\,,$$

by $s_f(g)(h) = d_f(h)(g) = f(g, h)$ (cf. II, p. 268, Cor. to Prop. 1, when G and H are commutative).

PROPOSITION 8. — *Suppose that* G *and* H *are commutative and annihilated by n. If* s_f *is bijective and* H *is finite, then* d_f *is bijective and we have* $\mathrm{Card}(\mathrm{G}) = \mathrm{Card}(\mathrm{H})$.

If s_f is bijective and H is finite, we have by Cor. 2 to Prop. 7 the relation $\mathrm{Card}(\mathrm{G}) = \mathrm{Card}(\mathrm{Hom}(\mathrm{H}, \mathrm{T})) = \mathrm{Card}(\mathrm{H})$, hence $\mathrm{Card}(\mathrm{G})$ is finite and by another application of the Corollary, $\mathrm{Card}(\mathrm{Hom}(\mathrm{G}, \mathrm{T})) = \mathrm{Card}(\mathrm{H})$. So it is enough to prove that d_f is injective. Now if $h \in \mathrm{Ker}(d_f)$, we have $f(g, h) = 1$ for all $g \in \mathrm{G}$, hence since s_f is bijective, $\varphi(h) = 1$ for all $\varphi \in \mathrm{Hom}(\mathrm{H}, \mathrm{T})$; by Prop. 7 this implies $\mathrm{Hom}(\mathrm{Ker}(d_f), \mathrm{T}) = \{1\}$, hence $\mathrm{Ker}(d_f) = \{1\}$ by Cor. 1 to Prop. 7.

8. Kummer theory

In this No. we denote by n an integer > 0, and we shall suppose that $\mu_n(\mathrm{K})$ has n elements ; by V, p. 78 this means also that n is prime to the characteristic exponent of K and that all the n-th roots of unity in an algebraic closure Ω of K lie in K.

We shall say that an extension L of K is *abelian of exponent dividing n* if it is abelian (V, p. 77, Def. 1) and its Galois group $\mathrm{Gal}(\mathrm{L}/\mathrm{K})$ is annihilated by n (V, p. 86).

Let A be a subset of K^* ; we denote by $\mathrm{K}(\mathrm{A}^{1/n})$ the subextension of Ω generated by all $\theta \in \Omega$ such that $\theta^n \in \mathrm{A}$.

Lemma 5. — $\mathrm{K}(\mathrm{A}^{1/n})$ *is an abelian extension of* K *of exponent dividing n.*

Since the polynomials $\mathrm{X}^n - a$, $a \in \mathrm{A}$ are separable over K, $\mathrm{L} = \mathrm{K}(\mathrm{A}^{1/n})$ is a separable, hence a Galois extension of K. Let $\sigma \in \mathrm{Gal}(\mathrm{L}/\mathrm{K})$ and let $\theta \in \Omega$ be such that $\theta^n \in \mathrm{A}$. We have $\sigma(\theta)^n = \theta^n$; hence there exists $\zeta \in \mu_n(\Omega) = \mu_n(\mathrm{K})$ such that $\sigma(\theta) = \zeta\theta$; this implies that $\sigma^n(\theta) = \zeta^n\theta = \theta$, whence $\sigma^n = 1$. If σ' is another element of $\mathrm{Gal}(\mathrm{L}/\mathrm{K})$ there exists $\zeta' \in \mu_n(\mathrm{K})$ such that $\sigma'(\theta) = \zeta'\theta$, whence $\sigma'\sigma(\theta) = \zeta\zeta'\theta = \sigma\sigma'(\theta)$ and so $\sigma'\sigma = \sigma\sigma'$.

Lemma 6. — *Let* L *be a Galois extension of* K. *There exists a unique mapping* $(\sigma, a) \mapsto \langle\sigma, a\rangle$ *of* $\mathrm{Gal}(\mathrm{L}/\mathrm{K}) \times ((\mathrm{L}^n \cap \mathrm{K}^*)/\mathrm{K}^{*n})$ *into* $\mu_n(\mathrm{K})$ *such that for every* $\sigma \in \mathrm{Gal}(\mathrm{L}/\mathrm{K})$ *and every element* $\theta \in \mathrm{L}^*$ *such that* $\theta^n \in \mathrm{K}$ *we have, on denoting by* $\bar{\theta}^n$ *the residue class of* θ^n mod K^{*n} :

$$\langle\sigma, \bar{\theta}^n\rangle = \sigma(\theta)/\theta . \tag{13}$$

This mapping is bimultiplicative.

In effect the right-hand side of (13) is an n-th root of unity which only depends on the residue class mod K^{*n} of θ^n ; this proves the first assertion. The second assertion is verified without difficulty.

For every Galois extension L of K let us write

$$k_{\mathrm{L}} : (\mathrm{L}^n \cap \mathrm{K}^*)/\mathrm{K}^{*n} \to \mathrm{Hom}(\mathrm{Gal}(\mathrm{L}/\mathrm{K}), \mu_n(\mathrm{K})),$$
$$k'_{\mathrm{L}} : \mathrm{Gal}(\mathrm{L}/\mathrm{K}) \to \mathrm{Hom}((\mathrm{L}^n \cap \mathrm{K}^*)/\mathrm{K}^{*n}, \mu_n(\mathrm{K})),$$

for the homomorphisms derived from the above bimultiplicative mapping (V, p. 87).

PROPOSITION 9. — *For every Galois extension of finite degree* L *of* K, *the homomorphism* k_{L} *is bijective.*

Let $\theta \in \mathrm{L}^*$ be such that $\theta^n \in \mathrm{K}$ and the residue class of θ^n mod K^{*n} belongs to the kernel of k_{L}. For every $\sigma \in \mathrm{Gal}(\mathrm{L}/\mathrm{K})$ we have by definition $\sigma(\theta) = \theta$; hence $\theta \in \mathrm{K}^*$ and $\theta^n \in \mathrm{K}^{*n}$. This proves the injectivity of k_{L}. Now let $f : \mathrm{Gal}(\mathrm{L}/\mathrm{K}) \to \mu_n(\mathrm{K})$ be a homomorphism ; for all $\sigma, \tau \in \mathrm{Gal}(\mathrm{L}/\mathrm{K})$ we have

$$f(\sigma\tau) = f(\sigma)\,f(\tau) = f(\sigma)\,.\,\sigma f(\tau), \quad f(\sigma)^n = 1 .$$

By V, p. 65, Cor. 1 there exists $\theta \in \mathrm{L}^*$ such that $f(\sigma) = \sigma(\theta)/\theta$ for all $\sigma \in \mathrm{Gal}(\mathrm{L}/\mathrm{K})$; since $f(\sigma)^n = 1$, we have $\sigma(\theta^n) = \theta^n$ for all $\sigma \in \mathrm{Gal}(\mathrm{L}/\mathrm{K})$, hence $\theta^n \in \mathrm{K}^*$; if a is the residue class of θ^n mod K^{*n}, we have by definition $f(\sigma) = \langle \sigma, a \rangle$ for $\sigma \in \mathrm{Gal}(\mathrm{L}/\mathrm{K})$, hence $f = k_{\mathrm{L}}(a)$.

COROLLARY. — *If* L *is a Galois extension of* K, *the homomorphism* k_{L} *is injective and its image is the group* $\mathrm{Hom}_c(\mathrm{Gal}(\mathrm{L}/\mathrm{K}), \mu_n(\mathrm{K}))$ *of continuous homomorphisms of the topological group* $\mathrm{Gal}(\mathrm{L}/\mathrm{K})$ *into the discrete group* $\mu_n(\mathrm{K})$.

This follows at once from what has been said, using the fact that L is an increasing directed union of Galois extensions L_i of finite degree and that the homomorphism of $\mathrm{Gal}(\mathrm{L}/\mathrm{K})$ into $\mu_n(\mathrm{K})$ is continuous if and only if it can be factored through one of the quotients $\mathrm{Gal}(\mathrm{L}_i/\mathrm{K})$ of $\mathrm{Gal}(\mathrm{L}/\mathrm{K})$.

THEOREM 4. — *a) The mapping* $\mathrm{H} \mapsto \mathrm{K}(\mathrm{H}^{1/n})$ *is an inclusion preserving bijection of the set of subgroups of* K^* *containing* K^{*n} *onto the set of abelian subextensions of exponent dividing* n *of* Ω. *The inverse mapping is* $\mathrm{L} \mapsto \mathrm{L}^n \cap \mathrm{K}^*$.

b) For every subgroup H *of* K^* *containing* K^{*n} *the homomorphism*

$$k' : \mathrm{Gal}(\mathrm{K}(\mathrm{H}^{1/n})/\mathrm{K}) \to \mathrm{Hom}(\mathrm{H}/\mathrm{K}^{*n}, \mu_n(\mathrm{K}))$$

is bijective, and it is a homeomorphism when the group $\mathrm{Hom}(\mathrm{H}/\mathrm{K}^{*n}, \mu_n(\mathrm{K}))$ *is equipped with the topology of simple convergence.*

c) Let H *be a subgroup of* K^* *containing* K^{*n}. *For every* $a \in \mathrm{H}/\mathrm{K}^{*n}$ *let* θ_a *be an element of* Ω *such that* θ_a^n *is a representative of* a *in* H. *Then the* θ_a, $a \in \mathrm{H}/\mathrm{K}^{*n}$, *form a basis of the vector* K*-space* $\mathrm{K}(\mathrm{H}^{1/n})$. *In particular,*

$$[\mathrm{K}(\mathrm{H}^{1/n}) : \mathrm{K}] = (\mathrm{H} : \mathrm{K}^{*n}) .$$

A) For every abelian extension L of K of exponent dividing n let us put $\mathrm{H}_{\mathrm{L}} = \mathrm{L}^n \cap \mathrm{K}^*$. If $[\mathrm{L} : \mathrm{K}]$ is finite, the homomorphism k'_{L} of $\mathrm{Gal}(\mathrm{L}/\mathrm{K})$ into $\mathrm{Hom}(\mathrm{H}_{\mathrm{L}}, \mu_n(\mathrm{K}))$ is bijective by Prop. 9 and V, p. 88, Prop. 8.

Since every abelian extension of K of exponent dividing n is an increasing directed union of abelian subextensions of finite degree of exponent dividing n, we deduce by passage to inverse limits that k'_L is a homeomorphism of topological groups for every abelian extension L of K dividing n.

B) Let L be an abelian extension of finite degree, of exponent dividing n, of K, and let $L' = K(H_L^{1/n})$; this is a subextension of L ; moreover $H_{L'}$ contains H_L and so is equal to it. Since the homomorphisms k'_L and $k'_{L'}$ are bijective by *A*) and $H_L = H_{L'}$, the groups $\mathrm{Gal}(L/K)$ and $\mathrm{Gal}(L'/K)$ have the same order and so are equal. This shows that $L' = L$, and hence that $L = K(H_L^{1/n})$. If L is an abelian extension of exponent dividing n of K, we have $K(H_L^{1/n}) = L$, because $K(H_L^{1/n})$ is a subextension of L which contains every subextension of finite degree of L.

C) Let H be a subgroup of K^* containing K^{*n} ; put $L = K(H^{1/n})$, then this is an abelian extension of K of exponent dividing n (V, p. 88, Lemma 5). We have $H \subset H_L$, whence we obtain an exact sequence of commutative groups annihilated by n

$$\{1\} \to H/K^{*n} \to H_L/K^{*n} \to H_L/H \to \{1\} .$$

From this we obtain an exact sequence

$$\{1\} \to \mathrm{Hom}(H_L/H, \mu_n(K)) \to \mathrm{Hom}(H_L/K^{*n}, \mu_n(K)) \xrightarrow{u} \mathrm{Hom}(H/K^{*n}, \mu_n(K)) ,$$

where u is the restriction homomorphism.

If we identify $\mathrm{Hom}(H_L/K^{*n}, \mu_n(K))$ with $\mathrm{Gal}(L/K)$ by means of the isomorphism k'_L, the kernel of u is identified with the set of $\sigma \in \mathrm{Gal}(L/K)$ such that $\sigma(\theta) = \theta$ for all $\theta \in H^{1/n}$. It follows that u is injective, hence $\mathrm{Hom}(H_L/H, \mu_n(K))$ is reduced to $\{1\}$; by Cor. 1 of V, p. 87 we thus have $H = H_L$. This completes the proof of *a*) and *b*).

D) Let us prove *c*). If $a, b \in H$, we have $\theta_a\theta_b/\theta_{ab} \in K$. It follows that the vector subspace of $K(H^{1/n})$ generated by the θ_a is stable under multiplication and so coincides with $K(H^{1/n})$. It only remains to show that the θ_a are linearly independent ; to do this we may evidently assume that H/K^{*n} is finite ; then $[K(H^{1/n}) : K] = (\mathrm{Gal}(K(H^{1/n})/K) : \{1\}) = (H : K^{*n})$ by *b*) and Cor. 2 of V, p. 87 ; since the number of θ_a is equal to $(H : K^{*n})$ and they generate the vector K-space $K(H^{1/n})$, they are linearly independent.

Examples. — 1) There exists a largest abelian extension of exponent dividing n of K, contained in Ω ; it is obtained by adjoining to K the n-th roots of all its elements ; its Galois group may be identified with $\mathrm{Hom}(K^*/K^{*n}, \mu_n(K))$, hence also with $\mathrm{Hom}(K^*, \mu_n(K))$.

* 2) Let us take $K = \mathbf{Q}$ and $n = 2$. Then $\mathbf{Q}^*/\mathbf{Q}^{*2}$ is a vector $\mathbf{F}_2$-space having as a basis the union of $\{-1\}$ and the set of all prime numbers. The largest abelian extension of exponent 2 of $\mathbf{Q}$ contained in $\mathbf{C}$ is thus the subfield $\mathbf{Q}(i, \sqrt{2}, \sqrt{3}, \sqrt{5}, \ldots)$ of $\mathbf{C}$. Its Galois group consists of all the automorphisms

obtained by multiplying each of the elements i, $\sqrt{2}$, $\sqrt{3}$, $\sqrt{5}$ etc. by ± 1. *

3) Let L be a cyclic extension of K of degree n ; then the group $(L^n \cap K^*)/K^{*n}$ is cyclic of order n. If $a \in K^*$ is such that the residue class of a mod K^{*n} is a generator of this group, then L is K-isomorphic to $K[X]/(X^n - a)$ and the group $\mathrm{Gal}(L/K)$ consists of the n automorphisms mapping X to ζX, $\zeta \in \mu_n(K)$.

4) Conversely let $a \in K^*$ and let r be the least integer > 0 such that $a^r \in K^{*n}$; then the subfield L of Ω generated by the roots of the polynomial $X^n - a$ is a cyclic extension of K of degree r. In particular, $X^n - a$ is irreducible if and only if $r = n$.

Remark. — Let $a \in K^*$ and let r be the least integer > 0 such that $a^r \in K^n$. Let B be the set of roots in K of the polynomial $X^{n/r} - a$; then we have

$$X^n - a = \prod_{b \in B} (X^r - b)\,, \tag{14}$$

by substitution of X^r for T in the relation $T^{n/r} - a = \Pi(T - b)$. By Example 4 each of the polynomials $X^r - b$ is irreducible, so that (14) is the decomposition of $X^n - a$ into irreducible polynomials in $K[X]$.

9. Artin-Schreier theory

In this No. we shall use p to denote a prime number and assume that K is of characteristic p. We denote by Ω an algebraic closure of K and by $\wp$ the endomorphism of the additive group of Ω defined by

$$\wp(x) = x^p - x\,.$$

By V, p. 93 the kernel of $\wp$ is the prime subfield $\mathbf{F}_p$ of K. For every subset A of K we denote by $K(\wp^{-1}(A))$ the subextension of Ω generated by all $x \in \Omega$ such that $\wp(x) \in A$.

Lemma 7. — $K(\wp^{-1}(A))$ *is an abelian extension of* K *of exponent dividing* p.

Since the polynomials $\wp - a = X^p - X - a$, $a \in A$ are separable over K, the extension $L = K(\wp^{-1}(A))$ is Galois. Let $\sigma \in \mathrm{Gal}(L/K)$ and $x \in \wp^{-1}(A)$; we have $\wp(\sigma(x)) = \wp(x)$, hence $\sigma(x) - x \in \mathbf{F}_p$, that is, $\sigma(x) = x + i$, $i \in \mathbf{F}_p$. This implies that $\sigma^p(x) = x + pi = x$, hence $\sigma^p = 1$; similarly if $\sigma' \in \mathrm{Gal}(L/K)$ and $\sigma'(x) = x + j$, then we have $\sigma \circ \sigma'(x) = x + i + j = \sigma' \circ \sigma(x)$, so $\sigma \circ \sigma' = \sigma' \circ \sigma$.

Lemma 8. — *Let* L *be a Galois extension of* K. *There exists a unique mapping* $(\sigma, a) \mapsto [\sigma, a\rangle$ *of* $\mathrm{Gal}(L/K) \times ((\wp(L) \cap K)/\wp(K))$ *into* $\mathbf{F}_p$ *such that for all* $\sigma \in \mathrm{Gal}(L/K)$ *and every element* $x \in L$ *such that* $\wp(x) \in K$ *we have, on denoting by* $\overline{\wp(x)}$ *the residue class of* $\wp(x)$ mod $\wp(K)$,

$$[\sigma, \overline{\wp(x)}\rangle = \sigma(x) - x\,. \tag{15}$$

This mapping is **Z**-*bilinear* (*for* $\sigma, \tau \in \mathrm{Gal}(\mathrm{L}/\mathrm{K})$, $a, b \in (\wp(\mathrm{L}) \cap \mathrm{K})/\wp(\mathrm{K})$, *we have* $[\sigma\tau, a\rangle = [\sigma, a\rangle + [\tau, a\rangle$, $[\sigma, a+b\rangle = [\sigma, a\rangle + [\sigma, b\rangle$).

For the right-hand side of (15) is an element of $\mathbf{F}_p$ which depends only on the residue class of $\wp(x) \bmod \wp(\mathrm{K})$; this proves the first assertion ; the second may be verified without difficulty.

For every Galois extension L of K let us write

$$a_{\mathrm{L}} : (\wp(\mathrm{L}) \cap \mathrm{K})/\wp(\mathrm{K}) \to \mathrm{Hom}(\mathrm{Gal}(\mathrm{L}/\mathrm{K}), \mathbf{F}_p)$$

$$a'_{\mathrm{L}} : \mathrm{Gal}(\mathrm{L}/\mathrm{K}) \to \mathrm{Hom}((\wp(\mathrm{L}) \cap \mathrm{K})/\wp(\mathrm{K}), \mathbf{F}_p)$$

for the homomorphisms obtained from the above **Z**-bilinear mapping (V, p. 87).

PROPOSITION 10. — *For every Galois extension* L *of finite degree of* K, *the homomorphism* a_{L} *is bijective.*

Let $x \in \mathrm{L}$ be such that $\wp(x) \in \mathrm{K}$ and the residue class of $\wp(x) \bmod \wp(\mathrm{K})$ lies in the kernel of a_{L}. For every $\sigma \in \mathrm{Gal}(\mathrm{L}/\mathrm{K})$ we have by definition $\sigma(x) = x$; hence $x \in \mathrm{K}$ and $\wp(x) \in \wp(\mathrm{K})$. This proves the injectivity of a_{L}. Now let $f : \mathrm{Gal}(\mathrm{L}/\mathrm{K}) \to \mathbf{F}_p$ be a homomorphism ; for all $\sigma, \tau \in \mathrm{Gal}(\mathrm{L}/\mathrm{K})$ we have

$$f(\sigma\tau) = f(\sigma) + \sigma(f(\tau)), \quad f(\sigma) \in \mathbf{F}_p .$$

By V, p. 65, Cor. 2 there exists $x \in \mathrm{L}$ such that $f(\sigma) = \sigma(x) - x$ for all $\sigma \in \mathrm{Gal}(\mathrm{L}/\mathrm{K})$. Since $f(\sigma) \in \mathbf{F}_p$ we have $\wp(\sigma(x)) = \wp(x)$, hence $\sigma(\wp(x)) = \wp(x)$ for all $\sigma \in \mathrm{Gal}(\mathrm{L}/\mathrm{K})$ and $\wp(x) \in \mathrm{K}$. If a is the residue class of $\wp(x) \bmod \wp(\mathrm{K})$, we have $f(\sigma) = [\sigma, a\rangle$, so $f = a_{\mathrm{L}}(a)$.

COROLLARY. — *If* L *is a Galois extension of* K, *the homomorphism* a_{L} *is injective and its image is the group* $\mathrm{Hom}_c(\mathrm{Gal}(\mathrm{L}/\mathrm{K}), \mathbf{F}_p)$ *of continuous homomorphisms of the topological group* $\mathrm{Gal}(\mathrm{L}/\mathrm{K})$ *into the discrete group* $\mathbf{F}_p$.

This is proved in the same way as the Cor. of Prop. 9, V, p. 89.

THEOREM 5. — *a*) *The mapping* $\mathrm{A} \mapsto \mathrm{K}(\wp^{-1}(\mathrm{A}))$ *is a bijection of the set of subgroups of* K *containing* $\wp(\mathrm{K})$ *onto the set of abelian subextensions of exponent dividing p in* Ω. *The inverse mapping is* $\mathrm{L} \mapsto \wp(\mathrm{L}) \cap \mathrm{K}$.

b) *For every subgroup* A *of* K *containing* $\wp(\mathrm{K})$, *the homomorphism*

$$a' : \mathrm{Gal}(\mathrm{K}(\wp^{-1}(\mathrm{A}))/\mathrm{K}) \to \mathrm{Hom}(\mathrm{A}/\wp(\mathrm{K}), \mathbf{F}_p)$$

is bijective, and it is a homeomorphism when $\mathrm{Hom}(\mathrm{A}/\wp(\mathrm{K}), \mathbf{F}_p)$ *is equipped with the topology of simple convergence.*

c) *Let* A *be a subgroup of* K *containing* $\wp(\mathrm{K})$ *and let* B *be a basis of the vector* $\mathbf{F}_p$-*space* $\mathrm{A}/\wp(\mathrm{K})$. *For each* $a \in \mathrm{B}$ *let* x_a *be an element of* Ω *such that* $\wp(x_a)$ *is a representative of a in* A. *Then the monomials* $x^\alpha = \prod_{a \in \mathrm{B}} x_a^{\alpha(a)}$ *with*

$\alpha = (\alpha(a))$ in $\mathbf{N}^{(B)}$ such that $0 \leqslant \alpha(a) < p$ for each $a \in B$ form a basis of the vector K-space $K(\wp^{-1}(A))$. In particular we have $[K(\wp^{-1}(A)) : K] = (A : \wp(K))$.

Th. 5 is proved in the same way as Th. 4 (V, p. 89) *mutatis mutandis*.

Examples. — 1) There exists a greatest abelian extension of K of exponent dividing p, contained in Ω ; it is $K(\wp^{-1}(K))$, its Galois group may be identified with $\mathrm{Hom}(K/\wp(K), \mathbf{F}_p)$.

2) Let L be a cyclic extension of K of degree p ; then the group $(\wp(L) \cap K)/\wp(K)$ is cyclic of order p. If $a \in K$ is such that the residue class of a mod $\wp(K)$ is a generator of this group, then L is K-isomorphic to $K[X]/(X^p - X - a)$ and the Galois group $\mathrm{Gal}(L/K)$ consists of the p automorphisms mapping X to $X + i$, $i \in \mathbf{F}_p$.

3) Conversely, if $a \in K - \wp(K)$, then the polynomial $X^p - X - a$ is irreducible and the subfield L of Ω generated by its roots is a cyclic extension of K of degree p. If $a \in \wp(K)$, then $X^p - X - a = \prod_{\alpha \in \wp^{-1}(a)} (X - \alpha)$.

§ 12. FINITE FIELDS

1. The structure of finite fields

PROPOSITION 1. — *Let K be a finite field with q elements.*

a) The characteristic of K is a prime number p, and there exists an integer $f \geqslant 1$ such that $q = p^f$.

b) The additive group of K is a direct sum of f cyclic groups of order p.

c) The multiplicative group of K is cyclic of order $q - 1$.

Since $\mathbf{Z}$ is infinite and K finite, the unique ring homomorphism $\varphi : \mathbf{Z} \to K$ is not injective and its kernel is a prime ideal of $\mathbf{Z}$ not equal to 0. Therefore the characteristic of K is a prime number and K is an algebra over the field $\mathbf{F}_p$ of p elements (V, p. 2). Let f be the degree of K over $\mathbf{F}_p$. If f were infinite, K would contain, for each integer $n \geqslant 0$ a subspace of dimension n over $\mathbf{F}_p$, whence $q \geqslant p^n$, which is absurd. Therefore f is finite. The additive group of K is thus isomorphic to $(\mathbf{F}_p)^f$, whence the assertions a) and b).

Assertion c) follows by Lemma 1 (V, p. 78).

PROPOSITION 2. — *Let K be a finite field of q elements. The field K is a splitting field of the polynomial $X^q - X$ of $\mathbf{F}_p[X]$ and it is the set of all roots of this polynomial.*

[1] In agreement with the conventions of this chapter we shall only be concerned here with finite *commutative* fields. In fact, every finite field is commutative, as we shall see in Chapter VIII (cf. V, p. 170, Exercise 14).

For every $x \neq 0$ in K we have $x^{q-1} = 1$ because K^* is a finite group of order $q-1$ (I, p. 52). It follows that $x^q = x$ for each x in K. The polynomial $X^q - X$ of $\mathbf{F}_p[X]$ is of degree q and it has q roots in K, whence

(1) $$X^q - X = \prod_{\xi \in K} (X - \xi).$$

Proposition 2 follows at once from this.

COROLLARY. — *Two finite fields of the same cardinal are isomorphic.*

Let K′ be a finite field of q elements; its characteristic is a prime number p' dividing $q = p^f$, whence $p' = p$. Therefore K′ is a splitting field of the polynomial $X^q - X$ in $\mathbf{F}_p[X]$ (Prop. 2), and so K and K′ are isomorphic (V, p. 22, Cor.).

When $K = \mathbf{F}_p$, Formula (1) reduces to the relation

(2) $$X^p - X \equiv \prod_{i=0}^{p-1} (X - i) \bmod p\mathbf{Z}[X]$$

in the polynomial ring $\mathbf{Z}[X]$.

Formula (2) may also be written

(3) $$X^{p-1} - 1 \equiv \prod_{i=1}^{p-1} (X - i) \bmod p\mathbf{Z}[X].$$

In particular for $X = 0$ we obtain (« Wilson's formula »)

(4) $$(p-1)! \equiv -1 \bmod p.$$

2. Algebraic extensions of a finite field

PROPOSITION 3. — *Let* K *be a finite field of* q *elements,* Ω *an algebraically closed extension of* K *and* m *an integer* $\geqslant 1$.

a) There exists a unique subextension K_m *of* Ω *which is of degree* m *over* K.

b) The field K_m *has* q^m *elements and is the set of fixed points of the automorphism* $x \mapsto x^{q^m}$ *of* Ω.

c) We have $K_m = K(\zeta)$ *for every generator* ζ *of the cyclic group* K_m^*.

Let p be the characteristic of K and f the degree of K over $\mathbf{F}_p$. We have $q^m = p^{fm}$ and the mapping $x \mapsto x^{q^m}$ is thus an automorphism of the perfect field Ω (V, p. 7, Prop. 4). Therefore the set K_m of roots of the polynomial $X^{q^m} - X$ of $K[X]$ is a subfield of Ω. Since the derivative of $X^{q^m} - X$ is equal to -1, all the roots of this polynomial are simple (IV, p. 17, Prop. 7) and so K_m has q^m elements. It follows that $[K_m : K] = m$.

Now let L be a subextension of Ω, of degree m over K. As vector space over K, L is isomorphic to K^m and so has q^m elements. We thus have $x^{q^m} = x$ for all $x \in L$ (Prop. 2), whence $L \subset K_m$. Since $[L:K] = [K_m:K] = m$, we finally have $L = K_m$.

We have thus proved Assertions *a)* and *b)*, and *c)* is trivial.

COROLLARY. — *Let* K *be a finite field and* Ω *an algebraically closed extension of* K. *The relative algebraic closure* $\bar{K}$ *of* K *in* Ω *consists of* 0 *and roots of unity and is an algebraic closure of* K.

We know $\bar{K}$ to be an algebraic closure of K (V, p. 22, Example 2) and it is clear that every root of unity in Ω belongs to $\bar{K}$. Further, given $x \neq 0$ in $\bar{K}$, of degree m over K, if K has q elements, then the field $K(x)$ has q^m, whence $x^{q^m - 1} = 1$, and so x is a root of unity in Ω.

Let p be a prime number and let $\mathbf{F}_p = \mathbf{Z}/p\mathbf{Z}$ be the field of p elements. We choose an algebraic closure Ω of $\mathbf{F}_p$, whose existence follows from Steinitz's theorem (V, p. 23, Th. 2). Let f be a positive integer and $q = p^f$. By Prop. 3 there exists a unique subfield of Ω which is of degree f over $\mathbf{F}_p$; we shall denote it by $\mathbf{F}_q(\Omega)$ or by abuse of notation $\mathbf{F}_q$. It is the unique subextension of Ω which has degree f over $\mathbf{F}_p$. It is the unique subfield of Ω of cardinal q, and every field of cardinal q is isomorphic (not canonically) to $\mathbf{F}_q$ (Cor. of Prop. 2). We note that $\mathbf{F}_q$ consists of those x in Ω for which $x^q = x$ and that $\mathbf{F}_q \subset \mathbf{F}_{q'}$ if and only if q' is a power of q.

PROPOSITION 4. — *Let* K *be a finite field of* q *elements and* K_m *an extension of finite degree* m *of* K.

a) The field K_m *is a Galois extension of* K *whose Galois group is the cyclic group of order* m *generated by the automorphism* $\sigma_q : x \mapsto x^q$.

b) For each $x \in K_m$ *the norm of* x *with respect to* K *is equal to* $x^{(q^m - 1)/(q - 1)}$.

c) Every element of K *is the trace* (resp. *norm*) *of an element of* K_m.

Let Γ be the cyclic group of automorphisms of K_m generated by σ_q. The field of invariants of Γ consists of the elements x of K_m such that $x^q = x$, hence is equal to K. Therefore K_m is a Galois extension of K with Galois group Γ, and this latter has order equal to $[K_m : K] = m$ (V, p. 66, Th. 3). Thus *a)* follows.

We have $\Gamma = \{1, \sigma_q, \sigma_q^2, \ldots, \sigma_q^{m-1}\}$; the norm of an element x of K_m with respect to K is thus $N(x) = \prod_{i=0}^{m-1} \sigma_q^i(x) = x^{1 + q + \cdots + q^{m-1}}$ and we have $1 + q + \cdots + q^{m-1} = \dfrac{q^m - 1}{q - 1}$. This proves *b)*. Let ζ be a generator of the cyclic group K_m^* ; the image of the norm $N : K_M^* \to K^*$ is the cyclic subgroup of K_i^* generated by the element $\xi = N(\zeta) = \zeta^{(q^m - 1)/(q - 1)}$; since ζ is of order $q^m - 1$, ξ is of order $q - 1$, and so generates K^*. This proves that every non-zero element of K is norm of a non-zero element of K_m ; moreover, we have $0 = N(0)$.

Finally since K_m is a separable algebraic extension of K, the trace is a non-zero linear form on the vector space K_m over K (V, p. 49, Cor.) ; so every element of K is the trace of an element of K_m.

3. The Galois group of the algebraic closure of a finite field

Let $S \neq \{1\}$ be a set of integers $\geqslant 1$ stable under multiplication ; we shall order it by the relation « m divides n ». When m divides n, we have $m\mathbf{Z} \supset n\mathbf{Z}$, hence there is a canonical homomorphism $\pi_{m,n}$ of the ring $\mathbf{Z}/n\mathbf{Z}$ onto the ring $\mathbf{Z}/m\mathbf{Z}$. We denote by $A(S)$ the inverse limit of the inverse system of rings $(\mathbf{Z}/m\mathbf{Z}, \pi_{m,n})$ indexed by S. Each finite set $\mathbf{Z}/m\mathbf{Z}$ is equipped with the discrete topology and $A(S)$ with the topology induced by that of the product $\prod_{n \in S} (\mathbf{Z}/n\mathbf{Z})$. Then $A(S)$ is a compact topological ring (Gen. Top. I, p. 64, Prop. 8). We see at once that the unique ring homomorphism φ of $\mathbf{Z}$ into $A(S)$ is injective with a dense image ; *we shall identify* $\mathbf{Z}$ *with its image under* φ *in* $A(S)$. For the topology induced on $\mathbf{Z}$ by that of $A(S)$, *the sets* $m\mathbf{Z}$ *(for* $m \in S$*) form a base of neighbourhoods of* 0.

When $S = \mathbf{N}^*$, $A(S)$ is written $\hat{\mathbf{Z}}$. When S consists of the powers of a prime number l, $A(S)$ is written $\mathbf{Z}_l$ and is called « ring of l-adic integers ». We thus have

$$\hat{\mathbf{Z}} = \varprojlim_{m \geqslant 1} \mathbf{Z}/m\mathbf{Z}\,, \quad \mathbf{Z}_l = \varprojlim_{n \geqslant 0} \mathbf{Z}/l^n\mathbf{Z}\,.$$

When S and T are two sets of integers stable under multiplication such that $S \supset T$, we have a natural projection $A(S) \to A(T)$ which is a continuous homomorphism of topological rings. In particular, for every prime number l we have a continuous homomorphism $\hat{\mathbf{Z}} \to \mathbf{Z}_l$. From it we obtain a continuous homomorphism

$$\hat{\mathbf{Z}} \to \prod_l \mathbf{Z}_l$$

(product extended over all prime numbers) ; this is an *isomorphism of topological rings,* as follows from the passage to the inverse limit in I, p. 112, Prop. 11.

Let K be a finite field of q elements and $\bar{K}$ an algebraic closure of K. For every integer $m \geqslant 1$ we denote by K_m the unique subfield of $\bar{K}$ which is of degree m over K (Prop. 3). We have $\bar{K} = \bigcup_{m \geqslant 1} K_m$. Further we denote by σ_q the automorphism $x \mapsto x^q$ of the perfect field $\bar{K}$; it is called the *Frobenius automorphism* of $\bar{K}$ (relative to K).

PROPOSITION 5. — *There exists a unique isomorphism of topological groups* $\pi_K : \hat{\mathbf{Z}} \to \mathrm{Gal}(\bar{K}/K)$ *such that* $\pi_K(1) = \sigma_q$.

Let Γ be the subgroup of $\mathrm{Gal}(\bar{K}/K)$ generated by σ_q. For every integer $m > 0$ we have $\sigma_q^m(x) = x^{q^m}$ for all $x \in \bar{K}$, hence the set of fixed points of σ_q^m is equal to K_m. Since $K_m \neq \bar{K}$, we have $\sigma_q^m \neq 1$. Hence there is an isomorphism π_0 of $\mathbf{Z}$ onto Γ which maps 1 to σ_q.

The field of invariants of Γ consists of the $x \in \bar{K}$ such that $x^q = x$, hence is equal to K. Therefore (V, p. 67, Lemma 2) the group Γ is *dense in* $\mathrm{Gal}(\bar{K}/K)$. Since every subextension of $\bar{K}$ of finite degree over K is one of the fields K_m, the subgroups $\mathrm{Gal}(\bar{K}/K_m)$ form a fundamental system of neighbourhoods of 1 in $\mathrm{Gal}(\bar{K}/K)$. It is clear that $\Gamma \cap \mathrm{Gal}(\bar{K}/K_m)$ is the cyclic group generated by σ_q^m, hence equal to $\pi_0(m\mathbf{Z})$.

From the above remarks about the topology of $\hat{\mathbf{Z}}$ the isomorphism $\pi_0 : \mathbf{Z} \to \Gamma$ extends in a unique fashion to an isomorphism of topological groups $\pi_K : \hat{\mathbf{Z}} \to \mathrm{Gal}(\bar{K}/K)$.

Let $m \geq 1$ be an integer ; it is clear that the Frobenius automorphism of $\bar{K}$ relative to K_m is σ_q^m. Hence we obtain the relation

$$\pi_{K_m}(a) = \pi_K(ma) \quad \text{for} \quad a \in \hat{\mathbf{Z}}\,. \tag{5}$$

4. Cyclotomic polynomials over a finite field

Let K be a finite field of q elements, $n \geq 1$ an integer not divisible by the characteristic p of K and R_n a cyclotomic extension of level n of K (V, p. 81). We know that the group $\mu_n(R_n) = \mu_n$ of n-th roots of unity in R_n is cyclic of order n, that $R_n = K(\mu_n)$ and that there exists an *injective* homomorphism

$$\varphi_n : \mathrm{Gal}(R_n/K) \to (\mathbf{Z}/n\mathbf{Z})^*$$

such that $\sigma(\zeta) = \zeta^j$ for $\sigma \in \mathrm{Gal}(R_n/K)$, $\zeta \in \mu_n$ and $j \in \varphi_n(\sigma)$.

Further, if f is the degree of R_n over K, then the Galois group of R_n over K is cyclic of order f, generated by the automorphism $\sigma_q : x \mapsto x^q$ (V, p. 95, Prop. 4). We have at once :

PROPOSITION 6. — *The image under* φ_n *of the Frobenius automorphism* σ_q *is the residue class of* q mod n.

Therefore, taking into account Prop. 6 of V, p. 84 :

COROLLARY. — *The degree of* R_n *over* K *is the least integer* $f \geq 1$ *such that* $q^f \equiv 1 \pmod{n}$. *For the cyclotomic polynomial* Φ_n *to be irreducible over* K *it is* *necessary and sufficient that the group* $(\mathbf{Z}/n\mathbf{Z})^*$ *should be generated by the residue class of* q *modulo* n.

Examples. — 1) The polynomial $\Phi_3(X) = X^2 + X + 1$ is irreducible in $\mathbf{F}_q[X]$ if and only if $q \equiv 2 \pmod 3$. Likewise $\Phi_4(X) = X^2 + 1$ is irreducible in $\mathbf{F}_q[X]$ if and only if $q \equiv 3 \pmod 4$ and for $\Phi_5 = X^4 + X^3 + X^2 + X + 1$, the irreducibility condition reads $q \equiv 2, 3 \pmod 5$.

2) We have $5^2 \equiv 1 \pmod{12}$, hence the residue class of 5 (mod 12) does not generate $(\mathbf{Z}/12\,\mathbf{Z})^*$. So the polynomial $\Phi_{12}(X) = X^4 - X^2 + 1$ is not irreducible in $\mathbf{F}_5[X]$; in fact we have

$$\Phi_5(X) = (X^2 + 2X - 1)(X^2 - 2X - 1)$$

in $\mathbf{F}_5[X]$.

§ 13. p-RADICAL EXTENSIONS OF HEIGHT $\leqslant 1$

Throughout this paragraph p *denotes a prime number. All the fields considered are of characteristic* p.

1. p-free subsets and p-bases

DEFINITION 1. — *Let* K *be a field and* L *a* p*-radical extension of height* $\leqslant 1$ *of* K. *A family* $(x_i)_{i \in I}$ *of elements of* L *is said to be* p*-free over* K (resp. *a* p*-basis of* L *over* K) *if* $x_i \notin K$ *for all* $i \in I$ *and the homomorphism of* $\bigotimes_{i \in I} K(x_i)$ *into* L *derived from the canonical injections* $u_i : K(x_i) \to L$ *is injective* (resp. *bijective*).

If a is an element of $L - K$, it is p-radical of height 1, and so its minimal polynomial over K is $X^p - a^p$ (V, p. 24, Prop. 1) ; therefore $\{1, a, \ldots, a^{p-1}\}$ is a basis of $K(a)$ over K. Let $(x_i)_{i \in I}$ be a family of elements of $L - K$ and Λ the subset of $\mathbf{N}^{(I)}$ consisting of all families of finite support $\alpha = (\alpha_i)_{i \in I}$ such that $\alpha_i < p$ for all $i \in I$; Prop. 9 (III, p. 471) shows that the elements $\bigotimes_{i \in I} x_i^{\alpha_i}$ as α runs over Λ form a basis of the vector space $\bigotimes_{i \in I} K(x_i)$ over K. Moreover, the canonical homomorphism of $\bigotimes_{i \in I} K(x_i)$ into L has as image $K(x_i)_{i \in I}$ (V, p. 18, Cor. 1), we thus have the following proposition :

PROPOSITION 1. — *Let* K *be a field,* L *a* p*-radical extension of height* $\leqslant 1$ *of* K *and* $\mathbf{x} = (x_i)_{i \in I}$ *a family of elements of* L. *Then the vector space* $K(x_i)_{i \in I}$ *over* K *is generated by the products* $\mathbf{x}^\alpha = \prod_{i \in I} x_i^{\alpha_i}$ *for* α *in* Λ. *For* $(x_i)_{i \in I}$ *to be* p*-free* (resp. *a* p*-basis*) *it is necessary and sufficient that the family* $(\mathbf{x}^\alpha)_{\alpha \in \Lambda}$ *should be free over* K (resp. *a basis of* L *over* K).

COROLLARY. — *Let* L' *be an extension of a field* K, *generated by two linearly disjoint subextensions* K' *and* L. *Suppose that* L *is p-radical of height* $\leqslant 1$ *over* K ; *then* L' *is p-radical of height* $\leqslant 1$ *over* K'. *Moreover, for a family of elements of* L *to be p-free over* K (resp. *a p-basis of* L *over* K) *it is necessary and sufficient that it should be p-free over* K' (resp. *a p-basis of* L' *over* K').

We have $L^p \subset K \subset K'$ and $L' = K'(L)$, whence $L'^p = K'^p(L^p) \subset K'$. In other words, L' is a p-radical extension of height $\leqslant 1$ of K'. The other assertions of the corollary follow at once from Prop. 1 and from V, p. 14, Prop. 5.

Remark. — Let K be a field, L a p-radical extension of height $\leqslant 1$ of K and $(x_i)_{i \in I}$ a family of elements of L. Let A be the polynomial algebra $K[(X_i)_{i \in I}]$, $\mathfrak{a}$ the ideal of A generated by the polynomials $X_i^p - x_i^p$ and $\varphi : A \to L$ the K-algebra homomorphism such that $\varphi(X_i) = x_i$ for each $i \in I$. The family $(x_i)_{i \in I}$ is p-free if and only if the kernel of φ is equal to $\mathfrak{a}$. For $K[(X_i)]/\mathfrak{a}$ may be identified with the algebra $\bigotimes_{i \in I} K[X_i]/(X_i^p - x_i^p)$.

Let L be a p-radical extension of height $\leqslant 1$ of the field K. A subset S of L is said to be p-free (resp. a p-basis) if the family defined by the identity mapping of S onto itself is p-free (resp. a p-basis). For a family $(x_i)_{i \in I}$ of elements of L to be p-free (resp. a p-basis) it is necessary and sufficient that the mapping $i \mapsto x_i$ should be a bijection of I on a p-free subset (resp. a p-basis) of L. By Prop. 1, every subset of a p-free set is p-free ; conversely, if S is a subset of L all of whose finite subsets are p-free, then S is p-free. Finally, a p-basis of L over K is a p-free subset B such that $L = K(B)$.

PROPOSITION 2. — *Let* K *be a field,* L *a p-radical extension of height* $\leqslant 1$ *of* K *and* S *a subset of* L. *For* S *to be p-free it is necessary and sufficient that* $K(T) \neq K(S)$ *for every subset* $T \neq S$ *of* S.

Suppose first that S is p-free and let $T \neq S$ be a subset of S. Denote by Λ the set of families with finite support $\alpha = (\alpha_s)_{s \in S}$ of integers between 0 and $p - 1$; let Λ' be the subset of Λ consisting of the families $\alpha = (\alpha_s)_{s \in S}$ such that $\alpha_s = 0$ for all s in $S - T$. We also put $u_\alpha = \prod_{s \in S} s^{\alpha_s}$ for $\alpha \in \Lambda$. Then (V, p. 98, Prop. 1) the family $(u_\alpha)_{\alpha \in \Lambda}$ is a basis of $K(S)$ over K and the subfamily $(u_\alpha)_{\alpha \in \Lambda'}$ is a basis of $K(T)$ over K. Since $\Lambda' \neq \Lambda$, we have $K(T) \neq K(S)$.

Suppose now that S is not p-free. Then there exists an integer $n \geqslant 1$ and a sequence of elements $x_1, \ldots, x_n$ of S such that $(x_1, \ldots, x_{n-1})$ is p-free but $(x_1, \ldots, x_n)$ is not. We have $[K(x_1, \ldots, x_{n-1}) : K] = p^{n-1}$ and $[K(x_1, \ldots, x_n) : K] < p^n$. Since $[K(x_1, \ldots, x_n) : K]$ is a multiple of $[K(x_1, \ldots, x_{n-1}) : K]$, we thus have

$$[K(x_1, \ldots, x_n) : K] = [K(x_1, \ldots, x_{n-1}) : K] .$$

whence $x_n \in K(x_1, \ldots, x_{n-1})$. This shows that $K(S - \{x_n\}) = K(S)$.

PROPOSITION 3. — *Let* K *be a field,* L *a p-radical extension of height* $\leqslant 1$ *of* K *and* S, T *two subsets of* L. *The following conditions are equivalent :*

a) The subset S *is p-free over* K *and* T *is p-free over* K(S).

b) $S \cap T = \varnothing$ *and* $S \cup T$ *is p-free over* K.

If T is *p*-free over K(S), we have $T \cap K(S) = \varnothing$ and *a fortiori* $S \cap T = \varnothing$. We may therefore assume S and T disjoint. Let Λ be the subset of $\mathbf{N}^{(S \cup T)}$ consisting of all families $\alpha = (\alpha_x)_{x \in S \cup T}$ with $\alpha_x < p$ for all $x \in S \cup T$, and define subsets Λ' of $\mathbf{N}^{(S)}$ and Λ'' of $\mathbf{N}^{(T)}$ in similar fashion. We can in a natural way, identify $\mathbf{N}^{(S \cup T)}$ with $\mathbf{N}^{(S)} \times \mathbf{N}^{(T)}$, and then Λ is identified with $\Lambda' \times \Lambda''$. For $\alpha \in \Lambda$ put $u_\alpha = \prod_{x \in S \cup T} x^{\alpha_x}$, and similarly define u'_β and u''_γ for $\beta \in \Lambda'$ and $\gamma \in \Lambda''$. We have $u_\alpha = u'_\beta u''_\gamma$ for $\alpha = (\beta, \gamma)$ in $\Lambda = \Lambda' \times \Lambda''$. Moreover $(u'_\beta)_{\beta \in \Lambda'}$ generates the vector K-space K(S) (V, p. 94, Prop. 1). For $S \cup T$ to be *p*-free over K it is necessary and sufficient that the family $(u'_\beta u''_\gamma)_{\beta \in \Lambda', \gamma \in \Lambda''}$ should be free over K (V, p. 98, Prop. 1) ; it comes to the same (II, p. 222) to suppose that the family $(u'_\beta)_{\beta \in \Lambda'}$ is free over K and the family $(u''_\gamma)_{\gamma \in \Lambda''}$ free over K(S). Now the equivalence of *a*) and *b*) follows from Prop. 1 (V, p. 98).

COROLLARY. — *Let* L *be a p-radical extension of height* $\leqslant 1$ *of* K *and* M *a subextension of* L. *Then* L *is p-radical of height* $\leqslant 1$ *over* M *and* M *is p-radical of height* $\leqslant 1$ *over* K. *Moreover, if* B *is a p-basis of* M *over* K *and* C *is a p-basis of* L *over* M, *then* $B \cap C = \varnothing$ *and* $B \cup C$ *is a p-basis of* L *over* K.

Let K be a field. A family $(x_i)_{i \in I}$ is said to be a (absolute) *p-basis* of K if it is a *p*-basis of K over K^p. For every integer $n \geqslant 1$ we denote by $\Lambda(n)$ the subset of $\mathbf{N}^{(I)}$ consisting of all $\alpha = (\alpha_i)_{i \in I}$ such that $\alpha_i < p^n$ for all $i \in I$.

PROPOSITION 4. — *Let* $\mathbf{x} = (x_i)_{i \in I}$ *be a p-basis of* K. *For every integer* $n \geqslant 1$, *the family* $(\mathbf{x}^\alpha)_{\alpha \in \Lambda(n)}$ *is a basis of* K *over* K^{p^n}.

For $n = 1$ the assertion reduces to Prop. 1 (V, p. 98). The set $\Lambda(n)$ consists of elements of $\mathbf{N}^{(I)}$ of the form $\alpha = \beta + p^{n-1}\gamma$ with $\beta \in \Lambda(n-1)$ and $\gamma \in \Lambda(1)$. Such a decomposition is unique and we have $\mathbf{x}^\alpha = \mathbf{x}^\beta (\mathbf{x}^\gamma)^{p^{n-1}}$. Moreover, the family $(x_i^{p^{n-1}})_{i \in I}$ is clearly a *p*-basis of $K^{p^{n-1}}$ over K^{p^n}, hence the family $(\mathbf{x}^\gamma)^{p^{n-1}}$ is a basis of $K^{p^{n-1}}$ over K^{p^n}. We thus obtain the conclusion by induction, using II, p. 222, Prop. 25.

2. Differentials and *p*-bases

Let K be a field (of characteristic p) and V a vector space over K. We recall (III, p. 552) that a derivation of K into V is a mapping D of K into V satisfying the relations

$$D(x + y) = D(x) + D(y) \tag{1}$$

$$D(xy) = xD(y) + yD(x) \tag{2}$$

for x, $y \in \mathrm{K}$. It follows that $\mathrm{D}(1) = 0$ and $\mathrm{D}(x^n) = nx^{n-1}\mathrm{D}(x)$ for all $x \neq 0$ and all $n \in \mathbf{Z}$ (III, p. 557 and 558). Since K is of characteristic p, we thus have for all $x \in \mathrm{K}$

$$\mathrm{D}(x^p) = 0 . \tag{3}$$

Moreover (III, p. 558), we have for x, $y \in \mathrm{K}$, $y \neq 0$,

$$\mathrm{D}(x/y) = (y\mathrm{D}(x) - x\mathrm{D}(y))/y^2 . \tag{4}$$

By the above formulae the kernel of D is a subfield E of K containing K^p. Let M be a subfield of K ; we say that D is an M-derivation if it is M-linear ; by (2) it comes to the same to assume that the restriction of D to M is zero.

We have (III, p. 570) defined the module $\Omega_{\mathrm{M}}(\mathrm{K})$ of M-differentials of K and the canonical M-derivation $d = d_{\mathrm{K/M}}$ of K into $\Omega_{\mathrm{M}}(\mathrm{K})$. The image of $d_{\mathrm{K/M}}$ generates the vector K-space $\Omega_{\mathrm{M}}(\mathrm{K})$. For a mapping D of K into V to be an M-derivation it is necessary and sufficient that there should exist a K-linear mapping $u : \Omega_{\mathrm{M}}(\mathrm{K}) \to \mathrm{V}$ such that $\mathrm{D} = u \circ d_{\mathrm{K/M}}$; this mapping u is uniquely determined.

PROPOSITION 5. — *Let* K *be a field and* L *an extension of* K *generated by an element* x *such that* $x \notin \mathrm{K}$, $x^p \in \mathrm{K}$. *Let* V *be a vector space over* L *and* Δ *a derivation of* K *into* V *such that* $\Delta(x^p) = 0$. *Then there exists a unique derivation* D *of* L *into* V *extending* Δ *and such that* $\mathrm{D}(x) = 0$.

By V, p. 24, Prop. 1, $\{1, x, \ldots, x^{p-1}\}$ is a basis of L over K. For every element $u = a_0 + a_1 x + \cdots + a_{p-1}x^{p-1}$ of L (with $a_0, \ldots, a_{p-1}$ in K) we put $\mathrm{D}(u) = \Delta(a_0) + x\Delta(a_1) + \cdots + x^{p-1}\Delta(a_{p-1})$. It is clear that D extends Δ and satisfies (1) ; so it is enough to establish the relation

$$\mathrm{D}(uv) = u \,.\, \mathrm{D}(v) + v \,.\, \mathrm{D}(u)$$

when u is of the form ax^i and v of the form bx^j with a, b in K, $0 \leqslant i < p$ and $0 \leqslant j < p$.

When $i + j < p$ we have $uv = x^{i+j} . ab$, whence $\mathrm{D}(uv) = x^{i+j}\Delta(ab)$. Otherwise we have $0 \leqslant i + j - p < p$ and so $uv = x^{i+j-p}(abx^p)$, with $abx^p \in \mathrm{K}$, but since $\Delta(x^p) = 0$, we have $\Delta(abx^p) = x^p\Delta(ab)$, whence again $\mathrm{D}(uv) = x^{i+j}\Delta(ab)$. So we have in all cases

$$\begin{aligned} \mathrm{D}(uv) &= x^{i+j}\Delta(ab) = x^i x^j (a \,.\, \Delta(b) + b \,.\, \Delta(a)) \\ &= (ax^i)\, x^j \,.\, \Delta(b) + (bx^j) \,.\, x^i \,.\, \Delta(a) = u \,.\, \mathrm{D}(v) + v \,.\, \mathrm{D}(u) . \end{aligned}$$

COROLLARY 1. — *Let* K *be a field,* L *a p-radical extension of height* $\leqslant 1$ *of* K *and* V *a vector space over* L. *Every derivation* Δ *of* K *into* V *vanishing on* L^p *extends to a derivation of* L *into* V.

By Zorn's lemma (E, III, p. 20) there exists a maximal extension of Δ to a derivation $\mathrm{D}_0 : \mathrm{L}_0 \to \mathrm{V}$, where L_0 is a subfield of L containing K. Let $x \in \mathrm{L}$; we have $x^p \in \mathrm{L}_0$ and $\Delta(x^p) = 0$, whence $\mathrm{D}_0(x^p) = 0$. By Prop. 5,

D_0 extends to a derivation defined on $L_0(x)$; by the maximal character of D_0 we thus have $L_0(x) = L_0$ whence $x \in L_0$. Since x was arbitrary we conclude that $L_0 = L$.

COROLLARY 2. — *Let* L *be a p-radical extension of height* $\leqslant 1$ *of a field* K *and let* E *be a subextension of* L. *Let* U *be the subspace of* $\Omega_K(L)$ *generated by the differentials of elements of* E. *Then* E *consists of the elements of* L *whose differential belongs to* U. *In particular we have* $d_{L/K}x \neq 0$ *for all* $x \in L - K$.

Let x be an element of L not belonging to E. Then $\{1, x, \ldots, x^{p-1}\}$ is a basis of $E(x)$ over E. Let Δ be the E-linear mapping of $E(x)$ into L such that $\Delta(x^i) = ix^i$ for $0 \leqslant i < p$. We verify at once that Δ is an E-derivation of $E(x)$ into L, and in particular a K-derivation. By Cor. 1 to Prop. 5 there exists a K-derivation D of L into L extending Δ. By the universal property of $\Omega_K(L)$, there exists a linear form u on this vector L-space such that $D = u \circ d_{L/K}$. We have $D(x) = x$ and $D | E = 0$; therefore we have $u(d_{L/K}x) \neq 0$ and $u | U = 0$, whence $d_{L/K}x \notin U$.

The last assertion is the particular case $E = K$; then we have $U = 0$.

THEOREM 1. — *Let* L *be a p-radical extension of height* $\leqslant 1$ *of a field* K *and* $(a_i)_{i \in I}$ *a family of elements of* L.

a) For $(x_i)_{i \in I}$ *to be p-free over* K *it is necessary and sufficient that the family* $(dx_i)_{i \in I}$ *should be free in the vector* L*-pace* $\Omega_K(L)$.

b) For $(x_i)_{i \in I}$ *to generate* L *over* K *it is necessary and sufficient that the family* $(dx_i)_{i \in I}$ *should generate the vector* L*-space* $\Omega_K(L)$.

c) For $(x_i)_{i \in I}$ *to be a p-basis of* L *over* K *it is necessary and sufficient that the family* $(dx_i)_{i \in I}$ *should be a basis of the vector* L*-space* $\Omega_K(L)$.

To begin with let us remark that the differential dx of an element of the field $K((x_i)_{i \in I})$ is a linear combination with coefficients in L of the differentials dx_i, $i \in I$. For the family $(x_i)_{i \in I}$ to be p-free it is necessary and sufficient that $x_i \notin K(x_j)_{j \in I - \{i\}}$ for all $i \in I$ (V, p. 99, Prop. 2). By Cor. 2 of Prop. 5 this means that dx_i is not a linear combination of the dx_j for $j \neq i$ in I. Assertion *a*) follows from this. Now *b*) follows at once from Cor. 2 to Prop. 5 and *c*) follows from *a*) and *b*).

The following corollary makes Cor. 1 of Prop. 5 more specific :

COROLLARY. — *Let* $(x_i)_{i \in I}$ *be a p-basis of* L *over* K. *Let* V *be a vector space over* L, Δ *a derivation of* K *into* V *which vanishes on* L^p *and* $(u_i)_{i \in I}$ *a family of elements of* V. *Then there exists a unique derivation* D *of* L *in* V *which extends* Δ *and maps* x_i *to* u_i *for all* $i \in I$.

By Cor. 1 of Prop. 5 there exists a derivation D_0 of L into V extending Δ. The derivations of L in V extending Δ are precisely the mappings of the form $D = D_0 + u \circ d_{L/K}$ where u is an L-linear mapping of $\Omega_K(L)$ into V. We have $D(x_i) = u_i$ if and only if the linear mapping u satisfies the conditions $u(dx_i) = u_i - D_0(x_i)$. Since the family dx_i is a basis of $\Omega_K(L)$, this determines u uniquely and the corollary follows.

Let L be a p-radical extension of height $\leqslant 1$ of K and $(x_i)_{i \in I}$ a p-basis of L over K. By the Cor. to Th. 1 there exists for each $i \in I$ a K-derivation D_i of L into L characterized by $D_i(x_j) = \delta_{ij}$ (Kronecker symbol) ; D_i is sometimes called the *partial derivation with respect to* x_i. When I is *finite*, the family $(D_i)_{i \in I}$ is a basis of the vector space over L consisting of the K-derivations of L into L.

Th. 1 allows the study of p-bases to be reduced to the study of bases of a vector space. For example we have the following result :

THEOREM 2. — *Let* L *be a p-radical extension of height* $\leqslant 1$ *of a field* K.

a) There exist p-bases of L *over* K. *More precisely, if* S *is a p-free subset of* L *over* K *and* T *a subset of* L *such that* $S \subset T$ *and* $L = K(T)$, *then there exists a p-basis* B *of* L *over* K *such that* $S \subset B \subset T$.

b) Two p-bases of L *over* K *have the same cardinal.*

c) For $[L:K]$ *to be finite it is necessary and sufficient that the vector space* $\Omega_K(L)$ *over* L *should be of finite dimension over* L *and then we have*

$$[L:K] = p^{[\Omega_K(L):L]}. \tag{5}$$

Assertions *a*) and *b*) are immediate (II, p. 292). If $[L:K]$ is finite, there exists by *a*) a finite p-basis $(x_1, \ldots, x_n)$ of L over K ; then the monomials $x_1^{\alpha_1} \ldots x_n^{\alpha_n}$ with $0 \leqslant \alpha_i < p$ for $1 \leqslant i \leqslant n$ form a basis of L over K and the differentials $dx_1, \ldots, dx_n$ form a basis of $\Omega_K(L)$ over L. We thus have $[L:K] = p^n$ and $[\Omega_K(L):L] = n$. Conversely if $\Omega_K(L)$ is of finite dimension over L, there exists a finite p-basis of L over K (V, p. 102, Th. 1) and $[L:K]$ is finite.

COROLLARY. — *For each* $x \in L - K$, *there exists a p-basis of* L *over* K *containing* x.

For $\{x\}$ is a p-free subset of L over K and so it is enough to apply Th. 2, *a*).

Let K be a field and L an extension of K. If D is a K-derivation of L with values in a vector L-space, then we have $D(x^p) = 0$ for all $x \in L$, and so D is a $K(L^p)$-derivation ; it follows that $\Omega_K(L) = \Omega_{K(L^p)}(L)$. Since L is a p-radical extension of height $\leqslant 1$ of $K(L^p)$, we can apply the above results. For example, from Th. 1, *c*) we deduce the following : let $(x_i)_{i \in I}$ be a family of elements of L ; for $(dx_i)_{i \in I}$ to be a basis of the vector L-space $\Omega_K(L)$ it is necessary and sufficient that $(x_i)_{i \in I}$ should be a p-basis of L over $K(L^p)$. Similarly, Cor. 2 of Prop. 5 implies :

PROPOSITION 6. — *Let* K *be a field (of characteristic p) and* L *an extension of* K.

a) Let $x \in L$; *the differential* $d_{L/K}x$ *vanishes if and only if* x *belongs to* $K(L^p)$.

b) $\Omega_K(L) = 0$ *if and only if* $L = K(L^p)$.

c) Suppose that L *is p-radical of finite height over* K ; *then* $\Omega_K(L) = 0$ *if and only if* $L = K$.

3. The correspondence between subfields and Lie algebras of derivations

We denote by E a field and by $\mathfrak{g}$ the set of all derivations of E into itself. We recall that $\mathfrak{g}$ is a vector space over E, with operations defined by

$$(6) \qquad (D + D')(x) = D(x) + D'(x), \quad (aD)(x) = a \cdot D(x)$$

for D, D′ in $\mathfrak{g}$ and a, x in E. Moreover, if D and D′ are derivations of E into E, the same is true of $[D, D'] = DD' - D'D$ (III, p. 554). Finally, Leibniz's formula (III, p. 556) gives

$$D^p(xy) = x \cdot D^p(y) + \sum_{j=1}^{p-1} \binom{p}{j} D^j(x) D^{p-j}(y) + D^p(x) \cdot y$$

(x, y in E) ; since the binomial coefficients $\binom{p}{j}$ for $1 \leqslant j \leqslant p-1$ are divisible by p (V, p. 4, Lemma 1), we see that D^p is a derivation of E into E. We immediately find the relation (for a, $a' \in E$)

$$(7) \qquad [aD, a'D'] = aa' \cdot [D, D'] + (aD(a')) \cdot D' - (a'D'(a)) \cdot D .$$

In particular, the mapping $(D, D') \mapsto [D, D']$ of $\mathfrak{g} \times \mathfrak{g}$ into $\mathfrak{g}$ is E^p-linear.

We denote by $\mathscr{C}$ the set of subfields K of E such that $E^p \subset K$ and $[E : K]$ is finite ; for each $K \in \mathscr{C}$ we denote by $\mathfrak{g}(K)$ the set of K-derivations of E. Further, we denote by $\mathscr{L}$ the set of vector subspaces $\mathfrak{h}$ of $\mathfrak{g}$ of finite dimension over E and such that $[D, D'] \in \mathfrak{h}$ and $D^p \in \mathfrak{h}$ for any D, D′ in $\mathfrak{h}$; for each $\mathfrak{h} \in \mathscr{L}$ we denote by $I(\mathfrak{h})$ the set of $x \in E$ such that $D(x) = 0$ for all $D \in \mathfrak{h}$.

THEOREM 3 (Jacobson). — *The mappings $K \mapsto \mathfrak{g}(K)$ and $\mathfrak{h} \mapsto I(\mathfrak{h})$ are bijections of $\mathscr{C}$ onto $\mathscr{L}$ and of $\mathscr{L}$ onto $\mathscr{C}$ respectively, which are mutually inverse. If $K \in \mathscr{C}$ and $\mathfrak{h} \in \mathscr{L}$ correspond to each other, then $[E : K] = p^{[\mathfrak{h} : E]}$.*

The proof requires several preliminary lemmas.

Lemma 1. — *Let L be a field, V a vector space over L and u an endomorphism of V such that $u^p = u$. For each $i \in \mathbf{F}_p$ let V_i be the kernel of $u - i$. Then we have*

$$V = \bigoplus_{i \in \mathbf{F}_p} V_i .$$

For each $i \in \mathbf{F}_p$ denote by $P_i(X)$ the polynomial $-\prod_{j \neq i} (X - j)$. We have

$$(8) \qquad (X - i) P_i(X) = X - X^p$$

by Formula (2) (V, p. 94). On deriving the quoted formula (2), we find

$$(9) \qquad \sum_{i \in \mathbf{F}_p} P_i(X) = 1 .$$

The formulae (8) and (9) show that the endomorphism $P_i(u)$ of V maps V into V_i and that $\sum_{i \in \mathbf{F}_p} P_i(u) = 1$, whence $V = \sum_{i \in \mathbf{F}_p} V_i$. It remains to show that the sum is direct. For each $i \in \mathbf{F}_p$ let $v_i \in V_i$; it is clear that $P_i(u)$ annihilates v_j for all $j \neq i$ and we have $P(u)\, v_i = av_i$ with $a = -\prod_{n \in \mathbf{F}_p^*} n \neq 0$. The relation $\sum_{i \in \mathbf{F}_p} v_i = 0$ thus implies $v_i = 0$ for all $i \in \mathbf{F}_p$ and so the sum is direct.

Lemma 2.— Let D *be a derivation of* E *such that* $D^p = D$ *and let* K *be the kernel of* D. *Suppose that there exists a non-zero* x *in* E *such that* $D(x) = x$. *Then* K *is a subfield of* E *containing* E^p *and we have* $[E : K] = p$.

It is clear that K is a subfield of E containing E^p. Denote by K_i the kernel of $D - i$ for $i \in \mathbf{F}_p$. We have $K_0 = K$ and Lemma 1 shows that $E = \bigoplus_{i \in \mathbf{F}_p} K_i$. Let $i \in \mathbf{F}_p$ and let u be in K_i ; we have

$$D(xu) = D(x) \,.\, u + x \,.\, D(u) = xu + x(iu) = (i+1)\, xu$$

whence $xu \in K_{i+1}$. Since x is non-zero, multiplication by x is an automorphism of the vector K-space E which maps K_i onto K_{i+1} for all $i \in \mathbf{F}_p$. Since $[K_0 : K] = 1$, we have $[K_i : K] = 1$ for all $i \in \mathbf{F}_p$ whence $[E : K] = p$.

Lemma 3. — Let $\mathfrak{h} \in \mathscr{L}$ *be of dimension* s *over* E. *Then* $I(\mathfrak{h})$ *belongs to* $\mathscr{C}$ *and we have* $[E : I(\mathfrak{h})] = p^s$.

It is clear that $I(\mathfrak{h})$ is a subfield of E containing E^p. For each $x \in E$ let f_x be the E-linear form $D \mapsto D(x)$ on $\mathfrak{h}$. Since the intersection of the kernels of these linear forms is equal to 0, they generate the vector space dual to $\mathfrak{h}$ (II, p. 301, Th. 7) ; hence there exist $x_1, \ldots, x_s$ in E such that the linear forms $f_{x_1}, \ldots, f_{x_s}$ form a basis of this dual. Let $(\Delta_1, \ldots, \Delta_s)$ be the basis of $\mathfrak{h}$ characterized by $\Delta_i(x_j) = f_{x_j}(\Delta_i) = \delta_{ij}$. Put $D_i = x_i \Delta_i$, then $(D_1, \ldots, D_s)$ is a basis of $\mathfrak{h}$ over E and we have $D_i(x_j) = x_i \delta_{ij}$. The derivations $D_i^p - D_i$ and $[D_i, D_j]$ for $i, j = 1, \ldots, s$ belong to $\mathfrak{h}$ and annihilate $x_1, \ldots, x_s$; we thus have

$$D_i^p = D_i \,, \quad [D_i, D_j] = 0 \,. \tag{10}$$

For i in the range 0 to s let us denote by K_i the intersections of the kernels of the derivations D_j for $1 \leqslant j \leqslant i$. Then K_i is a subfield of E and we have

$$E = K_0 \supset K_1 \supset \ldots \supset K_{s-1} \supset K_s = I(\mathfrak{h}) \,.$$

Let i be between 0 and $s - 1$; then K_i is stable under D_{i+1} because D_{i+1} commutes with $D_1, \ldots, D_i$. Moreover we have $D_{i+1}^p = D_{i+1}$, $D_{i+1}(x_{i+1}) = x_{i+1} \neq 0$ and $x_{i+1} \in K_i$. Hence Lemma 1 implies that $[K_i : K_{i+1}] = p$, whence finally $[E : K] = [K_0 : K_s] = p^s$.

We now come to the proof of the theorem. Let $\mathfrak{h} \in \mathscr{L}$ be of dimension s over E and put $K = I(\mathfrak{h})$; then $[E : K] = p^s$ by Lemma 3, whence $[\Omega_K(E) : E] = s$ by

Th. 2, *c*) (V, p. 103). By the universal property of the module of differentials the mapping $u \mapsto u \circ d_{E/K}$ is an isomorphism of the dual of $\Omega_K(E)$ onto $\mathfrak{g}(K)$, hence $[\mathfrak{g}(K):E] = s$. Now we have $[\mathfrak{h}:E] = s$ and $\mathfrak{h} \subset \mathfrak{g}(K)$, whence $\mathfrak{h} = \mathfrak{g}(K)$, that is $\mathfrak{h} = \mathfrak{g}(I(\mathfrak{h}))$.

Conversely, for every field $K \in \mathscr{C}$ it is clear that $\mathfrak{g}(K)$ belongs to $\mathscr{L}$ (V, p. 103, Th. 2, *c*)). If x belongs to $I(\mathfrak{g}(K))$, we have $u(d_{E/K}x) = 0$ for each linear form u on $\Omega_K(E)$, whence $d_{E/K}x = 0$ and so finally $x \in K$ by Cor. 2 of Prop. 5 (V, p. 102). Thus we have $K = I(\mathfrak{g}(K))$.

Remarks. — 1) The mutually inverse bijections $K \mapsto \mathfrak{g}(K)$ and $\mathfrak{h} \mapsto I(\mathfrak{h})$ are inclusion reversing; therefore $\mathfrak{h} \mapsto I(\mathfrak{h})$ is an isomorphism of the ordered set $\mathscr{L}$ onto the ordered set opposite to $\mathscr{C}$. Hence we obtain the relation $I(\mathfrak{h} \cap \mathfrak{h}') = E^p(I(\mathfrak{h}), I(\mathfrak{h}'))$ for $\mathfrak{h}, \mathfrak{h}'$ in $\mathscr{L}$, because $\mathfrak{h} \cap \mathfrak{h}'$ is the largest element of $\mathscr{L}$ contained both in $\mathfrak{h}$ and in $\mathfrak{h}'$.

2) It can be shown that every subspace of finite dimension of $\mathfrak{g}$ which is stable under the mapping $D \mapsto D^p$ is also stable under the bracket operation.

§ 14. TRANSCENDENTAL EXTENSIONS

1. Algebraically free families. Pure extensions

Let us recall (IV, p. 4) the following definition:

DEFINITION 1. — *Let* E *be an extension of a field* K; *a family* $\mathbf{x} = (x_i)_{i \in I}$ *of elements of* E *is said to be algebraically free over* K *if the monomials* $\mathbf{x}^\alpha = \prod_{i \in I} \mathbf{x}_i^{\alpha_i}$ *with respect to the* x_i (*for* $\alpha = (\alpha_i)_{i \in I}$ *in* $\mathbf{N}^{(I)}$) *are linearly independent over* K. *In the contrary case the family is said to be algebraically related over* K.

Definition 1 may also be expressed as follows:

PROPOSITION 1. — *For a family* $(x_i)_{i \in I}$ *of elements of an extension* E *of a field* K *to be algebraically free over* K *it is necessary and sufficient for the relation* $f((x_i)) = 0$, *where* f *is a polynomial in* $K[X_i]_{i \in I}$ *to imply* $f = 0$.

DEFINITION 2. — *Let* E *be an extension of a field* K. *A family* $(x_i)_{i \in I}$ *of elements of* E *is called a pure basis of* E (*over* K) *if it is algebraically free and* $E = K(x_i)_{i \in I}$. *The extension* E *of* K *is also called pure if it possesses a pure basis.*

The empty family is algebraically free, hence K is a pure extension of itself. With the notation of Def. 2, every element x_i is *transcendental* over K; if I is non-empty, E is thus a *transcendental* extension of K.

PROPOSITION 2. — *Let* E *and* E′ *be two fields and u an isomorphism of a subfield* K *of* E *onto a subfield* K′ *of* E′. *Let* $\mathbf{x} = (x_i)_{i \in I}$ (resp. $\mathbf{x}' = (x'_i)_{i \in I}$) *be a family of elements of* E (resp. E′) *which is algebraically free over* K (resp. K′). *Then there exists a unique isomorphism* v *of* $L = K(x_i)_{i \in I}$ *onto* $L' = K'(x'_i)_{i \in I}$ *which induces* u *on* K *and maps* x_i *to* x'_i *for each* $i \in I$.

The uniqueness of v is clear. Let us put $A = K[x_i]_{i \in I}$ and $A' = K'[x'_i]_{i \in I}$. By hypothesis the monomials $\mathbf{x}^\alpha = \prod_{i \in I} x_i^{\alpha_i}$ (for $\alpha = (\alpha_i)_{i \in I}$ in $\mathbf{N}^{(I)}$) form a basis of the vector K-space A and there is a similar property of A′. Hence there exists a ring isomorphism $w : A \to A'$ mapping each element $\sum_{\alpha \in \mathbf{N}^{(I)}} c_\alpha x^\alpha$ to $\sum_{\alpha \in \mathbf{N}^{(I)}} u(c_\alpha) x'^\alpha$. Since L is the field of fractions of A and L′ that of A′, the isomorphism w extends to an isomorphism v of the field L onto the field L′.

COROLLARY. — *For an extension* E *of a field* K *to be pure it is necessary and sufficient for* E *to be* K*-isomorphic to a field of rational fractions over* K. *More precisely, if the family* $(x_i)_{i \in I}$ *is a pure basis of* E, *then there exists a unique* K*-isomorphism of* $K(X_i)_{i \in I}$ *onto* E *which maps* X_i *to* x_i *for each* $i \in I$.

Remark. — It is clear that in an extension E of K an algebraically free family over K consists of *linearly independent* elements over K (hence pairwise distinct) ; in other words, it is also a free family for the *vector space* structure of E (with respect to K). But the converse is false, for if E is an algebraic extension of K, any non-empty family of elements of E (and *a fortiori* a non-empty family of linearly independent elements over K) is never algebraically free over K. When there is a risk of confusion, we shall say that a subset of an extension E of K which is free for the vector space structure of E with respect to K is *linearly free* over K.

Let E be an extension of a field K. A subset S of E is said to be *algebraically free* (over K) if the family defined by the identity mapping of S onto itself is algebraically free. The elements of an algebraically free subset of E are also called *algebraically independent.* If a subset of E is not algebraically free, it is said to be *algebraically related* and that its elements are *algebraically dependent.* For a family $(x_i)_{i \in I}$ of elements of E to be algebraically free it is necessary and sufficient that $i \mapsto x_i$ should be a bijection of I onto an algebraically free subset of E.

Every subset of an algebraically free set is algebraically free. Furthermore :

PROPOSITION 3. — *For a family* $(x_i)_{i \in I}$ *of elements of an extension* E *of a field* K *to be algebraically free over* K *it is necessary and sufficient that every finite subfamily of* $(x_i)_{i \in I}$ *should be algebraically free over* K.

This proposition follows immediately from Def. 1.

2. Transcendence bases

PROPOSITION 4. — *Let* E *be an extension of a field* K *and* S *and* T *two subsets of* E. *Then the following properties are equivalent :*

a) $S \cup T$ *is algebraically free over* K *and* $S \cap T = \varnothing$.

b) S *is algebraically free over* K *and* T *is algebraically free over* K(S).
c) T *is algebraically free over* K *and* S *is algebraically free over* K(T).
Evidently it is enough to prove that *a)* and *b)* are equivalent.

a) $\Rightarrow$ *b)* : Suppose that *a)* holds. Since S is contained in S $\cup$ T, it is algebraically free over K. If T is not algebraically free over K(S), there exists (Prop. 3) a finite family $(y_j)_{1 \leqslant j \leqslant n}$ of distinct elements of T which is algebraically related over K(S). Hence there is a non-zero polynomial f in the ring K(S) $[Y_1, \dots, Y_n]$ such that $f(y_1, \dots, y_n) = 0$; after multiplying f if necessary by a non-zero element of K[S] we may suppose that all the coefficients of f belong to K[S]. The coefficients of f are polynomials in a finite number of distinct elements x_i $(1 \leqslant i \leqslant m)$ of S, with coefficients in K. The elements $x_1, \dots, x_m, y_1, \dots, y_n$ are pairwise distinct because S $\cap$ T $= \varnothing$. The relation $f(y_1, \dots, y_n) = 0$ may thus be written

$$g(x_1, \dots, x_m ; y_1, \dots, y_n) = 0 ,$$

where g is a non-zero polynomial of $K[X_1, \dots, X_m, Y_1, \dots, Y_n]$, and such a relation contradicts the hypothesis that S $\cup$ T is algebraically free.

b) $\Rightarrow$ *a)* : Suppose that *b)* holds. In the first place it is clear that T $\cap$ K(S) $= \varnothing$ and *a fortiori* S $\cap$ T $= \varnothing$. It suffices to show that if x_i $(1 \leqslant i \leqslant m)$ are distinct elements of S, finite in number, and y_j $(1 \leqslant j \leqslant n)$ distinct elements of T finite in number, then the set of the x_i and y_j is algebraically free over K (Prop. 3). Consider a polynomial $f \in K[X_1, \dots, X_m, Y_1, \dots, Y_n]$ such that $f(x_1, \dots, x_m, y_1, \dots, y_n) = 0$ and put $f = \sum_{\alpha_1 \dots \alpha_n} \varphi_\alpha Y_1^{\alpha_1} \dots Y_n^{\alpha_n}$ with $\varphi_\alpha \in K[X_1, \dots, X_m]$ for all $\alpha = (\alpha_1, \dots, \alpha_n) \in \mathbf{N}^n$. Let $g = f(x_1, \dots, x_m, Y_1, \dots, Y_n)$; then g is a polynomial in the ring $K[S][Y_1, \dots, Y_n]$ and the relation $f(x_1, \dots, x_m, y_1, \dots, y_n) = 0$ may be written $g(y_1, \dots, y_n) = 0$. Since T is algebraically free over K(S), each of the coefficients $\varphi_\alpha(x_1, \dots, x_m)$ of g is zero ; since S is algebraically free over K, we have $\varphi_\alpha = 0$ for all $\alpha \in \mathbf{N}^n$, and hence $f = 0$.

COROLLARY. — *Let* E *be an extension of a field* K *and* S *a subset of* E *which is algebraically free over* K. *If* $x \in$ E *is transcendental over* K(S), *then* S $\cup$ $\{x\}$ *is algebraically free over* K.

PROPOSITION 5. — *Let* E *be an extension of the field* K. *For a subset* S *of* E *to be algebraically free over* K *it is necessary and sufficient that for each* $x \in$ S, *the element* x *should be transcendental over the field* K(S $-$ $\{x\}$).

The condition is necessary by Prop. 4.

To prove the sufficiency it is enough (Prop. 3) to show that every finite sequence $(x_1, \dots, x_n)$ of distinct elements of S is algebraically free. Now by hypothesis x_i is transcendental over $K(x_1, \dots, x_{i-1})$ for $1 \leqslant i \leqslant n$, and now our assertion follows by induction on n from the Cor. to Prop. 4.

PROPOSITION 6. — *Let* E *be an extension of a field* K *and* X *an algebraically free subset of* E *over* K. *If* K$'$ $\subset$ E *is an algebraic extension of* K *then* X *is algebraically free over* K$'$.

Arguing by contradiction, let us suppose that X is algebraically related over K'. By Prop. 5 there exists an element $x \in X$ which is algebraic over the field $K'(M)$, where $M = X - \{x\}$. Since $K'(M) = K(M)(K')$ and K' is algebraic over K, Cor. 2 of V, p. 18 shows that $K'(M)$ is algebraic over $K(M)$; since x is algebraic over $K'(M)$, it is therefore algebraic over $K(M) = K(X - \{x\})$, by Prop. 3 of V, p. 19. Now Prop. 5 shows X to be algebraically related over K and we have a contradiction.

DEFINITION 3. — *A subset* B *of an extension* E *of a field* K *is a transcendence basis of* E *(over* K*) if* B *is algebraically free over* K *and* E *is algebraic over* K(B).

Example. — A pure basis is a transcendence basis. On the other hand, if E is a pure extension of K, a transcendence basis of E over K is not always a pure basis of E. For example, in K(X), $\{X^2\}$ is a transcendence basis but does not generate K(X).

PROPOSITION 7. — *Let* E *be an extension of a field* K. *Every transcendence basis of* E *is a maximal element of the set (ordered by inclusion) of subsets of* E *algebraically free over* K. *Conversely, if* S *is a subset of* E *such that* E *is algebraic over* K(S), *every maximal algebraically free subset of* S *is a transcendence basis of* E.

Let B be a transcendence basis of E over K and $x \in E - B$. Then x is algebraic over K(B); by V, p. 107, Prop. 4, the subset $B \cup \{x\}$ of E is not algebraically free over K, whence the first part of the proposition. Secondly, if E is algebraic over K(S) and B is a maximal algebraically free part of S, it follows from the Cor. of Prop. 4 that every $x \in S$ is algebraic over K(B); hence (V, p. 18, Cor. 1) K(S) is algebraic over K(B), and so (V, p. 19, Prop. 3), E is algebraic over K(B).

THEOREM 1 (Steinitz). — *Every extension* E *of a field* K *admits a transcendence basis over* K. *In other words, every extension of a field* K *is an algebraic extension of a pure extension of* K.

By contrast, an extension is not always a pure extension of an algebraic extension (V, p. 171, Ex. 2).

This theorem is a consequence of the following more precise result:

THEOREM 2. — *Let* E *be an extension of a field* K, S *a subset of* E *such that* E *is algebraic over* K(S) *and* T *a subset of* S *which is algebraically free over* K; *then there exists a transcendence basis* B *of* E *over* K *such that* $T \subset B \subset S$.

For the set of algebraically free subsets of S, ordered by inclusion, is a set of finite character (E, III, p. 34) by Prop. 3. By Th. 1 of E, III, p. 35 it has a maximal element B containing T, and B is a transcendence basis of E over K, by Prop. 7.

COROLLARY (« Exchange Theorem »). — *Let* E *be an extension of* K, S *a subset of* E *such that* E *is algebraic over* K(S), T *a subset of* E *which is algebraically free over* K ; *then there exists a subset* S′ *of* S *such that* T ∪ S′ *is a transcendence basis of* E *over* K *and* T ∩ S′ = ∅ .

For E is algebraic over K(T ∪ S) and we have T ⊂ T ∪ S.

3. The transcendence degree of an extension

THEOREM 3. — *Let* E *be an extension of a field* K. *All transcendence bases of* E *over* K *have the same cardinal.*

It is enough to prove the inequality Card(B) ⩾ Card(B′), when B and B′ are two transcendence bases of E over K, and we may suppose that B′ is not empty. We first suppose that B is finite and use induction on its cardinal n ; for $n = 0$, E is algebraic over K and B′ is empty. Suppose now that $n \geqslant 1$; given $x \in \mathrm{B}'$, the exchange theorem provides a subset C of B such that $x \notin \mathrm{C}$ and $\{x\} \cup \mathrm{C}$ is a transcendence basis of E over K ; we have C ≠ B by Prop. 7 whence Card C < n. Put $\mathrm{K}_1 = \mathrm{K}(x)$ and $\mathrm{C}' = \mathrm{B}' - \{x\}$; then C and C′ are algebraically free over the field K_1 (V, p. 107, Prop. 4) and E is algebraic both over $\mathrm{K}_1(\mathrm{C}) = \mathrm{K}(\mathrm{C} \cup \{x\})$ and $\mathrm{K}_1(\mathrm{C}') = \mathrm{K}(\mathrm{B}')$. In other words, C and C′ are two transcendence bases of E over K_1. Since Card(C) < n, the induction hypothesis implies the inequality $\mathrm{Card}(\mathrm{C}') \leqslant \mathrm{Card}(\mathrm{C}) \leqslant n - 1$, whence $\mathrm{Card}(\mathrm{B}') \leqslant n = \mathrm{Card}(\mathrm{B})$.

Suppose now that B is infinite. Every $x \in \mathrm{B}$ is algebraic over K(B′) and so there exists a *finite* subset $\mathrm{S}(x)$ of B′ such that x is algebraic over $\mathrm{K}(\mathrm{S}(x))$. Write $\mathrm{S} = \bigcup_{x \in \mathrm{B}} \mathrm{S}(x)$, then S ⊂ B′ and since B is infinite, we have Card(S) ⩽ Card(B) (E, III, p. 49, Cor. 3). But since every element of B is algebraic over K(S) and E is algebraic over K(B), we conclude that E is algebraic over K(S) (V, p. 19, Prop. 3). Now Prop. 7 implies that S = B′, whence the desired inequality Card(B′) ⩽ Card(B).

DEFINITION 4. — *Let* E *be an extension of a field* K. *The cardinal of every transcendence basis of* E *over* K *is called the transcendence degree of* E *over* K *and written* $\mathrm{tr\,.\,deg}_{\mathrm{K}}\mathrm{E}$.

Th. 2 and 3 and Def. 4 imply the following corollaries, where E denotes an extension of a field K, of *finite transcendence* degree n.

COROLLARY 1. — *Let* S *be a subset of* E *such that* E *is algebraic over* K(S). *Then* Card(S) ⩾ n ; *if the cardinal of* S *is equal to* n, *then* S *is algebraically free over* K (*hence it is then a transcendence basis of* E *over* K).

COROLLARY 2. — *Suppose that* $\mathrm{E} = \mathrm{K}(x_1, \ldots, x_m)$, *then* $m \geqslant n$; *if moreover* $m = n$, *then* $(x_1, \ldots, x_m)$ *is a pure basis of* E *over* K, *and* E *is then a pure extension of* K.

COROLLARY 3. — *Every subset of* E *which is algebraically free over* K *has at most* n *elements, and if it has exactly* n *elements, it is a transcendence basis of* E *over* K.

THEOREM 4. — *Let* K, E *and* F *be three fields such that* $K \subset E \subset F$. *If* S *is a transcendence basis of* E *over* K *and* T *a transcendence basis of* F *over* E, *then* $S \cap T$ *is empty and* $S \cup T$ *is a transcendence basis of* F *over* K.

For F is algebraic over E(T); moreover, E(T) is algebraic over the field $K(S \cup T) = K(S)(T)$, because E is algebraic over K(S) (V, p. 18, Cor. 2); therefore (V, p. 19, Prop. 3) F is algebraic over $K(S \cup T)$. On the other hand, T being algebraically free over E is so *a fortiori* over K(S), hence (V, p. 107, Prop. 4) $S \cup T$ is algebraicaliy free over K and $S \cap T = \emptyset$.

COROLLARY. — *Let* K, E *and* F *be three fields such that* $K \subset E \subset F$. *Then we have*

$$\text{tr} . \deg_K F = \text{tr} . \deg_K E + \text{tr} . \deg_E F . \tag{1}$$

4. Extension of isomorphisms

PROPOSITION 8. — *Let* Ω *be an algebraically closed extension of a field* K, E *and* F *two subextensions of* Ω *and* u *a* K-*isomorphism of* E *onto* F. *For a* K-*automorphism* v *of* Ω *to exist which extends* u, *it is necessary and sufficient for* Ω *to have the same transcendence degree over* E *and* F.

The condition is clearly necessary.

Suppose then that Ω has the same transcendence degree over E and F and let us choose a transcendence basis B of Ω over E and a transcendence basis C of Ω over F. Since B and C are equipotent, Prop. 2 (V, p. 106) shows that u extends to a K-isomorphism u' of E(B) onto F(C). Since Ω is an algebraic closure of E(B) and F(C), the Cor. of V, p. 23 shows that u' extends to an automorphism v of Ω.

COROLLARY 1. — *Let* Ω *be an algebraically closed extension of a field* K *and* E *a subextension of* Ω. *Every* K-*automorphism of* E *extends to a* K-*automorphism of* Ω.

COROLLARY 2. — *Let* Ω *be an algebraically closed extension of a field* K, E *and* F *two subextensions of* Ω *and* u *a* K-*isomorphism of* E *onto* F. *If the transcendence degree of* E *over* K *is finite (in particular if* E *is algebraic over* K*) then there exists a* K-*automorphism of* Ω *extending* u.

Let us denote by n, $d(E)$ and $d(F)$ the transcendence degrees of E over K, of Ω over E and of Ω over F respectively. The existence of the K-isomorphism u shows that the transcendence degree of F over K is equal to n. By the Cor. of Th. 4 the transcendence degree of Ω over K is equal to $d(E) + n$ and also to $d(F) + n$. Therefore (E, III, p. 28, Prop. 8) we have $d(E) = d(F)$ and we can apply Prop. 8.

PROPOSITION 9. — *Let* K *be a field and* Ω *an algebraically closed extension of* K. *Suppose that* Ω *is not algebraic over* K ; *then the set of elements of* Ω *which are transcendental over* K *is infinite. Moreover, if* x *and* y *are two elements of* Ω *which are transcendental over* K, *then there exists an automorphism* u *of* Ω *over* K *such that* $u(x) = y$.

Since Ω is not algebraic over K, there exists an element x of Ω transcendental over K ; then the elements x^n $(n \in \mathbf{N})$ are distinct and transcendental over K. Suppose that x and y are distinct and transcendental over K ; by Prop. 2 (V, p. 106) there exists a K-isomorphism $\tilde{u}$ of $K(x)$ onto $K(y)$ such that $\tilde{u}(x) = y$; since $K(x)$ is of transcendence degree 1 over K, Cor. 2 of Prop. 8 shows that $\tilde{u}$ extends to a K-automorphism u of Ω.

PROPOSITION 10. — *Let* K *be a field,* Ω *an algebraically closed extension of* K *and* G *the group of* K-*automorphisms of* Ω. *Let* $x \in \Omega$.

a) For x *to be algebraic over* K *it is necessary and sufficient for the set of elements* $u(x)$ *as* u *runs over* G *to be finite,*

b) For x *to be* p-*radical over* K *it is necessary and sufficient that* $u(x) = x$ *for all* $u \in G$.

In particular if K *is perfect, the set of invariants of the group* G *is equal to* K.

Suppose first that x is transcendental. By Prop. 9 the set T of elements of Ω transcendental over K is infinite, and for each $y \in T$ there exists $u \in G$ such that $u(x) = y$. Hence the set of elements $u(x)$ as u runs over G is infinite.

Suppose now that x is algebraic over K and denote by f its minimal polynomial over K ; the set of roots of f in Ω is finite and for each $u \in G$ we have $f(u(x)) = u(f(x)) = 0$. Hence the set of elements $u(x)$ as u runs over G is finite. This proves *a*).

Let L be the set of elements y of Ω such that $u(y) = y$ for all $u \in G$, and let $\bar{K}$ be the relative algebraic closure of K in Ω. By what has been said, L is a subextension of $\bar{K}$ over K. Further (Cor. 1 to Prop. 8) every K-automorphism of $\bar{K}$ is the restriction to $\bar{K}$ of an element of G. Assertion *b*) of Prop. 10 now follows from Cor. 3 of V, p. 53.

5. Algebraically disjoint extensions

DEFINITION 5. — *Let* L *be an extension of a field* K, *and* E *and* F *two subextensions of* L. *Then* E *and* F *are said to be algebraically disjoint* (*over* K) *and* E *is said to be algebraically disjoint from* F *over* K, *if for every subset* A (resp. B) *of* E (resp. F) *algebraically free over* K, A *and* B *are disjoint and* $A \cup B$ *is algebraically free over* K.

Remarks. — 1) If E is a subextension of L which is algebraic over K, then it is algebraically disjoint from every subextension F of L. For an extension of K to be algebraic it is necessary and sufficient that it should be algebraically disjoint from itself.

2) It may happen that E is algebraically disjoint from F over K, but not over a subfield K_0 of K. * For example **C** is algebraically disjoint from itself over **R** but not over **Q**. *

3) It is clear that if E is algebraically disjoint from F over K, when E and F are considered as subextensions of L, the same is true when they are considered as subextensions of $K(E \cup F)$ and conversely.

PROPOSITION 11. — *If* E *and* F *are algebraically disjoint over* K *then* $E \cap F$ *is algebraic over* K.

This follows from Def. 5.

PROPOSITION 12. — *Let* L *be an extension of a field* K *and* E, F *subextensions of* L. *Then the following conditions are equivalent :*

a) E *and* F *are algebraically disjoint ;*

b) *there exists a transcendence basis of* E *over* K *which is algebraically free over* F ;

c) *every subset of* E *which is algebraically free over* K *is algebraically free over* F.

Let us introduce the following conditions :

b') there exists a transcendence basis of F over K which is algebraically free over E ;

c') every subset of F which is algebraically free over K is algebraically free over E.

a) ⇒ *b'*) : Suppose that E and F are algebraically disjoint. Let B (resp. C) be a transcendence basis of E (resp. F) over K. Then $B \cap C = \varnothing$ and $B \cup C$ is algebraically free over K, hence C is algebraically free over K(B) (V, p. 107, Prop. 4) ; since E is algebraic over K(B), Prop. 6 of V, p. 108 shows C to be algebraically free over E.

b') ⇒ *c*) : Suppose that there exists a transcendence basis C of F over K which is algebraically free over E. Let A be a subset of E algebraically free over K. Then C is algebraically free over K(A), hence A is algebraically free over K(C) (V, p. 107, Prop. 4) and therefore over F (V, p. 108, Prop. 6), because F is algebraic over K(C).

c) ⇒ *a*) : This follows at once from Prop. 4 (V, p. 107).

The implications *a*) ⇒ *b*) ⇒ *c'*) ⇒ *a*) may be proved in the same way.

COROLLARY. — *Let* E *and* F *be algebraically disjoint over* K. *Let* E' *be the relative algebraic closure of* E *in* L *and* F' *that of* F (V, p. 19). *Then* E' *and* F' *are algebraically disjoint over* K.

Let B be a transcendence basis of E over K ; this is also a transcendence basis of E' over K. Since E is algebraically disjoint from F over K, B is algebraically free over F, hence over F' (V, p. 108, Prop. 6) ; now we can apply Prop. 12.

PROPOSITION 13. — *Let* L *be an extension of a field* K *and* E, F *two subextensions of* L.

a) *We have* $\operatorname{tr.deg}_F F(E) \leq \operatorname{tr.deg}_K E$. *When* E *and* F *are algebraically disjoint over* K, *then every transcendence basis of* E *over* K *is a transcendence basis of*

F(E) *over* F *and we have* $\operatorname{tr.deg}_F F(E) = \operatorname{tr.deg}_K E$. *Conversely, this equality implies that* E *and* F *are algebraically disjoint over* K *when* $\operatorname{tr.deg}_K E$ *is finite.*

b) *We have* $\operatorname{tr.deg}_K K(E \cup F) \leqslant \operatorname{tr.deg}_K E + \operatorname{tr.deg}_K F$. *When* E *and* F *are algebraically disjoint over* K, *we have* $\operatorname{tr.deg}_K K(E \cup F) = \operatorname{tr.deg}_K E + \operatorname{tr.deg}_K F$. *Conversely, this equality implies that* E *and* F *are algebraically disjoint over* K *when* E *and* F *are of finite transcendence degree over* K.

a) Let B be a transcendence basis of E over K ; then E is algebraic over K(B), and Cor. 2 of V, p. 18 shows that F(E) is algebraic over F(K(B)) = F(B). By Th. 2 (V, p. 109) B contains a transcendence basis of F(E) over F ; when E is algebraically disjoint from F over K, B is algebraically free over F (Prop. 12) and this is then a transcendence basis of F(E) over F. The three first assertions of *a*) follow from this. Suppose now that E is of finite transcendence degree over K, equal to that of F(E) over F ; since F(E) is algebraic over F(B) and Card B = $\operatorname{tr.deg}_F F(E)$, Cor. 1 of V, p. 110 shows B to be algebraically free over F, so E is algebraically disjoint from F over K (Prop. 12).

b) We have $K(E \cup F) = F(E)$ and so the Cor. of V, p. 110 implies the equality :

$$\operatorname{tr.deg}_K K(E \cup F) = \operatorname{tr.deg}_F F(E) + \operatorname{tr.deg}_K F .$$

Now *b*) follows at once from *a*) and this equality.

PROPOSITION 14. — *Let* L *be an extension of a field* K, E *and* F *two subextensions of* L *and* B *a transcendence basis of* E *over* K. *For* E *and* F *to be algebraically disjoint over* K *it is necessary and sufficient that* K(B) *and* F *should be linearly disjoint over* K.

For E and F to be algebraically disjoint over K it is necessary and sufficient for B to be algebraically free over F (Prop. 12), that is, for the monomials in the elements of B to be linearly independent over F. Since these monomials form a basis of the vector K-space K[B], it comes to the same to say that K[B] and F are linearly disjoint over K. Finally, since K(B) is the field of fractions of K[B], Prop. 6 of V, p. 14 shows that K[B] and F are linearly disjoint if and only if this is so for K(B) and F.

COROLLARY 1. — *If* E *and* F *are linearly disjoint, then* E *is algebraically disjoint from* F *over* K. *Conversely, if* E *is a pure extension of* K *and is algebraically disjoint from* F *over* K, *then* E *and* F *are linearly disjoint over* K.

COROLLARY 2. — *Every pure extension of* K *is linearly disjoint from every algebraic extension of* K ; *in particular,* K *is relatively algebraically closed in every pure extension of* K.

6. Algebraically free families of extensions

DEFINITION 6. — *Let* L *be an extension of a field* K *and* $(E_i)_{i \in I}$ *a family of subextensions of* L. *The family* $(E_i)_{i \in I}$ *is said to be algebraically free if the following condition is satisfied :*

(AF) *For each* $i \in I$ *let* A_i *be a subset of* E_i *which is algebraically free over* K. *Then* $A_i \cap A_j = \varnothing$ *for* $i \neq j$ *and* $\bigcup_{i \in I} A_i$ *is algebraically free over* K.

Remark. — By Prop. 3 (V, p. 107) it is enough to verify the condition (AF) for finite subsets A_i. We thus obtain the following result : if $(E_i)_{i \in I}$ is an algebraically free family, the same is true of $(E_i')_{i \in I}$ if E_i' is a subextension of E_i for each $i \in I$; conversely, if every family $(E_i')_{i \in I}$, where E_i' is a *finitely generated* subextension of E for each $i \in I$ is algebraically free, then $(E_i)_{i \in I}$ is algebraically free. On the other hand, for $(E_i)_{i \in I}$ to be algebraically free it is necessary and sufficient that $(E_i)_{i \in J}$ should be algebraically free for every finite subset J of I. Speaking intuitively we may say that the algebraic independence of extensions is a property « of finite character ».

PROPOSITION 15. — *Let* $(E_i)_{i \in I}$ *be a family of subextensions of a given extension* L *of a field* K. *Then the following conditions are equivalent :*

a) *The family* $(E_i)_{i \in I}$ *is algebraically free.*

b) *For each* $i \in I$ *the extension* E_i *is algebraically disjoint over* K *from the extension* F_i *generated by the* E_j *for* $j \neq i$.

c) *There exists a family* $(B_i)_{i \in I}$ *of disjoint subsets of* L, *such that* B_i *is a transcendence basis of* E_i *over* K *for each* $i \in I$, *and* $B = \bigcup_{i \in I} B_i$ *is algebraically free over* K.

It is clear that *a*) implies *c*).

Assuming *c*), let us choose i in I ; put $C_i = \bigcup_{j \neq i} B_j$. For each $j \neq i$ every element of E_j is algebraic over $K(B_j)$ and *a fortiori* over $K(C_i)$. By Cor. 1 of V, p. 18, the field F_i is therefore algebraic over $K(C_i)$. Further, we have $B_i \cap C_i = \varnothing$ and $B = B_i \cup C_i$ is algebraically free over K ; therefore B_i is algebraically free over $K(C_i)$ (V, p. 107, Prop. 4), hence also over F_i (which is algebraic over $K(C_i)$) by Prop. 6 of V, p. 108. Thus we have proved that E_i is algebraically disjoint from F_i over K (V, p. 113, Prop. 12), hence *c*) implies *b*).

Let us now assume *b*) and prove *a*). It is enough to show that if $i_1, \ldots, i_n$ are distinct elements of I, then the family of extensions $(E_{i_1}, \ldots, E_{i_n})$ is algebraically free ; we argue by induction on n, the case $n = 1$ being trivial. Suppose then that $n > 1$ and that the family $(E_{i_1}, \ldots, E_{i_{n-1}})$ is algebraically free ; for $1 \leqslant k \leqslant n$ choose a subset A_k of E_{i_k} algebraically free over K and put

$B = A_1 \cup \ldots \cup A_{n-1}$. By the induction hypothesis the subsets $A_1, \ldots, A_{n-1}$ are pairwise disjoint and B is algebraically free over K ; by *b*) E_{i_n} is algebraically disjoint from F_{i_n} and since B is contained in F_{i_n} we have $B \cap A_n = \varnothing$ and $B \cup A_n = A_1 \cup \ldots \cup A_n$ is algebraically free over K. We have thus shown that the family $(E_{i_1}, \ldots, E_{i_n})$ is algebraically free.

The following proposition generalizes part *b*) of Prop. 13 (V, p. 113).

PROPOSITION 16. — *Let* $(E_i)_{i \in I}$ *be a family of subextensions of an extension of a field* K, *and let* E *be the field generated by* $\bigcup_{i \in I} E_i$.

a) *We have* $\operatorname{tr.deg}_K E \leqslant \sum_{i \in I} \operatorname{tr.deg}_K E_i$, *with equality when the family* $(E_i)_{i \in I}$ *is algebraically free over* K.

b) *Conversely, assume that* $\operatorname{tr.deg}_K E = \sum_{i \in I} \operatorname{tr.deg}_K E_i$ *and that* $\operatorname{tr.deg}_K E$ *is finite ; then the family* $(E_i)_{i \in I}$ *is algebraically free over* K.

For each $i \in I$ let B_i be a transcendence basis of E_i over K and put $B = \bigcup_{i \in I} B_i$. For each $i \in I$, each element of E_i is algebraic over $K(B_i)$, hence over $K(B)$. Now Cor. 1 of V, p. 18 shows that E is algebraic over $K(B)$; by V, p. 109, Th. 2, B thus contains a transcendence basis of E over K. If moreover the family $(E_i)_{i \in I}$ is algebraically free over K, then the B_i are disjoint and the set B is algebraically free over K. This establishes *a*) (*Set Theory*, III, p. 160, Cor. of Prop. 4).

Under the hypotheses of *b*), E is algebraic over $K(B)$ and of finite transcendence degree over K, and we have $\operatorname{Card}(B) \leqslant \operatorname{tr.deg}_K E$. By Cor. 1 of V, p. 110, B is algebraically free over K and the B_i are disjoint. Now Prop. 15 shows that the family $(E_i)_{i \in I}$ is algebraically free over K.

Before stating the next theorem let us remark that there exist algebraically closed extensions of K of arbitrary transcendence degree, for example an algebraic closure of an appropriate field of rational fractions.

THEOREM 5. — *Let* $(E_i)_{i \in I}$ *be a family of extensions of a field* K *and* Ω *an algebraically closed extension of* K. *Suppose that the inequality*

$$\operatorname{tr.deg}_K \Omega \geqslant \sum_{i \in I} \operatorname{tr.deg}_K E_i \tag{2}$$

holds. Then there exists an algebraically free family $(F_i)_{i \in I}$ *of subextensions of* Ω *such that* F_i *is* K-*isomorphic to* E_i *for all* $i \in I$.

For each $i \in I$ let B_i be a transcendence basis of E_i over K. Let B be a transcendence basis of Ω over K. By (2) we have $\operatorname{Card} B \geqslant \sum_{i \in I} \operatorname{Card} B_i$; hence there exists a family $(B_i')_{i \in I}$ of pairwise disjoint subsets of B and bijections $u_i : B_i \to B_i'$ (for $i \in I$). By Prop. 2 of V, p. 106, u_i extends to a K-isomorphism v_i of $K(B_i)$ onto $K(B_i')$; since Ω is algebraically closed and E_i algebraic over

$K(B_i)$, the Cor. (V, p. 23) shows that v_i extends to a K-isomorphism of E_i onto a subextension F_i of Ω. By construction B_i' is a transcendence basis of F_i over K, and Prop. 15 (V, p. 115) shows that the family $(F_i)_{i \in I}$ of subextensions of Ω is algebraically free over K.

COROLLARY 1. — *Let* E *and* Ω *be two extensions of a field* K. *Suppose that* Ω *is algebraically closed, of transcendence degree at least equal to that of* E. *Then* E *is* K-*isomorphic to a subextension of* Ω.

COROLLARY 2. — *Let* Ω *be an algebraically closed field of infinite transcendence degree over its prime subfield. Then every field of the same characteristic as* Ω *is the ascending directed union of fields isomorphic to subfields of* Ω.

For each field is the ascending directed union of finitely generated subfields over its prime field, and now it is enough to apply Cor. 1.

* *Example.* — This applies particularly in characteristic 0 in taking $\Omega = \mathbf{C}$ (« Lefschetz' principle »). *

7. Finitely generated extensions

PROPOSITION 17. — *Let* E *be an extension of a field* K *and* B *a transcendence basis of* E *over* K. *For* E *to be finitely generated over* K (V, p. 11, Def. 2) *it is necessary and sufficient that* B *be finite and that the degree* $[E : K(B)]$ *be finite.*

Suppose that E is finitely generated over K and let S be a finite subset of E such that $E = K(S)$. By Th. 2 (V, p. 109) S contains a transcendence basis B' of E over K and this has the same cardinal as B (V, p. 110, Th. 3). Therefore B is finite. Put $K' = K(B)$; then E is algebraic over K' and we have $E = K'(S)$; since S is finite, Th. 2 of V, p. 18 shows that $[E : K']$ is finite.

Conversely, suppose that B is finite and $[E : K(B)]$ is finite. If C is a (finite) basis of the vector space E over $K(B)$, we have $E = K(B)(C) = K(B \cup C)$ and so E is a finitely generated extension of K.

COROLLARY 1. — *Suppose that* E *is a finitely generated extension of* K, *and let* K' *be the relative algebraic closure of* K *in* E (V, p. 19). *Then* K' *is of finite degree over* K.

Let B be a transcendence basis of E over K. By Cor. 2 of V, p. 114, K' is linearly disjoint from $K(B)$ over K, whence $[K' : K] = [K'(B) : K(B)] \leq [E : K(B)]$ and now the finiteness of $[K' : K]$ follows from that of $[E : K(B)]$.

COROLLARY 2. — *A field that is finitely generated over its prime subfield contains only finitely many roots of unity.*

By Cor. 1 we are reduced to proving that a field L which is an extension of finite degree of its prime subfield possesses only a finite number of roots of unity. This is clear when L is of characteristic $\neq 0$ because it is then finite. If L is of characteristic 0 and contains infinitely many roots of unity, then it contains primitive roots of unity of arbitrarily high order. By V, p. 84, Th. 2 there exist therefore an infinity of integers $n > 0$ such that $\varphi(n) \leqslant [L : \mathbf{Q}]$, which is absurd (V, p. 80, Formulae (2) and (3)).

COROLLARY 3. — *If* E *is a finitely generated extension of a field* K, *then every subextension* E′ *of* E *is finitely generated.*

Let B′ be a transcendence basis of E′ over K. By V, p. 109, Th. 2, B′ is contained in a transcendence basis B of E over K and so is finite, by Prop. 17. Since E′ is algebraic over K(B′) and E is a finitely generated extension of K(B′), Cor. 1 shows [E′ : K(B′)] to be finite. Now Prop. 17 shows that E′ is finitely generated over K.

Prop. 17 may be paraphrased by saying that a finitely generated extension of K is an algebraic extension of finite degree of a purely transcendental extension $K(x_1, \dots, x_n)$.

§ 15. SEPARABLE EXTENSIONS

1. Characterization of the nilpotent elements of a ring

PROPOSITION 1. — *Let* A *be a commutative ring and* x *an element of* A. *For* x *to be nilpotent it is necessary and sufficient that* $1 - x\mathrm{T}$ *be invertible in the ring* A[T].

We note that A[T] is a subring of the formal power series ring A[[T]] and that $1 - x\mathrm{T}$ has in A[[T]] the inverse $\sum_{n=0}^{\infty} x^n \mathrm{T}^n$ (IV, p. 30, Prop. 5). For $1 - x\mathrm{T}$ to be invertible in A[T] it is necessary and sufficient for $\sum_{n=0}^{\infty} x^n \mathrm{T}^n$ to be a polynomial, that is, for x to be nilpotent.

PROPOSITION 2. — *Let* A *be a commutative ring. The set of nilpotent elements of* A *is an ideal of* A *equal to the intersection of the set of all prime ideals of* A.

Let x be a nilpotent element of A and $\mathfrak{p}$ a prime ideal of A. The residue class of x mod $\mathfrak{p}$ is a nilpotent element of the integral domain $A/\mathfrak{p}$, hence is zero ; thus we have $x \in \mathfrak{p}$.

Let x be a non-nilpotent element of A. By Prop. 1 the principal ideal $(1 - x\mathrm{T})$ of A[T] is distinct from A[T]. By Krull's theorem (I, p. 104) there exists a maximal ideal $\mathfrak{m}$ of A[T] containing $1 - x\mathrm{T}$. Then $\mathfrak{m}$ is a prime ideal of A[T], hence $\mathfrak{p} = A \cap \mathfrak{m}$ is a prime ideal of A. We have $1 \notin \mathfrak{m}$ and $1 - \mathrm{T}x \in \mathfrak{m}$, hence $\mathrm{T}x \notin \mathfrak{m}$ and *a fortiori* $x \notin \mathfrak{p}$.

We have thus shown that the set $\mathfrak{n}$ of nilpotent elements of A is the intersection of the set of all prime ideals of A ; since every intersection of ideals is an ideal, $\mathfrak{n}$ is an ideal.

COROLLARY. — *For a commutative ring to be reduced* (V, p. 34, Def. 2) *it is necessary and sufficient for it to be isomorphic to a subring of a product of fields.*

The condition is clearly sufficient.

Let A be reduced. By Prop. 2 the intersection $\mathfrak{n}$ of the set of prime ideals of A is reduced to 0. For every prime ideal $\mathfrak{p}$ of A, let $k(\mathfrak{p})$ be the field of fractions of $A/\mathfrak{p}$ and $\varphi_\mathfrak{p}$ the canonical homomorphism of A into $k(\mathfrak{p})$. Let φ be the homomorphism of A into $\prod_\mathfrak{p} k(\mathfrak{p})$ whose component of index $\mathfrak{p}$ is $\varphi_\mathfrak{p}$. The kernel of $\varphi_\mathfrak{p}$ is $\mathfrak{p}$, hence that of φ is $\mathfrak{n} = 0$; therefore φ is an isomorphism of A onto a subring of $\prod_\mathfrak{p} k(\mathfrak{p})$.

2. Separable algebras

DEFINITION 1. — *Let* A *be a commutative algebra over a field* K. *Then* A *is said to be separable over* K, *or also a separable* K*-algebra if the ring* $L \otimes_K A$ *is reduced for every extension* L *of* K.

Every separable algebra is clearly reduced. For a partial converse see Th. 3 (V, p. 125).

Examples. — 1) Let A be a polynomial algebra $K[X_i]_{i \in I}$. For every extension L of K the ring $L \otimes_K A$ is isomorphic to $L[X_i]_{i \in I}$ (III, p. 449, Remark 2) and hence is an integral domain (IV, p. 9, Prop. 8). In other words, every polynomial algebra over a field K is a separable K-algebra.

2) Let A be a commutative algebra of finite degree over a field K. For A to be separable it is necessary and sufficient for it to be etale (V, p. 34, Th. 4).

3) Let E be an algebraic extension of a field K. If L is an extension of K, then the ring $L \otimes_K E$ is a union of the subrings $L \otimes_K F$ where F ranges over the set of all subextensions of finite degree of E ; therefore the ring $L \otimes_K E$ is reduced if and only if this is true of $L \otimes_K F$ for every subextension F of E, of finite degree over K. Taking Example 2 into account we see that E is a separable algebra in the sense of the above Def. 1 if and only if it is a separable algebraic extension in the sense of Def. 1 of V, p. 36.

PROPOSITION 3. — *Let* K *be a field.*

a) Every subalgebra of a separable K*-algebra is separable.*

b) Every direct limit of separable K*-algebras is separable.*

c) Every product of separable K*-algebras is separable.*

d) Let A *be a* K*-algebra and* K' *an extension of* K. *For* A *to be separable it is necessary and sufficient for the* K'*-algebra* $A_{(K')}$ *obtained from* A *by extension of scalars to be separable.*

Let L be an extension of K. Let A be a separable algebra over K and B a subalgebra of A ; then the ring $L \otimes_K A$ is reduced, and $L \otimes_K B$ is isomorphic to a subring of $L \otimes_K A$, hence is reduced. Thus B is separable and *a*) is proved. In the same way *b*) may be proved, using the canonical isomorphism of $L \otimes_K \varinjlim A_i$ with $\varinjlim L \otimes_K A_i$ (II, p. 290, Prop. 7) and *c*) is proved by remarking that $L \otimes_K \left(\prod_{i \in I} A_i\right)$ is isomorphic to a subring of $\prod_{i \in I} (L \otimes_K A_i)$ (II, p. 306, Prop. 15).

We shall use the notation of *d*). For every extension L′ of K′ the rings $L' \otimes_{K'} A_{(K')}$ and $L' \otimes_K A$ are isomorphic (II, p. 278, Prop. 2). We conclude that if A is a separable K-algebra, then $A_{(K')}$ is a separable K′-algebra. Conversely suppose that $A_{(K')}$ is a separable K′-algebra ; the preceding remark shows that $L' \otimes_K A$ is reduced for every extension L′ of K containing K′ as subextension. Let L be an extension of K ; by the Scholium (V, p. 13) there exists an extension L′ of K containing K′ as a subextension and a K-homomorphism of L into L′ ; the ring $L \otimes_K A$ is thus isomorphic to a subring of $L' \otimes_K A$ and hence reduced. This proves that A is a separable K-algebra.

PROPOSITION 4. — *Let* A *be a separable algebra over a field* K *and let* B *be the total ring of fractions of* A ; *then the* K-*algebra* B *is separable.*

Let S be the set of cancellable elements of A. We have identified A with a subring of B (I, p. 113) ; further, every element of S is invertible in B and every element of B has the form as^{-1} with $a \in A$ and $s \in S$. Let *L* be an extension of K and x a nilpotent element of $L \otimes_K B$. Then x may be written in the form $x = \sum_{i=1}^{n} y_i \otimes a_i s_i^{-1}$ with $y_i \in L$, $a_i \in A$, $s_i \in S$ for $1 \leqslant i \leqslant n$. If we put $s = s_1 \dots s_n$ then $x(1 \otimes s)$ belongs to the subring $L \otimes_K A$ of $L \otimes_K B$; since $x(1 \otimes s)$ is nilpotent and A separable over K, we have $x(1 \otimes s) = 0$, and since s is invertible in B we find that $x = 0$. This proves the ring $L \otimes_K B$ to be reduced, whence the proposition.

PROPOSITION 5. — *Let* K *be a field and* A, B *commutative* K-*algebras. If* A *is reduced and* B *separable then* $A \otimes_K B$ *is reduced.*

By the Cor. of Prop. 2 (V, p. 119) A is isomorphic to a subalgebra of a product $\prod_{i \in I} L_i$ where L_i is an extension of K for $i \in I$. Therefore $A \otimes_K B$ is isomorphic to a subring of $\left(\prod_{i \in I} L_i\right) \otimes_K B$ and this latter ring is isomorphic to a subring of $\prod_{i \in I} (L_i \otimes_K B)$ (II, p. 306, Prop. 15). Since B is separable, each of the rings $L_i \otimes_K B$ is reduced, and so the same is true of $\prod_{i \in I} (L_i \otimes_K B)$, and *a fortiori* of $A \otimes_K B$.

COROLLARY 1. — *Let* K *be a field,* L *a separable extension of* K *and* f *a polynomial in* K[X]. *If* f *has no multiple factors in* K[X], *it has also no multiple factors in* L[X].

If f has no multiple factor in K[X], the quotient ring $K[X]/(f)$ is reduced ; for f is a product of irreducible polynomials f_i relatively prime in pairs, hence

$K[X]/(f)$ is isomorphic, by I, p. 110, Prop. 9, to the product of fields $K[X]/(f_i)$. By Prop. 5 the ring $L[X]/(f)$, being isomorphic to $L \otimes_K K[X]/(f)$, is reduced ; if g is a non-constant polynomial of $L[X]$ such that g^2 divides f, then the residue class of fg^{-1} in $L[X]/(f)$ is a non-zero nilpotent element ; hence f has no multiple factors in $L[X]$.

COROLLARY 2. — *Let* A *and* B *be two commutative* K*-algebras. If* A *and* B *are separable, the same is true of* $A \otimes_K B$.

Let L be an extension of K. The ring $L \otimes_K A$ is reduced because A is separable ; now Prop. 5 shows the ring $(L \otimes_K A) \otimes_K B$ (isomorphic to $L \otimes_K (A \otimes_K B)$) to be reduced, whence the Corollary.

3. Separable extensions

Let K be a field. Since any extension of K is a K-algebra, the notion of separability introduced in Def. 1 (V, p. 119) applies in particular to the case of extensions of K. By example 3 (V, p. 119) that definition of separability agrees in the case of algebraic extensions with that of § 7 (V, p. 36, Def. 1).

PROPOSITION 6. — *Every pure extension of a field* K *is separable.*

This follows at once from Example 1 (V, p. 119) and Prop. 4 (V, p. 120).

PROPOSITION 7. — *Let* E *be a field,* G *a group of automorphisms of* E *and* K *the subfield of* E *consisting of the invariants of* G. *Then* E *is a separable extension of* K.

Let L be an extension of K ; there exists an algebraically closed extension Ω of L whose transcendence degree over K is at least equal to that of E over K. By Cor. 1 (V, p. 117) there exists a K-homomorphism u of E into Ω. We denote by v the Ω-algebra homomorphism of $A = \Omega \otimes_K E$ into Ω which maps $\lambda \otimes x$ to $\lambda . u(x)$ for $\lambda \in \Omega$ and $x \in \Omega$; we write $\mathfrak{a}$ for the kernel of v.

For every $s \in G$ let h_s be the automorphism $\mathrm{Id}_\Omega \otimes s$ of the Ω-algebra A ; the kernel of the homomorphism $v \circ h_s$ of A into Ω is the prime ideal $\mathfrak{a}_s = h_s^{-1}(\mathfrak{a})$ of A. It is clear that the ideal $\mathfrak{b} = \bigcap_{s \in G} \mathfrak{a}_s$ of A is stable under the automorphisms h_s. Therefore (V, p. 63, Cor.) the ideal $\mathfrak{b}$ is of the form $\mathfrak{c} \otimes_K E$, where $\mathfrak{c}$ is an ideal of Ω. Now $\mathfrak{b} \subset \mathfrak{a} \neq A$, and so $\mathfrak{c} \neq \Omega$; since Ω is a field, we have $\mathfrak{c} = 0$, hence $\mathfrak{b} = 0$.

The family $(\mathfrak{a}_s)_{s \in G}$ of prime ideals of A thus has zero intersection. By Prop. 2 (V, p. 118) the ring is reduced, and *a fortiori* the same holds of the subring $L \otimes_K E$ of A. Since L was an arbitrary extension of K, this proves that E is separable over K.

PROPOSITION 8. — *Let* L *be an extension of a field* K. *If* L *is separable over* K, *every subextension of* L *is separable over* K. *Conversely, if every finitely generated subextension of* L *is separable over* K, *then* L *is separable over* K.

This follows at once from Prop. 3, *a*) and *b*) (V, p. 119).

Thus separability may be called a property of « finite character ».

PROPOSITION 9. — *Let* L *be an extension of a field* K *and* M *a commutative* L-*algebra* (for example, an extension of L). *If* M *is separable over* L *and* L *is separable over* K, *then* M *is separable over* K.

Let K′ be an extension of K. Since L is a separable extension of K, the ring $K' \otimes_K L$ is reduced ; since M is a separable L-algebra, Prop. 5 (V, p. 120) shows the ring $(K' \otimes_K L) \otimes_L M$ to be reduced. Now the ring $K' \otimes_K M$ is isomorphic to $(K' \otimes_K L) \otimes_L M$ (II, p. 278, Prop. 2), hence is reduced. This proves that M is separable over K.

If the extension M is separable over K, it is not necessarily separable over L (cf. however V, p. 124, Cor. 3). For example, if p is a prime number, the field $\mathbf{F}_p(X)$ of rational fractions in one indeterminate X over $\mathbf{F}_p$ is separable over $\mathbf{F}_p$ (V, p. 121, Prop. 6) but it is a p-radical algebraic extension of $\mathbf{F}_p(X^p)$; in particular $\mathbf{F}_p(X)$ is not separable over $\mathbf{F}_p(X^p)$.

Later (V, p. 137 Prop. 5) we shall study the separability of composite extensions.

4. Mac Lane's separability criterion

THEOREM 1. — *Let* K *be a field of characteristic* 0. *Every reduced* K-*algebra and in particular every extension of* K *is separable over* K.

We first show that every extension L of K is separable. Let B be a transcendence basis of L over K (V, p. 109, Th. 1) and let $L_1 = K(B)$. Then L_1 is separable over K (V, p. 121, Prop. 6). Further, L is an algebraic extension of L_1 and L_1 is a field of characteristic 0 ; therefore L is separable over L_1 (V, p. 37, Cor.). By Prop. 9, L is thus separable over K.

Now let A be a reduced algebra over the field K. By the Cor. of Prop. 2 (V, p. 119) there exists a family $(L_i)_{i \in I}$ of extensions of K such that A is isomorphic to a subalgebra of $\prod_{i \in I} L_i$. Each of the algebras L_i is separable over K by what has been said and so A has the same property, by Prop. 3 *a*) and *c*) (V, p. 119).

THEOREM 2. — *Let* K *be a field of characteristic* $p \neq 0$, $K^{p^{-\infty}}$ *a perfect closure of* K *and* A *a commutative* K-*algebra. The following properties are equivalent :*

a) A *is separable.*

b) *There exists an extension* K′ *of* K *such that* K′ *is perfect and* $K' \otimes_K A$ *is reduced.*

c) *The ring* $K^{p^{-\infty}} \otimes_K A$ *is reduced.*

d) The ring $K' \otimes_K A$ *is reduced for every extension* K' *of* K *which is of finite degree and p-radical of height* $\leqslant 1$.

e) For every family $(a_i)_{i \in I}$ *of elements of* A *linearly free over* K, *the family* $(a_i^p)_{i \in I}$ *is linearly free over* K.

f) There exists a basis $(a_i)_{i \in I}$ *of the vector* K*-space* A *such that the family* $(a_i^p)_{i \in I}$ *is linearly free over* K.

If an extension K' of K is a perfect field, it contains a subextension K-isomorphic to $K^{p^{-\infty}}$ (V, p. 6, Prop. 3) ; moreover, every p-radical extension of K is isomorphic to a subextension of $K^{p^{-\infty}}$ (V, p. 26, Prop. 3). These remarks show the implications $a) \Rightarrow b) \Rightarrow c) \Rightarrow d)$.

Let us prove that $d)$ implies $e)$. Let $(a_i)_{i \in I}$ be a linearly free family in A and let $(\lambda_i)_{i \in I}$ be a family of finite support in K such that $\sum \lambda_i a_i^p = 0$. Let K' be the subextension of $K^{p^{-\infty}}$ generated by the elements $\lambda_i^{p^{-1}}$; it is of finite degree and height $\leqslant 1$. Put $x = \sum_{i \in I} \lambda_i^{p^{-1}} \otimes a_i$ in $K' \otimes_K A$; we have

$$x^p = \sum_{i \in I} \lambda_i \otimes a_i^p = 1 \otimes \sum_{i \in I} \lambda_i a_i^p = 0 .$$

By the hypothesis $d)$ we have $x = 0$, whence $\lambda_i = 0$ for all $i \in I$.

Clearly $e)$ implies $f)$ and it remains to show that $f)$ implies $a)$. Thus let $(a_i)_{i \in I}$ be a basis of A over K such that the family $(a_i^p)_{i \in I}$ is linearly free over K. Let L be an extension of K and let x be an element of $L \otimes_K A$ such that $x^p = 0$. Write $x = \sum_{i \in I} \lambda_i \otimes a_i$ with $\lambda_i \in L$ for all $i \in I$. We have $x^p = \sum_{i \in I} \lambda_i^p \otimes a_i^p = 0$ and since the family $(a_i^p)_{i \in I}$ is linearly free over K, we have $\lambda_i^p = 0$, whence $\lambda_i = 0$ for all $i \in I$; it follows that $x = 0$. So we have proved that $x^p = 0$ implies $x = 0$ in $L \otimes_K A$, from which it follows at once that $L \otimes_K A$ is reduced.

COROLLARY 1 (Mac Lane). — *Let* K *be a field of characteristic exponent* p, Ω *a perfect extension of* K *and* L *a subextension of* Ω. *Then the following conditions are equivalent :*

a) L *is separable over* K.

b) L *is linearly disjoint from* $K^{p^{-\infty}}$ *over* K.

c) L *is linearly disjoint over* K *from every p-radical extension of* K *contained in* Ω, *of finite degree and height* $\leqslant 1$.

The case where K is of characteristic 0 is trivial because then L is separable over K (Th. 1) and $K^{p^{-\infty}} = K$ by convention. Suppose then that $p \neq 1$, and let us first show that $a)$ implies $b)$. Assume that L is separable over K and let $(a_i)_{i \in I}$ be a basis of L over K. Let $(\lambda_i)_{i \in I}$ be a family of finite support of elements of $K^{p^{-\infty}}$ such that $\sum_{i \in I} \lambda_i a_i = 0$; there exists an integer $f \geqslant 0$ and elements

μ_i of K such that $\lambda_i = \mu_i^{p^{-f}}$. We have

$$\sum_{i \in I} \mu_i a_i^{p^f} = \Big(\sum_{i \in I} \lambda_i a_i\Big)^{p^f} = 0$$

and Th. 2 implies, by induction on f, that the family $(a_i^{p^f})_{i \in I}$ is linearly free over K. We thus have $\mu_i = 0$, whence $\lambda_i = 0$ for all $i \in I$. Finally L is linearly disjoint from $K^{p^{-\infty}}$ over K.

It is clear that *b*) implies *c*). Lastly suppose that *c*) holds and let K′ be an extension of K of finite degree and p-radical of height $\leqslant 1$. The ring $K' \otimes_K L$ is isomorphic to a subring of Ω, hence is reduced. Now it follows from Th. 2 that L is separable over K.

COROLLARY 2. — *Let* K *be a field of characteristic exponent* p, $K^{p^{-\infty}}$ *a perfect closure of* K *and* L *a separable extension of* K. *Then the ring* $L \otimes_K K^{p^{-\infty}}$ *is a field. If moreover* L *is algebraic over* K, *then* $L \otimes_K K^{p^{-\infty}}$ *is a perfect closure of* L.

The case $p = 1$ is trivial, so let us suppose $p \neq 1$. Let Ω be a perfect closure of L. By Cor. 1 there exists a K-algebra homomorphism of $L \otimes_K K^{p^{-\infty}}$ onto $L[K^{p^{-\infty}}]$ which maps $x \otimes y$ to xy for $x \in L$ and $y \in K^{p^{-\infty}}$. Since $K^{p^{-\infty}}$ is algebraic over K, the ring $L[K^{p^{-\infty}}]$ is a subfield of Ω (V, p. 18, Cor. 1). Suppose further that L is algebraic over K, then $L[K^{p^{-\infty}}]$ is an algebraic extension of the perfect field $K^{p^{-\infty}}$, hence it is a perfect field (V, p. 43, Cor. 1) ; finally, since the field $L[K^{p^{-\infty}}]$ is a p-radical extension of L (V, p. 25, Cor.), it is a perfect closure of L.

COROLLARY 3.— *Let* L *be an algebraic extension of* K *and* M *a commutative* L-*algebra* (for example an extension of L). *If* M *is separable over* K, *it is separable over* L.

The algebra L is K-isomorphic to a subalgebra of M, hence L is a separable extension of K. Therefore (Cor. 2) there exists an L-isomorphism of $L^{p^{-\infty}}$ onto $K^{p^{-\infty}} \otimes_K L$. The ring $L^{p^{-\infty}} \otimes_L M$ is thus isomorphic to $(K^{p^{-\infty}} \otimes_K L) \otimes_L M$, and so to $K^{p^{-\infty}} \otimes_K M$ (II, p. 278, Prop. 2) and this latter ring is reduced because M is separable over K. Thus the ring $L^{p^{-\infty}} \otimes_L M$ is reduced, which proves that M is separable over L (V, p. 122, Th. 2).

Remark. — Mac Lane's criterion may be formulated without introducing any extensions of K other than L. For by *c*) of Cor. 1, L is separable over K if and only if L and $K^{p^{-1}}$ are linearly disjoint over K. Since the mapping $x \mapsto x^p$ is an isomorphism of L onto the subfield L^p, we obtain the following criterion (cf. V, p. 177, Ex. 11 for an analogous criterion for algebras) :

L *is separable over* K *if and only if the subfields* L^p *and* K *of* L *are linearly disjoint over* K^p.

5. Extensions of a perfect field

For ease of reference we summarize the principal properties of extensions of perfect fields :

THEOREM 3. — *Let* K *be a perfect field.*

a) *Every algebraic extension of* K *is a perfect field.*

b) *Every extension of* K *is separable.*

c) *For a* K*-algebra to be separable it is necessary and sufficient that it should be reduced.*

d) *Let* A *and* B *be two reduced* K*-algebras. Then* $A \otimes_K B$ *is reduced.*

e) *If* E *and* F *are two extensions of* K, *then the ring* $E \otimes_K F$ *is reduced.*

Assertion *a*) is just Cor. 1 of Prop. 11 (V, p. 43).

Assertion *b*) follows from Th. 1 (V, p. 122) when K is of characteristic 0 and from Cor. 1 (V, p. 123) when K is of characteristic $p \neq 0$.

Let us prove *c*). The case where K is of characteristic 0 follows from Th. 1 (V, p. 122). So it is enough to show that if K is perfect of characteristic $p \neq 0$ and A is a reduced K-algebra, then A is separable over K. But this follows from the equivalence of the conditions *a*) and *b*) of Th. 2 (V, p. 122 ; take $K' = K$ in *b*)).

Finally, *d*) follows from *c*) and Prop. 5 (V, p. 120) and *e*) is a particular case of *d*).

6. The characterization of separability by automorphisms

THEOREM 4. — *Let* Ω *be an algebraically closed extension of a field* K *and* L *a subextension of* Ω. *Then the following conditions are equivalent :*

a) L *is separable over* K.

b) *The intersection of the kernels of the* Ω*-algebra homomorphisms of* $\Omega \otimes_K L$ *into* Ω *is reduced to* 0.

c) *For any elements* $a_1, \ldots, a_n$ *of* L *linearly independent over* K *there exist* K*-automorphisms* $\sigma_1, \ldots, \sigma_n$ *of* Ω *such that* $\det(\sigma_i(a_j)) \neq 0$.

d) *Let* V *be a vector sub-*K*-space of* L *of finite dimension. Every* K*-linear mapping of* V *into* Ω *is a linear combination (with coefficients in* Ω*) of the restrictions to* V *of* K*-automorphisms of* Ω.

d) $\Rightarrow$ *c*) : Let $a_1, \ldots, a_n$ be elements of L linearly independent over K and let V be the vector sub-K-space of L generated by $a_1, \ldots, a_n$. The mapping $f \mapsto (f(a_1), \ldots, f(a_n))$ is an Ω-linear bijection of $\mathrm{Hom}_K(V, \Omega)$ onto Ω^n. Suppose that *d*) holds, then there exist K-automorphisms $\sigma_1, \ldots, \sigma_n$ of Ω such that the elements $(\sigma_i(a_1), \ldots, \sigma_i(a_n))$ of Ω^n (for $1 \leqslant i \leqslant n$) form a basis of Ω^n. We thus have $\det(\sigma_i(a_j)) \neq 0$, so *d*) implies *c*).

c) $\Rightarrow$ *b)* : Assume that *c)* holds and let x be in $\Omega \otimes_K L$. We write x in the form $\sum_{j=1}^{n} x_j \otimes a_j$ with $x_1, \ldots, x_n$ in Ω and the elements $a_1, \ldots, a_n$ of L linearly independent over K. Choose K-automorphisms $\sigma_1, \ldots, \sigma_n$ of Ω such that det $\sigma_i(a_j) \neq 0$; let $\chi_1, \ldots, \chi_n$ be the Ω-homomorphisms of $\Omega \otimes_K L$ into Ω such that $\chi_i(a \otimes b) = a \,.\, \sigma_i(b)$ for $a \in \Omega$ and $b \in L$. Suppose that $\chi_i(x) = 0$ for $1 \leqslant i \leqslant n$, in other words, that $\sum_{j=1}^{n} x_j \,.\, \sigma_i(a_j) = 0$ for $1 \leqslant i \leqslant n$. Since we have assumed that the matrix $(\sigma_i(a_j))$ has a non-zero determinant, we have $x_i = 0$ for $1 \leqslant i \leqslant n$ and so $x = 0$.

b) $\Rightarrow$ *a)* : Since every extension of a field of characteristic 0 is separable (V, p. 122, Th. 1) it suffices to examine the case where K is of characteristic $p \neq 0$. Let X be the set of all Ω-algebra homomorphisms of $\Omega \otimes_K L$ into Ω and f the homomorphism of $\Omega \otimes_K L$ into Ω^X defined by $f(u) = (\chi(u))_{\chi \in X}$ for $u \in \Omega \otimes_K L$. Condition *b)* means that f is injective and so implies that the ring $\Omega \otimes_K L$ is reduced. Hence condition *b)* of Th. 2 (V, p. 122) is satisfied with $K' = \Omega$, so L is separable over K.

a) $\Rightarrow$ *d)* : Suppose that L is separable over K. Let V be a vector sub-K-space of finite dimension of L, $V_{(\Omega)} = \Omega \otimes_K V$ the vector Ω-space derived from V by extension of scalars and f_0 the linear form on $V_{(\Omega)}$ such that $f_0(x \otimes y) = xy$ for $x \in \Omega$, $y \in V$. Denote by G the group of K-automorphisms of Ω ; for $\sigma \in G$ we put $\sigma_V = \sigma \otimes \mathrm{Id}_V$ and $g_\sigma = \sigma \circ f_0 \circ \sigma_V^{-1}$.

For each $\sigma \in G$ the mapping g_σ of $V_{(\Omega)}$ into Ω is Ω-linear and maps $x \otimes y$ to $x \,.\, \sigma(y)$ for $x \in \Omega$, $y \in V$. The kernel N_σ of g_σ is therefore a vector subspace of $V_{(\Omega)}$, and the same holds for $N = \bigcap_{\sigma \in G} N_\sigma$. If p is the characteristic exponent of K, the field of invariants of G in Ω is equal to $K^{p^{-\infty}}$ (V, p. 112, Prop. 10). We clearly have $\sigma_V(N) = N$ for all $\sigma \in G$; therefore (V, p. 63, Cor.) the vector Ω-space N is generated by $N_0 = N \cap (K^{p^{-\infty}} \otimes V)$. Since L is separable over K, the fields $K^{p^{-\infty}}$ and L are linearly disjoint over K (V, p. 123, Cor. 1) ; we have $K^{p^{-\infty}} \otimes_K V \subset K^{p^{-\infty}} \otimes_K L$ and $f_0(x \otimes y) = xy$ for $x \in \Omega$ and $y \in V$. It follows that the restriction of f_0 to $K^{p^{-\infty}} \otimes_K V$ is injective. Now $f_0 = g_1$ is zero on N and *a fortiori* on $N_0 \subset K^{p^{-\infty}} \otimes_K V$. We thus have $N_0 = 0$, whence $N = 0$. Since V is of finite dimension over K, $V_{(\Omega)}$ is of finite dimension over Ω ; the intersection of the kernels of the linear forms g_σ is zero, hence (II, p. 301, Th. 7) the family $(g_\sigma)_{\sigma \in G}$ generates the dual of $V_{(\Omega)}$. Now let u be a K-linear mapping of V into Ω ; let $\tilde{u}$ be the linear form on $V_{(\Omega)}$ which maps $x \otimes y$ to $xu(y)$ for $x \in \Omega$ and $y \in V$. By what has been said there exist elements $\sigma_1, \ldots, \sigma_n$ of G and $\lambda_1, \ldots, \lambda_n$ of Ω such that $\tilde{u} = \sum_{i=1}^{n} \lambda_i g_{\sigma_i}$, whence $u(y) = \sum_{i \in 1}^{n} \lambda_i \sigma_i(y)$ for all $y \in V$. So we have shown that *a)* implies *d)*.

§ 16. DIFFERENTIAL CRITERIA OF SEPARABILITY

1. Extension of derivations: the case of rings

Let K be a commutative ring, A a commutative K-algebra and $\mathbf{x} = (x_i)_{i \in I}$ a family of elements of A. Further, let Δ be a derivation of K into an A-module M, in other words (III, p. 553) a **Z**-linear mapping of K into M satisfying the relation $\Delta(cc') = c \cdot \Delta(c') + c' \cdot \Delta(c)$ for c, c' in K. For each $i \in I$ let D_i be the partial derivation with respect to X_i in the polynomial ring $K[X_i]_{i \in I}$; this is the unique derivation of that ring into itself which is zero on K and on X_j for $j \in I - \{i\}$, and takes the value 1 on X_i (IV, p. 6). For every polynomial $f = \sum_{\alpha \in \mathbf{N}^{(I)}} c_\alpha \cdot X^\alpha$ in $K[X_i]_{i \in I}$ we denote by $f^\Delta(\mathbf{x})$ the element $\sum_{\alpha \in \mathbf{N}^{(I)}} \mathbf{x}^\alpha \cdot \Delta(c_\alpha)$ of M.

PROPOSITION 1. — *Suppose that the family $\mathbf{x} = (x_i)_{i \in I}$ generates the K-algebra A. Let $(m_i)_{i \in I}$ be a family of elements of M and $(f_\lambda)_{\lambda \in \Lambda}$ a family of polynomials generating the ideal $\mathfrak{a}$ of all polynomials $f \in K[X_i]_{i \in I}$ such that $f(\mathbf{x}) = 0$. For a derivation D of A into M to exist such that $D(c \cdot 1) = \Delta(c)$ for all $c \in K$ and $D(x_i) = m_i$ for all $i \in I$ it is necessary and sufficient that*

$$(1) \qquad f_\lambda^\Delta(\mathbf{x}) + \sum_{i \in I} D_i f_\lambda(\mathbf{x}) \cdot m_i = 0 \quad \textit{for all} \quad \lambda \in \Lambda .$$

If this is so, the derivation D is unique and satisfies

$$(2) \qquad D(f(\mathbf{x})) = f^\Delta(\mathbf{x}) + \sum_{i \in I} D_i f(\mathbf{x}) \cdot m_i \quad \textit{for all} \quad f \in K[X_i]_{i \in I} .$$

Put $E = K[X_i]_{i \in I}$ and write φ for the K-homomorphism of E onto A which maps X_i to x_i for all $i \in I$; we thus have $\varphi(f) = f(\mathbf{x})$ for all $f \in E$. We consider M as an E-module by means of the homomorphism $\varphi : E \to A$ and define a mapping D' of E into M by $D'(f) = f^\Delta(\mathbf{x}) + \sum_{i \in I} D_i f(\mathbf{x}) \cdot m_i$ (note that the family $(D_i f)_{i \in I}$ has finite support for each $f \in E$). It is clear that D' is the unique derivation of E into M extending Δ and mapping X_i to m_i for each $i \in I$.

Let D be a derivation of A into M such that $D(c \cdot 1) = \Delta(c)$ for all $c \in K$ and $D(x_i) = m_i$ for all $i \in I$. Then $D \circ \varphi$ is a derivation of E into M extending Δ and mapping X_i to m_i for all $i \in I$. We thus have $D \circ \varphi = D'$, that is, the relation (2) holds. This proves the uniqueness of D; moreover (1) is a consequence of $f_\lambda(\mathbf{x}) = 0$ and (2).

Conversely assume that (1) holds ; in other words, we have $D'(f_\lambda) = 0$ for all $\lambda \in \Lambda$. Let $f \in \mathfrak{a}$; there exists a family with finite support $(q_\lambda)_{\lambda \in \Lambda}$ in E such that $f = \sum_{\lambda \in \Lambda} q_\lambda \cdot f_\lambda$. We have

$$D'(f) = \sum_{\lambda \in \Lambda} [f_\lambda(\mathbf{x}) \cdot D'(q_\lambda) + q_\lambda(\mathbf{x}) \cdot D'(f_\lambda)]$$

whence $D'(f) = 0$ because $f_\lambda(\mathbf{x})$ and $D'(f_\lambda)$ are zero for all $\lambda \in \Lambda$. Since D' vanishes on $\mathfrak{a}$, there exists a **Z**-linear mapping D of A into M such that $D' = D \circ \varphi$; clearly D is the required derivation of A into M.

2. Extension of derivations : the case of fields

Let K, L and M be fields such that $K \subset L \subset M$. By Prop. 21 (III, p. 572) we have an exact sequence of vector M-spaces

$$(E_{K,L,M}) \qquad \Omega_K(L) \otimes_L M \xrightarrow{\alpha} \Omega_K(M) \xrightarrow{\beta} \Omega_L(M) \to 0 ;$$

the M-linear mappings α and β are characterized by the relations

$$\alpha(d_{L/K}x \otimes 1) = d_{M/K}x \quad \text{for} \quad x \in L \tag{3}$$

$$\beta(d_{M/K}y) = d_{M/L}y \quad \text{for} \quad y \in M . \tag{4}$$

Let V be a vector M-space, denote by $D_K(M, V)$ the vector space of K-derivations of M with values in V and similarly introduce $D_K(L, V)$ and $D_L(M, V)$. By III, p. 571, Diagram (42) and III, p. 572, Diagram (44), we have a commutative diagram of homomorphisms of vector spaces

$$\begin{array}{ccccc}
0 \to \mathrm{Hom}_M(\Omega_L(M), V) & \xrightarrow{\mathrm{Hom}(\beta, 1)} & \mathrm{Hom}_M(\Omega_K(M), V) & \xrightarrow{\mathrm{Hom}(\alpha, 1)} & \mathrm{Hom}_M(\Omega_K(L) \otimes_L M, V) \\
\downarrow & & \downarrow & & \downarrow \\
0 \to \mathrm{Der}_L(M, V) & \xrightarrow{i_V} & \mathrm{Der}_K(M, V) & \xrightarrow{r_V} & \mathrm{Der}_K(L, V) ,
\end{array}$$

where the vertical arrows are isomorphisms, i_V is the canonical injection and r_V the restriction mapping.

We thus obtain from II, p. 299, Th. 5 and II, p. 301, Prop. 10 the following proposition :

PROPOSITION 2. — *The following conditions are equivalent :*

a) The mapping α is injective.

b) Every K*-derivation of* L *into* M *extends to a* K*-derivation of* M *into* M.

c) Every K*-derivation of* L *into a vector* M*-space* V *extends to a* K*-derivation of* M *into* V.

PROPOSITION 3. — *Let* K *be a field,* L *a pure extension of* K *and* $(x_i)_{i \in I}$ *a pure basis of* L (V, p. 106, Def. 2).

a) Let V *be a vector space over* L, Δ *a derivation of* K *into* V *and* $(v_i)_{i \in I}$ *a family of elements of* V. *There exists a unique derivation* D *of* L *into* V, *extending* Δ *and such that* $D(x_i) = v_i$ *for all* $i \in I$.

b) The vector L-*space* $\Omega_K(L)$ *of* K-*differentials of* L *has the family* $(dx_i)_{i \in I}$ *as basis.*

Assertion *b*) has been proved in IV, p. 23 and Assertion *a*) follows directly from this.

COROLLARY. — *Let* P *be a subfield of* K. *The canonical mapping* α *of* $\Omega_P(K) \otimes_K L$ *into* $\Omega_P(L)$ *is injective and the family* $(d_{L/P}x_i)_{i \in I}$ *is a basis (over* L) *of a subspace of* $\Omega_P(L)$ *supplementary to the image of* α: $\Omega_P(K) \otimes_K L \to \Omega_P(L)$.

Prop. 3, *a*) shows that every P-derivation of K into a vector space V over L extends to a P-derivation of L into V ; the injectivity of α then follows by Prop. 2. The second assertion of the corollary follows from the exactness of the sequence $(E_{P,K,L})$ and Prop. 3, *b*) (taking Formula (4) into account).

PROPOSITION 4. — *Let* K *be a field,* L *a separable algebraic extension of* K *and* V *a vector space over* L.

a) Every K-*derivation of* L *into* V *is zero.*

b) If Δ *is a derivation of* K *into* V, *there exists a unique derivation* D *of* L *into* V *which extends* Δ.

Let D be a K-derivation of L into V. If E is a subextension of L of finite degree over K, then the K-algebra E is etale, and so $\Omega_K(E) = 0$ (V, p. 33, Th. 3), whence $D \mid E = 0$ by the universal property of $\Omega_K(E)$ (III, p. 569). Since L is the union of a family of subextensions of finite degree over K, we have $D = 0$, whence *a*).

Let Δ be a derivation of K into V. If D′ and D″ are two extensions of Δ to a derivation of L into V, then the difference $D' - D''$ is a K-derivation of L into V ; hence it is zero by *a*) and so $D' = D''$.

It remains to prove the *existence* of an extension of Δ. Zorn's Lemma (E, III, p. 20) implies the existence of a *maximal* extension D_0 of Δ to a derivation defined on a subfield L_0 of L containing K.

Let x be in L and g the minimal polynomial of x over L_0. Since L is algebraic and separable over K, x is algebraic and separable over L_0 (V, p. 39, Prop. 6 and p. 39, Cor. 2) ; therefore x is a simple root of g (V, p. 39, Prop. 5), whence $g'(x) \neq 0$ (IV, p. 17, Prop. 7). If we define $g^{D_0}(x)$ as in V, p. 127, there exists thus an element u of V such that $g^{D_0}(x) + g'(x) . u = 0$; by Prop. 1 (V, p. 127) there exists a derivation D of $L_0(x)$ into V extending D_0 and such that $D(x) = u$. In view of the maximal character of (L_0, D_0) we thus have $L_0(x) = L_0$, whence $x \in L_0$. Since x was arbitrary we conclude that $L_0 = L$.

COROLLARY 1. — *We have* $\Omega_K(L) = 0$ *if* L *is algebraic and separable over* K.

The corollary follows from Prop. 4, *a*) because the vector L-space $\Omega_K(L)$ is generated by the image of the canonical K-derivation $d_{L/K} : L \to \Omega_K(L)$.

COROLLARY 2. — *If* L *is an algebraic and separable extension of a field* K, *the canonical mapping* $\alpha : \Omega_P(K) \otimes_K L \to \Omega_P(L)$ *is an isomorphism for every subfield* P *of* K.

The mapping α is injective by Prop. 2 (V, p. 128) and Prop. 4, *b*) ; since $\Omega_K(L)$ reduces to 0 (Cor. 1) the exactness the sequence $(E_{P,K,L})$ implies that α is surjective.

COROLLARY 3. — *Let* E *be an extension of a field* K *and* D *a derivation of* E *into* E, *mapping* K *into* K. *If* L *is a subextension of* E *which is algebraic and separable over* K, *then* $D(L) \subset L$.

Let Δ be the derivation of K into L which agrees with D on K. By Prop. 4 (V, p. 129) there exists a derivation D′ of L into L extending Δ. Now we may consider D′ and the restriction D″ of D to L as derivations of L into E ; since they agree on K, we have $D' = D''$ by Prop. 4, whence

$$D(L) = D''(L) = D'(L) \subset L .$$

Remarks. — 1) Later (V, p. 131, Cor. 3 and p. 135, Cor. 2) we shall prove a converse to Cor. 1 of Prop. 4.

2) Every algebraic extension of a prime field is separable (V, p. 37, Cor.). Since every derivation of a prime field is zero, every derivation of an algebraic extension of a prime field is zero (Prop. 4).

3. Derivations in fields of characteristic zero

THEOREM 1. — *Let* K *be a field of characteristic* 0, L *an extension of* K *and* V *a vector space over* L. *Let* Δ *be a derivation of* K *into* V, $(x_i)_{i \in I}$ *a transcendence basis of* L *over* K *and* $(u_i)_{i \in I}$ *a family of elements of* V. *Then there exists a unique derivation* D *of* L *into* V *extending* Δ *and such that* $D(x_i) = u_i$ *for all* $i \in I$.

Put $E = K(x_i)_{i \in I}$; Prop. 3 (V, p. 128) shows that Δ extends to a unique derivation D_0 of E into V such that $D_0(x_i) = u_i$ for all $i \in I$. The field L is an algebraic extension of E and since L is of characteristic 0, L is separable over E (V, p. 37, Cor.). Therefore (V, p. 129, Prop. 4) D_0 extends to a unique derivation D of L into V.

COROLLARY 1. — *Every derivation of* K *into* V *extends to a derivation of* L *into* V.

COROLLARY 2. — *Let* E *be a subextension of* L *and* U *the vector sub-*L*-space of* $\Omega_K(L)$ *generated by the differentials of elements of* E. *For an element x to be algebraic over* E *it is necessary and sufficient that* $dx \in U$.

For each $y \in L$ let $D(y)$ be the residue class of dy mod U. Then D is an E-derivation of L into $\Omega_K(L)/U$. Since K is of characteristic 0, every algebraic extension of E is separable (V, p. 37, Cor.) ; if $x \in L$ is algebraic over E, we have $Dx = 0$ by Prop. 4 (V, p. 129), that is, $dx \in U$.

If $x \in L$ is transcendental over E, there exists an E-derivation Δ of $E(x)$ into L such that $\Delta(x) = 1$ (V, p. 128, Prop. 3) ; by Th. 1, Δ extends to an E-derivation D of L into L. Let φ be the linear form on $\Omega_K(L)$ such that $D = \varphi \circ d$; we have $\varphi(dy) = 0$ for $y \in E$ and $\varphi(x) = 1$, whence $dx \notin U$.

COROLLARY 3. — *For an element x of* L *to be algebraic over* K *it is necessary and sufficient that* $dx = 0$ *in* $\Omega_K(L)$. *In particular, for* L *to be an algebraic extension of* K *it is necessary and sufficient that* $\Omega_K(L) = 0$.

This follows immediately from Cor. 2 on taking E = K.

COROLLARY 4. — *Let* K, L *and* M *be fields of characteristic* 0 *such that* $K \subset L \subset M$; *then the canonical mapping* $\alpha : \Omega_K(L) \otimes_L M \to \Omega_K(M)$ *is injective.*

This follows at once from Cor. 1 and V, p. 128, Prop. 2.

THEOREM 2.— *Let* K *be a field of characteristic* 0, L *an extension of* K *and* $(x_i)_{i \in I}$ *a family of elements of* L.

a) For $(x_i)_{i \in I}$ *to be algebraically free over* K *it is necessary and sufficient that the differentials* dx_i *(for* $i \in I$*) in* $\Omega_K(L)$ *should be linearly free over* L.

b) For L *to be algebraic over* $K(x_i)_{i \in I}$ *it is necessary and sufficient that the differentials* dx_i *for* $i \in I$ *generate the vector space* $\Omega_K(L)$ *over* L.

c) For $(x_i)_{i \in I}$ *to be a transcendence basis of* L *over* K, *it is necessary and sufficient that the family* $(dx_i)_{i \in I}$ *should be a basis of* $\Omega_K(L)$ *over* L.

For every $i \in I$ let E_i be the subfield $K(x_j)_{j \in I - \{i\}}$ of L. For the family $(x_i)_{i \in I}$ to be algebraically free over K it is necessary and sufficient (V, p. 108, Prop. 5) that x_i should be transcendental over E_i for each $i \in I$. By Cor. 2 of Th. 1 this means that for each $i \in I$ the differential dx_i is not a linear combination with coefficients in L of the differentials dx_j with $j \neq i$ in I ; this latter condition just means that the family $(dx_i)_{i \in I}$ is free over L, whence *a*).

Assertion *b*) follows at once from Cor. 2 of Th. 1 and *c*) is a consequence of *a*) and *b*).

COROLLARY. — *We have* $[\Omega_K(L) : L] = \mathrm{tr.deg}_K L$ *when* K *is of characteristic* 0.

4. Derivations in separable extensions

We have seen (V, p. 122, Th. 1) that every extension L of a field K of characteristic 0 is separable ; moreover, every derivation of K into a vector space over L extends then to a derivation of L (V, p. 130, Cor. 1). More generally we have the following statement :

THEOREM 3. — *Let* K *be a field and* L *an extension of* K. *For* L *to be separable over* K *it is necessary and sufficient that every derivation of* K *into a vector* L*-space should extend to a derivation of* L.

We may assume that K has characteristic $\neq 0$. Suppose first that L is separable over K. Let V be a vector space over L and Δ a derivation of K into V. By Mac

Lane's criterion (V, p. 124, Remark), the fields L^p and K are linearly disjoint over K^p. Since Δ is a K^p-linear mapping of K into V, it extends in a unique way to an L^p-linear mapping Δ' of $K[L^p] = K(L^p)$ into V. It is clear that Δ' is a derivation of $K(L^p)$ into V which vanishes on L^p ; thus it extends (V, p. 101, Cor. 1) to a derivation of L into V.

Conversely, suppose that every derivation of K with values in L extends to a derivation of L into L. Let B be a p-basis of K (V, p. 99) and let Λ be the set of families $(\alpha_b)_{b \in B}$ with finite support consisting of integers between 0 and $p-1$. For each $b \in B$ there exists a derivation Δ_b of K into K characterized by $\Delta_b(b') = \delta_{bb'}$ (Kronecker symbol) for each $b' \in B$ (V, p. 103). By hypothesis there exists for each $b \in B$ a derivation D_b of L into L extending Δ_b. We have $D_b(b') = \delta_{bb'}$ for b, b' in B, which proves that in $\Omega_{L^p}(L)$ the differentials db (for $b \in B$) are linearly independent over L. Therefore (V, p 102, Th. 1), B is p-free over L^p. It follows (V, p. 98, Prop. 1 and p. 124, Remark) that the extension L of K is separable.

COROLLARY. — *If* L *is a separable extension of* K, *the canonical mapping* $\alpha_P : \Omega_P(K) \otimes_K L \to \Omega_P(L)$ *is injective for every subfield* P *of* K. *Conversely, if there exists a perfect subfield* P *of* K (for example the prime subfield of K) *such that the mapping* α_P *is injective, then* L *is separable over* K.

The first assertion follows from Prop. 2 (V, p. 128) and Th. 3. Conversely, let P be a perfect subfield of K ; we have $P = P^p \subset K^p$, hence every derivation of K into a vector L-space is a P-derivation ; the second assertion of the corollary now follows from Cor. 2 (V, p. 130) and Th. 3.

5. The index of a linear mapping

Let L be a field [1], let U and V be two vector spaces [2] over L and $f : U \to V$ an L-linear mapping ; f is said to have *finite index* if the kernel N and cokernel C of f have finite dimension, and the integer

$$\chi(f) = [C : L] - [N : L]$$

is called the *index* of f.

Lemma 1. — *Let* U *and* V *be two vector spaces of finite dimension over a field* L. *Every linear mapping* $f : U \to V$ *possesses an index equal to* $[V : L] - [U : L]$.

Let N be the kernel, I the image and C the cokernel of f. We have $C = V/I$ and I is isomorphic to U/N ; thus we have $[U : L] = [N : L] + [I : L]$ and $[V : L] = [C : L] + [I : L]$, whence the result.

[1] Not necessarily commutative.
[2] On the left.

Lemma 2. — *Let* $f : \mathrm{U} \to \mathrm{V}$ *and* $g : \mathrm{V} \to \mathrm{W}$ *be two* L-*linear mappings. If* f *and* g *each have an index, the same is true of* $g \circ f$ *and we have*

$$\chi(g \circ f) = \chi(f) + \chi(g) .$$

Put $h = g \circ f$, and denote by N, N′, N″ the kernels of f, g, h respectively and by C, C′, C″ their cokernels. We have $\mathrm{N} \subset \mathrm{N}'' \subset \mathrm{U}$ and $f(\mathrm{N}'') = f(\mathrm{U}) \cap \mathrm{N}'$; hence there exists a linear mapping $\bar{f} : \mathrm{N}'' \to \mathrm{N}'$ agreeing with f on N″ and with kernel N. The canonical mapping π of V onto $\mathrm{C} = \mathrm{V}/f(\mathrm{U})$ induces a mapping π' of N′ into C whose kernel is $f(\mathrm{U}) \cap \mathrm{N}' = \bar{f}(\mathrm{N}'')$. By passage to quotients g defines a mapping $\bar{g}$ of $\mathrm{C} = \mathrm{V}/f(\mathrm{U})$ into $\mathrm{C}'' = \mathrm{W}/g(f(\mathrm{U}))$ whose kernel is clearly $(\mathrm{N}' + f(\mathrm{U}))/f(\mathrm{U}) = \pi'(\mathrm{N}')$. Finally the canonical mapping ρ of $\mathrm{C}'' = \mathrm{W}/g(f(\mathrm{U}))$ onto $\mathrm{C}' = \mathrm{W}/g(\mathrm{V})$ has the kernel $g(\mathrm{V})/g(f(\mathrm{U})) = \bar{g}(\mathrm{C})$. To sum up, we have established the exactness of the sequence

$$0 \to \mathrm{N} \xrightarrow{i} \mathrm{N}'' \xrightarrow{\bar{f}} \mathrm{N}' \xrightarrow{\pi'} \mathrm{C} \xrightarrow{\bar{g}} \mathrm{C}'' \xrightarrow{\rho} \mathrm{C}' \to 0$$

(where i is the canonical injection of N in N″).

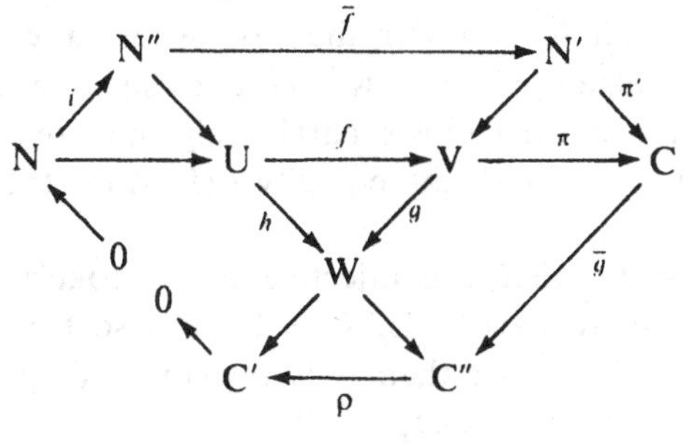

FIG. 1

By hypothesis N, N′, C and C′ are of finite dimension ; hence the same is true of N″ and C″. By Cor. 2 (II, p. 295) we then have

$$[\mathrm{N} : \mathrm{L}] - [\mathrm{N}'' : \mathrm{L}] + [\mathrm{N}' : \mathrm{L}] - [\mathrm{C} : \mathrm{L}] + [\mathrm{C}'' : \mathrm{L}] - [\mathrm{C}' : \mathrm{L}] = 0$$

whence $\chi(h) = \chi(f) + \chi(g)$.

6. Differential properties of finitely generated extensions

THEOREM 4. — *Let* P *be a perfect field,* L *an extension of* P *and* K *a subextension of* L ; *we suppose that* L *is a finitely generated extension of* K. *Then the canonical* L-*linear mapping* $\alpha : \Omega_{\mathrm{P}}(\mathrm{K}) \otimes_{\mathrm{K}} \mathrm{L} \to \Omega_{\mathrm{P}}(\mathrm{L})$ *has index equal to the transcendence degree of* L *over* K.

If E and F are two subextensions of L such that $E \subset F$, we denote by $\alpha(F/E)$ the canonical F-linear mapping of $\Omega_P(E) \otimes_E F$ into $\Omega_P(F)$ and when it is defined, the index of $\alpha(F/E)$ will be written $d(F/E)$. If E, F and G are three subextensions of L such that $E \subset F \subset G$, we have a commutative diagram

$$\begin{array}{ccc} \Omega_P(F) \otimes_F G & \xrightarrow{\alpha(G/F)} & \Omega_P(G) \\ \uparrow{\scriptstyle \alpha(F/E) \otimes_F \mathrm{Id}_G} & & \uparrow{\scriptstyle \alpha(G/E)} \\ (\Omega_P(E) \otimes_E F) \otimes_F G & \xrightarrow{\beta} & \Omega_P(E) \otimes_E G \end{array}$$

where β is the canonical isomorphism defined in Prop. 2 (II, p. 278). Since the index is clearly invariant under extension by scalars and the index of an isomorphism is zero, Lemma 2 (V, p. 133) shows that the index $d(G/E)$ is defined when $d(F/E)$ and $d(G/F)$ are and then

$$d(G/E) = d(G/F) + d(F/E)\,. \tag{5}$$

Since L is a finitely generated extension of K, Formula (5) and the Cor. of V, p. 111 shows that it is enough to consider the case where x exists such that $L = K(x)$; further if x is algebraic over K, there exists a power q of the characteristic exponent of L such that x^q is separable algebraic over K (V, p. 44, Prop. 13). So it is enough to establish the equality $d(L/K) = \mathrm{tr}\,.\,\deg_K L$ in the three special cases below :

a) x is transcendental over K : then α is injective and its cokernel is of rank 1 over L (V, p. 129, Cor.) ; so we have $d(L/K) = 1$ and also $\mathrm{tr}\,.\,\deg_K L = 1$.

b) x is separable algebraic over K : then α is bijective (V, p. 130, Cor. 2), whence $d(L/K) = 0$; clearly also $\mathrm{tr}\,.\,\deg_K L = 0$.

c) The field L *is of characteristic $p \neq 0$, $x \notin$* K *and $x^p = a$ is in* K : the cokernel C of α is isomorphic to $\Omega_K(L)$, and since $\{x\}$ is a p-basis of L over K, the space C is of dimension 1 over L (V, p. 102, Th. 1). Since a is a p-th power in L, we have $d_{L/P}a = 0$ and the kernel of α contains the subspace R of $U = \Omega_P(K) \otimes_K L$ generated by $d_{K/P}a \otimes 1$. For each $y \in K$ let $\Delta(y)$ be the residue class of $d_{K/P}y \otimes 1$ mod R ; then Δ is a P-derivation of K into U/R such that $\Delta(a) = 0$. Prop. 5 (V, p. 101) shows that Δ extends to a P-derivation D of L into U/R. Hence there exists an L-linear mapping $\beta : \Omega_P(L) \to U/R$ such that $D = \beta \circ d_{L/P}$ and $\beta \circ \alpha$ is the canonical mapping of U onto U/R. This proves R to be the kernel of α. Since P is perfect, we have $P(K^p) = K^p$, whence $a \notin P(K^p)$ and finally $d_{K/P}a \neq 0$ (V, p. 103, Prop. 6). The kernel and cokernel of α are thus of dimension 1, whence $d(L/K) = 0$, and we also have $\mathrm{tr}\,.\,\deg_K L = 0$.

COROLLARY 1. — *Let* L *be a finitely generated extension of a field* K *of transcendence degree s. The vector space $\Omega_K(L)$ is of dimension $\geqslant s$ over* L, *with equality if and only if* L *is separable over* K.

Let P be the prime subfield of K. Let N be the kernel of α, then by exactness of the sequence $(E_{P,K,L})$ (V, p. 128) and Th. 4, we have $[\Omega_K(L):L] = s + [N:L]$; by V, p. 132, Cor., the extension L of K is separable if and only if $N = 0$. The corollary now follows.

COROLLARY 2. — *Let* L *be a finitely generated extension of a field* K. *For* L *to be algebraic and separable over* K *it is necessary and sufficient that* $\Omega_K(L) = 0$.

This follows immediately from Cor. 1.

COROLLARY 3. — *Let* K *be a field of characteristic* $p \neq 0$ *and* L *a finitely generated extension of* K *of transcendence degree* s. *If* $[K:K^p]$ *is finite, the same is true of* $[L:L^p]$ *and we have* $[L:L^p] = p^s . [K:K^p]$.

Let P be the prime subfield of K. If $[K:K^p]$ is finite, then the vector space $\Omega_P(K) = \Omega_{K^p}(K)$ is of finite dimension m over K, and we have $[K:K^p] = p^m$ (V, p. 103, Th. 2); the vector space $\Omega_P(K) \otimes_K L$ is then of finite dimension m over L. Further, since K is of finite degree over K^p, the field $K(L^p)$ is of finite degree over $K^p(L^p) = L^p$; since the field L is a finitely generated extension of K and is algebraic over $K(L^p)$, it is an extension of finite degree of $K(L^p)$ (V, p. 18, Th. 2); we thus conclude that L is of finite degree over L^p (V, p. 10, Th. 1). Then $\Omega_P(L)$ is a vector space of finite dimension n over L and we have $[L:L^p] = p^n$ (V, p. 103, Th. 2). By Lemma 1 (V, p. 132) the L-linear mapping $\alpha : \Omega_P(K) \otimes_K L \to \Omega_P(L)$ therefore has index $n - m$, whence $n - m = s$ by Th. 4 (V, p. 133) and $p^n = p^s . p^m$.

Remarks. — 1) Let K be a field of characteristic $p \neq 0$ and L an extension of K. We have $\Omega_K(L) = 0$ if and only $L = K(L^p)$ (V, p. 103, Prop. 6). Therefore if L is finitely generated over K, then it is an algebraic and separable extension if and only if we have $L = K(L^p)$. When L is not finitely generated over K, this result no longer holds generally as is shown by the case where L is the perfect closure of K.

2) Let K be a field, $F_1, \ldots, F_m$ polynomials in $K[X_1, \ldots, X_n]$ and L an extension of K generated by the elements $x_1, \ldots, x_n$. Suppose that the polynomials $F_1, \ldots, F_m$ generate the ideal of $K[X_1, \ldots, X_n]$ consisting of all polynomials F such that $F(x_1, \ldots, x_n) = 0$. From Prop. 1 (V, p. 127) and the universal property of the module of differentials (III, p. 569) we easily deduce the following result: the vector space $\Omega_K(L)$ over L is generated by $dx_1, \ldots, dx_n$; we have the relations

$$(6) \qquad \sum_{i=1}^{n} D_iF_j(x_1, \ldots, x_n) . dx_i = 0 \quad (\text{for } 1 \leqslant j \leqslant m);$$

finally if $u_1, \ldots, u_n$ are elements of L such that $\sum_{i=1}^{n} u_i . dx_i = 0$, there exist elements $v_1, \ldots, v_m$ of L such that $u_i = \sum_{j=1}^{m} D_iF_j(x_1, \ldots, x_n)\, v_j$ for $1 \leqslant i \leqslant n$. Let us denote by r the rank of the matrix $(D_iF_j(x_1, \ldots, x_n))$ with n rows and m columns; let s be the transcendence degree of L over K. Then we have $[\Omega_K(L):L] = n - r$. Therefore the extension L of K is separable if and only if $r + s = n$ (Cor. 1), and it is algebraic and separable if and only if $r = n$ (Cor. 2).

7. Separating transcendence bases

DEFINITION 1. — *Let* L *be an extension of a field* K. *A transcendence basis* B *of* L *over* K *is said to be separating if the algebraic extension* L *of* K(B) *is separable.*

If K has characteristic 0, every transcendence basis of L over K is separating because every algebraic extension of a field of characteristic 0 is separable (V, p. 37, Cor.). If an extension admits a separating transcendence basis, it is separable (V, p. 121, Prop. 6 and p. 122, Prop. 9). The following theorem shows that every *finitely generated* separable extension admits a separating transcendence basis ; this restriction is essential (V, p. 177, Ex. 1).

THEOREM 5. — *Let* K *be a field,* L *an extension of* K *and* $(x_i)_{i \in I}$ *a family of elements of* L. *If the family* $(x_i)_{i \in I}$ *is a separating transcendence basis of* L *over* K, *then the family* $(dx_i)_{i \in I}$ *is a basis of the vector space* $\Omega_K(L)$ *over* L. *The converse holds if* L *is a finitely generated separable extension of* K.

Put $M = K(x_i)_{i \in I}$ and denote by α the canonical mapping of $\Omega_K(M) \otimes_M L$ into $\Omega_K(L)$. If $(x_i)_{i \in I}$ is a separating transcendence basis of L over K, the family $(d_{M/K}x_i)_{i \in I}$ is a basis of the vector M-space $\Omega_K(M)$ (V, p. 128, Prop. 3) and α is an isomorphism of vector L-spaces since L is algebraic and separable over M (V, p. 130, Cor. 2). Since $\alpha(d_{M/K}x_i \otimes 1) = d_{L/K}x_i$, the family $(d_{L/K}x_i)_{i \in I}$ is thus a basis of $\Omega_K(L)$ over L.

Conversely, suppose that L is a separable finitely generated extension of K and that the family $(d_{L/K}x_i)_{i \in I}$ is a basis of the vector space $\Omega_K(L)$ over L. By Cor. 1 of V, p. 134, the transcendence degree of L over K is equal to the dimension of $\Omega_K(L)$ over L, hence to the cardinal of I. From the exact sequence $(E_{K,M,L})$ (V, p. 128) we have $\Omega_M(L) = 0$; since L is a finitely generated extension of M, Cor. 2 of V, p. 135 shows that L is algebraic and separable over $M = K(x_i)_{i \in I}$; since the transcendence degree of L over K is finite and equal to the cardinal of I, the family $(x_i)_{i \in I}$ is a transcendence basis of L over K (V, p. 110, Cor. 1).

COROLLARY. — *Let* L *be a separable finitely generated extension of* K *and let* S *be a subset of* L *such that* L = K(S). *Then there exists a separating transcendence basis* B *of* L *over* K, *contained in* S.

Since $\Omega_K(L)$ is generated by the differentials of the elements of S, there exist s elements $x_1, \ldots, x_s$ of S such that $(dx_1, \ldots, dx_s)$ is a basis of $\Omega_K(L)$ over L. Now it suffices to apply Th. 5.

Remark. — Let L be a separable finitely generated extension of a field K of characteristic $p \neq 0$; there may exist transcendence bases of L which are not separating. It is enough to observe that $\{X^p\}$ is a transcendence basis of K(X) but K(X) is a p-radical extension of degree p of $K(X^p)$.

PROPOSITION 5. — *Let* L *and* M *be two extensions of a field* K *contained in a given extension and algebraically disjoint over* K. *If* M *is separable over* K, *then* L(M) *is separable over* L.

It is enough to show that for every finite subset S of M, L(S) is separable over L (V, p. 122, Prop. 8). Let S be a finite subset of M. Since the field K(S) is separable over K, it has a separating transcendence basis B (Cor. of Th. 5). Since L and M are algebraically disjoint over K, B is a transcendence basis of L(B) over L (V, p. 113, Prop. 12). Further, every element of S is algebraic and separable over K(B), hence over L(B) (V, p. 39, Cor. 2). We deduce (V, p. 39, Prop. 6) that L(S) = L(B)(S) is algebraic and separable over L(B), hence L(S), is separable over L.

COROLLARY. — *If* L *and* M *are separable extensions, algebraically disjoint over* K, *then the field* K(L ∪ M) *is separable over* K.

For K(L ∪ M) = L(M) is separable over L by Prop. 5 (because M is separable over K) and L is separable over K, whence the corollary (V, p. 122, Prop. 9).

The hypothesis that the extensions L and M are algebraically disjoint is indispensable in Prop. 5 and its corollary. For let K be an imperfect field of characteristic $p \neq 0$ and E an extension of the form $K(x, a)$ with x transcendental over K and a p-radical of height 1 over K ; put $L = K(x)$ and $M = K(x + a)$. Then $x + a$ is transcendental over K (if not, $x = (x + a) - a$ would be algebraic over K) and the fields L and M are separable over K. However, $K(L \cup M) = K(x, a)$ is p-radical of degree p over $L = K(x)$ and is not separable over L, nor over K.

§ 17. REGULAR EXTENSIONS

1. Complements on the relative separable algebraic closure

THEOREM 1 (Zariski). — *Let* K *be a field,* L *an extension of* K, K_1 *the relative separable algebraic closure of* K *in* L (V, p. 44), K_2 *the relative algebraic closure of* K *in* L (V, p. 19) *and* $(X_i)_{i \in I}$ *a family of indeterminates. Then* $K_1(X_i)_{i \in I}$ *is the relative separable algebraic closure of* $K(X_i)_{i \in I}$ *in* $L(X_i)_{i \in I}$ *and* $K_2(X_i)_{i \in I}$ *is the relative algebraic closure of* $K(X_i)_{i \in I}$ *in* $L(X_i)_{i \in I}$.

A) Suppose that E is a field and F an extension field of E and that every element of F which is separable algebraic over E belongs to E. Let u be an element of F(X) which is separable algebraic over E(X) ; we shall show that u belongs to E(X). There exist in F(X) two relatively prime polynomials P and Q such that $u = P/Q$ and we may take Q to be monic. Let S be the *finite* subset of F consisting of the coefficients of P and Q, $F_0 = E(S)$ and let Δ be an E-derivation of F_0 into F_0. Let D be the derivation of $F_0(X)$ into itself which agrees with Δ on F_0 and maps X to 0 (V, p. 128, Prop. 3).

Since $u \in F_0(X)$ is separable algebraic over $E(X)$ and D is zero on $E(X)$, we have $D(u) = 0$ (V, p. 129, Prop. 4), whence $D(P) . Q = P . D(Q)$. Since P and Q are relatively prime, we conclude that Q divides $D(Q)$ (IV, p. 13, Cor. 4). Now Q may be written in the form

$$Q(X) = X^n + a_1X^{n-1} + \cdots + a_{n-1}X + a_n$$

with $a_1, \ldots, a_n$ in F_0 ; since $D(x) = 0$, we thus have

$$D(Q) = \Delta(a_1)X^{n-1} + \cdots + \Delta(a_{n-1})X + \Delta(a_n) \tag{1}$$

whence $\deg D(Q) < \deg Q$. Since Q divides $D(Q)$, this is possible only if $D(Q) = 0$. But then $D(P) = 0$ because $D(P) . Q = P . D(Q)$. Now (1) and a similar formula for P show that Δ annihilates the set S of coefficients of P and Q, whence $\Delta = 0$, because $F_0 = E(S)$. By V, p. 135, Cor. 2, the finitely generated extension F_0 of E is thus separable algebraic ; by the hypotheses on E and F we have $F_0 = E$, so finally $u \in E(X)$.

B) Suppose now that E is relatively algebraically closed in the extension field F and denote by p the characteristic exponent of E. Let u be an element of $F(X)$ which is algebraic over $E(X)$. There exists an integer $f \geqslant 0$ such that $v = u^{p^f}$ is separable algebraic over $E(X)$ (V, p. 44, Prop. 13). By *A*) we thus have $v \in E(X)$. There exists a unique representation of u in the form P/Q with relatively prime polynomials P and Q in $F[X]$ and Q monic ; we have a similar decomposition $v = P_1/Q_1$ with P_1 and Q_1 relatively prime in $E[X]$ and Q_1 monic. It follows that $P_1/Q_1 = P^{p^f}/Q^{p^f}$; the polynomials P^{p^f} and Q^{p^f} are relatively prime in $F[X]$ (IV, p. 13, Cor. 6), like P_1 and Q_1, and Q^{p^f} is monic. It follows that $P^{p^f} = P_1 \in E[X]$ and $Q^{p^f} = Q_1 \in E[X]$. Therefore the coefficients of P and Q are p-radical over E, and hence belong to E because E is relatively algebraically closed in F. We thus have $P \in E[X]$, $Q \in E[X]$ and finally $u \in E(X)$. This proves $E(X)$ to be relatively algebraically closed in $F(X)$.

C) We use the notation of Theorem 1. Since K_1 is a separable algebraic extension of K, the extension $K_1(X_i)_{i \in I}$ of $K(X_i)_{i \in I}$ is algebraic and separable (V, p. 39, Prop. 6). Moreover, every element of L which is algebraic and separable over K_1 belongs to K_1 (V, p. 44, Prop. 13, *a*)). Let J be a finite subset of I ; by an immediate induction on the cardinal of J we deduce from *A*) that every element of $L(X_i)_{i \in J}$ which is algebraic and separable over $K(X_i)_{i \in J}$ belongs to $K_1(X_i)_{i \in J}$. Finally let u be an element of $L(X_i)_{i \in I}$ algebraic and separable over $K(X_i)_{i \in I}$; there exists a finite subset J of I such that u belongs to $L(X_i)_{i \in J}$ and is algebraic and separable over $K(X_i)_{i \in J}$; by what has been said, u belongs to $K_1(X_i)_{i \in J}$ and *a fortiori* to $K_1(X_i)_{i \in I}$.

We have thus deduced from *A*) that $K_1(X_i)_{i \in I}$ is the relative separable closure of $K(X_i)_{i \in I}$ in $L(X_i)_{i \in I}$; in the same way it follows from *B*) that $K_2(X_i)_{i \in I}$ is the relative algebraic closure of $K(X_i)_{i \in I}$ in $L(X_i)_{i \in I}$.

2. The tensor product of extensions

PROPOSITION 1. — *Let* Ω *be an extension of a field* K *and let* L, M *be two subextensions of* Ω *which are algebraically disjoint over* K. *Suppose that the relative separable algebraic closure of* K *in* L *is equal to* K [1]. *Let* φ *be the* K*-algebra homomorphism of* $L \otimes_K M$ *into* Ω *mapping* $x \otimes y$ *to* xy *for* $x \in L$, $y \in M$, *and let* $\mathfrak{p}$ *be the kernel of* φ. *Then* $\mathfrak{p}$ *is the set of nilpotent elements of* $L \otimes_K M$ *and this is the smallest prime ideal of* $L \otimes_K M$.

On replacing Ω by an algebraic closure if necessary, we may suppose Ω algebraically closed. Let B be a transcendence basis of M over K, N the relative algebraic closure of K(B) in Ω and N_s (resp. N_r) the set of elements of N which are separable (resp. p-radical) over K(B). We remark that M is algebraic over K(B), so that N is the relative algebraic closure of M in Ω.

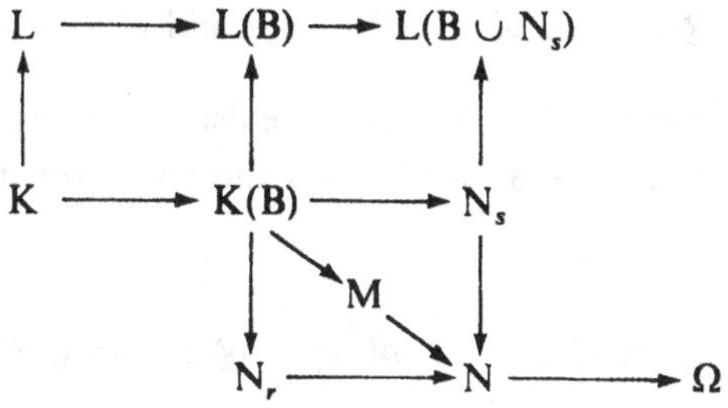

FIG. 1.

Let us define the following chain of homomorphisms :

$$L \otimes_K M \xrightarrow{\alpha} L \otimes_K N \xrightarrow{\beta} (L \otimes_K K(B)) \otimes_{K(B)} N \xrightarrow{\gamma} L(B) \otimes_{K(B)} N$$

$$\rightarrow L(B) \otimes_{K(B)} N_s \otimes_{K(B)} N_r \xrightarrow{\varepsilon} L(B \cup N_s) \otimes_{K(B)} N_r \xrightarrow{\zeta} \Omega\,.$$

We have $\alpha = \mathrm{Id}_L \otimes u$ where u is the canonical injection of M into N, hence α is *injective*. The mapping β is the *isomorphism* of commutative groups which maps $x \otimes y$ to $(x \otimes 1) \otimes y$ (II, p. 278, Prop. 2) for $x \in L$ and $y \in N$. We have $\gamma = v \otimes \mathrm{Id}_N$, where v is the K-algebra homomorphism of $L \otimes_K K(B)$ into L(B) which maps $x \otimes y$ to xy, for $x \in L$ and $y \in K(B)$; since L and M are algebraically disjoint over K, Prop. 14 (V, p. 114) shows that L and K(B) are linearly disjoint over K, in other words, v is injective, hence γ is *injective*. Since N is a quasi-Galois extension of K(B), there exists (V, p. 76, Prop. 13) a K(B)-algebra isomorphism w of $N_s \otimes_{K(B)} N_r$ onto N mapping $x \otimes y$ to

[1] This hypothesis is sometimes expressed by saying that L is a *primary* extension of K.

xy for $x \in \mathrm{N}_s$ and $y \in \mathrm{N}_r$; we denote by δ the *isomorphism* $\mathrm{Id}_{\mathrm{L(B)}} \otimes w^{-1}$. By Th. 1 (V, p. 137) and the hypothesis on the extension L of K, every element of L(B) which is algebraic and separable over K(B) belongs to K(B); in particular, we have $\mathrm{L(B)} \cap \mathrm{N}_s = \mathrm{K(B)}$. Since N_s is a Galois extension of K(B), Th. 5 (V, p. 71) shows that there exists a K(B)-algebra isomorphism w' of $\mathrm{L(B)} \otimes_{\mathrm{K(B)}} \mathrm{N}_s$ onto $\mathrm{L(B} \cup \mathrm{N}_s)$ mapping $x \otimes y$ to xy for $x \in \mathrm{L(B)}$ and $y \in \mathrm{N}_s$; we denote by ε the *isomorphism* $w' \otimes \mathrm{Id}_{\mathrm{N}_r}$. Finally, ζ is the K-algebra homomorphism mapping $x \otimes y$ to xy for $x \in \mathrm{L(B} \cup \mathrm{N}_s)$ and $y \in \mathrm{N}_r$.

What has been said shows that $\eta = \varepsilon\delta\gamma\beta\alpha$ is an injective K-algebra homomorphism of $\mathrm{L} \otimes_{\mathrm{K}} \mathrm{M}$ into $\mathrm{L(B} \cup \mathrm{N}_s) \otimes_{\mathrm{K(B)}} \mathrm{N}_r$. Further, every element of M is of the form $\sum_{i=1}^{n} a_i b_i$ with $a_i \in \mathrm{N}_s$ and $b_i \in \mathrm{N}_r$ for $1 \leqslant i \leqslant n$; hence we obtain $\varphi = \zeta\eta$.

The kernel $\mathfrak{p}$ of φ is a prime ideal of $\mathrm{L} \otimes_{\mathrm{K}} \mathrm{M}$, hence every nilpotent element of $\mathrm{L} \otimes_{\mathrm{K}} \mathrm{M}$ belongs to $\mathfrak{p}$, by Prop. 2 (V, p. 118). Conversely let a be an element of $\mathfrak{p}$; put $\eta(a) = \sum_{i=1}^{s} b_i \otimes c_i$ with $b_i \in \mathrm{L(B} \cup \mathrm{N}_s)$ and $c_i \in \mathrm{N}_r$ for $1 \leqslant i \leqslant s$. Since N_r is a p-radical extension of K(B), there exists an integer $f \geqslant 0$ such that $c_i^{p^f}$ belongs to K(B) for $1 \leqslant i \leqslant s$ (where p is the characteristic exponent of K). But we have

$$\eta(a^{p^f}) = \sum_{i=1}^{s} b_i^{p^f} \otimes c_i^{p^f} = \left(\sum_{i=1}^{s} b_i^{p^f} c_i^{p^f}\right) \otimes 1 = \zeta\eta(a)^{p^f} \otimes 1 = 0$$

and since η is injective we finally have $a^{p^f} = 0$. We have thus shown $\mathfrak{p}$ to be the set of all nilpotent elements of $\mathrm{L} \otimes_{\mathrm{K}} \mathrm{M}$. Now every prime ideal of $\mathrm{L} \otimes_{\mathrm{K}} \mathrm{M}$ contains $\mathfrak{p}$ by Prop. 2 (V, p. 118).

COROLLARY. — *Let* L *and* M *be two extensions of a field* K. *Suppose that the relative separable algebraic closure of* K *in* L *is equal to* K. *Then the set* $\mathfrak{p}$ *of nilpotent elements of* $\mathrm{L} \otimes_{\mathrm{K}} \mathrm{M}$ *is a prime ideal. If moreover* L *or* M *is separable over* K, *then* $\mathrm{L} \otimes_{\mathrm{K}} \mathrm{M}$ *is an integral domain.*

We may assume that L and M are algebraically disjoint subextensions of an extension Ω of K (V, p. 116, Th. 5); then $\mathfrak{p}$ is a prime ideal by Prop. 1 (V, p. 139). If moreover L or M is separable over K, then $\mathrm{L} \otimes_{\mathrm{K}} \mathrm{M}$ is a reduced ring by the definition of separable extension (V, p. 119, Def. 1); so $\mathfrak{p} = 0$ and $\mathrm{L} \otimes_{\mathrm{K}} \mathrm{M}$ is an integral domain because $\mathfrak{p}$ was prime.

3. Regular algebras

DEFINITION 1. — *Let* K *be a field. An algebra* A *over* K *is said to be regular if* $\mathrm{L} \otimes_{\mathrm{K}} \mathrm{A}$ *is an integral domain for every extension* L *of* K.

A regular algebra is in particular an integral domain, hence commutative.

PROPOSITION 2. — *Let* A *and* B *be two algebras over a field* K. *If* A *is an integral domain and* B *is regular then* $A \otimes_K B$ *is an integral domain.*

Let L be the field of fractions of A. Since B is a regular K-algebra, $L \otimes_K B$ is an integral domain, hence so is $A \otimes_K B$, since it is isomorphic to a subring of $L \otimes_K B$.

PROPOSITION 3. — *Let* K *be a field.*

a) Every subalgebra of a regular K*-algebra is itself regular.*

b) The tensor product of two regular K*-algebras is again regular.*

c) Let A *be a* K*-algebra and* K' *an extension of* K. *For* A *to be regular it is necessary and sufficient that the* K'*-algebra* $A_{(K')}$ *derived from* A *by extension of scalars should be regular.*

The proof of *a*) (resp. *c*)) is identical to that of part *a*) (resp. *d*)) of Prop. 3, V, p. 119, after replacing everywhere « reduced ring » by « integral domain » and « separable algebra » by « regular algebra ». Let us prove *b*).

Let A and B be two regular K-algebras. Let L be an extension of K. Since A is regular, the ring $L \otimes_K A$ is an integral domain. By Prop. 2, the ring $(L \otimes_K A) \otimes_K B$ is an integral domain, because B is regular. Finally, the ring $L \otimes_K (A \otimes_K B)$ is isomorphic to $(L \otimes_K A) \otimes_K B$, and hence is an integral domain. This shows $A \otimes_K B$ to be a regular K-algebra.

4. Regular extensions

DEFINITION 2. — *An extension of a field* K *is said to be regular if it is regular as* K*-algebra.*

PROPOSITION 4. — *Let* A *be an algebra over field* K *which is an integral domain, and* E *its field of fractions. Let* L *be an extension of* K ; *if the ring* $L \otimes_K A$ *is an integral domain the same is true of* $L \otimes_K E$.

If $L \otimes_K A$ is an integral domain, it may be embedded in its field of fractions F. Write $u(x) = x \otimes 1$ for $x \in L$ and denote by v the K-homomorphism of E into F which extends the injective homomorphism $y \mapsto 1 \otimes y$ of A into F. By Prop. 6 (V, p. 14) the subfields $u(L)$ and $v(E)$ of F are linearly disjoint over K ; therefore the homomorphism $u * v$ of $L \otimes_K E$ into F (V, p. 12) is injective. This shows $L \otimes_K E$ to be an integral domain.

COROLLARY. — *For* A *to be a regular* K*-algebra it is necessary and sufficient that its field of fractions should be a regular extension of* K.

The condition is necessary by Prop. 4 and sufficient by Prop. 3, *a*).

PROPOSITION 5. — *Every pure extension of a field* K *is regular.*

By the preceding corollary it is enough to prove that every polynomial algebra $A = K[X_i]_{i \in I}$ is a regular K-algebra. Let L be an extension of K ; the ring $L \otimes_K A$ is isomorphic to $L[X_i]_{i \in I}$ (III, p. 449, Remark 2), and hence an integral domain (IV, p. 9, Prop. 8).

PROPOSITION 6. — *Let* L *be an extension of a field* K. *If* L *is regular, then every subextension of* L *is regular. Conversely, if every finitely generated subextension of* L *is regular, then* L *is regular.*

The first assertion follows from Prop. 3, *a*).

Let M be an extension of K and let $\mathcal{U}$ be the set of all finitely generated subextensions of L. For each $E \in \mathcal{U}$, the ring $M \otimes_K E$ may be identified with a subring of $M \otimes_K L$ and we thus have an increasing directed family of subrings of $M \otimes_K L$ whose union is $M \otimes_K L$. Now the second assertion follows immediately.

PROPOSITION 7. — *Let* L *be an extension of a field* K *and* M *an* L*-algebra* (for example, an extension of L). *If* L *is regular over* K *and* M *is a regular* L*-algebra, then* M *is regular as* K*-algebra.*

Let E be an extension of K ; since L is regular over K, $E \otimes_K L$ is an integral domain. By Prop. 2 (V, p. 141) the ring $(E \otimes_K L) \otimes_L M$ is thus an integral domain and the same is true of the ring $E \otimes_K M$ isomorphic to it (II, p. 278, Prop. 2). Hence the result.

PROPOSITION 8. — *Let* L *and* M *be two extensions of a field* K.

a) *If* M *is regular over* K, *then the field of fractions of the integral domain* $L \otimes_K M$ *is a regular extension of* L.

b) *If* L *and* M *are regular extensions of* K, *the same is true of the field of fractions of* $L \otimes_K M$.

Assertion *a*) follows from Prop. 3, *c*) (V, p. 141) and the Cor. of V, p. 141; Assertion *b*) follows from Prop. 3, *b*) (V, p. 141) and the Cor. of V, p. 141.

5. Characterization of regular extensions

PROPOSITION 9. — *Let* K *be a field,* $\bar{K}$ *an algebraic closure of* K *and* L *an extension of* K. *Then the following conditions are equivalent :*

a) L *is separable over* K *and* K *is relatively algebraically closed in* L.

b) L *is a regular extension of* K.

c) *The ring* $\bar{K} \otimes_K L$ *is an integral domain.*

d) *Let* $\bar{L}$ *be an algebraic closure of* L. *Then* L *is linearly disjoint over* K *from the relative algebraic closure of* K *in* $\bar{L}$.

Moreover, when these conditions hold, then $\bar{K} \otimes_K L$ *is a field.*

a) $\Rightarrow$ *b*) : Let M be an extension of K. Under the hypotheses of *a*), the ring $M \otimes_K L$ is an integral domain, by V, p. 140, Cor.

b) $\Rightarrow$ *c*) : This follows from Def. 2.

c) $\Rightarrow$ *d*) : With the notation of *d*) we can identify $\bar{K}$ with the relative algebraic closure of K in $\bar{L}$ (V, p. 22, Ex. 2). Suppose that the ring $A = \bar{K} \otimes_K L$ is an integral domain. Let E be a subextension of $\bar{K}$ of finite degree over K ; the subring

$E \otimes_K L$ of A is an integral domain and hence is an algebra of finite degree over L ; by the Cor. of V, p. 10 it is a field. Since $\bar{K}$ is the union the increasing directed set of extensions E of the above type, A is a field (V, p. 11, Prop. 3). The canonical homomorphism of A into $\bar{L}$ mapping $x \otimes y$ to xy (for $x \in \bar{K}$ and $y \in L$) is therefore injective, so L and $\bar{K}$ are linearly disjoint over K.

d) $\Rightarrow$ *a*) : Under the hypotheses of *d*) we have $L \cap \bar{K} = K$, so K is relatively algebraically closed in L ; further if p is the characteristic exponent of K, the field L is linearly disjoint from $K^{p^{-\infty}}$ over K, hence L is separable over K (V, p. 123, Cor. 1).

COROLLARY 1. — *Let* A *be an algebra over a field* K. *For* A *to be a regular* K-*algebra it is necessary and sufficient for the ring* $\bar{K} \otimes_K A$ *to be an integral domain.*

The stated condition is clearly necessary. Conversely, assume that $\bar{K} \otimes_K A$ is an integral domain and denote by E the field of fractions of A. By Prop. 4 (V, p. 141) the ring $\bar{K} \otimes_K E$ is an integral domain, hence E is a regular extension of K, by Prop. 9 ; by V, p. 141, Cor., we conclude that A is regular as K-algebra.

COROLLARY 2. — *Let* K *be an algebraically closed field. Every* K-*algebra which is an integral domain is a regular* K-*algebra. In particular, every extension of* K *is regular.*

This follows from Cor. 1.

COROLLARY 3. — *Let* K *be an algebraically closed field. If* A *and* B *are two* K-*algebras which are integral domains, then the same is true of* $A \otimes_K B$.

By Cor. 2, A and B are regular K-algebras, and it suffices to apply Prop. 2 (V, p. 141).

6. Application to composite extensions

PROPOSITION 10. — *Let* L *and* M *be two extensions of a field* K *and* (E, u, v) *a composite extension of* L *and* M (V, p. 12). *Suppose that the ring* $L \otimes_K M$ *is an integral domain and that the subextensions* $u(L)$ *and* $v(M)$ *of* E *are algebraically disjoint over* K. *Then* $u(L)$ *and* $v(M)$ *are linearly disjoint over* K.

Put $w = u * v$ (V, p. 12), denote by F the field of fractions of the integral domain $L \otimes_K M$ and identify L (resp. M) with a subfield of F by means of the mapping $x \mapsto x \otimes 1$ (resp. $y \mapsto 1 \otimes y$) ; then the restriction of w to L (resp. M) is u (resp. v). Let B be a transcendence basis of M over K (V, p. 109, Th. 1).

By hypothesis $u(L)$ and $v(M)$ are algebraically disjoint over K ; therefore (V, p. 114, Prop. 14), $u(L)$ and $v(K(B))$ are linearly disjoint over K. Thus there exists a K-homomorphism $u' : L(B) \to E$ which agrees with u on L and with v ou K(B). By construction L and M are linearly disjoint over K in F ; by Prop. 8 (V, p. 15) the subfields L(B) and M of F are linearly disjoint over K(B). It follows that there exists a K-homomorphism w' : $M[L(B)] \to E$ which agrees with u' on $L(B)$ and with v on M. But the field F is generated by $M \cup L(B)$ and M is algebraic over $K(B)$; we thus have $M[L(B)] = F$ (V, p. 18, Cor. 2). Hence we conclude that w' is a K-isomorphism of F onto E whose restriction to L (resp. M) is u (resp. v). This shows $u(L)$ and $v(M)$ to be linearly disjoint over K.

COROLLARY 1. — *Let Ω be an extension of a field* K *and* L *a subextension of* Ω *which is regular over* K. *Every subextension* M *of* Ω *which is algebraically disjoint from* L *over* K *is linearly disjoint.*

The ring $L \otimes_K M$ is an integral domain by definition of regular extension and it suffices now to apply Prop. 10.

COROLLARY 2. — *Let Ω be an extension of a field* K *and* L, M *two subextensions of* Ω. *Suppose that* L *is separable over* K *and that the relative separable closure of* K *in* M *is equal to* K. *If* L *and* M *are algebraically disjoint over* K, *then they are linearly disjoint over* K.

By Prop. 10 it is enough to remark that the ring $L \otimes_K M$ is an integral domain (V, p. 140, Cor.).

Exercises

§ 1

1) Let A be a not necessarily commutative ring, not zero and without divisors of zero (I, p. 98). Show that if there exists an integer $m \geqslant 1$ such that $mA = 0$, then the set of all integers having this property is an ideal $p\mathbf{Z}$, where p is a prime number, and A is a ring of characteristic p.

2) Let m, n be integers such that $0 \leqslant n \leqslant m$. Let p be a prime number and let $m = \alpha_0 + \alpha_1 p + \cdots + \alpha_r p^r$ and $n = \beta_0 + \beta_1 p + \cdots + \beta_r p^r$ be the expansions of m and n in base p, where we thus have $0 \leqslant \alpha_i < p$, $0 \leqslant \beta_i < p$ for $0 \leqslant i \leqslant r$; suppose that $\alpha_r \neq 0$. Show that $\binom{m}{n} \equiv 0 \pmod{p}$ if $\beta_i > \alpha_i$ for at least one index i. If on the other hand, $\beta_i \leqslant \alpha_i$ for $0 \leqslant i \leqslant r$, then

$$\binom{m}{n} \equiv \binom{\alpha_0}{\beta_0} \cdots \binom{\alpha_r}{\beta_r} \pmod{p} .$$

(In $(\mathbf{Z}/p\mathbf{Z})[X]$ we have $(1+X)^m = (1+X)^{\alpha_0}(1+X^p)^{\alpha_1} \ldots (1+X^{p^r})^{\alpha_r}$.)

3) Let A be a commutative ring. For every polynomial $f \in A[X]$ we write in the polynomial ring $A[X, Y]$ of polynomials in two indeterminates over A,

$$f(X+Y) = \sum_{m=0}^{\infty} \Delta_m f(X)\, Y^m ,$$

where Δ_m is thus an A-linear mapping of $A[X]$ into itself.

a) If Z is an indeterminate and τ_Z the A-linear mapping of $A[X]$ into $A[X, Z]$ such that $\tau_Z f = f(X+Z)$, show that $\tau_Z \circ \Delta_m = \Delta_m \circ \tau_Z$ (where on the right Δ_m is the mapping defined above in the ring $B[X]$, where $B = A[Z]$). Deduce that in $A[X]$ we have

$$\Delta_m \Delta_n = \Delta_n \Delta_m = \binom{m+n}{m} \Delta_{m+n}$$

for integers $m, n \geqslant 0$ (cf. IV, p. 8).

b) Suppose that A is a ring of characteristic $p > 0$ and for each integer $k \geqslant 0$ put $D_k = \Delta_{p^k}$. Show that if $f \in A[X]$ and $g \in A[X^{p^k}]$ then $D_k(fg) = g \,.\, D_k f + f \,.\, D_k g$.

Every polynomial $f \in A[X]$ can be uniquely written in the form $f(X) = \sum_{m=0}^{\infty} X^{mp^k} g_m(X)$ where $\deg(g_m) < p^k$; show that

$$D_k f(X) = \sum_{m=0}^{\infty} m X^{(m-1)p^k} g_m(X)$$

and deduce that $D_k^p = 0$.

c) For every integer $m > 0$ let $m = \alpha_0 + \alpha_1 p + \cdots + \alpha_k p^k$ be the expansion of m in base p (*Set Theory*, III, p. 177), where thus $0 \leqslant \alpha_j < p$ for $0 \leqslant j \leqslant k$; suppose that $\alpha_k \neq 0$. Show that

$$\Delta_m = \left(\frac{1}{\alpha_0!} D_0^{\alpha_0}\right)\left(\frac{1}{\alpha_1!} D_1^{\alpha_1}\right)\cdots\left(\frac{1}{\alpha_k!} D_k^{\alpha_k}\right).$$

(If Δ'_m is the right-hand side of this relation, note that if we introduce the expansion $f(X+Y) = \sum_{q=0}^{\infty} f_q(Y) X^q$, we have $\Delta'_m f(X+Y) = \sum_{q=0}^{\infty} f_q(Y) \Delta'_m(X^q)$ and deduce that this may be written $\Delta'_m f(X+Y) = f_m(X) + Y g(X, Y)$, where g is a polynomial.)

4) Let G be a commutative group, written additively, and let p be a prime number. Denote by $G\left[\frac{1}{p}\right]$ the commutative group $G \otimes_{\mathbf{Z}} \mathbf{Z}\left[\frac{1}{p}\right]$. Show that if K is a field of characteristic p, then the perfect closure of the group algebra K[G] is the group algebra $K^{p^{-\infty}}\left[G\left[\frac{1}{p}\right]\right]$.

$_*$5) Let $\mathfrak{P}$ be the set of prime numbers and put $A = \prod_{p \in \mathfrak{P}} \mathbf{F}_p$. For every filter $\mathfrak{F}$ on $\mathfrak{P}$ (*Gen. Top.*, I, p. 57) denote by $m_{\mathfrak{F}} \subset A$ the set of $u = (u_p)_{p \in \mathfrak{P}}$ such that the set of $p \in \mathfrak{P}$ with $u_p = 0$ is a member of $\mathfrak{F}$. Show that $m_{\mathfrak{F}}$ is an ideal and that we thus obtain a bijection between the ideals of A distinct from A and the filters on $\mathfrak{P}$. Let $\mathfrak{F}$ be an ultrafilter (*Gen. Top.*, I, p. 60) on $\mathfrak{P}$ which does not possess a limit in the discrete topology. Show that $A/m_{\mathfrak{F}}$ is a field of characteristic 0, isomorphic to a subfield of **C**. $_*$

§ 2

1) *a*) Show that for every integer $a \neq 0$ the polynomial $X^4 - aX - 1$ is irreducible in **Q**[X].
b) Let α be a root of $X^4 - aX - 1$ in an extension of **Q**, and let $K = \mathbf{Q}(\alpha)$, which is of degree 4 over **Q**. Show that there is an infinity of values of $a \in \mathbf{Z}$ such that K contains no subfield F distinct from **Q** and K.

$_*$2) The polynomial $X^3 - 2$ of **Q**[X] is irreducible ; let E be a splitting field of this polynomial (V, p. 21) and let F be the field generated by a single one of the roots of this polynomial in E. Show that two composite extensions of E and F are always isomorphic qua extensions of E but not qua composite extensions. $_*$

3) Let A be an algebra of finite degree over a commutative field K, where A is not assumed unital. Show that if $a \in A$ is cancellable on the left, then there exists $e \in A$ such that $ex = x$ for all $x \in A$ and there exists $b \in A$ with $ab = e$. If a is also cancellable on the right, e is the unit element of A and a is invertible in A. The case of commutative A.

4) Let K be a field, Ω an extension of K and E, F linearly disjoint subextensions over K. Let C be the K-algebra $K[E \cup F]$ isomorphic to $E \otimes_K F$.
a) Let E′ (resp. F′) be a subextension of E (resp. F) and put $C' = K[E' \cup F']$. If an element a of C′ is invertible in C then $a^{-1} \in C'$. (Reduce to the case $F' = F$. Let

E″ be a supplement of E′ in the vector E′-space E, so that C is isomorphic to $C' \oplus (E'' \otimes_K F)$. Express the matrix of multiplication by a in this decomposition.)
* *b*) Show that if C is a field, then E or F is algebraic over K. (If $x \in E$ and $y \in F$ are transcendental over K, take $E' = K(x)$, $F' = K(y)$ and $a = x + y$ in *a*).) *

§ 3

1) Let E be an extension of a field K and x and y two distinct roots in E of the same irreducible polynomial of K[X]. Show that the extensions $K(x)$ and $K(y)$ are not linearly disjoint over K (use Th. 2). * Give an example where $K = \mathbf{Q}$ and $K(x) \cap K(y) = K$ (cf. V, p. 161, Ex. 2). *

2) Let $(E_i)_{i \in I}$ be a family of extensions of a field K, contained in an extension L of K. If F_i is the relative algebraic closure of K in E_i, show that the relative algebraic closure of K in $E = \bigcap_{i \in I} E_i$ is $F = \bigcap_{i \in I} F_i$.

3) Show that every algebraic extension E of a field K is equipotent to a subset of $K \times \mathbf{N}$ (consider the mapping which associates with each element of E its minimal polynomial over K). In particular, every algebraic extension of a finite field is countable and every algebraic extension of an infinite field K is equipotent to K. * Deduce that in the field **R** of real numbers there exist numbers that are transcendental over the prime subfield **Q** and that the set of all these numbers has the power of the continuum. *

4) Let F be a transcendental extension of a field E. Show that

$$[F : E] \geqslant \operatorname{Sup}(\operatorname{Card}(E), \operatorname{Card}(\mathbf{N}))$$

with equality if the extension F can be generated by a countable family of elements. (Use Ex. 7 of IV, p. 89.)

5) Let A be a commutative algebra and let S be a subset of A generating A as K-algebra. Suppose that $\operatorname{Card}(S) < \operatorname{Card}(K)$. Show that if $\mathfrak{M}$ is a maximal ideal of A, then the field $A/\mathfrak{M}$ is an algebraic extension of K. (Use Ex. 15 of V, p. 174 when $\operatorname{Card}(K) \leqslant \operatorname{Card}(\mathbf{N})$ and Ex. 4 when

$$\operatorname{Card}(K) > \operatorname{Card}(\mathbf{N}).)$$

* Deduce that $A/\mathfrak{M} = K$ when K is algebraically closed. *

* 6) Let C be an uncountable algebraically closed field and $P = C[(X_i)_{i \in I}]$ a polynomial algebra over C. Given two families $(f_\alpha)_{\alpha \in A}$, $(g_\beta)_{\beta \in B}$ of elements of P suppose that :
a) B and I are countable.
b) For every finite subset A′ of A and every finite subset B′ of B there exists $(x_i)_{i \in I} \in C^I$ such that

$$f_\alpha(x_i) = 0 \quad \text{and} \quad g_\beta(x_i) \neq 0 \quad \text{for all} \quad \alpha \in A', \beta \in B'.$$

Show that then there exists $(x_i)_{i \in I} \in C^I$ such that

$$f_\alpha(x_i) = 0 \quad \text{and} \quad g_\beta(x_i) \neq 0 \quad \text{for all} \quad \alpha \in A, \beta \in B .$$

(Let $(Y_\beta)_{\beta \in B}$ be a family of indeterminates indexed by B and let Q be the polynomial algebra $C[(X_i)_{i \in I}, (Y_\beta)_{\beta \in B}]$. Let Λ be the quotient of Q by the ideal generated by the f_α and the $1 - g_\alpha Y_\alpha$. Show that $\Lambda \neq 0$ thanks to *a*) and *b*) and apply Ex. 5 to a maximal ideal of Λ.) *

7) Let L be a field, A a subring of L and $K \subset L$ the field of fractions of A.
a) Show that if L is a finitely generated A-module (II, p. 206) then necessarily $A = K$ (by considering a supplementary subspace of K in L, qua vector space over K, show that K is a finitely generated A-module, hence necessarily of the form $s^{-1}A$, where $s \in A$; finally show that s must be invertible in A).
b) Show that if there exist a finite number of elements x_j $(1 \leqslant j \leqslant n)$ of L, *algebraic* over K, such that $L = A[x_1, \ldots, x_n]$, then there exists an element $b \neq 0$ in A such that $K = A[b^{-1}]$ (show that there exists $b \neq 0$ in A such that L is a finitely generated $A[b^{-1}]$-module). Show that either b is invertible (and $K = A$) or every non-zero ideal of A contains a principal ideal Ab^m. Converse. The ring $k[[X]]$ of formal power series over a field k is a ring of this type.

* 8) If K is a field, show that K is algebraically closed in the field K(X) of rational fractions in one indeterminate over K (use the fact that K[X] is a principal ideal domain). *

9) Let K be a field, E an extension of K and S a subset of E – K.
a) Show that the set of extensions $L \subset E$ of K such that $L \cap S = \varnothing$, ordered by inclusion, has a maximal element M.
b) Show that if S is finite, E is an algebraic extension of M.

* 10) Let K be a field and P a polynomial in the ring K[X, Y] in two indeterminates. Suppose that there exists a polynomial $H \in K[X]$ and two polynomials $Q, R \in K[X, Y]$ such that $H(X) P(X, Y) = Q(X, Y) R(X, Y)$ and that the coefficients of $Q(X, Y)$, qua polynomial in Y, are polynomials in K[X] without non-constant common factor. Show that all the coefficients of $R(X, Y)$, qua polynomial in Y, are divisible by $H(X)$. (Consider H, P, Q, R as polynomials in X over the field K(Y) and decompose them into factors of the first degree in $\Omega[X]$ where Ω is an algebraic closure of K(Y); observe that if $H(X)$ and $Q(X, Y)$ had a factor of the first degree in common, this factor would divide all the coefficients of $Q(X, Y)$, qua polynomial in Y, and use the fact (IV, p. 12) that if K′ is an extension of K, the GCD in K′[X] of two polynomials of K[X] belongs to K[X].) *

11) Let E be an extension of a field K, $x \in E$ a transcendental element over K, so that $K(x)$ is isomorphic to the field K(T) of rational fractions in an indeterminate over K. Thus every $y \in K(x)$ may be written $y = g(x)/h(x)$, where g and h are two polynomials of K[X] that are relatively prime (IV, p. 12), determined up to a non-zero common factor in K; the larger of the degrees of g and h is called the *height* of y with respect to x.
a) Show that in the polynomial ring $K(y)[X]$ the polynomial $g(X) - yh(X)$ is irreducible if $y \notin K$. Deduce that if y has height $n > 0$ with respect to x, then $K(x)$ is an algebraic extension of degree n of $K(y)$.

Let $P(X) = \sum_{i=0}^{n} y_i X^i$ be the minimal polynomial of x over $K(y)$. Show that for $0 \leqslant i \leqslant n$ we have either $y_i \in K$ or $K(y_i) = K(y)$.

b) Deduce from *a*) that every $y \in K(x)$ such that $K(y) = K(x)$ is of the form $(ax+b)/(cx+d)$ where a, b, c, d are elements of K such that $ad - bc \neq 0$; converse. Find all K-automorphisms of $K(x)$.

c) Show that if $y \in K(x)$ is of height n with respect to x, and $z \in K(y)$ of height m with respect to y, then z is of height mn with respect to x.

* *d*) Let F be an extension of K such that $K \subset F \subset K(x)$ and $F \neq K$; let y be an element of F whose height m with respect to x has the least possible value ; show that $F = K(y)$ (« Lüroth's theorem »). (Let P be the minimal polynomial of x in $F[T]$; show that $P(T) = Q(T, x)/R(x)$, where Q is a polynomial of $K[T, X]$ of degree $\geqslant m$ in X and is not divisible by any non-constant polynomial of $K[X]$, and R is a polynomial of $K[X]$. If $y = g(x)/h(x)$, where g and h are relatively prime in $K[X]$, observe that by *a*) the polynomial $g(T)h(X) - g(X)h(T)$ is not divisible by any non-constant polynomial of $K[T]$ or of $K[X]$; deduce that necessarily

$$g(T)h(X) - g(X)h(T) = c \cdot Q(T, X), \quad \text{where } c \in K,$$

using Ex. 10.) *

* 12) *a*) With the notation and hypotheses as in Ex. 11, show that for an extension F of K such that $K \subset F \subset K(x)$ to be such that $F \cap K[x] \neq K$, it is necessary and sufficient that $F = K(y)$ where $y = g(x)$ for a non-constant polynomial $g \in K[T]$; then we have $K(y) \cap K[x] = K[y]$. (Write $F = K(y)$ with $y = g(x)/h(x)$, g and h being relatively prime polynomials of $K[T]$. If there exists a non-constant polynomial $P \in K[T]$ such that $P(x) \in F$, we can write $P(x) = Q(y)/R(y)$, where Q and R are two relatively prime monic polynomials of $K[T]$, not both constant. Distinguish two cases, according as $R = 1$ or R is non-constant ; in the second case, decompose Q and R into factors of the first degree in an algebraic closure Ω of K and observe that if a, b are distinct elements of Ω, then $g - ah$ and $g - bh$ are relatively prime in $\Omega[T]$.)

b) If $F = K(y)$ with $y = g(x)$, where $g \in K[T]$ is non-constant, we may suppose that $g(T) = Tg_1(T)$, where $g_1 \in K[T]$. Show that if $K(xg(x)) \cap K[y] \neq K$, we necessarily have $g_1(T) = aT^n$ with $a \in K$ and $n > 0$. (Observe that by hypothesis and *a*) there exist two non-constant polynomials P, Q in $K[T]$ such that $P(xg(x)) = Q(g(x)) \notin K$; deduce that there are two elements a, b in K and an integer m such that $g(T)$ divides $a - bT^m$.)

c) Let y, z be two elements of $K[x]$ such that $K(y) \cap K(z) \neq K$; show that then $K[y] \cap K[z] \neq K$.

d) Find $y \in K(x)$ such that $K(y) = K(x)$ but $K[y] \cap K[x] = K$. *

* 13) Let K be a field, $P(T, X) = a_0(T)X^n + a_1(T)X^{n-1} + \dots + a_n(T)$ a polynomial in $K[T, X]$; suppose that : 1° the polynomial a_0 is not divisible by T ; 2° each of the polynomials $a_1, \dots, a_n$ is divisible by T ; 3° the polynomial a_n is not divisible by T^2. Show that under these conditions P, as polynomial of $K(T)[X]$, is irreducible (argue by contradiction, using Ex. 10). *

* 14) With the notation as in Ex. 11, show that if we take $g(X) = X$ and $h(X) = X^n + X + 1$ $(n \geqslant 2)$, the polynomial $(g(X)h(x) - h(X)g(x))/(X - x)$ of $K(x)[X]$ is irreducible (use Ex. 13). *

* 15) Give an example of a field K containing two subfields K_1, K_2 such that K is algebraic over K_1 and over K_2 but not over $K_1 \cap K_2$. (Take $K = \mathbf{Q}(T)$, $K_1 = \mathbf{Q}(T^2)$, and $K_2 = \mathbf{Q}(T^2 - T)$. Verify that $[K : K_1] = 2$, $[K : K_2] = 2$ and $K_1 \cap K_2 = \mathbf{Q}$. To do this note that an element $f(T)$ of $K_1 \cap K_2$ satisfies the conditions $f(-T) = f(T)$ and $f(1 - T) = f(T)$.)

§ 4

1) Let K be a field and $(E_i)_{i \in I}$ any family of extensions of K. Show that there exists an extension E of K and for each $i \in I$, a K-isomorphism u_i of E_i into E such that E is generated by the union of the $u_i(E_i)$ (argue as in Prop. 4 (V, p. 22)).

* 2) Let K be an algebraically closed field of characteristic 0, $F = K((X^{1/n!}))$ the field of formal power series in $X^{1/n!}$, and $E = K((X))$, where $X = (X^{1/n!})^{n!}$, the subfield of F consisting of formal power series in X. For every formal power series $f \in K((X))$ we denote by $\omega(f)$ its order.

a) Let $P(Y) = a_0 Y^n + a_1 Y^{n-1} + \cdots + a_n$ be a polynomial of $E[Y]$ of degree n, such that $a_n \neq 0$. Show that there exists a strictly increasing sequence $(i_k)_{0 \leqslant k \leqslant r}$ of integers in the interval $(0, n)$ such that: 1° $i_0 = 0$, $i_r = n$; 2° $\omega(a_{i_k})$ is finite for $0 \leqslant k \leqslant r$; 3° for each index j such that $0 \leqslant j \leqslant n$, distinct from the i_k and such that $\omega(a_j)$ is finite, the point $(j, \omega(a_j)) \in \mathbf{R}^2$ is above the straight line passing through the points $(i_k, \omega(a_{i_k}))$, $(i_{k-1}, \omega(a_{i_{k-1}}))$ and strictly above it if $j < i_{k-1}$ or $j > i_k$ $(1 \leqslant k \leqslant r)$. The union of the segments joining the points $(i_{k-1}, \omega(a_{i_{k-1}}))$ and $(i_k, \omega(a_{i_k}))$ for $1 \leqslant k \leqslant r$ is called the *Newton polygon* of P, the preceding segments are called its *sides* and the points $(i_k, \omega(a_{i_k}))$ its *vertices*.

b) Put $\rho_k = i_k - i_{k-1}$, $\sigma_k = (\omega(a_{i_k}) - \omega(a_{i_{k-1}}))/\rho_k$; show that P has exactly ρ_k zeros (counted with the right multiplicity) belonging to the field $K((X^{1/(\rho_k!)}))$ and whose orders are equal to σ_k (determine by recursion the coefficients of a formal power series of this type satisfying the equation $P(z) = 0$).

c) Conclude that the union of the fields $K((X^{1/n}))$ for all integers $n > 0$ is an algebraic closure of the field $E = K((X))$ (« Puiseux's theorem »).

d) If K′ is a field of characteristic 2 and $E' = K'((X))$, then the polynomial $Y^2 + XY + X$ of $E'[Y]$ does not have a root in the union of the fields $K'((X^{1/n!}))$. *

§ 5

1) *a*) Let E be an extension of a field K, $x \in E$ a transcendental element over K and let $F = K(x^n)$ for an integer $n > 1$. Show that if the intersection $F \cap K[(x-1)(x^n - 1)]$ is distinct from K, then K is of characteristic $p > 0$ and $n = p^e$, so that $K(x)$ is a p-radical extension of F. (Writing $y = x - 1$ and using Ex. 12, *b*) of V, p. 149, show that $(y+1)^n - 1 = ay^n$ for an $a \in K$.)

b) Show that if $K \subset F' \subset K(x)$ and if $K(x)$ is not a p-radical extension of F′ and if $K \neq F' \neq K(x)$, then there exists $z \in K[x]$ such that $z \notin K$ and $F' \cap K[z] = K$. (Argue by contradiction, using *a*) and Ex. 12, *b*) of V, p. 149.)

c) Under the same hypotheses as in *b*), suppose that $F' = K(y)$, where $y \in K[x]$; show that there exists $u \in K[x]$ such that $u \notin K$ and $F' \cap K(u) = F' \cap K[u] = K$. (Observe that if this were false, we should have $F' \cap K[z] \neq K$ for all $z \in K[x]$ such that $z \notin K$, using Ex. 12, *c*) of V, p. 149.)

2) Let K be a field of characteristic $p > 0$ and f an irreducible monic polynomial of $K[X]$. Show that in $K[X]$ the polynomial $f(X^p)$ is either irreducible or the p-th power of an irreducible polynomial, depending on whether or not there exists a coefficient of f not belonging to K^p (decompose $f(X^p)$ into factors of the first degree in $\Omega[X]$, where Ω is an algebraically closed extension of K).

3) Let K be a field of characteristic $p > 2$ and let F be the field of rational fractions $K(X, Y)$. Let $E = F(\theta)$ be an extension of F generated by a root θ of the polynomial

$$f(Z) = Z^{2p} + XZ^p + Y$$

of $F[Z]$. Show that $[E : F(\theta^p)] = p$, but that E contains no p-radical element over F not in F. (Firstly remark that f is irreducible in $F[Z]$; if $\beta \in E$ existed such that $\beta^p \in F$, $\beta \notin F$, f would be reducible in $F(\beta)[Z]$; using Ex. 2, show that then $X^{1/p}$ and $Y^{1/p}$ would belong to E and we would have $[E : F] \geqslant p^2$.) (See Ex. 1, V, § 7).

§ 6

1) Let K be a field and A a commutative K-algebra. For A to be etale it is necessary and sufficient that there should exist a finite subset S of A generating the K-algebra A and such that for all $x \in S$ the subalgebra $K[x]$ generated by x is etale. (If S generates A, the K-algebra A is isomorphic to a quotient of the tensor product of the $K[x]$, where x runs over S.)

2) Let Γ be a finite commutative monoid. Consider the following conditions E_0 and E_p where p denotes a prime number.
(E_0) If $x \in \Gamma$ and if e, d are integers $\geqslant 0$, the relation $x^{e+d} = x^d$ implies the relation $x^{e+1} = x$.
(E_p) If $x, y \in \Gamma$, the relation $x^p = y^p$ implies $x = y$.
Let K be a field of characteristic p ($\geqslant 0$) and let A be the K-algebra $K^{(\Gamma)}$ of the monoid Γ.
a) For A to be etale it is necessary and sufficient that the condition (E_p) be satisfied. (Use Ex. 1 for the sufficiency. For the necessity observe for $p = 0$, that $x^{e+1} - x$ is nilpotent if $x^{e+d} = x^d$; for $p > 0$ use the Cor. of V, p. 35.)
b) Let X be the set of homomorphisms of Γ into the multiplicative monoid of K. We have $\text{Card } X \leqslant \text{Card } \Gamma$. If K is algebraically closed, then $\text{Card } X = \text{Card } \Gamma$ holds if and only if condition (E_p) is satisfied.
c) If Γ is a group, condition (E_0) always holds ; if $p > 0$, condition (E_p) is equivalent to saying that each element of Γ has order prime to p, or also that $\text{Card } \Gamma$ is not divisible by p.

§ 7

1) *a*) Let K be a field of characteristic $p > 0$. An algebraic extension F of K of finite degree is said to be *exceptional* if it is not separable over K and if it contains no element

$x \notin K$ which is p-radical over K, in other words if $F^p \cap K = K^p$ (V, p. 151, Ex. 3). Show that if F is exceptional and if $E = F_s$, $z \in F$, $z \notin E$ and $z^p \in E$, then for each $t \in K$ we have $z^p \notin E^p(t)$ (observe that otherwise we should have $E^p(t) = E^p(t^p)$ and so $t \in F^p$).

b) Conversely, let E be a separable algebraic extension of K of finite degree, and let $x \in E$ be such that for each $t \in K$ we have $x \notin E^p(t)$. Let $F = E(x^{1/p})$, so that $E = F_s$. Show that F is an exceptional extension of K. (If there existed $y \in F$ such that $y \notin E$ and $y^p = z \in K$, show that we would have $z \in E^p(x)$; this would imply $z \in E^p$, hence $y \in E$, which is a contradiction.)

A separable extension E of K of finite degree is called *strongly separable* if the union of the fields $E^p(t)$, where t ranges over K, is distinct from E; it thus comes to the same to say that $E = F_s$, where F is a inseparable exceptional extension of K. A strongly separable extension of K cannot be a perfect field.

c) Show that if E is a strongly separable extension of K then it is impossible that $[E : E^p] = p$. (Argue by contradiction, showing that $[E : E^p] = p$ would imply $E^p(t) = E$ for all $t \in K$ such that $t \notin K^p$; observe on the other hand that if $K \subset E^p$, E cannot be strongly separable over K unless it is perfect.)

d) Show that if $[K : K^p] = p$, then there is no strongly separable extension of K, and so neither is there an exceptional extension of K.

2) Let k be an imperfect field of characteristic p, $a \in k$ an element such that $a \notin k^p$, F the field $k(X)$ of rational fractions in one indeterminate over k; put $y = X^{p^2}/(X^p + a)$ and $K = k(y)$.

a) Show that $[F : K] = p^2$, so that $T^{p^2} - yT^p - ay$ is the minimal polynomial of X in K[T]; deduce that the relative separable closure of K in F is $E = k(X^p)$ (cf. V, p. 148, Ex. 11).

b) Show that F is an exceptional extension (Ex. 1) of K. (Argue by contradiction, assuming $b \in F$ such that $b \notin K$ and $b^p \in K$. Observe that $K(b) = k(z)$, where $z = r(X)/s(X)$ with $\deg r = p$, $\deg s \leqslant p$ and r, s being relatively prime polynomials in $k[X]$; show that $z^p \in K$ and that the minimal polynomial of X over K is $r(T)^p - z^p s(T)^p$ up to a factor from K. Conclude, using *a*) that we would have $a \in k \cap F^p = k^p$ contrary to the hypothesis.)

3) Let K be a field of characteristic $p > 0$, E an extension of finite degree of K, E_r and E_s the largest p-radical extension and separable extension of K contained in E. Show that $E_s \otimes_K E_r$ is isomorphic to the relative separable closure of E_r in E, with which it will be identified; if it is not equal to E, then E is an exceptional extension of E_r and $E_s \otimes_K E_r$ is a strongly separable extension of E_r.

Conversely, let E be an exceptional extension of a field L and let K be a subfield of L such that L is p-radical over K and $[L : K] < +\infty$; then L is the largest p-radical extension of K contained in E.

4) Let K be a field, $E = K(x)$ a separable algebraic extension of K and f the minimal polynomial of x in K[X].

a) Let F be an extension of K; in the ring F[X] the polynomial f decomposes into a product $f_1 f_2 \dots f_r$ of irreducible and separable pairwise distinct polynomials. Show that the algebra $E \otimes_K F$ is isomorphic to the product of fields $F[X]/(f_j)$ $(1 \leqslant j \leqslant r)$.

b) Suppose that $F = K(y)$ is a separable algebraic extension of K; let g be the minimal polynomial of y in K[X], and let $g = g_1 g_2 \dots g_s$ be the decomposition of g into a product of irreducible polynomials in E[X]. Show that $r = s$ and that if $m = \deg f$, $n = \deg g$, $m_j = \deg f_j$, $n_j = \deg g_j$ we can, after permuting the g_j if necessary, suppose that $m_j/n_j = m/n$ for $1 \leqslant j \leqslant r$.

5) Let K be a finite field with q elements, and let A be the K-algebra K^n. For A to be generated by a single element, it is necessary and sufficient that $n \leqslant q$.

§ 8

1) Let K be a commutative ring, $f(X)$ a monic polynomial in K[X], A the K-algebra $K[X]/(f)$ and x the residue class of X in A. For each $a \in K$ we have $f(a) = N_{A/K}(a - x)$.

2) Let K be a field and A a K-algebra of finite degree n. For each integer m relatively prime to n we have $A^{*m} \cap K^* = K^{*m}$; in other words, if $x \in K^*$ and if there exists an element y of A such that $x = y^m$, then there exists an element z in K such that $x = z^m$.

§ 9

1) The polynomial $X^2 - 2$ is irreducible in $\mathbf{Q}[X]$ (cf. I, p. 51, Th. 7) ; let α be one of its roots in an algebraic closure Ω of $\mathbf{Q}$. Show that the polynomial $X^2 - \alpha$ is irreducible over $E = \mathbf{Q}(\alpha)$; let β be one of the roots of this polynomial in Ω and let $F = E(\beta) = \mathbf{Q}(\beta)$. Show that F is not a quasi-Galois extension of $\mathbf{Q}$; what is the quasi-Galois extension of $\mathbf{Q}$ generated by F ?

* 2) Let E be the relative algebraic closure of $\mathbf{Q}$ in $\mathbf{R}$ (« the field of real algebraic numbers ») ; if u is an automorphism of E, then $u(x) = x$ for all $x \in \mathbf{Q}$; show that if $y \in E$ is such that $y > 0$, then also $u(y) > 0$. Deduce that $u(y) = y$ for all $y \in E$. *

3) Show that every algebraic extension of a field K, generated by a set of elements of degree 2 over K is quasi-Galois over K.

4) Let K be a field, f an irreducible polynomial in K[X], separable over K and of degree n and let α_i $(1 \leqslant i \leqslant n)$ be its roots in an algebraic closure Ω of K. Let g be a polynomial in K[X], and h an irreducible factor in K[X] of the polynomial $f(g(X))$. Show that the degree of h is a multiple rn of n and that h has exactly r roots in common with each of the polynomials $g(X) - \alpha_i$ $(1 \leqslant i \leqslant n)$ of $\Omega[X]$ (consider the conjugates over K of a root of h).

5) *a*) Let K be a field, Ω an algebraic closure of K and E, F two subextensions of Ω. Show that every composite extension of E and F (V, p. 12) is isomorphic to a composite extension of the form $(L_w, 1, w)$, where w denotes a K-isomorphism of F onto a subfield of Ω, 1 is the identity mapping of E and L_w is the subfield of Ω generated by $E \cup w(F)$.
b) Deduce from *a*) that if F is of finite degree n over K, there are at most n composite extensions of E and F that are pairwise non-isomorphic.
c) Suppose that E is a quasi-Galois extension of K and F a subextension of E. Show that every composite extension of E and F is isomorphic to a composite extension of the form $(E, 1, v)$, where v is a K-isomorphism of F onto a subfield of E.

6) *a*) Let K be a field of characteristic exponent p, L an extension of K, Ω an algebraic closure of L and $\bar{K}$ the relative algebraic closure of K in Ω. Show that $\bar{K}$ and $L^{p^{-\infty}}$ are linearly disjoint over $E = \bar{K} \cap L^{p^{-\infty}}$. (Let (a_λ) be a basis of the vector space $L^{p^{-\infty}}$ over E ; consider a relation $\sum_\lambda c_\lambda a_\lambda = 0$, where $c_\lambda \in \bar{K}$ and the number of non-zero c_λ is as small as possible, and apply an L-automorphism of Ω to this relation.)
b) Deduce from *a*) that if E and F are two extensions of K such that E is algebraic over K and K is relatively algebraically closed in F, then any two composite extensions of E and F are isomorphic.

7) Let K be a field of characteristic 0, α, β, ξ three elements of an algebraic closure of K such that $[K(\alpha):K] = 2$, $[K(\beta):K] = 3$, $[K(\xi):K] = n > 1$; denote by α' (resp. β', β'') the unique conjugate $\neq \alpha$ of α (resp. the conjugates $\neq \beta$ of β).
a) Show that $[K(\alpha, \xi):K] = n$ if $\alpha \in K(\xi)$, $[K(\alpha, \xi):K] = 2n$ otherwise.
b) Show that $[K(\beta, \xi):K] = n$ if $\beta \in K(\xi)$, $[K(\beta, \xi):K] = 2n$ if $\beta \in K(\xi)$ and just one of β', β'' belongs to $K(\xi)$ and finally $[K(\beta, \xi):K] = 3n$ if none of the three elements β, β', β'' belongs to $K(\xi)$.
c) If d is the discriminant of the minimal polynomial of β, show that $K(\beta, \beta') = K(\beta, \sqrt{d})$.
d) Show that $K(\beta, \beta') = K(\beta - \beta')$.
e) If $n = 2$ and $\alpha \notin K(\xi)$, show that $K(\alpha, \xi) = K(\alpha + \xi)$.
f) Show that $K(\alpha, \beta) = K(\alpha + \beta)$.
g) Show that $K(\alpha, \beta) = K(\alpha\beta)$ except when $\alpha = aj$ and $\beta^3 = b$, where a, b are elements of K and $j^3 = 1$.

8) Let K be an infinite field, E an algebraic extension of K of finite degree and τ an automorphism of the vector K-space E such that for all $x \in E$, $\tau(x)$ is conjugate to x over K. Show that τ is a K-automorphism of the field E. (Let N be the quasi-Galois extension of K generated by E and let Γ be the group of K-automorphisms of N, Δ the subgroup of Γ consisting of all E-automorphisms and σ_j $(1 \leqslant j \leqslant r)$ a system of representatives of the left cosets of Δ. If $(u_i)_{1 \leqslant i \leqslant n}$ is a basis of the vector K-space E, consider the polynomial

$$F(X_1, \ldots, X_n) = \prod_{j=1}^{r} \left(\sum_{i=1}^{n} (\tau(u_i) - \sigma_j(u_i)) X_i \right) .)$$

§ 10

1) *a*) Let K be a field, f a separable polynomial in K[X] and N a splitting field of f. Show that for the group Gal(N/K) to operate transitively on the set of roots of f in N it is necessary and sufficient for f to be irreducible in K[X].
b) Suppose further that f is irreducible in K[X] ; let x be a root of f in N. For there to exist no field E such that $K \subset E \subset K(x)$, distinct from K and $K(x)$ it is necessary and sufficient that Gal(N/K), qua transitive permutation group of the set of roots of f in N, should be primitive (I, p. 142, Ex. 13).

2) Let K be a field, f_0 an irreducible and separable polynomial in K[X], of degree n ; let α_i $(1 \leqslant i \leqslant n)$ be its roots in a splitting field N of f_0. Let F be the field of rational fractions $N(X_1, X_2, \ldots, X_n)$, E the subfield $K(X_1, X_2, \ldots, X_n)$ of F ; F is a Galois extension of E and Gal(E/F) is canonically isomorphic to Gal(N/K) (V, p. 71, Th. 5 ; V, p. 154, Ex. 8). Show that if $\theta = \alpha_1 X_1 + \alpha_2 X_2 + \cdots + \alpha_n X_n$ then $F = E(\theta)$ and the minimal polynomial g_1 of θ in E[T] is an irreducible factor of the polynomial

$$f(T) = \prod_{\sigma \in \mathfrak{S}_n} (T - \alpha_1 X_{\sigma(1)} - \alpha_2 X_{\sigma(2)} - \cdots - \alpha_n X_{\sigma(n)}) .$$

Each permutation $\sigma \in \mathfrak{S}_n$ defines a K-automorphism of E, again denoted by σ, by the condition $\sigma(X_i) = X_{\sigma(i)}$ for $1 \leqslant i \leqslant n$; show that Gal(N/K) is isomorphic to the subgroup of $\mathfrak{S}_n$ consisting of permutations σ such that $\sigma(g_1) = g_1$. Further, if $f = g_1 g_2 \ldots g_r$ is a decomposition of f into irreducible factors in E[T], show that for every index j such that $1 \leqslant j \leqslant r$ there exists a permutation $\sigma_j \in \mathfrak{S}_n$ such that $\sigma_j(g_1) = g_j$.

3) Let N be a Galois extension of a field K of infinite degree over K. Show that the cardinal of the Galois group Gal(N/K) is greater than that of the set of subsets of N (use V, p. 61 and 62). Deduce that there exist non-closed subgroups of Gal(N/K).

4) Let N be a Galois extension of a field K, $(E_i)_{i \in I}$ a family of subextensions of N, Galois over K, such that: 1° for each index i, if F_i is the field generated by the E_j for $j \neq i$, then $E_i \cap F_i = K$; 2° N is generated by the union of the E_i.
a) Show that N is isomorphic to the tensor product $\bigotimes_{i \in I} E_i$ (III, p. 470) (define an isomorphism of this tensor product onto N, using Th. 5 of V, p. 71).
b) Show that Gal(N/K) is isomorphic to the product of the topological groups $Gal(E_i/K)$.

5) Show that every compact totally disconnected group is isomorphic to the group Gal(N/K) of a Galois extension of a field (using Gen. Top. III, p. 325, Ex. 3, reduce to the case where the group is a product of finite groups and use Ex. 4 as well as V, p. 59, Example 5).

6) Let N be a Galois extension of a field K, of infinite degree, and let ρ, σ be two distinct elements of Gal(N/K). If J is the set of $x \in N$ such that $\rho(x) \neq \sigma(x)$, show that K(J) is an extension of K of infinite degree. (Observe that a field cannot be the union of two of its subfields unless it is equal to one of them.)

7) Let N be a Galois extension of a field K such that Gal(N/K) is isomorphic to the symmetric group $\mathfrak{S}_n$ (V, p. 59, Example 5), n being a non-prime integer ; identify Gal(N/K) and $\mathfrak{S}_n$. Let Γ_1 be the subgroup of $\mathfrak{S}_n$ of order $(n-1)!$ leaving the number 1 invariant and let Γ_2 be the cyclic subgroup of $\mathfrak{S}_n$ of order n generated by the cycle $(1, 2, \ldots, n)$ (I, p. 142, Ex. 12). Let E_1, E_2 be the field of invariants of the subgroups Γ_1, Γ_2 respectively. Show that E_1 and E_2 are linearly disjoint over K, that there exists no field F other than K and E_1 such that $K \subset F \subset E_1$ but that there exist fields F' distinct from E_2 and N such that $E_2 \subset F' \subset N$ (use Ex. 13 of I, p. 142 and Ex. 14, I, p. 143).

8) Let K be a field and N a Galois extension of K of finite degree n.

a) Let p be a prime number dividing n and let p^r be the highest power of p dividing n. Show that there exist subextensions L_i of N for $0 \leqslant i \leqslant r$ such that for $0 \leqslant i \leqslant r-1$, L_i contains L_{i+1}, $L_0 = N$, L_i is a Galois extension of degree p of L_{i+1} and $[L_r : K]$ is not divisible by p (cf. I, p. 77, Th. 1 and I, p. 78, Th. 2).

b) Let $x \in N$ be such that N is generated by x and its conjugates. Show that there exists a subextension E of N such that $x \notin E$ and $[N : E] = p^r$. (Consider the Sylow p-groups H_j of Gal(N/K) and the (normal) subgroup H generated by their union ; if F_j (resp. F) is the subfield of invariants of H_j (resp. H) show that $x \notin F$ and deduce that there is an index j such that $x \notin F_j$; then we can take $E = F_j$.) Conclude that there exists a subextension L of N such that $L(x)$ is a Galois extension of degree p of L (use I, p. 77, Prop. 12).

c) Conversely, let Ω be an algebraic closure of N and suppose that there exists a subextension L of Ω such that $L(x)$ is Galois of degree p over L. Show that then p divides n. (Observe that $N \cap L(x) = (N \cap L)(x)$ and that $(N \cap L)(x)$ is a Galois extension of degree p of $N \cap L$.)

9) Let K be a field of characteristic p, Ω an algebraically closed extension of K, x an element of Ω such that $x \notin K$ and M a maximal element of the set of subextensions of Ω not containing x (V, p. 148, Ex. 9), so that Ω is an algebraic extension of M (*loc. cit.*).

a) Show that if M is not a perfect field, Ω is a p-radical extension of M, union of the fields $M(x^{p^{-n}})$ for $n \geqslant 0$ (observe that for every element $y \in \Omega$ p-radical over M, the field $M(y)$ contains $M(x)$; deduce that we necessarily have $[M(x) : M] = p$, because every element of Ω separable over M is necessarily in M).

b) Show that if M is perfect, $M(x)$ is a Galois extension of M, of prime degree q, and for every subextension N of Ω which is Galois of finite degree over M, the degree $[N : M]$ is a power of q. (Take for q a prime number dividing $[M(x) : M]$; note that every extension of M distinct from M and contained in Ω contains $M(x)$, and use Ex. 8, *a*).)

10) Let K be a perfect field such that for every integer $n > 1$ there exists a Galois extension N of K such that Gal(N/K) is isomorphic to the alternating group $\mathfrak{A}_n$; for example, for every field k of characteristic 0 the field $K = k(X_n)_{n \in \mathbf{N}}$ of rational fractions in an infinity of indeterminates has this property (V, p. 59, Example 5).

a) Let A be a finite set of integers > 1 and let Ω be an algebraic closure of K. An element $x \in \Omega$ is said to be *A-constructible* if there exists an increasing sequence $K = E_0 \subset E_1 \subset \cdots \subset E_r \subset \Omega$ of extensions of K such that each of the degrees $[E_j : E_{j-1}]$ belongs to A for $1 \leqslant j \leqslant r$ and $x \in E_r$. Show that there exists a maximal element L in the set of subextensions $E \subset \Omega$ such that each $x \in E$ is A-constructible. For each $x \notin L$ show that the degree $[L(x) : L]$ belongs to $\mathbf{N} - A$.

b) Let a be the greatest element of A and let n be an integer such that $n \geqslant \max(a+1, 5)$. Show that there exists an extension M of L such that $[M : L] = n$. (Observe firstly that there exists an irreducible polynomial $f \in K[K]$ of degree n such that if N is the splitting field of f, Gal(N/K) is isomorphic to $\mathfrak{A}_n$; let $y \in \Omega$ be a root of f in Ω and let $E = K(y)$. Show that we necessarily have $E \cap L = K$ and therefore $L(E)$ is the desired field. Argue by contradiction, noting that the elements of $E \cap L$ are A-constructible and that the relation $E \cap L \neq K$ would imply the existence of a normal subgroup Γ of $\mathfrak{A}_n$ such that $1 < (\mathfrak{A}_n : \Gamma) < n$, which would contradict the simplicity of $\mathfrak{A}_n$ (I, p. 143, Ex. 16).)

11) *a*) Let E be a separable extension of a field K of finite degree and N the Galois extension of K generated by E. If α is an element of N whose conjugates form a normal basis of N over K and if $\beta = \mathrm{Tr}_{N/E}(\alpha)$, show that $E = K(\beta)$.
b) Let N be a Galois extension of K of finite degree and let α be an element of N whose conjugates form a normal basis over K. If L is a subextension of N which is Galois over K and if $\beta = \mathrm{Tr}_{N/L}(\alpha)$, show that the conjugates of β form a normal basis of L over K.
c) Let L, M be two Galois extensions of K of finite degree, such that $L \cap M = K$; let α (resp. β) be an element of L (resp. M) whose conjugates over K form a normal basis of L (resp. M) over K ; show that the conjugates over K of $\alpha\beta$ form a normal basis of $K(L \cup M)$.

12) Let K be an infinite field, N a Galois extension of degree n of K, $f \in K[X]$ the minimal polynomial of an element θ of N such that $N = K(\theta)$. The group $\Gamma = \mathrm{Gal}(N/K)$ is allowed to operate on the polynomials in N[X] by operating on the coefficients of these polynomials.
a) Let $g \in N[X]$ be the polynomial of degree $\leqslant n-1$ such that $g(\theta) = 1$, $g(\sigma(\theta)) = 0$ for all $\sigma \neq 1$ in Γ. For every subgroup Δ of Γ let $p_\Delta \in N[X]$ be the polynomial $\det(\sigma\tau(g))_{\sigma \in \Delta, \tau \in \Delta}$. Show that

$$p_\Delta^2 \equiv \sum_{\sigma \in \Delta} \sigma(g) \mod f$$

and deduce that $p_\Delta \neq 0$.
b) Deduce that there exists an element $\alpha \in K$ such that for *every* subgroup Δ of Γ the conjugates of $g(\alpha)$ over the subfield $k(\Delta)$ form a normal basis of N over $k(\Delta)$.

13) Let E be an algebraic extension of a field K, E_s the largest separable extension and E_r the largest p-radical extension of K contained in E. In an algebraic closure of K let N be the quasi-Galois extension of K generated by E ; then the largest separable extension of K contained in N is the quasi-Galois extension N_s generated by E_s. For E_r to be the largest p-radical extension of K contained in N, it is necessary and sufficient that E should be separable over E_r (cf. V, p. 152, Ex. 3).

14) Let E be an algebraic extension of a field K, Γ the group of K-automorphisms of E and $S \subset E$ the field of invariants of Γ.
a) Show that for E to be quasi-Galois over K it is necessary and sufficient for S to be a p-radical extension of K.
b) Let S_s be the largest separable extension of K contained in S. Show that S_s is the smallest of the fields F such that $K \subset F \subset E$ and E is a quasi-Galois extension of F.
c) Let E_s be the largest separable extension of K contained in E ; show that $E = S(E_s)$ (observe that no K-automorphism of E other than the identity leaves all the elements of $S(E_s)$ invariant).

15) Let E be a Galois extension of finite degree n of a field K whose Galois group G is *nilpotent*. Let P be the set of prime numbers dividing n. Then there exist Galois subextensions E_p of E ($p \in P$) such that $\mathrm{Gal}(E_p/K)$ is a p-group for each $p \in P$, and such that the canonical homomorphism $\bigotimes_{p \in P} E_p \to E$ is an isomorphism.

16) Let a be an integer and let x_1, x_2, x_3, x_4 be the roots of $P(X) = X^4 - aX - 1$ in an algebraic closure of $\mathbf{Q}$. We identify the Galois group G of P with a group of permutations of $\{x_1, \ldots, x_4\}$. Suppose that $\mathbf{Q}(x_1)$ contains no quadratic subextension (cf. Ex. 1 of V, p. 154).

a) Show that $G = \mathfrak{S}_4$ or $\mathfrak{A}_4$. (Use Ex. 1 of V, p. 154 : P is irreducible, hence G acts transitively ; further $\mathbf{Q}(x_1)$ contains no quadratic subextensions, hence the stabilizer H of x_1 is not contained in a subgroup of index 2 of G, whereas $[G : H] = 4$; deduce that G is not a 2-group, hence its order is divisible by 3.)

b) Let $\delta = \prod_{1 \leqslant i < j \leqslant 4} (x_i - x_j)$ and $d = \delta^2$. Show that $d = 4^4 - 3^3 a^4$; further $s(\delta) = \varepsilon_s \delta$ for all $s \in G$. Deduce that $G = \mathfrak{A}_4$ if $\delta \in \mathbf{Q}$, and $G = \mathfrak{S}_4$ otherwise. Show that δ is not rational. (We have thus constructed an infinite family of equations of degree 4 over $\mathbf{Q}$ with Galois group $\mathfrak{S}_4$.)

* 17) Let f be an irreducible polynomial in $\mathbf{Q}[X]$ of prime degree p, having exactly two roots in $\mathbf{C} - \mathbf{R}$. Then the Galois group G of f is isomorphic to $\mathfrak{S}_p$. (G contains an element of order p and a transposition.) *

18) Let K be a field and F the field of rational fractions K(T). The group $GL_2(K)$ operates on F by the law $(\gamma f)(T) = f\left(\frac{aT + b}{cT + d}\right)$, for $\gamma = \begin{pmatrix} a & b \\ c & d \end{pmatrix} \in GL_2(K)$ and $f \in F$. The centre $K^* . I$ of $GL_2(K)$ operates trivially, whence we obtain, by passage to quotients, a homomorphism of $PGL_2(K) = GL_2(K)/K^* . I$ into the automorphism group of the extension F of K. This homomorphism is bijective (V, p. 148, Ex. 11).

a) Let H be a subgroup of $PGL_2(K)$ of finite order h, and let E be the field of invariants of H, so that $F = E(T)$ is a Galois extension of E with Galois group H (V, p. 66, Th. 3). Let

$$H(X) = \prod_{s \in H} (X - s(T)) = \sum_{i=0}^{h} (-1)^i S_i(T) X^{h-i}$$

be the minimal polynomial of T over E. For each i such that $0 \leqslant i \leqslant h$, either $S_i(T) \in K$ or $E = K(S_i(T))$. (Use Ex. 11, *a*) of V, p. 157.)

b) For each $s \in H$ let $\begin{pmatrix} a_s & b_s \\ c_s & d_s \end{pmatrix}$ be a representative in $GL_2(K)$ and let H_0 be the subgroup of H consisting of those s for which $c_s = 0$; let $h_0 = \operatorname{Card} H_0$ and put $R^H(T) = S_{h_0}(T)$. Show that $E = K(R^H(T))$ and that $R^H(T) = P(T)/Q(T)^{h_0}$ where P is a polynomial of degree h and $Q(T) = \prod_s (c_s T + d_s)$ where s runs over a system of representatives of the cosets $H_0 s$ different from H_0 ; the degree of Q is thus $(h/h_0) - 1$.

c) Suppose that K^* contains a subgroup μ_n of finite order n. Let C_n be the image in $PGL_2(K)$ of the subgroup of $GL_2(K)$ consisting of the matrices $\begin{pmatrix} a & 0 \\ 0 & 1 \end{pmatrix}$, $a \in \mu_n$. Then $C_n(X) = X^n - T^n$ and $R^{C_n}(T) = (-1)^{n+1} T^n$.

d) Let w be the element of $PGL_2(K)$ represented by $\begin{pmatrix} 0 & 1 \\ 1 & 0 \end{pmatrix}$. Then $w^2 = 1$ and $wsw^{-1} = s^{-1}$ if s is represented by a diagonal matrix. The subgroup D_n of $PGL_2(K)$ generated by C_n and w is therefore of order $2n$. We have $D_n(X) = (X^n - T^n)\left(X^n - \left(\frac{1}{T}\right)^n\right) = X^{2n} + (-1) R^{D_n}(T) + 1$ where $R^{D_n}(T) = -((-T)^n + (-T)^{-n})$.

e) Let S be the group of order 3 in $PGL_2(K)$ generated by the image of the matrix $\begin{pmatrix} 0 & -1 \\ 1 & 1 \end{pmatrix}$. We have $S(X) = (X - T)\left(X - \frac{1}{1-T}\right)\left(X - \frac{T-1}{T}\right)$ and $R^S(T) = \frac{T^3 - 3T + 1}{T^2 - 1}$.

f) Suppose that K is a finite field with q elements. Let U be the image in $PGL_2(K)$ of the group of all matrices $\begin{pmatrix} 1 & 0 \\ b & 1 \end{pmatrix}$, $b \in K$. Then

$$U(X) = (X - T)^q - (X - T) = X^q - X - (T^q - T) \quad \text{and} \quad R^U(T) = T^q - T .$$

Let B be the group generated by C_{q-1} and U ; it is of order $q(q-1)$. We have

$$B(X) = (X^q - X)^{q-1} - (T^q - T)^{q-1} \quad \text{and} \quad R^H(T) = (T^q - T)^{q-1} .$$

19) Let E be a Galois extension of a field K with Galois group G. For each $x \in E$ denote by Gx the set of conjugates of x in E ; this is a finite G-set. Let x, y be two elements of E. For the G-sets Gx and Gy to be isomorphic (that is, for a bijection $Gx \to Gy$ to exist, compatible with the operation of G on the two sets), it is necessary and sufficient that the extensions $K(x)$ and $K(y)$ should be conjugate.

20) Let E be an algebraic extension of a field K such that every non-constant polynomial in K[X] has at least one root in E. Show that E is algebraically closed. (In an algebraic closure of E, hence of K, let F be an extension of finite degree of K. To show that $F \subset E$ reduce to the case where F is quasi-Galois and then, by means of Prop. 13 of V, p. 76, to the case where F is either Galois or p-radical.)

21) Let $f(X) = X^n + a_1X^{n-1} + \cdots + a_n$ be a polynomial of $\mathbf{Z}[X]$ and p a prime number. Suppose that $a_i \equiv 0 \pmod p$, $1 \leqslant i \leqslant n$ and that $a_n \not\equiv 0 \pmod{p^2}$. Show that $f(X)$ is irreducible in $\mathbf{Q}[X]$. (Let $f(X) = P(X)\,Q(X)$ be a decomposition in $\mathbf{Q}[X]$. Show that if P and Q are monic, they have integer coefficients. Reduce mod p.)

* 22) Let m be an even integer > 0, r an integer $\geqslant 3$ and $n_1 < \cdots < n_{r-2}$, $r-2$ even integers. Put $g(X) = (X^2 + m)(X - n_1) \ldots (X - n_{r-2})$.

a) Put $f(X) = g(X) - 2$. Show that $f(X)$ has at least $r-2$ real roots. Calculate the sum of the squares of the roots of f in $\mathbf{C}$. Deduce that for $m \geqslant \sum n_i^2/2$, $f(X)$ is a polynomial with integer coefficients having $r-2$ real roots and two non-real conjugate roots.

b) Show that $f(X)$ is irreducible in $\mathbf{Q}[X]$. (Write $f(X) = X^r + \sum_{k=1}^{r} a_k X^{r-k}$; show that a_k is even and that a_r is not divisible by 4 ; apply Ex. 21.)

c) Show that when r is prime and $m \geqslant \sum n_i^2/2$, then the Galois group G of the splitting field of f is isomorphic to $\mathfrak{S}_r$ (cf. V, p. 158, Ex. 17).

d) Show that when $r = 4$ and $m \geqslant \sum n_i^2/2$, the degree over $\mathbf{Q}$ of the splitting field is equal to 8 or 24, and that it is equal to 8 if and only if $n_1 = -n_2$ or also if and only if $f(X)$ is of the form $P(X^2)$, where P is a polynomial of degree 2. *

23) Let k be a field of characteristic 2.

a) Let $A_1, \ldots, A_n$ be indeterminates, K the field $k(A_1, \ldots, A_n)$, E a splitting field of the polynomial $P = X^n + A_1X^{n-1} + \cdots + A_n \in K[X]$ and $X_1, \ldots, X_n$ the roots of P in E, so that $E = k(X_1, \ldots, X_n)$ (V, p. 21). We identify the Galois group of E/K with the symmetric group $\mathfrak{S}_n$ acting by permutations of the variables $X_1, \ldots, X_n$ (V, p. 59). Put

$$\beta(X_1, \ldots, X_n) = \sum_{i<j} X_i/(X_i + X_j) \in E \text{ and } C(A_1, \ldots, A_n) = \sum_{i<j} (X_iX_j)/(X_i^2 + X_j^2) \in K.$$

Then $\beta \notin K$ and

$$\beta^2 + \beta + C = 0 .$$

We have $\sigma(\beta) = \beta$ for $\sigma \in \mathfrak{A}_n$ and $\sigma(\beta) = \beta + 1$ for $\sigma \in \mathfrak{S}_n - \mathfrak{A}_n$. Deduce that $K(\beta)$ is the unique quadratic subextension of E/K.

b) Let $D \in \mathbf{Z}[T_1, \ldots, T_n]$ be the discriminant of the polynomial $X^n + T_1X^{n-1} + \cdots + T_n$. Show that there exist Q, R, $S \in \mathbf{Z}[T_1, \ldots, T_n]$ with

$$D = Q^2 - 4R + 8S$$

and

$$C = C(A_1, \ldots, A_n) = \frac{R(A_1, \ldots, A_n)}{Q^2(A_1, \ldots, A_n)} = \frac{R(A_1, \ldots, A_n)}{D(A_1, \ldots, A_n)} .$$

For every triple of elements U, V, W of $\mathbf{Z}[T_1, \ldots, T_n]$ such that $D = U^2 - 4\,V + 8\,W$ there exists $M \in \mathbf{Z}[A_1, \ldots, A_n]$ such that

$$C = \frac{V(A_1, \ldots, A_n)}{D(A_1, \ldots, A_n)} + \left(\frac{M}{D(A_1, \ldots, A_n)}\right)^2 + \frac{M}{D(A_1, \ldots, A_n)} .$$

c) Let $f = X^n + a_1X^{n-1} + \cdots + a_n$ be a *separable* polynomial of $k[X]$ and G the Galois group of a splitting extension. Show that $(a_1, \ldots, a_n)$ is substitutable in C and that the following conditions are equivalent

(i) every element of G induces an *even* permutation of the roots of f;

(ii) there exists $\alpha \in k$ such that $C(a_1, \ldots, a_n) = \alpha^2 + \alpha$.

* *d*) Henceforth assume that k is *finite* and choose the elements U, V, W of $\mathbf{Z}[T_1, \ldots, T_n]$ arbitrarily to satisfy the condition stated in *b*).

Show that (i) and (ii) of *c*) are equivalent to the following conditions :

(iii) $$\mathrm{Tr}_{k/\mathbf{F}_2}(C) = 0 ;$$

(iv) $$\mathrm{Tr}_{k/\mathbf{F}_2}\left(\frac{V(a_1, \ldots, a_n)}{D(a_1, \ldots, a_n)}\right) = 0 .$$

(v) The number of irreducible factors of f is congruent to n modulo 2 (use Ex. 22 of V, p. 165). *

§ 11

1) Let K be a field of characteristic p, n an integer not a multiple of p, $R_n(K)$ the field of n-th roots of unity and E a cyclic extension of $R_n(K)$ of degree n over K ; E is thus the root

field of a polynomial $X^n - \alpha \in R_n(K)[X]$ (V, p. 89, Th. 4). For E to be a Galois extension of K it is necessary and sufficient that for every automorphism $\tau \in \mathrm{Gal}(R_n(K)/K)$ there exists an integer $r > 0$ and an element $\gamma \in R_n(K)$ such that $\tau(\alpha) = \gamma^n \alpha^r$.

2) *a*) Let E be a splitting extension of the polynomial $X^3 - 2 \in \mathbf{Q}[X]$. The field E is a cyclic extension of degree 3 of $R_3(\mathbf{Q})$ but contains no cyclic extension of degree 3 of $\mathbf{Q}$.
b) The field $E = R_9(\mathbf{Q})$ is a cyclic extension of degree 3 of $R_3(\mathbf{Q})$ and contains a cyclic extension of degree 3 of $\mathbf{Q}$.
c) With the hypothesis and notation as in V, p. 154, Ex. 7, show that for $K(\alpha, \beta)$ to be Galois over K it is necessary and sufficient for d to be a square in $K(\alpha)$; for $K(\beta)$ to be cyclic over K it is necessary and sufficient for d to be a square in K.

* 3) Show that for every integer $n > 1$ there exists a cyclic extension of degree n of the field $\mathbf{Q}$. (Reduce to the case where n is a prime number q and use the structure of the multiplicative group $(\mathbf{Z}/q^m\mathbf{Z})^*$ (VII, p. 13). *

4) Let K be a field of characteristic p, q a prime number $\neq p$ such that K contains the q-th roots of unity and let ζ be a primitive q-th root of unity.
a) Let E be a cyclic extension of K of degree q^e and σ a K-automorphism of E generating $\mathrm{Gal}(E/K)$. Let F be the field of degree q^{e-1} over K such that $K \subset F \subset E$; there exists $\theta \in E$, a root of an irreducible polynomial $X^q - \alpha$ in F[X] such that $E = F(\theta)$ and $\theta^{\sigma^m} = \zeta\theta$ (where $m = q^{e-1}$). Show that also $E = K(\theta)$ and $\theta^\sigma = \beta\theta$ where $\beta \in F$ is such that $\alpha^{\sigma - 1} = \beta^q$ and $N_{F/K}(\beta) = \zeta$ (observe that $\theta^\sigma = \beta\theta^k$ with $0 < k < q$ and $\beta \in F$, by Ex. 1 ; by calculating $\theta^{\sigma^{mq}}$, show that $k^{mq} - 1 \equiv 0 \pmod q$ and deduce that $k = 1$).
b) Conversely, let F be a cyclic extension of degree q^{e-1} of K with $e \geqslant 2$ and let σ be a K-automorphism of F generating $\mathrm{Gal}(F/K)$. If there exists an element $\beta \in F$ such that $N_{F/K}(\beta) = \zeta$ and if $\alpha \in F$ is such that $\alpha^{\sigma - 1} = \beta^q$, show that for every $c \in K^*$ the polynomial $X^q - c\alpha$ is irreducible in F[X]. If θ is one of the roots of this polynomial in an algebraic closure Ω of F, show that there exists a K-homomorphism $\bar\sigma$ of $E = F(\theta)$ into Ω, extending σ and such that $\theta^{\bar\sigma} = \beta\theta$; deduce that E is a cyclic extension of K of degree q^e, that $\bar\sigma$ generates $\mathrm{Gal}(E/K)$ and that $E = K(\theta)$. Finally show that every cyclic extension of degree q^e of K, containing F, is the root field of a polynomial $X^q - c\alpha$ in F[X], for an appropriate $c \in K^*$ (use Th. 3 of V, p. 85).
c) Take K to be the field $\mathbf{Q}$ of rational numbers. The polynomial $X^2 + 1$ is irreducible in $\mathbf{Q}[X]$; if i is one of its roots, then $F = \mathbf{Q}(i)$ is a cyclic extension of $\mathbf{Q}$ of degree 2 but there exists no cyclic extension of degree 4 of $\mathbf{Q}$ containing F.

5) Let K be a field of characteristic p and n an integer not a multiple of p. For a polynomial of K[X] of the form $X^n - a$ to be irreducible it is necessary and sufficient that for every prime factor q of n, a should not equal the q-th power of an element of K and further when $n \equiv 0 \pmod 4$, a should not be of the form $-4c^4$ with $c \in K$. (To prove the sufficiency of the condition, reduce to the case where $n = q^e$ with q prime and use Ex. 4 of V, p. 153. Argue by induction an e, determining, with the help of Ex. 4 of V, p. 153, the form of the constant term of each irreducible factor of $X^{q^e} - a$.)

6) Let K be a field of characteristic p, n an integer not a multiple of p and N a splitting field of the polynomial $X^n - a$ of K[X].

a) Show that the group Gal(N/K) is isomorphic to a subgroup of the group Γ of matrices $\begin{pmatrix} x & y \\ 0 & 1 \end{pmatrix}$, where $x \in (\mathbf{Z}/n\mathbf{Z})^*$ and $y \in \mathbf{Z}/n\mathbf{Z}$. If n is a prime number, the group Gal(N/K) is commutative only if $R_n(K) = K$ or if a is the n-th power of an element of K.
b) If $K = \mathbf{Q}$ and if n is a prime number, show that if a is not the n-th power of a rational number, then Gal(N/K) is isomorphic to the whole of Γ.
c) If $K = \mathbf{Q}$, determine Gal(N/K) when a is a square-free integer.

7) An extension E of a field K is called a *Kummer extension* if there exists an integer $n > 0$ such that K contains a primitive n-th root of unity and E is an abelian extension with Galois group annihilated by n.
a) Let K be a field and N a Galois extension of degree m of K with Galois group Γ. For N to be a Kummer extension of K it is necessary and sufficient that there should exist a basis $(\theta_\sigma)_{\sigma \in \Gamma}$ of N over K such that the elements $\gamma_{\sigma\tau} = \theta_\sigma^{1-\tau}$ belong to K for each pair (σ, τ). (For the necessity of the condition reduce to the case where N is cyclic of prime power degree. For the sufficiency observe that if $\tau \in \Gamma$ is of order n, then there exists $\sigma \in \Gamma$ such that $\gamma_{\sigma\tau}$ is a primitive n-th root of unity.)
b) If $(\theta_\sigma)_{\sigma \in \Gamma}$ is a basis satisfying condition *a*), show that for an element $x = \sum_\sigma a_\sigma \theta_\sigma$ of N to be such that the conjugates of x form a normal basis of N over K, it is necessary and sufficient that $a_\sigma \neq 0$ for all $\sigma \in \Gamma$.

* 8) Let N be a Kummer extension of a field K.
a) Let E be a subextension of N such that $[E : K] = q$ is a prime number. Show that there exists a decomposition $N = N_1 \otimes N_2 \otimes \cdots \otimes N_r$ of N into cyclic extensions of K such that $E \subset N_j$ for an index j.
b) Deduce from *a*) that if $x \in N$ is such that the conjugates of x over K form a normal basis of N over K, then for each field E such that $K \subset E \subset N$, the conjugates of x over E form a normal basis of N over E (reduce to the case where $[E : K]$ is prime and use Ex. 7, *b*)). *

9) Let K be a field of characteristic $p > 0$.
a) Let E be a cyclic extension of degree p^e of K and let σ be a K-automorphism generating the group Gal(E/K). Let F be the field of degree p^{e-1} over K such that $K \subset F \subset E$; if $m = p^{e-1}$, show that there exists $\theta \in E$, root of an irreducible polynomial $X^p - X - \alpha$ of F[X] such that $E = F(\theta)$ and $\sigma^m(\theta) = \theta + 1$. Show that we also have $E = K(\theta)$ and $\sigma(\theta) = \theta + \beta$ where $\beta \in F$ is such that $\sigma(\alpha) - \alpha = \beta^p - \beta$ and $\mathrm{Tr}_{F/K}(\beta) = 1$.
b) Conversely, let F be a cyclic extension of degree p^{e-1} of K (with $e > 1$), σ a K-automorphism of F generating Gal(F/K). Show that there exist two elements α, β of F such that $\mathrm{Tr}_{F/K}(\beta) = 1$ and $\sigma(\alpha) - \alpha = \beta^p - \beta$ (use Th. 3 of V, p. 85 and the Cor. of Prop. 1 of V, p. 49). Deduce that for all $c \in K$ the polynomial $X^p - X - \alpha - c$ is irreducible in F[X]; if θ is a root of this polynomial in an algebraic closure Ω of F, show that there exists a K-homomorphism $\bar{\sigma}$ of $E = F(\theta)$ in Ω extending σ and such that $\bar{\sigma}(\theta) = \theta + \beta$; conclude that E is a cyclic extension of degree p^e of K, that $\bar{\sigma}$ generates Gal(E/K) and that $E = K(\theta)$. Finally show that every cyclic extension of degree p^e of K containing F is the splitting field of a polynomial $X^p - X - \alpha - c$ of F[X] for a suitable $c \in K$.

10) *a*) Let K be a field whose algebraic closure is an extension of prime degree q of K. Such that K is perfect and that q must be distinct from the characteristic of K (use Ex. 9).

b) Show that K contains the q-th roots of unity and that Ω is the splitting field of an irreducible polynomial $X^q - a$ of $K[X]$; deduce that $q = 2$ (in the contrary case deduce from Ex. 5 that the polynomial $X^{q^2} - a$ would be irreducible). Show further that $-a$ is a square in K (*loc. cit.*), that -1 is not a square in K, and deduce that $\Omega = K(i)$, where $i^2 = -1$.

11) Let K be a field whose algebraic closure Ω is an extension of finite degree > 1 of K. Show that $\Omega = K(i)$ where $i^2 = -1$. (If we had $\Omega \neq K(i)$, show with the help of Galois theory and Sylow's theorem that there would then exist a field E such that $K(i) \subset E \subset \Omega$ and Ω is an extension of prime degree of E ; now apply Ex. 10.)

12) Let K be a field of characteristic q, Ω an algebraically closed extension of K, x an element of Ω such that $x \notin K$ and M a maximal element of the set of subextensions of Ω not containing x (V, p. 156, Ex. 9) ; suppose further that M is perfect so that $M(x)$ is a Galois extension of prime degree p of M (*loc. cit.*).
a) Show that either $p = 2$ and $\Omega = M(i)$ or for each integer $r \geqslant 1$, there exists a single extension of degree p^r of M ; this extension M_r is cyclic over M, Ω is the union of the M_r and the M_r are the only extensions of finite degree of M. (Note that if a p-group Γ of order p^r has only one subgroup of order p^{r-1}, it is necessarily cyclic, arguing by induction on r and using I, p. 133, Ex. 10, and I, p. 77, Cor. of Prop. 11).
b) Suppose that $\Omega \neq M(i)$ and that $p \neq q$. Show that M contains the p-th roots of unity ; we thus have $M(x) = M_1 = M(\alpha)$ with $\alpha^p \in M$. Show that if α is the p-th power of an element of M_1, then $M_1 = M(\zeta)$, where ζ is p^2-th root of unity (if $\alpha = \beta^p$ for $\beta \in M_1$, consider the minimal polynomial of β in $M[X]$).
c) Under the hypotheses of *b*) show that if α is not a p-th power of an element of M_1 then M_r is the splitting field of the polynomial $X^{p^{r-1}} - \alpha$ of $M_1[X]$ (use *b*)).
d) Under the hypotheses of *b*), show that if α is the p-th power of an element of M_1, there exists $\gamma \in M_1$ such that $\gamma \notin M$ and that M_2 is the splitting field of $X^p - \gamma$; show that γ is not the p-th power of an element of M_2 and deduce that M_r is the splitting field of the polynomial $X^{p^{r-1}} - \gamma$ of $M_1[X]$.

13) A field K is called a *Moriya field* if every separable algebraic extension of finite degree over K is cyclic. * Every finite field is a Moriya field (V, p. 95, Prop. 4). * The field M of Ex. 12 is a Moriya field.
a) For K to be a Moriya field it is necessary and sufficient that for each integer $n > 0$ there should exist at most one separable extension of degree n over K. (Use the following fact : if G is a finite group and for each integer $m \geqslant 1$ dividing the order of G there exists at most one subgroup of G of order m, then G is cyclic ; this follows from the same property for p-groups (Ex. 12) and from I, p. 80, Th. 4).
b) If K is a Moriya field, show that every algebraic extension E of K is a Moriya field and that if F is an extension of E of degree m over E, then there exists an extension of K of degree m over K.

14) *a*) Let K be a field such that there exist algebraic extensions of K of arbitrarily large finite degree, and that all these extensions have as degree a power of the same prime number p. Show that under these conditions, for each power p^h ($h \geqslant 0$) there exists an extension of K of degree p^h (use I, p. 77, Th. 1).

b) Let p be a prime number and K a field such that there exists an algebraic extension of K of degree divisible by p if $p \neq 2$ and divisible by 4 if $p = 2$. Show that there exists an algebraic extension E of K such that the set of degrees of algebraic extensions of E is the set of powers p^h of p. (If K is perfect or of characteristic $\neq p$, let K_1 be the perfect closure of K ; consider a maximal algebraic extension of K_1, among all those whose elements are of degree not divisible by p over K_1 ; use *a*) and Ex. 11.)

15) Let K be a Moriya field (Ex. 13) and P the set of prime numbers dividing the degree of an algebraic extension of finite degree of K. Show that the set of degrees of algebraic extensions of finite degree of K is either equal to the set N_P of integers all of whose prime factors are in P or to the subset of N_P consisting of integers not divisible by 4.

16) For every prime number p there exists a Moriya field of characteristic p such that the set of degrees of algebraic extensions of finite degree of this field is the set N^* of integers > 0 [1]. Show that for every set P of prime numbers, there exists a Moriya field K of any characteristic, such that the set of degrees of algebraic extensions of finite degree of K is the set N_P (argue as in Ex. 14, *b*)).

* 17) Let P_n be the set of monic polynomials $X^n + a_1X^{n-1} + \cdots + a_n$ of $\mathbf{Z}[X]$ whose roots in $\mathbf{C}$ all have absolute value $\leqslant 1$.

a) Show that the set P_n is finite. Deduce that for every polynomial $F \in P_n$, if F_h denotes the polynomial whose roots are the h-th powers of the roots of F, then there exist two distinct integers h, k such that $F_h = F_k$.

b) Conclude from *a*) that all the roots of a polynomial $F \in P_n$ are either zero or roots of unity (« Kroneckers's theorem »). *

18) Let $p \in \mathbf{N}$ be a prime number, C_p the cyclotomic extension of level p of $\mathbf{Q}$, $\zeta \in C_p$ a primitive p-th root of unity and n an integer. Write

$$X_n(\zeta) = \sum_{\alpha \in \mathbf{Z}/p\mathbf{Z}} \zeta^{\alpha^n}.$$

a) Calculate $d = [\mathbf{Q}(X_n(\zeta)) : \mathbf{Q}]$. Determine the Galois group of the extension $\mathbf{Q}(X_n(\zeta))$ of $\mathbf{Q}$ and the conjugates of $X_n(\zeta)$.

b) Let $T^d + a_1T^{d-1} + \cdots + a_d$ be the minimal polynomial of $X_n(\zeta)$ over $\mathbf{Q}$. Show that $a_1 = 0$.

c) Suppose that $n > 1$ and $p \equiv 1 \pmod{n}$. Show that $a_2 = -\dfrac{n(n-1)p}{2}$ if $\dfrac{p-1}{n}$ is even and $a_2 = \dfrac{np}{2}$ if $\dfrac{p-1}{n}$ is odd.

d) Suppose that $p \equiv 1 \pmod{n}$. Let $\beta \in \mathbf{F}_p^*$ be an element whose residue class in $\mathbf{F}_p^*/\mathbf{F}_p^{*n}$ is a generator. Let μ be the number of solutions in $\mathbf{F}_p^n$ of the equation $\sum_{i=1}^{n} \beta^{i-1}X_i^n = 0$. Show that $a_n = \dfrac{(-1)^n p(\mu - p^{n-1})}{p-1}$.

e) Calculate a_3 when $n = 3$ and $p = 7$, 13, 19.

* [1] The largest totally ramified extension of the field $\mathbf{Q}_p$ of p-adic numbers in its algebraic closure is a field of characteristic 0 with the same property. *

* 19) *a*) Assume the prime number theorem; if $\pi(x)$ is the number of primes $p \leqslant x$, then $\pi(x) \sim x/\log x$ as x tends to $+\infty$. Deduce that for every odd integer m, there exists an increasing sequence of prime numbers $p_1 < p_2 < \cdots < p_m$ such that $p_m < p_1 + p_2$ (consider the prime numbers between x and $3x/2$).
b) With the finite sequence of prime numbers p_j $(1 \leqslant j \leqslant m)$ determined as in *a*), put $n = p_1 p_2 \ldots p_m$. Show that in the cyclotomic polynomial $\Phi_n(X)$ the terms of degree $< p_m + 1$ are the same as those of the polynomial

$$(1 - X)^{-1} \prod_{j=1}^{m} (1 - X^{p_j}) .$$

Deduce that the coefficient of X^{p_m} in $\Phi_n(X)$ is $1 - m$ (« I. Schur's theorem »). *

* 20) Let p be a prime number and G a commutative p-torsion group (VII, p. 10).
a) Let $G_i \subset G$ be the set of elements $x \in G$ such that for each l the equation $p^l y = x$ has a solution. Show that G_i is a direct factor of G.
b) Suppose that the equation $px = 0$ has only a finite number of solutions in G. Show that for each integer l, the kernel ${}_lG$ of the multiplication by p^l in G is a finite group and that the rank $r(l)$ of the vector $\mathbf{F}_p$-space ${}_lG/{}_{(l-1)}G$ is a decreasing function of l, hence constant for $l > l_0$, l_0 large enough. (Use the theorem of elementary divisors (VII, p. 24). Show that G/G_i is a finite group annihilated by p^{l_0}.)
c) Suppose further that $G = G_i$. Show that the mapping $l \mapsto r(l)$ is constant. Writing $H = (\mathbf{Z}[1/p]/\mathbf{Z})^{r(1)}$, show that for each l there exists an isomorphism ${}_lG \to {}_lH$ and that every isomorphism ${}_lG \to {}_lH$ extends to an isomorphism ${}_{(l+1)}G \to {}_{(l+1)}H$. Deduce that G is isomorphic to H.
d) Under the hypotheses of *b*) show that there exists an integer n and an application $i \to n_i$ of $\mathbf{N} - \{0\}$ into $\mathbf{N}$, with finite support, such that G is isomorphic to $(\mathbf{Z}[1/p]/\mathbf{Z})^n \oplus \bigoplus_i (\mathbf{Z}/p^i\mathbf{Z})^{n_i}$. Show that the integer n and the mapping $i \mapsto n_i$ are uniquely determined by the mapping $l \mapsto \operatorname{Card}({}_lG)$. *

* 21) Let G and H be two commutative torsion groups such that for every integer n, the equation $nx = 0$ has the same *finite* number of solutions in G and in H. Show that G is isomorphic to H. (Decompose G and H into direct sums of p-torsion groups and apply Ex. 20.) Let $G = \bigoplus_{i \in \mathbf{N}} \mathbf{Z}/p^i\mathbf{Z}$ and $H = \bigoplus_{i \in \mathbf{N}} \mathbf{Z}/p^{2i}\mathbf{Z}$. Show that G and H are not isomorphic and that for each integer n the cardinal of the set of solutions of the equation $nx = 0$ is the same in G and in H. *

22) *a*) Let k be a field, f a separable monic polynomial in $k[X]$, D a splitting extension of f, $x_1, \ldots, x_n$ the roots of f in D and G the Galois group of D over k. Assume that G is *cyclic*. Show that for every element of G to induce an *even* permutation of the x_i it is necessary and sufficient that the number of irreducible factors of f should be congruent to the degree of f mod 2 (consider a generator σ of G and the decomposition into cycles of the permutation of the x_i induced by σ).
b) Suppose further that k is of characteristic $\neq 2$. Show that the two above conditions are equivalent to the fact that the discriminant of f should be a square in k (« Stickelberger's theorem »).

23) Let n and p be odd and distinct prime numbers.
a) Let $Q \in \mathbf{F}_p[X]$ be the image of the cyclotomic polynomial Φ_n. Show that the number i of irreducible factors of Q is even if and only if p is a square mod n. (Let K be the splitting field of Q. Show that $\mathrm{Gal}(K/\mathbf{F}_p)$ may be identified with the subgroup of $(\mathbf{F}_n)^*$ generated by the image of p. Study the parity of the index $((\mathbf{F}_n)^* : \mathrm{Gal}(K/\mathbf{F}_p))$.)
b) Show that the discriminant of $X^n - 1 \in \mathbf{F}_p[X]$ is

$$(-1)^{\frac{n(n-1)}{2}+n-1} n^n .$$

c) Show that i is even if and only if $(-1)^{\frac{n(n-1)}{2}} n$ is a square mod p (apply Ex. 22 to the polynomial $X^n - 1$).
d) Put $\left(\frac{n}{p}\right) = 1$ if n is a square mod p and $\left(\frac{n}{p}\right) = -1$ otherwise. Deduce from *c*) that $\left(\frac{n}{p}\right)\left(\frac{p}{n}\right) = (-1)^{\frac{(n-1)(p-1)}{4}}$ (« Law of quadratic reciprocity »).

§ 12

¶ 1) *a*) Let q be a power of a prime number. Show that for every integer $n > 1$, if h_i $(1 \leq i \leq r)$ are the distinct prime divisors of the integer n, then the number of elements $\zeta \in \mathbf{F}_{q^n}$ such that $\mathbf{F}_{q^n} = \mathbf{F}_q(\zeta)$ is equal to

$$\nu = q^n - \sum_i q^{n/h_i} + \sum_{i<j} q^{n/h_i h_j} - \sum_{i<j<k} q^{n/h_i h_j h_k} + \cdots + (-1)^r q^{n/h_1 h_2 \ldots h_r}$$

(observe that such an element is characterized by the property of not belonging to any of the fields $\mathbf{F}_{q^{n/h_i}}$). If $h_1 \leq \cdots \leq h_r$, we have

$$q^n - \sum_i q^{n/h_i} \leq \nu \leq q^n - q^{n/h_1} .$$

Examine the case where n is a prime power.
b) Let $b_n(q)$ be the number of monic irreducible polynomials of degree n over $\mathbf{F}_q$. Deduce from *a*) an expression for $b_n(q)$. In particular, if $q \geq 3$, show that the number of polynomials (not necessarily monic) of degree n in $\mathbf{F}_q[X]$ which are irreducible is at least equal to $q^n(q-2)/n \geq q^{n+1}/3n$.
c) Prove that $\sum_{d \mid n} d b_d(q) = q^n$ (establish the identity $1/(1 - qT) = \prod_n (1 - T^n)^{-b_n(q)}$).
* Using the Möbius inversion formula (*Lie*, II, p. 176, Appendix), show that

$$n b_n(q) = \sum_{d \mid n} \mu(n/d)\, q^d .$$

Compare this result with that obtained in *b*). *
d) Let $a_n(q)$ be the number of monic polynomials without multiple factors in $\mathbf{F}_q[X]$. Show that $a_0(q) = 1$, $a_1(q) = q$, $a_n(q) = q^n - q^{n-1}$ for $n > 1$ (establish the identity $q^n = \sum_i a_{n-2i}(q)\, q^i$).

2) Show that the number of monic polynomials of degree n of $\mathbf{F}_q[X]$ which have no root in $\mathbf{F}_q$ is $q^{n-q}(q-1)^q$ for $n \geqslant q$ (observe that every function $\mathbf{F}_q \to \mathbf{F}_q$ is a polynomial function).

3) Let K be a finite field of q elements ; for every prime number l let N_l be the union of the extensions of K (contained in an algebraic closure Ω of K) whose degree is a power of l. Show that N_l is an abelian extension of K whose Galois group is isomorphic to the group $\mathbf{Z}_l$ of l-adic integers. Deduce that Ω is isomorphic to the tensor product $\bigotimes_l N_l$, and hence, that $\mathrm{Gal}(\Omega/K)$ is isomorphic to the product $\prod_l \mathbf{Z}_l$.

* 4) Decompose the polynomial $X^4 + 1 \in \mathbf{F}_p[X]$ into irreducible factors. *

5) Let K be a finite field of q elements where q is not a power of 2, and let a_1, a_2, b be three elements of K such that $a_1a_2 \neq 0$. Show that the number ν of solutions $(x_1, x_2) \in K^2$ of the equation $a_1x_1^2 + a_2x_2^2 = b$ is given by the following formulae :
1° if $b = 0$ and $-a_1a_2$ is not a square in K, $\nu = 1$;
2° if $b \neq 0$ and $-a_1a_2$ is not a square in K, $\nu = q + 1$;
3° if $b = 0$ and $-a_1a_2$ is a square in K, $\nu = 2q - 1$;
4° if $b \neq 0$ and $-a_1a_2$ is a square in K, $\nu = q - 1$.
(When $-a_1a_2$ is not a square, adjoin to K a root of $X^2 + a_1a_2$ and use the fact that the multiplicative group of a finite field is cyclic.)

6) Let K be a finite field of q elements where q is not a power of 2.
a) Let $a_1, a_2, \dots, a_{2m}, b$ be elements of K such that $a_1a_2 \dots a_{2m} \neq 0$. Show that the number ν of solutions $(x_1, \dots, x_{2m}) \in K^{2m}$ of the equation $a_1x_1^2 + \cdots + a_{2m}x_{2m}^2 = b$ is given by the following formulae :
1° if $b = 0$ and $(-1)^m a_1a_2 \dots a_{2m}$ is not a square in K, $\nu = q^{2m-1} - q^m + q^{m-1}$;
2° if $b \neq 0$ and $(-1)^m a_1a_2 \dots a_{2m}$ is not a square in K, $\nu = q^{2m-1} + q^{m-1}$;
3° if $b = 0$ and $(-1)^m a_1a_2 \dots a_{2m}$ is a square in K, $\nu = q^{2m-1} + q^m - q^{m-1}$;
4° if $b \neq 0$ and $(-1)^m a_1a_2 \dots a_{2m}$ is a square in K, $\nu = q^{2m-1} - q^{m-1}$.
(Argue by induction on m, using Ex. 5.)
b) Let $a_1, a_2, \dots, a_{2m+1}, b$ be elements of K such that $a_1a_2 \dots a_{2m+1} \neq 0$. Show that the number ν of solutions $(x_1, x_2, \dots, x_{2m+1}) \in K^{2m+1}$ of the equation

$$a_1x_1^2 + \cdots + a_{2m+1}x_{2m+1}^2 = b$$

is given by the following formulae :
1° if $b = 0$, $\nu = q^{2m}$;
2° if $b \neq 0$ and $(-1)^m a_1a_2 \dots a_{2m+1}b$ is not a square in K, $\nu = q^{2m} - q^m$;
3° if $b \neq 0$ and $(-1)^m a_1a_2 \dots a_{2m+1}b$ is a square in K, $\nu = q^{2m} + q^m$. (Reduce to the case *a*).)

7) Let K be a finite field, E an algebraic extension of degree n of K and $(\alpha_1, \dots, \alpha_n)$ a normal basis of E over K ; suppose that the cyclic Galois group $\mathrm{Gal}(E/K)$ is generated by the cyclic permutation $\sigma = (\alpha_1\alpha_2 \dots \alpha_n)$. Let τ be an automorphism of the vector K-space E such that for each $x \in E$, $\tau(x)$ is conjugate to x over E (cf. V, p. 154, Ex. 8).
a) Show that if K has at least 3 elements, τ is a K-automorphism of the field E (reduce to the case where $\tau(\alpha_1) = \alpha_1$ and consider $\tau(\alpha_1 + c\alpha_j)$ for $c \in K$).

b) If $K = F_2$, show that for each element $\rho \in \mathrm{Gal}(E/K)$, the relations $\rho(\alpha_r) = \alpha_{r'}$ and $\rho(\alpha_s) = \alpha_{s'}$ imply $r' - r \equiv s' - s \pmod{n}$. Deduce that if $n > 5$, τ is a K-automorphism of the field E. (Note that if $\tau(\alpha_1) = \alpha_1$ and $\tau(\alpha_1 + \alpha_h) = \alpha_1 + \alpha_k$ with $k \neq h$, we must have $h + k = n + 2$; deduce from this that necessarily $\tau(\alpha_2) = \alpha_n$ and $\tau(\alpha_4) = \alpha_4$ or $\tau(\alpha_4) = \alpha_{n-2}$ which leads to a contradiction.)
c) For $K = F_2$ and $3 \leqslant n \leqslant 5$ the following automorphisms of the vector K-space E are not field automorphisms of E :

$$\text{for } n = 3, \tau : (\alpha_1, \alpha_2, \alpha_3) \mapsto (\alpha_1, \alpha_3, \alpha_2)$$
$$\text{for } n = 4, \tau : (\alpha_1, \alpha_2, \alpha_3, \alpha_4) \mapsto (\alpha_1, \alpha_4, \alpha_3, \alpha_2)$$
$$\text{for } n = 5, \tau : (\alpha_1, \alpha_2, \alpha_3, \alpha_4, \alpha_5) \mapsto (\alpha_1, \alpha_5, \alpha_4, \alpha_3, \alpha_2)$$

but are such that $\tau(x)$ is conjugate to x for all $x \in E$.

8) Let K be a finite field of q elements and

$$P(X) = a_0 + a_1 X + \cdots + a_{q-2} X^{q-2}$$

a polynomial of K[X] of degree $q - 2$. Let r be the rank of the matrix

$$A = \begin{pmatrix} a_0 & a_1 & \cdots & a_{q-2} \\ a_1 & a_2 & \cdots & a_0 \\ \cdots & \cdots & \cdots & \cdots \\ a_{q-2} & a_0 & \cdots & a_{q-3} \end{pmatrix}.$$

Show that the number of distinct roots of P in K, other than 0, is $q - 1 - r$. (Consider the product VA, where V is the Vandermonde matrix of elements $x_1, x_2, \ldots, x_{q-1}$ of K^*.)

9) Let K be a finite field of q elements and let $r = q^n$ for an integer $n \geqslant 1$. For every polynomial $P(X) = \sum_{j=0}^{m} a_j X^j$ of K[X], put

$$\hat{P}(X) = \sum_j a_j X^{(r^j - 1)/(r-1)}$$

and

$$\tilde{P}(X) = X\hat{P}(X^{r-1}) = \sum_j a_j X^{r^j}.$$

a) Show that if P, Q are two polynomials in K[X] and if $R = PQ$, then we have $\tilde{R}(X) = \tilde{P}(\tilde{Q}(X))$. Deduce that for two polynomials F, G of K[X] to be such that F divides G it is necessary and sufficient that $\hat{F}$ should divide $\hat{G}$. (To see that the condition is sufficient, consider the Euclidean division of G by F.)
b) Let F be an irreducible polynomial in K[X] and let G be any polynomial in K[X]. If there exists an integer $d \geqslant 1$ such that $\hat{F}(X^d)$ and $\hat{G}(X^d)$ have a common root, show that F divides G. (If H is the GCD of $\hat{F}(X^d)$ and $\hat{G}(X^d)$ consider the set of polynomials $P \in K[X]$ such that H divides $\hat{P}(X^d)$.)
c) Every polynomial $F \neq 0$ of K[X] divides a polynomial of the form $X^N - 1$. Show that if $r - 1 = de$ and if α is a root of $\hat{F}(X^d)$, then α belongs to the extension K of degree nN and

deduce that the degree of every irreducible factor of $\hat{F}(X^d)$ in $K[X]$ divides nN (use *a*)). If further e and dN are relatively prime and if m is the degree of α over the extension of K of degree n, show that m divides N (use the fact that $(r^N - 1)/e \equiv dN \pmod{e}$).

d) With the notation as in *c*), suppose that F is irreducible and that N is the least of the integers h such that F divides $X^h - 1$. If e and dN are relatively prime and if every prime factor of n divides N, show that every irreducible factor of $\hat{F}(X^d)$ is of degree nN. (Note that in virtue of *b*) we then have $m = N$.)

e) Deduce in particular from *d*) that if $P(X) = \sum_j a_j X^j$ is an irreducible polynomial in $K[X]$ and if N is the least of the integers h such that $P(X)$ divides $X^h - 1$, then every irreducible factor of $\sum a_j X^{q^j - 1}$ is of degree N.

10) Let t, n, k be three integers $\geqslant 1$ such that $t \leqslant n$ and n divides $2^k - 1$. Let S be the smallest set of integers such that $[1, t] \subset S \subset [1, n]$ and the condition « $i \in S$, $2i \equiv j \pmod{n}$ and $1 \leqslant j \leqslant n$ » implies $j \in S$. Put $m = \operatorname{Card} S$. Let ξ be an element of order n in $\mathbf{F}_{2^k}^*$ and put $P(X) = \prod_{j \in S} (X - \xi^j)$.

a) Prove that $P(X) \in \mathbf{F}_2[X]$.

b) Show that if $R(X) \in \mathbf{F}_2[X]$ is of degree $< n$ and is a non-zero multiple of $P(X)$, then there are strictly more than t non-zero coefficients.

c) Deduce the construction of a mapping $\varphi : \mathbf{F}_2^n \to \mathbf{F}_2^m$ such that if $u, v \in \mathbf{F}_2^n$ are distinct and have at most t distinct coordinates then $\varphi(u) \equiv \varphi(v)$.

d) Calculate S, m and P when $t = 6$, $n = 15$, $k = 4$.

11) Let f be a monic polynomial in $\mathbf{Z}[X]$, irreducible in $\mathbf{Q}[X]$ and let p be a prime number such that the canonical image $\bar{f}$ of f in $\mathbf{F}_p[X]$ is a polynomial without multiple roots. If Γ (resp. $\bar{\Gamma}$) is the Galois group of the polynomial f (resp. $\bar{f}$) over $\mathbf{Q}$ (resp. $\mathbf{F}_p$), define an isomorphism of a subgroup of Γ onto $\bar{\Gamma}$. Deduce that if $\bar{f} = g_1 g_2 \dots g_r$ where the g_j are irreducible in $\mathbf{F}_p[X]$ and if g_j is of degree n_j (so that f has degree $n = n_1 + n_2 + \cdots + n_r$), then the group Γ, qua subgroup of $\mathfrak{S}_n$ contains a product $\sigma_1 \sigma_2 \dots \sigma_r$ of cycles with disjoint supports, σ_j being of order n_j for $1 \leqslant j \leqslant r$ (cf. I, p. 62).

Generalize to the case where f is not monic, but where its leading coefficient is not divisible by p.

12) Let f be a polynomial of $\mathbf{Z}[X]$ of degree n and let p_1, p_2, p_3 be three distinct prime numbers not dividing the leading coefficient of f; let f_1, f_2, f_3 be the canonical images of f in $\mathbf{F}_{p_1}[X]$, $\mathbf{F}_{p_2}[X]$, $\mathbf{F}_{p_3}[X]$ respectively. Suppose that f_1 is the product of a linear factor and an irreducible factor of degree $n - 1$, f_2 is the product of an irreducible factor of the second degree and irreducible factors of odd degrees, and finally f_3 is irreducible. Show that under these conditions f is irreducible in $\mathbf{Q}[X]$ and that the Galois group of the polynomial f over $\mathbf{Q}$ is isomorphic to the symmetric group $\mathfrak{S}_n$ (« Dedekind's criterion » ; use Ex. 11 as well as I, p. 63, Prop. 9).

13) Let $p_1, p_2, \dots, p_m$ be distinct odd prime numbers and $P = p_1 p_2 \dots p_m$ $(m \geqslant 3)$.

a) Given an integer $n \geqslant 1$, let E be the set of polynomials in $\mathbf{Z}[X]$ of degree $\leqslant n$, whose coefficients lie in the interval $[0, P[$ of $\mathbf{N}$; we thus have $\operatorname{Card}(E) = P^{n+1}$. Let k be a number such that $0 < k < 1/3n$; show that among the polynomials $f \in E$ there are at least $k^m P^{n+1}$ such that for *every* index j such that $1 \leqslant j \leqslant m$, the canonical image f_j of f in $\mathbf{F}_{p_j}[X]$ is of degree n and irreducible (use Ex. 1, *b*)). Suppose moreover that $k < \dfrac{1}{18(n-2)}$ and $n > 2$; arguing in the same way and using Dedekind's criterion (Ex. 12) show that there exists a set of polynomials $f \in E$, of cardinal at least equal to $(1 - 3(1-k)^m) P^{n+1}$ which are of degree n, irreducible and whose Galois group over $\mathbf{Q}$ is isomorphic to $\mathfrak{S}_n$.

b) For every integer N let $L_{n,N}$ be the set of polynomials of $\mathbf{Z}[X]$ of degree $\leqslant n$, whose coefficients belong to the interval $(-N, N)$; we thus have $\mathrm{Card}(L_{n,N}) = (2N+1)^{n+1}$. The integer n being fixed, show that for every ε such that $0 < \varepsilon < 1$, there exists an integer N_0 such that for $N \geqslant N_0$, there are among the polynomials $f \in L_{n,N}$ at least $(1-\varepsilon)(2N+1)^{n+1}$ which are irreducible of degree n, and whose Galois group over $\mathbf{Q}$ is isomorphic to $\mathfrak{S}_n$ (use *a*)).

* 14) Let K be a finite field, *commutative or not*, let Z be its centre, q the number of elements of Z and n the rank $[K:Z]$.

a) If E is a subfield of K containing Z, show that $[E:Z]$ divides n.

b) Let $x \in K$ be an element not belonging to Z; show that the number of distinct conjugates yxy^{-1} of x in K^* is of the form $(q^n - 1)/(q^d - 1)$, where d divides n and is distinct from n (consider in K the set of elements permutable with x, and use *a*)).

c) Deduce from *b*) that $q - 1$ is divisible by the integer $\Phi_n(q)$ (partition the group K^* into conjugacy classes and use the relation (6) of V, p. 82).

d) Show that if $n > 1$, then $\Phi_n(q) > (q-1)^{\varphi(n)}$ (decompose $\Phi_n(X)$ in the field $\mathbf{C}$ of complex numbers). Deduce that we must have $K = Z$, in other words, that K is necessarily commutative (Wedderburn's theorem). *

§ 13

1) Let K be a field of characteristic $p > 0$ and E an extension of K; E is a p-radical extension of height $\leqslant 1$ of $K(E^p)$. The cardinal of a p-basis of E over $K(E^p)$ (cf. V, p. 103, Th. 2) is called the *degree of imperfection* of E over K. If $K(E^p) = E$, E is said to be *relatively perfect* over K; if E is perfect, it is relatively perfect over each of its subfields. The (absolute) *degree of imperfection* of K is the degree of imperfection of K over its prime subfield $\mathbf{F}_p$, or also the cardinal of a (absolute) p-basis of K.

a) If B is a p-basis of E over $K(E^p)$, show that for every integer $k > 0$ we have $E = K(E^{p^k})(B)$.

b) Suppose that $E \subset K^{p^{-n}}$ for an integer $n > 0$. Show that for the degree $[E:K]$ to be finite it is necessary and sufficient that the degree of imperfection m_0 of E over K should be finite; then m_0 is the least cardinal of the sets S such that $E = K(S)$. Show that if m_k is the degree of imperfection of $K(E^{p^k})$ over K, then $m_{k+1} \leqslant m_k$ for all k, and if $f = \sum_k m_k$, then $[E:K] = p^f$.

2) *a*) Let E be an extension of a field K of characteristic $p > 0$ and F an extension of E. Show that if B is a p-basis of E over $K(E^p)$ and C a p-basis of F over $E(F^p)$, then there exists a p-basis of F over $K(F^p)$ contained in $B \cup C$.

b) Suppose that B is finite and that F is an algebraic extension of K of finite degree. Show that $[E : K(E^p)] \geqslant [F : K(F^p)]$. If moreover F is separable over E, show that $[E : K(E^p)] = [F : K(F^p)]$.

3) Let K be an imperfect field and E an algebraic extension of K of finite degree. For an element $x \in E$ to exist such that $E = K(x)$ it is necessary and sufficient that the degree of imperfection of E over K (Ex. 1) should be 0 or 1. (To see that the condition is sufficient, note that if E_s is the relative separable closure of K in E, we have $E = E_s(\alpha)$ and $E_s = K(\beta)$ and if $\alpha \notin E_s$, consider the conjugates of $\alpha + c\beta$ over K, for $c \in K$.)

4) Let K be a field of characteristic $p > 0$ and E an algebraic extension of K of finite degree. If $r > 0$ is the degree of imperfection of E over K (Ex. 1), show that r is the least cardinal of sets S such that $E = K(S)$. (To see that E can be generated by r of its elements note that if E_s is the relative separable closure of K in E, then there exist r elements a_i $(1 \leqslant i \leqslant r)$ of E such that $E = E_s(a_1, \ldots, a_r)$, and use Ex. 3.)

5) Let K be a field and E an algebraic extension of K of finite degree.
a) Suppose that K is of characteristic $p > 0$. Show that if the degree of imperfection of E over K (Ex. 1) is > 1, there exist an infinity of distinct fields F such that $K \subset F \subset E$ (reduce to the case where $K(E^p) = K$, and consider the fields $K(a + \lambda b)$, where the set $\{a, b\}$ is p-free over K and $\lambda \in K$).
b) Conclude that for there to exist only a finite number of fields F such that $K \subset F \subset E$ it is necessary and sufficient that the degree of imperfection of E over K be $\leqslant 1$.

6) Let K be a field of characteristic $p > 0$, E an algebraic extension of K of finite degree and N the quasi-Galois extension of K generated by E.
a) Let E_r and E_s (resp. N_r and N_s) be the largest p-radical extension and the largest separable extension of K contained in E (resp. in N). Show that for $E = E_r(E_s)$ (in which case E is isomorphic to $E_s \otimes_K E_r$) it is necessary and sufficient that $E_r = N_r$.
b) If the degree of imperfection of E over K is 1, then a necessary and sufficient condition for $E = E_r(E_s)$ is that the degree of imperfection of N over K be 1. Deduce that if N is a quasi-Galois extension of K, whose degree of imperfection over K is equal to 1, then $E = E_r(E_s)$ for every subextension E of N (cf. Ex. 2, *b*)).
c) Conversely if N is a quasi-Galois extension of K of finite degree whose degree of imperfection over K is > 1, show that if $N_r \neq N$, then there exists $t \in N$ such that for the extension $E = K(t)$ of K we have $E \neq E_r(E_s)$ (if a, b are two elements of a p-basis of N over $K(N^p)$ and if $x \in N$ is separable over K and $x \notin K$, take $t = a + bx$).

§ 14

1) Let E be an extension of a field K and let B be a transcendence basis of E over K. Show that E is equipotent to $K \times B$ if one of the sets K, B is infinite, and countable otherwise. * Deduce in particular that every transcendence basis of the field **R** of real numbers over the field **Q** of rational numbers has the power of the continuum. *

2) Let K be the field **Q**(X) of rational fractions in one indeterminate over the field **Q** of rational numbers. Show that in the ring K[Y] the polynomial $Y^2 + X^2 + 1$ is irreducible and if E is the extension of K generated by a root of this polynomial (in an algebraic closure of

K), then **Q** is relatively algebraically closed in E but E is not a pure extension of **Q**. (To see that there is in E no element $a \notin \mathbf{Q}$ algebraic over **Q** note that the existence of such an element would imply the relation $E = \mathbf{Q}(a)(X)$, and observe that $-(X^2+1)$ is not a square in $\Omega(X)$, where Ω is an algebraic closure of **Q**.) Show that if i is a root of X^2+1, then $E(i)$ is a pure transcendental extension of $\mathbf{Q}(i)$.

* 3) Let K be the field $\mathbf{C}(X)$ of rational fractions in one indeterminate over the (algebraically closed) field **C** of complex numbers. Show that in the ring K[Y], the polynomial $Y^3 + X^3 + 1$ is irreducible ; let E be the extension of K generated by a root of this polynomial (in an algebraic closure of K). Show that E is not a pure extension of **C**, even though **C** is algebraically closed. (Show that it is impossible for a relation $u^3 + v^3 + w^3 = 0$ to exist between three polynomials u, v, w of $\mathbf{C}[X]$, pairwise relatively prime and not all three constant. Argue by contradiction : if r is the largest of the degrees of u, v, w and if for example $\deg(w) = r$, deduce from the relation

$$w^3 = -(u+v)(u+jv)(u+j^2v)$$

where j is a primitive cube root of unity, the existence of three polynomials u_1, v_1, w_1 in $\mathbf{C}[X]$, pairwise relatively prime, not all three constant, of degrees $< r$ and such that $u_1^3 + v_1^3 + w_1^3 = 0$.) *

4) Let Ω be an algebraically closed extension of a field K, of infinite transcendence degree over K.

a) Show that there exist an infinity of K-endomorphisms of Ω having as images subfields of Ω distinct from Ω, and with respect to which Ω may have a transcendence degree equal to any cardinal up to at most $\operatorname{tr.deg}_K \Omega$. * In particular, there exist an infinity of distinct **Q**-isomorphisms of the field **C** of complex numbers onto subfields of **C** distinct from **C**. *

b) Show that for every subextension E of Ω such that $\operatorname{tr.deg}_K E < \operatorname{tr.deg}_K \Omega$, every K-isomorphism of E onto a subfield of Ω may be extended to a K-automorphism of Ω. Give examples where $\operatorname{tr.deg}_K E = \operatorname{tr.deg}_K \Omega$ and where there exists a K-isomorphism of E onto a subfield of Ω which cannot be extended to a K-isomorphism of any extension F of E, contained in Ω and distinct from E, onto a subfield of Ω.

5) Let Ω be an algebraically closed extension of a field K. Show that for every subextension E of Ω, which is transcendental over K, there exist an infinity of K-isomorphisms of E onto a subfield of Ω (if $x \in E$ is transcendental over K, consider a subextension F of E such that x is transcendental over F and E algebraic over $F(x)$, and show that there exist an infinity of F-isomorphisms of E onto a subfield of Ω).

6) Let Ω be an algebraically closed extension of a field K, and N a subextension of Ω. Show that if N is quasi-Galois over K and if E and E′ are two conjugate extensions of K contained in Ω, then N(E) and N(E′) are conjugate over K. Conversely, if N has this property and if $\operatorname{tr.deg}_K N$ is finite and $< \operatorname{tr.deg}_K \Omega$, then it is necessarily algebraic and quasi-Galois over K.

7) Let E be an extension of a field K and $\operatorname{Aut}_K(E)$ the group of all K-automorphisms of E.

a) For every finitely generated subextension L of E over K, $\operatorname{Aut}_L(E)$ is a subgroup of $\operatorname{Aut}_K(E)$. Show that when L ranges over the set of subextensions of E finitely generated over K, the groups $\operatorname{Aut}_L(E)$ form a fundamental system of neighbourhoods of the neutral

element of $\mathrm{Aut}_K(E)$ for a topology compatible with the group structure of $\mathrm{Aut}_K(E)$ and for which $\mathrm{Aut}_K(E)$ is a separated and totally disconnected group.
b) When E is equipped with the discrete topology, the topology of $\mathrm{Aut}_K(E)$ is the coarsest in which the mapping $(u, x) \to u(x)$ of $\mathrm{Aut}_K(E) \times E$ into E is continuous.
* *c*) The group $\mathrm{Aut}_{\mathbf{Q}}(\mathbf{R})$ is reduced to the neutral element (V, p. 153, Ex. 2 of § 9). *
d) Let $K_0 \supset K$ be the subextension of E consisting of the elements invariant under all K-automorphisms of E. Show that for E to be algebraic over K_0 it is necessary and sufficient that $\mathrm{Aut}_K(E)$ should be compact.
e) Assume that K is infinite. If $E = K(X)$, the field of fractions in one indeterminate over K, show that $\mathrm{Aut}_K(E)$ is an infinite discrete group (V, p. 148, Ex. 11). If K is of characteristic 0, the subgroup Γ of $\mathrm{Aut}_K(E)$ generated by the automorphism $\sigma : X \mapsto X + 1$ is closed in $\mathrm{Aut}_K(E)$ and distinct from the latter, but has the same field of invariants.

8) Let E be an extension of a field K.
a) Let x be an element of $E - K$. Show that there exists a pure extension $L \subset E$ of K such that E is algebraic over L and $x \notin L$ (consider the set of subextensions L of E which are pure transcendental over K and such that $x \notin L$).
b) Let $x \in E$ be a transcendental element over K and p an integer $\geqslant 2$. Show that there exists a pure transcendental extension $L \subset E$ of K such that E is algebraic over L, $x \notin L$ and $x^p \in L$ (same method).
c) Let F be a subextension of E and let $x \in E$ be an algebraic element over F and $P \in F[X]$ its minimal polynomial. Show that there exists a pure extension $L \subset E$ of K such that E is algebraic over L and P is irreducible over the field $F(L)$.

9) An extension E of a field K is called a *Dedekind extension* of K if for *every* subextension F of E, F is identical with the subfield of elements of E invariant under the group $\mathrm{Aut}_F(E)$ of all F-automorphisms of E.
a) For an algebraic extension E of K to be Dedekind over K it is necessary and sufficient that E should be a Galois extension of K.
b) Suppose that K is of characteristic $p > 0$. Show that a Dedekind extension of K is necessarily algebraic over K (hence Galois). (Use Ex. 8, *b*).)
c) If E is a Dedekind extension of K and L a pure transcendental subextension of E, distinct from K and such that E is an algebraic extension of L, show that E has infinite degree over L.

10) Let E be an extension of a field K such that the mapping $F \mapsto \mathrm{Aut}_F(E)$ is a *bijection* of the set of subextensions of E onto the set of subgroups of $\mathrm{Aut}_K(E)$. Show that E is then a Galois extension of K of finite degree. (Using Ex. 9, *c*), show first that E is necessarily algebraic over K, then use Ex. 9, *a*).)

11) Let E be an extension of a field K and Ω an algebraically closed extension of E. Show that the following conditions are equivalent :
α) E is a p-radical extension of K ;
β) for every extension F of K, $E \otimes_K F$ has only a single prime ideal ;
γ) for every extension F of K, any two composite extensions of E and F are isomorphic ;
δ) $E \otimes_K \Omega$ has only a single prime ideal ;
ε) any two composite extensions of E and Ω are isomorphic.

12) *a*) Let K be a field, Ω an algebraically closed extension of K, $E \subset \Omega$ and F two extensions of K ; assume that $\mathrm{tr.deg}_E \Omega \geqslant \mathrm{tr.deg}_K F$. Show that every composite extension of E and F is isomorphic to a composite of the form $(L_w, 1, w)$, where w is a K-isomorphism of F onto a subfield of Ω and L_w is the subfield of Ω generated by E and $w(F)$.
b) Deduce from *a*) that if F is algebraic of finite degree n over K, then there are at most n composite extensions of E and F that are pairwise non-isomorphic.

13) In an algebraic closure Ω of $\mathbf{Q}(X)$ consider two pure transcendental extensions $E = \mathbf{Q}(X)$ and $F = \mathbf{Q}(X + i)$ (where $i^2 = -1$) of the field $\mathbf{Q}$. Show that $E \cap F = \mathbf{Q}$ but that E and F are not algebraically disjoint over $\mathbf{Q}$.

14) Let E, F, G be three extensions of a field K contained in an extension Ω of K and such that $F \subset G$. Show that for E and G to be algebraically disjoint over K it is necessary and sufficient that E and F should be algebraically disjoint over K and E(F) and G algebraically disjoint over F.

15) *a*) Let K be a field and L a subfield of K such that K is a *finitely generated* L-*algebra,* in other words $K = L[a_1, a_2, \ldots, a_n]$ for elements in K. Show that the a_j are *algebraic* over L. (Argue by contradiction : if $a_1, \ldots, a_m$ $(m \geqslant 1)$ form a maximal algebraically free subfamily of $(a_j)_{1 \leqslant j \leqslant n}$, observe that in the ring $A = L[a_1, \ldots, a_m]$ the intersection of all non-zero ideals is reduced to zero, and use Ex. 5 of V, p. 147.)
b) Deduce from *a*) that in the polynomial algebra $K[X_1, \ldots, X_n]$ every maximal ideal is of *finite* codimension over K. Conclude that if A is a commutative finitely generated algebra over K, then every subfield of A containing K is of finite degree over K.

16) *a*) Let K be an algebraically closed field, Ω an algebraically closed extension of K, E a subextension of Ω and x an element of Ω transcendental over E. Let $\overline{K(x)}$ be the relative algebraic closure of $K(x)$ in Ω and let $M = E(\overline{K(x)})$; show that M is an algebraic extension of infinite degree of $E(x)$. (Observe that for every integer $m > 1$ the polynomial $X^m - x$ is irreducible in $E(x)[X]$ (V, p. 91, Remark).
b) Let E be an algebraically closed field and F a finitely generated extension of E. Show that every algebraically closed subfield of F is necessarily contained in E. (Otherwise there would exist a subfield $L \subset F$ which is algebraically closed and an element $x \in L$ which is transcendental over E. Apply *a*), taking for K the relative algebraic closure in E of the prime subfield of E, noting that M should be finitely generated over E.)

17) Let K be a field, Ω an algebraically closed extension of K and x an element of Ω transcendental over K. Show that for every prime number q there exists a subextension M of Ω such that $M(x)$ is a Galois extension of degree q of M. (Consider the element $y = x^q$ if q is not the characteristic of K, $y = x^q - x$ in the opposite case, and a subextension L of Ω generated by a transcendence basis of Ω over K containing y and besides the q-th roots of unity if $y = x^q$; use Ex. 12 of V, p. 163.)

18) Let K be a field and $K((X))$ the field of formal power series in one indeterminate over K (IV, p. 38). Show that $\mathrm{tr.deg}_K K((X)) = \mathrm{Card}(K^{\mathbf{N}})$. Distinguish two cases :
a) If $\mathrm{Card}(K) < \mathrm{Card}(K^{\mathbf{N}})$, note that $\mathrm{Card}(K((X))) = \mathrm{Card}(K^{\mathbf{N}})$.
b) If $\mathrm{Card}(K) = \mathrm{Card}(K^{\mathbf{N}})$, let P be the prime subfield of K, S an infinite set of elements of $K((X))$ which is algebraically independent over K and T the set of coefficients of all the formal power series belonging to S. Let L be the relative algebraic closure in

K((X)) of the field K(S) and u an element of L ; there is an equation $g(s_1, \dots, s_m, u) = 0$ where $g \neq 0$ is a polynomial in $K[X_1, \dots, X_m, X_{m+1}]$ and $s_1, \dots, s_m$ elements of S. Let A be the set of coefficients of the polynomial g and $C(u)$ the set of coefficients of the formal power series u ; show that the field $P(T)(A \cup C(u))$ is algebraic over $P(T \cup A)$ (denoting by Ω an algebraic closure of K, show that otherwise there would exist an infinity of formal power series $v \in \Omega((X))$ satisfying the equation $g(s_1, \dots, s_m, v) = 0$). Using Ex. 1, show that if $\mathrm{Card}(S) < \mathrm{Card}(K)$, then the transcendence degree of K over P(T) is infinite and deduce that in this case L is distinct from K((X)) (if $(t_n)_{n \geqslant 0}$ is an infinite sequence of elements of K algebraically independent over P(T), consider the formal power series $\sum_{n=0}^{\infty} t_n X^n$).

19) Let E and F be two transcendental extensions of a field K, linearly disjoint over K. Show that $K(E \cup F)$ is distinct from the ring C (isomorphic to $E \otimes_K F$) generated by $E \cup F$. (Reduce to the case where the transcendence degrees of E and F over K are equal to 1 ; if $x \in E$ and $y \in F$ are transcendental over K show that we cannot have $(x+y)^{-1} \in C$; arguing by contradiction, show that in the opposite case, if p is the characteristic exponent of K, there would exist an integer $r \geqslant 0$ such that $(x+y)^{-p^r}$ belonged to the subring of C generated by $K(x) \cup K(y)$.)

* 20) Let n be an integer, K a field of characteristic p prime to n, possessing primitive n-th roots of unity, $\mu_n \subset K^*$ the subgroup of n-th roots of unity, G a finite commutative group annihilated by n and $A = K^{(G)}$ the group algebra of G (III, p. 446).

a) Show that A is a diagonalizable algebra (V, p. 28). (Decompose G into a direct sum of cyclic groups (VII, p. 22).) Deduce that every A-module is a direct sum of modules which are vector K-spaces of dimension 1.

b) Let V be an A-module of finite dimension over K, S the symmetric algebra of the vector K-space V and F the field of fractions of S. The group G operates on V, hence on S and F. Assume that G acts *faithfully* on V. Show that G operates faithfully on F. Calculate $[F : F^G]$.

c) Let $(x_1, \dots, x_n)$ be a basis of the vector K-space V consisting of eigenvectors for the action of G. Let Γ be the subgroup of F^* generated by $x_1, \dots, x_n$. Show that Γ is a free commutative group and that the sub-K-algebra B of F generated by Γ may be identified with $K^{(\Gamma)}$ and is stable under the operation of G.

d) For all $x \in \Gamma$, the mapping $\gamma \mapsto \frac{\gamma x}{x}$ is a homomorphism of G into μ_n, giving rise to a homomorphism $\theta : \Gamma \to \mathrm{Hom}(G, \mu_n)$. Show that θ is surjective. Put $\Gamma_1 = \mathrm{Ker}(\theta)$. Using the elementary divisor theorem (VII, p. 24), show that B is a free $K^{(\Gamma_1)}$-module of rank Card(G).

e) By passing to the field of fractions deduce from *d*) and *a*) that F^G is the field of fractions of $K^{(\Gamma_1)}$. Deduce that F^G is a pure transcendental extension of K. *

§ 15

1) Let K be a field of characteristic $p > 0$, E an extension of K and B a p-basis of E over $K(E^p)$ (V, p. 98).

a) Assume that $E \subset K^{p^{-1}}$; let K_0 be a subfield of K such that E is *separable* over K_0. Show that the set B^p is a p-independent subset over $K_0(K^p)$ (observe that if

(a_λ) is a basis of the vector K_0-space K, then (a_λ^p) is a basis of $K_0(K^p)$ over K_0). Let C be a subset of K, disjoint from B^p and such that $B^p \cup C$ is a p-basis of K over $K_0(K^p)$; show that $B \cup C$ is a p-basis of E over $K_0(E^p)$.

b) Assume that $E \subset K^{p^{-n}}$ and that E is separable over a subfield K_0 of K. Show that if the degree of imperfection of K over K_0 is finite (V, p. 170, Ex. 1), then it is equal to the degree of imperfection of E over K_0 (use *a*)).

2) Let K be a field of characteristic $p > 0$, E an extension of K and F a *separable* extension of E. Show that any two of the following three properties imply the third :

a) B is a p-basis of E over $K(E^p)$;

b) C is a p-basis of F over $E(F^p)$;

c) $B \subset E$, $B \cup C$ is a p-basis of F over $K(F^p)$ and $B \cap C = \varnothing$. (Use the fact that if (c_μ) is a basis of the vector E-space F, then (c_μ^p) is a basis of $K(E^p)[F^p]$ over $K(E^p)$ and of $E[F^p]$ over E.)

3) Let K be a field of characteristic $p > 0$, E an extension of K, and F a finitely generated extension of E. Show that the degree of imperfection of F over K is at least equal to that of E over K (reduce to the following two cases : 1° F is separable over E ; 2° $F = E(x)$, where $x^p \in E$ (cf. V, p. 170, Ex. 2)).

4) Let F be a separable extension of a field K of characteristic $p > 0$. Show that if E is a subextension of F, relatively perfect over K (V, p. 170, Ex. 1), F is separable over E.

5) Let K be a field of characteristic $p > 0$. Show that if E is an algebraic extension of K or a relatively perfect extension of K, then $E^{p^{-\infty}} = E(K^{p^{-\infty}})$.

6) Let K be a field of characteristic $p > 0$, E a *separable* extension of K and B a p-basis of E over $K(E^p)$.

a) Show that B is algebraically free over K (consider an algebraic relation of least degree between the elements of B and put the degrees of the variables which occur in the form $kp + h$ with $0 \leqslant h \leqslant p - 1$). Deduce that if $\operatorname{tr.deg}_K E < +\infty$, then the degree of imperfection of E over K is at most equal to $\operatorname{tr.deg}_K E$.

b) Show that E is separable and relatively perfect over K(B).

7) Show that a relatively perfect transcendental extension E of a field K of characteristic $p > 0$ is not finitely generated over K. (In the opposite case show that for every transcendence basis B of E over K, E would be algebraic and separable over K(B).)

8) If E is a separable algebraic extension of a field K of characteristic $p > 0$, then the largest perfect subextension E^{p^∞} of E (the intersection of the fields E^{p^n} for $n \geqslant 0$) is algebraic over the largest perfect subfield K^{p^∞} of K. (If $x \in E^{p^\infty}$, show that for every integer $n > 0$ we have $x^{p^{-n}} \in K(x)$ and deduce that the minimal polynomial of x over K^{p^n} is the same as its minimal polynomial over K.)

9) Let E be a field, G a group of automorphisms of E and K the subfield of E consisting of the elements invariant under G.

a) Show that for every subfield L of E which is stable for the elements of G, L and K are linearly disjoint over $L \cap K$. (Consider a linear relation $\sum_i \lambda_i x_i = 0$ between the elements

$x_j \in K$ which are linearly independent over $L \cap K$, with $\lambda_j \in L$ not all zero, the number of $\lambda_j \neq 0$ being the least possible.)

b) If $L \cap K$ is the field of invariants of a group of automorphisms of K, show that L is the field of invariants of a group of automorphisms of L(K).

10) Let E be an extension of a field K, let G be a compact subgroup of the topological group $\mathrm{Aut}_K(E)$ and let F be the field of invariants of G. Show that E is a Galois extension of F and that G is canonically isomorphic to the topological group $\mathrm{Aut}_F(E)$. (Observe that G operates in a continuous fashion on the discrete space E to establish that an element $x \in E$ has a finite orbit under the action of G.)

11) Let K be a field of characteristic $p > 0$. For every commutative K-algebra A, raising to the p-th power makes the ring A into an A-algebra, hence also a K-algebra which is denoted by $A^{p^{-1}}$. Let A be a commutative K-algebra. Show that the following conditions are equivalent :

(i) A is a separable K-algebra.

(ii) The unique K-homomorphism $\Phi : A \otimes_K K^{p^{-1}} \to A^{p^{-1}}$ such that $\Phi(a \otimes \lambda) = a^p . \lambda$ $(a \in A, \lambda \in K)$ is injective.

(iii) For every family $(b_i)_{i \in I}$ of K which is linearly free over K^p and every family $(a_i)_{i \in I}$ of A the equality $\sum_i a_i^p b_i = 0$ implies that $a_i = 0$ for all i.

(It may be verified that (ii) $\Leftrightarrow$ (iii) and that (ii) is a consequence of V, p. 123, Th. 2, *e*).)

12) Let K be a field and $(X_i)_{i \in I}$ a family of indeterminates. Show that the algebra of formal power series $K[[(X_i)_{i \in I}]]$ is a separable K-algebra. (Apply Ex. 11 ; it may also be verified that for every *finite* extension K′ of K, $K[[(X_i)_{i \in I}]] \otimes_K K'$ is isomorphic to $K'[[(X_i)_{i \in I}]]$ and then Th. 2, *d*) of V, p. 123 applied.) Deduce that the field of fractions of $K[[(X_i)_{i \in I}]]$ is a separable extension of K (V, p. 120, Prop. 4).

§ 16

1) Let K be a field of characteristic $p > 0$, E a *separable* extension of K, of *finite* transcendence degree over K.

a) If there exists a transcendence basis B_0 of E over K and an integer $m \geq 0$ such that $K(E^{p^m})$ is separable over $K(B_0)$, show that for every transcendence basis B of E over K there exists an integer $n \geq 0$ such that $K(E^{p^n})$ is separable over K(B).

b) Deduce that if the condition in *a*) holds, E admits a separating transcendence basis over K (if S is a p-basis of E over $K(E^p)$ and B a transcendence basis of E over K containing S (V, p. 176, Ex. 6), show that B is a separating transcendence basis of E over K, using Ex. 6 of V, p. 176).

c) Let K be a field of characteristic $p > 0$ and x an element transcendental over K (in an algebraically closed extension Ω of K). Show that the union E of the extensions $K(x^{p^{-n}})$ of K is a separable extension of K such that $\mathrm{tr.deg}_K E = 1$ but which does not admit a separating transcendence basis.

d) Let K be a perfect field. Show that the perfect closure of the field of rational fractions K(T) is an extension of K of transcendence degree 1 which does not admit a separating transcendence basis.

2) Let K be field of characteristic $p > 0$ and E an extension of *finite* transcendence degree over K. For E to admit a separating transcendence basis over K it is necessary and sufficient that E should be separable over K and the degree of imperfection of E over K equal to its transcendence degree over K.

3) Let E be an extension of a field K and F an extension of E.
a) If E admits a separating transcendence basis over K and if F admits a separating transcendence basis over E, then F admits a separating transcendence basis over K.
b) If F admits a separating transcendence basis over K and if $\operatorname{tr.deg}_K E < +\infty$, then E admits a separating transcendence basis over K (reduce to the case where $\operatorname{tr.deg}_K F < +\infty$ and apply Ex. 1).
c) If F admits a finite separating transcendence basis over K and is separable over E then F admits a separating transcendence basis over E (use Ex. 1).

4) Let K be a field of characteristic $p > 0$ whose absolute degree of imperfection (V, p. 170, Ex. 1) is equal to 1. Show that for an extension E of K to be separable over K it is necessary and sufficient that E should contain no element p-radical over K and not belonging to K (if $a \in K$ forms a p-basis of K over K^p show that $K^{p^{-1}} = K(a^{p^{-1}})$ and E are linearly disjoint over K).

5) *a*) Let K be a field of characteristic $p > 0$ and E an extension of K admitting a separating transcendence basis over K. Show that the intersection L of the fields $K(E^{p^n})$, where n runs over $\mathbf{N}$, is an algebraic extension of K. (Consider first the case where $\operatorname{tr.deg}_K E < +\infty$; show that L is relatively perfect over K, hence E is separable over L. Now use Ex. 2 and Ex. 6 of V, p. 176. In the general case, if B is a separating transcendence basis of E over K, for each $x \in L$, there is a finite subset B_0 of B such that the minimal polynomial of x over each of $K(B_0^{p^n})$ is the same as its minimal polynomial over $K(B)$; now deduce that if $F = K(B_0)(x)$, then $x \in K(F^{p^n})$ for all n.)
b) Show that the largest perfect field E^{p^∞} (intersection of the E^{p^n}) contained in E is an algebraic extension of K^{p^∞} (use *a*) and Ex. 8 of V, p. 176). Give an example where E is a finitely generated extension of a perfect field K but E^{p^∞} is not contained in K.

6) Let K be a field of characteristic $p > 0$, E an extension of K such that there exists an integer $h \geq 0$ for which $K(E^{p^h})$ is separable over K (which is always the case when E is a finitely generated extension of K).
a) If C is a p-basis of E over $K(E^p)$, there is a subset B of C such that B^{p^h} is a p-basis of $K(E^{p^h})$ over $K(E^{p^{h+1}})$; B is an algebraically free set over K. Show that the extension $F = K(E^{p^h})(B)$ is separable over K and E is an algebraic extension of F.
b) Show that B is a p-basis of F over $K(F^p)$. Show that if $x \in E$ and if m is the least integer such that $x^{p^m} \in F$, then $x^{p^m} \in K(F^{p^m})$; therefore E is a subfield of the field $K^{p^{-\infty}}(F)$ (isomorphic to $K^{p^{-\infty}} \otimes_K K$). A subextension F of E is said to be *distinguished* if it is separable over K and if E is a subfield of $K^{p^{-\infty}}(F)$; F is then a *maximal* separable extension of K in E.
c) Let P be a perfect field of characteristic $p > 0$, $K = P(X, Y)$ the field of rational fractions in two indeterminates and Ω an algebraically closed extension of the field of

rational fractions K(Z) over K. If $u \in \Omega$ is such that $u^p = X + YZ^p$, show that $E = K(Z, u)$ is an extension of K such that K(Z) and K(u) are two distinguished subextensions of E, so that there exists no *larger* separable subextension of E. If $v \in \Omega$ is such that $v^{p^2} = X + YZ^p$, show that in the extension $E' = K(Z, v)$, K(Z) is a maximal separable subextension of E but is not a distinguished subextension.

7) Assume that the hypotheses of Ex. 6 hold, and further that the degree of imperfection of E over K (V, p. 170, Ex. 1) is *finite* (which is always the case if E is a finitely generated extension of K).

a) Let B be a p-basis of F over $K(F^p)$ which is also a separating transcendence basis of F over K (and a transcendence basis of E over K). Show that for every integer $k > 0$, B^{p^k} is a p-independent subset of $K(E^{p^k})$ over $K(E^{p^{k+1}})$. Let $q = \text{tr.deg}_K E$; if $m_k + q$ is the degree of imperfection of $K(E^{p^k})$ over $K(E^{p^{k+1}})$, show that p^f, where $f = \sum_{k=0}^{h-1} m_k$, is equal to the degree $[E:F]$ and is therefore independent of the choice of the distinguished subextension F of E. The integer p^f is called the *order of inseparability* of E over K and is written $[E:K]_i$; this number coincides with the inseparable degree denoted by $[E:K]_i$ in V, p. 46 when E is of finite degree.

b) For $0 \leqslant k \leqslant h-1$, $[K^{p^{-k}}(E) : K^{p^{-k}}]_i$ is equal to p^{f_k}, where $f_k = \sum_{j=k}^{h-1} m_j$.

c) Let L be a subextension of E, separable over K and such that E is p-radical and of finite degree over L ; show that $[E:L] \geqslant [E:K]_i$ (if $[L(E^{p^k}) : K(E^{p^k})] = p^{r_k}$, show that $r_{k+1} \leqslant r_k + q$, by considering a basis of $L(E^{p^k})$ over $K(E^{p^k})$ and a basis of L over $K(L^p)$).

d) Let K′ be an extension of K such that K′ and E are contained in the same algebraically closed extension Ω of K. Show that if K′(F) is separable over K′, it is a distinguished subextension of the extension K′(E) of K′. This is so when K′ and E are algebraically disjoint over K and then we have $[K'(E) : K']_i \leqslant [E:K]_i$; equality holds when K′ and E are linearly disjoint over K or when K′ is a separable algebraic extension of K.

e) Show that $[E:K]_i$ is the least value of $[E:K(B)]_i$ for all transcendence bases B of E over K ; if $\bar{K}$ is the relative algebraic closure of K in Ω (which is an algebraic closure of K), then $[E:K(B)]_i = [E:K]_i [\bar{K}(E) : \bar{K}(B)]_i$. If L is a subextension of E such that E is an algebraic extension of L and $\bar{K}(E)$ a separable (algebraic) extension of $\bar{K}(L)$, then the largest separable extension F of L contained in E is distinguished.

f) Suppose that E is a finitely generated extension of K ; show that for every subextension L of E we have $[L:K]_i \leqslant [E:K]_i \leqslant [L:K]_i [E:L]_i$. Give an example where

$$[L:K]_i = [E:K]_i \quad \text{and} \quad [E:L]_i > 1 .$$

8) Let k be a field and L an extension of k. Show that the following conditions are equivalent :

(i) L is a separable algebraic extension of a finitely generated pure transcendental extension.

(ii) L is a separable extension of k and

$$[\Omega_{L/k} : L] = \text{tr.deg}_k L < +\infty .$$

(To prove (ii) ⇒ (i) we may proceed as follows: let $x_1, \dots, x_n$ in L be such that $dx_1, \dots, dx_n$ form a basis of $\Omega_k(L)$ and put $K = k(x_1, \dots, x_n)$. One first shows that $\text{tr.deg}_k(K) = n$ and hence that K is a pure transcendental extension of k. Let L′ be a finite algebraic extension of K contained in L. By comparing the canonical mapping $u : \Omega_k(K) \otimes_K L' \to \Omega_k(L')$ with the canonical mapping $\Omega_k(K) \otimes_K L \to \Omega_k(L)$, show that u is injective. Calculate $[\Omega_k(L') : L']$ (V, p. 134, Cor. 1). Deduce that u is bijective and hence that $\Omega_K(L') = 0$. Show that L′ is separable over K ; deduce that L is separable over K.)

9) With the notation of Ex. 8 let $M \subset L$ be a subextension. Put $m = \text{tr.deg}_k M$, $n = \text{tr.deg}_k L$, $r = n - m$. Let $x_1, \dots, x_n$ in L be such that L is a separable algebraic extension of $k(x_1, \dots, x_n) = K$.

a) Show that, after renumbering the x_i if necessary, we may suppose that L is algebraic over $M(x_1, \dots, x_r)$.

b) Show that there exists a finitely generated subextension $N \subset M$ such that:

α) L is algebraic over $N(x_1, \dots, x_s)$;

β) $K' = K(N(x_1, \dots, x_r))$ is a finite extension of $N(x_1, \dots, x_r)$;

γ) the canonical homomorphism $K' \otimes_{N(x_1, \dots, x_r)} M(x_1, \dots, x_r) \to K(M(x_1, \dots, x_r)) = K''$ is bijective.

c) Let N be a subextension as in *b*). Show successively that K″ is a separable algebraic extension of K′ (because L is separable over K), that $M(x_1, \dots, x_r)$ is a separable algebraic extension of $N(x_1, \dots, x_r)$ (use *b*), γ)) and that M is a separable algebraic extension of N.

d) Show that M is a separable algebraic extension of a pure transcendental extension (use *c*) and the fact that N is a finitely generated separable extension of k).

§ 17

1) Let K_0 be a field of characteristic $p > 0$, $K = K_0(X, Y)$ the field of rational fractions in two indeterminates over K_0, U and V two indeterminates, Ω an algebraically closed extension of K(U, V), $E = K(U, u)$, where $u^p = X + YU^p$ and $F = K(V, v)$, where $v^p = X + YV^p$; E and F are not separable over K (V, p. 178, Ex. 6, *c*)).

a) Show that K is relatively algebraically closed in E and in F (if $x \in E$ is algebraic over K, show that we must have $x^p \in K$; if $x \notin K$, then $E = K(U, x)$, deduce that then $X^{1/p}$ and $Y^{1/p}$ would lie in $K(x)$).

b) Show that E and F are linearly disjoint over K but that K is not relatively algebraically closed in K(E ∪ F). (Prove that $X^{1/p} \in K(E \cup F)$; deduce that v cannot belong to E(V), and conclude from this that E and F are linearly disjoint over K.)

2) Let K be a field of characteristic $p > 0$, Ω an extension of K and E, F two subextensions of Ω, algebraically disjoint over K.

a) Let F_s be the relative separable closure of K in F. Show that $E(F_s)$ is the relative separable closure of E in E(F). (Reduce to the case where $F_s = K$; let B be a transcendence basis of E over K and K′ the relative algebraic closure of K in F, which is radical over F ; then (V, p. 137) K′(B) is the relative algebraic closure of K(B) in F(B) and is p-radical over K(B). If $x \in E(F)$ is algebraic over E, there exists an integer $r \geqslant 0$ such that $x^{p^r} \in M$, where M is a separable extension of finite degree of F(B) having a basis $(u_i)_{1 \leqslant i \leqslant n}$ over F(B) consisting of elements of E. If $y = \sum_i b_i u_i$ with $b_i \in F(B)$, show by considering the F(B)-isomorphisms of M into an algebraic closure of E(F) that the b_i belong to K′(B) and deduce that y is p-radical over E.)

b) Suppose that E and F are linearly disjoint over K, that K is relatively algebraically closed in F and that E is a separable extension of K. Show that E is relatively algebraically closed in E(F). (Reduce to the case where E is a separable algebraic extension of K, with the help of V, p. 138. Deduce from *a*) that the relative algebraic closure of E in E(F) is p-radical over E and then use the fact that if (a_λ) is a basis of E over K, the $a_\lambda^{p^r}$ also form a basis of E over K.)

3) Let E be a field, G a group of automorphisms of E and K the field of invariants of G.
a) Show that if there exists no normal subgroup of finite index in G other than G, then E is a regular extension of K.
b) Deduce from *a*) that if G is isomorphic to a finite product of infinite simple groups, then E is a regular extension of K.

4) Let K be a field and E, F two extensions of K.
a) Suppose that K is relatively algebraically closed in F. Show that if (L, u, v), (L', u', v') are two composite extensions of E and F such that $u(E)$ and $v(F)$ (resp $u'(E)$ and $v'(F)$) are algebraically disjoint over K, then these two composite extensions are isomorphic. (Consider first the case where E is a pure transcendental extension of K ; then use Ex. 2, *a*), as well as Ex. 6, *b*) of V, p. 154)
b) Let E_s, F_s be the relative separable closures of K in E and F respectively. Suppose that all the composite extensions of E_s and F_s are isomorphic. Show that if (L, u, v) and (L', u', v') are two composite extensions of E and F such that $u(E)$ and $v(F)$ (resp. $u'(E)$ and $v'(F)$) are algebraically disjoint over K, then these two composite extensions are isomorphic (use *a*) and Ex. 11 of V, p. 173).

5) Let K be a field and p its characteristic exponent. For every integer $n > 0$ and prime to p, denote by $\varphi_K(n)$ the degree over K of $R_n(K) = K(\mu_n)$, where μ_n denotes the group of n-th roots of unity in an algebraic closure of K. For example $\varphi_Q(n) = \varphi(n)$ (V, p. 84, Th. 2) ; if K is finite with q elements, $\varphi_K(n)$ is the order of the residue class of q in $(\mathbf{Z}/n\mathbf{Z})^*$ (V, p. 97, Cor.) ; if K is algebraically closed, $\varphi_K(n) = 1$ for all n.
a) Let E be the relative algebraic closure of the prime field K_0 in K. Show that $\varphi_K(n) = \varphi_E(n)$.
b) Suppose that K is a finitely generated extension of K_0. Let m be an integer > 0 ; then there exists an integer $N > 0$ divisible by every n such that $\varphi_K(n) \leqslant m$.
c) Let s be an element of $GL_m(K)$ of finite order. Show that s^N has order a power of p. (We have $s = tu = ut$, where u has order a power of p and t has order n prime to p. The element t is root of a separable polynomial $X^n - 1$ as well as of its characteristic polynomial ; deduce that the subalgebra A of $M_m(K)$ generated by t is etale, of degree $\leqslant m$, isomorphic to a product of extensions $R_d(K)$, where d runs over certain divisors of n including n itself. Therefore $\varphi_K(n) \leqslant m$; use *b*).)

Historical Note

(Chapters IV and V)

(Numbers in brackets refer to the bibliography at the end of this Note.)

The theory of commutative fields — and the theory of polynomials which is closely related to it — derives directly from what constituted until the middle of the 19th century the principal object of classical Algebra : the solution of algebraic equations, and of the problems of geometrical constructions equivalent to them.

Once one seeks to solve an algebraic equation of degree > 1, one finds oneself in the presence of quite new computational difficulties, as it is no longer possible to determine the unknown by performing « rational » calculations on the data. This difficulty must have been perceived at a very early stage ; it must be reckoned as one of the most important contributions of the Babylonians that they were able to reduce the solution of quadratic and biquadratic equations to a single new algebraic operation, the extraction of square roots (as is shown by the many numerical equations which have been found solved in this way in the texts that have come down to us ([1], p. 183-193)). As far as the formal calculation is concerned, Antiquity never passes beyond this point in the problem of the solution of algebraic equations ; the Greeks of the classical epoch in effect confine themselves to restating the Babylonian formulae in geometrical terms, and their use in algebraic form is not documented before Hero (100 AD) and Diophantus.

It is in quite a different direction that the Greeks made decisive progress. We have little information about the way in which the Babylonians conceived and calculated the square roots of integers which were not perfect squares * ; in the rare texts on this question which have come down to us, they seem to have contented themselves with fairly coarse methods of approximation ([1], p. 33-38). The Pythagorean School, which had quite rigorously determined the concept of commensurable magnitudes and attached to it a quasi-religious character, was not able to keep to this point of view ; possibly it was the failure of repeated attempts to express $\sqrt{2}$ rationally that led them finally to demonstrate that this number is irrational **.

* In all the examples of quadratic and biquadratic equations in Babylonian texts the data are always so chosen that the radicals apply to perfect squares.

** A recent author has made the ingenious remark that the construction of the regular pentagram, known to the Pythagoreans (for whom it was one of their mystic symbols),

We have said elsewhere (cf. Hist. Note to Book III, Chapter IV) how this discovery, which marks a watershed in the history of Mathematics, reacted profoundly on the conception of « number » among the Greeks, and led them to create an algebra of exclusively geometrical character, in order to find a mode of representation (or perhaps an « existence » proof) for the incommensurable ratios, which they refused to consider as numbers. Most frequently an algebraic problem was reduced by them to the intersection of two auxiliary plane curves, appropriately chosen, or to several successive determinations of such intersections. Late traditions of little authority ascribe to Plato the introduction of a first classification of these constructions, destined to a long and brilliant career : for reasons that are philosophical rather than mathematical, it seems, that what are known as « ruler-and-compass-constructions », that is those where the only auxiliary curves occurring are straight lines and circles, have been put on one side *. In any case Euclid in his Elements [2] limits himself exclusively to treating problems that can be solved in this fashion (without however characterizing them by a particular name) ; a circumstance which undoubtedly contributed no little to fix the attention of mathematicians of succeeding centuries on these problems. But we know today ** that the algebraic equations that may be solved « by ruler and compasses » are of a very special type ; in particular, an irreducible equation

leads immediately to a demonstration of the irrationality of $\sqrt{5}$, and he has put forward the hypothesis (which unfortunately is not supported by any text) that it was in this fashion that the Pythagoreans discovered irrational numbers (K. von FRITZ, *Ann. of. Math.*, vol. 46 (1945), p. 242).

* In connection with this principle one also attributes to Plato the classification of plane curves into « plane loci » (straight line and circle), « solid loci » (the conics, obtained by plane section of a solid body, the cone), all the other curves being grouped together under the name « τόποι γραμμικοι ». It is curious to see the influence of this classification extending even to Descartes, who in his Geometry arranges in the same « genus » the equations of degree $2n - 1$ and $2n$, no doubt because those of degree 1 or 2 are solved by intersections of « plane loci » and those of degree 3 or 4 by intersections of « solid loci ».

** The determination of the points of intersection of a straight line and a circle (or of two circles) is equivalent to finding the solution of an equation of the second degree whose coefficients are rational functions of the coefficients of the equations of the straight line and circle (or the two circles) considered. It follows easily that the coordinates of a point constructed « by ruler and compasses » from given points belong to an extension L of the field **Q** of rational numbers, obtained as follows : if K is the field obtained by adjoining to **Q** the coordinates of the given points, then there exists an increasing sequence $(L_i)_{0 \leqslant i \leqslant n}$ of fields intermediate between K and L, satisfying the conditions $K = L_0$, $L = L_n$, $[L_i : L_{i-1}] = 2$ for $1 \leqslant i \leqslant n$. By induction on n we deduce that the degree over K of the Galois extension N generated by L is a *power of* 2 ; conversely it may be shown that if this condition holds, then there exists a sequence (L_i) of fields intermediate between K and L with the above properties, and therefore the problem posed can be solved by ruler and compasses (cf. N. TSCHEBOTARÖW, *Grundzüge der Galoisschen Theorie* (transl. H. Schwerdtfeger), Groningen (P. Noordhoff), 1950, p. 351).

of the 3rd degree (over the field of rational numbers) cannot be solved in this way, and the Greeks had encountered quite early problems of the 3rd degree which have remained famous, such as the duplication of the cube (solution of $x^3 = 2$) and the trisection of an angle ; on the other hand the quadrature of the circle placed them in the presence of a transcendental problem. We find that to solve these problems they introduce numerous curves, both algebraic (conics, cissoid of Diocles, conchoid of Nicomedes) and transcendental (quadratrix of Dinostratus, Archimedes' spiral) ; all this inevitably led them to conduct an independent study of these various curves, thus paving the way to the future developments of Analytic Geometry, Algebraic Geometry and of Infinitesimal Calculus. But these methods contribute no progress to the solution of algebraic equations *, and the only work of Antiquity which has made a notable contribution to this question and exerted a lasting influence on the algebraists of the Middle Ages and the Renaissance is Book X of Euclid's Elements [2] ; in this book (of which the principal results are ascribed by certain historians to Theaetetus), he considers the expressions obtained by the combination of several radicals such as $\sqrt{\sqrt{a} \pm \sqrt{b}}$ (a and b rational), gives conditions under which these expressions are irrational, classifies them in numerous categories (which he shows to be distinct) and studies the algebraic relations between these various irrationals, such as those which we would write today as

$$\sqrt{\sqrt{p} + \sqrt{q}} = \sqrt{\frac{1}{2}(\sqrt{p} + \sqrt{p-q})} + \sqrt{\frac{1}{2}(\sqrt{p} - \sqrt{p-q})} ;$$

all expressed in the usual geometric language of the *Elements,* which makes the exposition particularly dense and awkward.

After the decline of classical Greek Mathematics the notions relating to algebraic equations underwent modification. There can be no doubt that during the whole of the classical period the Greeks possessed methods of indefinite approximation of square roots, about which we are unfortunately ill informed **.

* In the absence of a manageable algebraic method of calculation, one finds no trace among the Greeks of any attempt at a classification of the problems they were unable to solve by ruler and compasses ; the Arabs are the first to reduce numerous problems of this kind (for example the construction of regular polygons of 7 or 9 sides) to equations of the 3rd degree.

** For example, Archimedes' method for the approximate calculation of the number π necessitates a knowledge of several square roots to a fairly good approximation, but we are ignorant of the procedure employed by Archimedes to obtain these values. The method of approximation of $\sqrt{2}$ which provides the expansion of this number as « continued fraction » is known (in geometric form) from a text by Theon of Smyrna (2nd century AD), but perhaps goes back to the first Pythagoreans. As regards the method of indefinite approximation of square roots which is still in use today in elementary teaching, this is not documented before Theon of Alexandria (4th century AD), even though it was no doubt known to Ptolemy. We note finally that an approximate calculation of a cube root is found in Hero (ca. 100 AD) (cf. G. ENESTRÖM, *Bibl. Math.* (3), vol. VIII (1907), p. 412).

With the Hindus, later the Arabs and their Western emulators of the Middle Ages, the extraction of roots of all orders becomes an operation which tends to be considered as fundamental with the same right as the rational operations of algebra, and to be denoted, like the latter, by symbols which are more and more easily handled in calculations *. The theory of the equation of the second degree which was perfected during the Middle Ages (number of roots, negative roots, case of impossibility, double root), and that of biquadratic equations, provided models of formulae for the solution of equations « by radicals », on which algebraists will attempt for centuries to model analogous formulae for the solution of equations of higher degree, in the first instance for the equation of the 3rd degree. Leonardo of Pisa, who was chiefly responsible for introducing Arab science to the West in the 13th century, recognized that in any case the irrationals classified by Euclid in his tenth book cannot serve for this purpose (a new impossibility proof, in a theory which abounds with them), and we see him trying already the analogous calculations for cube roots, obtaining relations such as

$$\sqrt[3]{16} + \sqrt[3]{54} = \sqrt[3]{250},$$

analogous to Euclid's formulae for square roots (and of which we actually find earlier examples among the Arabs). But three centuries of fruitless efforts are still to pass before Scipio del Ferro, at the beginning of the 16th century, finally arrives at the formula for the solution of the equation $x^3 + ax = b$:

$$(1) \qquad x = \sqrt[3]{\frac{b}{2} + \sqrt{\left(\frac{b}{2}\right)^2 + \left(\frac{a}{3}\right)^3}} + \sqrt[3]{\frac{b}{2} - \sqrt{\left(\frac{b}{2}\right)^2 + \left(\frac{a}{3}\right)^3}}\,.$$

We refrain from describing here the picturesque side of this sensational discovery — the quarrels it provoked between Tartaglia on the one hand and Cardano and his school on the other — as well as the often engaging figures of the scholars who were the protagonists. But we must note the decisive progress ensuing in the theory of equations at the hands of Cardano and his pupils. Cardano, who has less aversion than the majority of his contemporaries to the use of negative numbers, observes thus that the equations of third degree may have three roots, and the biquadratic equations four ([3] vol. IV, p. 259), and he remarks that the sum of the three roots of $x^3 + bx = ax^2 + c$ (an equation in which he knows how to make the term in x^2 vanish) is always equal to a (*ibid.*). Guided no doubt by this relation and the intuition of its general character, he has the first idea of the notion of the multiplicity of a root ; above all he is emboldened (not without oratory precautions), to calculate formally with expressions containing

* The irrationality of $\sqrt[n]{a}$ when a is an integer which is not a perfect n-th power, is neither mentioned nor demonstrated before Stifel (16th century) ; the proof by the latter is modelled on that of Euclid for $n = 2$ and it is very unlikely that this easy generalization was not noticed earlier.

square roots of negative numbers. It is probable that he was led to this by the fact that such expressions arise naturally in the use of the formula (1) when $\left(\frac{b}{2}\right)^2 + \left(\frac{a}{3}\right)^3 < 0$ (the so-called « irreducible case », in which Cardano recognized the existence of three real roots) ; in any case this is brought out clearly by his disciple R. Bombelli who in his Algebra ([4], p. 293) proves the relation

$$\sqrt[3]{2 + \sqrt{-121}} = 2 + \sqrt{-1}$$

and who takes care to give explicitly the rules of calculating with complex numbers in a form which is already very close to modern expositions *. Finally in 1545 another student of Cardano, L. Ferrari, succeeds in solving the general equation of the 4th degree, by means of an auxiliary equation of the 3rd degree **.

After such rapid progress, the period that followed, until the middle of the 18th century, barely develops the new ideas introduced by the Italian School. Thanks to the essential progress which he brings to algebraic notation, Vieta is able to express in general fashion the relations between the coefficients and the roots of an algebraic equation, at least when the roots are all positive *** ([5], p. 158). More daring, A. Girard [6] does not hesitate to affirm (of course without proof) that an equation of degree n has exactly n roots, provided that the « impossible roots » are counted, each with its degree of multiplicity, and that these roots satisfy the relations given by Vieta ; he also obtains, for the first time, the expression for the sums of equal powers of the roots, up to exponent 4.

But the spirit of the 17th century is turned towards other directions and it is only indirectly that Algebra reaps some benefit from the new discoveries of Analytic Geometry and Infinitesimal Calculus. Thus Descartes' method of obtaining the

* Bombelli ([4], p. 169 and 190) considers complex numbers as « linear combinations » with positive coefficients, of four basis elements : « più » (+ 1), « meno » (– 1), « piu de meno » $(+i)$ and « meno de meno » $(-i)$; in particular he postulates an axiom that « piu » and « piu de meno » cannot be added, the first appearance of the notion of linear independence.

** The equation being reduced to the form $x^4 = ax^2 + bx + c$, we determine a number z so that the right-hand side of the equation

$$(x^2 + z)^2 = (a + 2z)\,x^2 + bx + (c + z^2)$$

is a perfect square, which gives an equation of the third degree for z.

*** Vieta, a passionate admirer of the Ancients, refrains systematically from introducing negative numbers into his arguments ; he is nonetheless able on occasion to express in his language the relations between coefficients and roots when certain of the latter are negative ; for example, if the equation $x^3 + b = ax$ has two positive roots x_1, x_2 $(a > 0, b > 0)$, Vieta shows that $x_1^2 + x_2^2 + x_1x_2 = a$ and $x_1x_2(x_1 + x_2) = b$ ([5], p. 106).

tangents to algebraic curves (cf. Hist. Note to Book IV, Chap. I-II-III, p. 46) is related to the criterion for the multiplicity of a root of an algebraic equation, stated by his disciple Hudde ([7], p. 433 and 507-509). It is no doubt also to Descartes' influence that the distinction between algebraic functions and transcendental functions should be ascribed, parallel to that which he introduces in his *Géométrie* between the « geometric » curves and the « mechanical » curves (cf. Hist. Note to Book IV, Chap. I-II-III, p. 46 and 61). In any case this distinction is made perfectly clear by J. Gregory who in 1667, seeks even to prove that the area of a circular sector cannot be an algebraic function of the chord and the radius *. The expression « transcendental » is due to Leibniz, whom these questions of classification never cease to interest all through his career and who, about 1682 discovers a simple proof of the result Gregory was pursuing, by proving that $\sin x$ is not an algebraic function of x ([8], vol. V, p. 97-98) **. With his friend Tschirnhaus, Leibniz is actually one of the few mathematicians of his time to interest himself still in the problem of the solution « by radicals » of algebraic equations. At his beginnings we find him studying the « irreducible case » of the equation of the 3rd degree, and convince himself (as it happens, with insufficient proof) that it is impossible in this case to eliminate imaginary quantities from the formulae of the solution ([9], p. 547-564). At about the same time he attacks, also without success, the solution by radicals of the equation of the 5th degree ; and when later Tschirnhaus claims to solve the problem by making all the terms of the equation disappear except the two extremes, by a transformation of the form $y = P(x)$, where P is a suitably chosen polynomial of the 4th degree, Leibniz notices at once that the equations which determine the coefficients of $P(x)$ are of degree > 5, and regards the method as doomed to failure ([9], p. 402-403).

It seems to have been the needs of the new Analysis which gradually rekindled the interest in algebra. The integration of rational fractions, effected by Leibniz and Johann Bernoulli, and the question of imaginary logarithms closely related to it, provided an opportunity to develop the calculations with imaginary numbers in greater depth and to take up again the question of the decomposition of a polynomial into factors of the first degree (« fundamental theorem of alge-

* J. GREGORY, *Vera Circuli et Hyperbolae Quadratura*..., Pataviae, 1667 ; cf. G. HEINRICH, *Bibl. Math* (3), vol. II (1901), p. 77-85.

** The definition given by Leibniz of « transcendental quantities » ([8], vol. V, p. 228 ; see also *ibid.*, p. 120) seems to apply to functions rather than numbers (in modern language, what he does amounts to defining transcendental elements over the field obtained by adjoining to the field of rational numbers the data of the problem) ; however it is likely that he had a fairly clear notion of transcendental numbers (even though the latter do not seem to have been defined in precise fashion before the end of the 18th century) ; at any rate he observes explicitly that a transcendental function can assume rational values for rational values of the variable, and hence that his proof of the transcendence of $\sin x$ is not sufficient to prove that π is irrational ([8], vol. V, p. 97 and 124-126).

bra ») *. Right at the beginning of the 18th century, the solution of the binomial equation $x^n - 1 = 0$ is reduced by Cotes and de Moivre to the division of the circle into n equal parts ; to obtain the expressions of its roots « by radicals » it is therefore enough to know how to do it when n is an odd prime, and de Moivre remarks that the substitution $y = x + \frac{1}{x}$ then reduces the problem to the solution « by radicals » of an equation of degree $(n-1)/2$. As far as the « fundamental theorem » is concerned, after the repeated failures of a general solution « by radicals » (including several attempts by Euler [10]), efforts are beginning to tend towards *a priori* proofs, not using explicit formulae of solution. Without entering into the details of the methods proposed (which were to bear fruit in the proofs of Lagrange and Gauss ; cf. Hist. Notes to Book II, Chap. VI-VII and Book III, Chap. VIII), it is appropriate to note here the point of view from which the problem was seen in the middle of the 18th century ; it is admitted (without any justification other than a vague feeling of generality, coming no doubt, as with A. Girard, from the existence of relations between the coefficients and the roots) that an equation of degree n always has n « ideal » roots, an which calculations may be performed as on numbers, *without knowing whether they are numbers* (real or complex) ; what has to be proved (using if necessary calculations on the ideal roots), is that at least one of these roots is an ordinary complex number **. In this defective form the first germ of the general idea of « formal adjunction » may be recognized, which in spite of Gauss's objections ([13], vol. III, p. 1) was to become the basis of the modern theory of commutative fields.

With the fundamental memoirs of Lagrange ([11], *a*) and Vandermonde [12] the year 1770 sees the opening of a new and decisive period in the history of algebraic equations. The empiricism of more or less lucky attempts to find formulae of solution, which until then had reigned undivided, was followed by a systematic analysis of the problems posed and of the methods capable of solving them, an analysis which in sixty years led to the definitive results of Galois. Both Lagrange and Vandermonde started out from the ambiguity introduced by the multiple determinations of the radicals in the formulae for the solution of equations of degree $\leqslant 4$; this fact had already attracted the attention of Euler ([10], *a*), who had shown among other things how in del Ferro's formula one has to associate the radicals occurring there in such a way as to obtain 3 roots and not 9. Lagrange remarked that each of the cube roots in del Ferro's formula may be

* One obtains a good idea of the rudimentary state in which calculations with complex numbers were at that epoch when one sees Leibniz (yet one of the most practised in that technique, among the mathematicians of his time) expressing himself as if it were not possible to decompose $x^4 + 1$ into two real factors of the second degree ([8], vol. V, p. 359-360).

** It is to be noted that what the mathematicians of the 18th century call « imaginary roots » are frequently only the above « ideal » roots ; and they try to prove that these roots are of the form $a + b\sqrt{-1}$ (see for example [11], *b*).

written in the form $\frac{1}{3}(x_1 + \omega x_2 + \omega^2 x_3)$, where ω is a cube root of unity and x_1, x_2, x_3 the roots of the proposed equation, taken in a certain order, and he made the fundamental observation that the function $(x_1 + \omega x_2 + \omega^2 x_3)^3$ of the three roots can only take *two* distinct values for every *permutation* of the three roots, which explains *a priori* the success of the methods of solution of this equation. A similar analysis of the methods of solution of the equation of the 4th degree leads him to the function $x_1x_2 + x_3x_4$ of the four roots, which only takes *three* distinct values for every permutation of the roots and is therefore root of an equation of the third degree with coefficients which are rational functions in those of the given equation * ; these facts constitute as Lagrange says « *the true principles and, so to speak, the metaphysics* ** *of the solution of equations of the* 3th *and* 4th *degree* » ([11], *a*), p. 357). Guided by these examples he plans to study in general, for an equation of degree n the number ν of values *** which a rational function V of the n roots may assume when the latter are permuted arbitrarily ; he thus in reality inaugurates (in this terminology still closely adapted to the theory of equations) the theory of groups and that of fields, of which he already obtains several fundamental results by making use of the same principles as those that are employed today. For example, he shows that the number ν is a divisor of $n!$, by the reasoning which today serves to prove that the order of a subgroup of a finite group divides the order of this group. Still more remarkable is the theorem in which he shows that if V_1 and V_2 are two rational functions of the roots such that V_1 and V_2 remain invariant under the same permutations, then either is a rational function of the other and the coefficients of the equation (a particular case of the theorem of Galois characterizing a subextension of a Galois extension as field of invariants of its Galois group) : « *This problem* » he says « *seems to me one of the most important of the theory of equations, and the general solution we shall give will serve to throw new light on this part of Algebra* » ([11], *a*), p. 374).

All these researches are naturally, in Lagrange's mind, preliminaries to the analysis of possible methods of solution of algebraic equations by successive reduction to equations of lower degree, such a method being related, as he shows, to the formation of rational functions of the roots taking fewer than n values under permutation of the roots. Guided no doubt by results on the equation of the 3rd degree, he introduces in general the « Lagrange resolvents » $y_k = \sum_{h=1}^{n} \omega_k^h x_h$, where

* Waring also makes this observation in his *Meditationes algebraicae*, published in that same year 1770, but he is far from drawing the same conclusions as Lagrange.

** In this word, flowing so frequently from the pen of 18th century authors, may be discerned a first (still very vague) glimpse of the modern concept of *structure*.

*** Lagrange already makes a distinction between the different *rational fractions* obtained from V by permuting the indeterminates x_i $(1 \leq i \leq n)$, and the different *values* these fractions take when the x_i are roots of an algebraic equation with given numerical coefficients ; but there still subsists in his exposition a certain hesitation on this subject, and it is only with Galois that the distinction becomes clear-cut.

ω_k is an n-th root of unity ($1 \leqslant k \leqslant n$), shows clearly how a knowledge of these n numbers entails that of the roots x_k, and examines in general the degree of the equation satisfied by the y_k ; he shows for example that if n is prime, then the y_k^n are the roots of an equation of degree $n-1$ whose coefficients are rational functions of a root of an equation of degree $(n-2)!$ with coefficients which may be expressed rationally by means of the coefficients of the given equation. « *Here, unless I am mistaken* » he concludes, « *are the true principles of the solution of equations, and the analysis most appropriate to lead us there ; as we see, everything reduces to a kind of calculus of combinations, by which one finds* a priori *the results one should expect* » ([11], *a*), p. 403).

As regards the memoir of Vandermonde, though independent of that of Lagrange, it makes contact in a number of points with the latter, notably in the idea of studying rational functions of the roots taking as few distinct values as possible under permutations of the roots *, and the study of « Lagrange resolvents » which he also introduces for this purpose. His work is far from possessing the clarity and generality of that of Lagrange ; however on one point he goes decidedly further, in applying the same ideas to the equation of the division of the circle $x^n - 1 = 0$, for n an odd prime. While Lagrange contents himself with recalling that this equation reduces to an equation of degree $m = (n-1)/2$ with rational coefficients, without trying to solve it when $n \geqslant 11$, Vandermonde affirms that the m-th powers of the Lagrange resolvents of this latter equation are rational, by reason of the relations between the various roots of $x^n - 1 = 0$; but he confines himself to verifying the wellfoundedness of this assertion for $n = 11$, without justifying it generally.

It is only 30 years later that the result announced by Vandermonde was completely proved by C. F. Gauss **. His decisive results on the equation $x^n - 1 = 0$ (n an odd prime) fit into the general programme of his memorable arithmetical researches ([13], vol. I, p. 413 *et seq.*), and they illustrate especially his mastery in the manipulation of what we now call the theory of cyclic groups. After having proved that the polynomial $\Phi_n(x) = (x^n - 1)/(x-1)$ is irreducible for odd prime n ***, he has the idea of writing his $n-1$ roots in the form

* In this research (which in fact he only develops for equations of the 5th degree) there appears for the first time the notion of *imprimitivity* ([12], p. 390-391). It is also tempting to compare the methods of Lagrange and Vandermonde with their contemporary work on determinants, which was to make them familiar with the idea of a permutation and all that appertains to it.

** Gauss does not quote Vandermonde in his *Disquisitiones*, but it is probable that he had read the memoir of the latter (cf. [13], vol. X_2, Abh. 4, p. 58).

*** The notion of an irreducible polynomial (with rational coefficients) goes back to the 17th century, and Newton and Leibniz had already given procedures which allowed (at least theoretically) to determine the irreducible factors of a polynomial with explicit rational coefficients ([8], vol. IV, p. 329 and 355) ; but Gauss's proof is the first irreducibility proof applying to a whole class of polynomials of arbitrarily high degree.

$\zeta^{g^k} = \zeta_k$ $(0 \leqslant k \leqslant n-2)$, where g is a primitive root of the congruence $z^{n-1} \equiv 1 \pmod{n}$ (which, in modern terms, comes to putting in evidence the fact that the group Γ of the equation $\Phi_n(x) = 0$ is cyclic). With every divisor e of $n-1$ he associates the $f = (n-1)/e$ « periods » $\eta_\nu = \zeta_\nu + \zeta_{\nu+e} + \zeta_{\nu+2e} + \cdots + \zeta_{\nu+(f-1)e}$ $(1 \leqslant \nu \leqslant f)$ and he shows in essence that the linear combinations with rational coefficients of the η_ν form a field, generated by any one of the f periods η_ν, and of degree f over the field of rational numbers (this field of course corresponds to the subgroup of Γ of order e). We cannot here enter into the details of this analysis, and of the important arithmetic consequences which it entails ; let us merely note that it gave him in particular the celebrated theorem on the possibility of constructing « by ruler and compasses » the regular polygons having a number of sides equal to a prime number of the form $2^{2^k} + 1$ *. As regards the solutions by radicals of the equation $\Phi_n(x) = 0$, this follows easily from the theory of periods, applied to the f-th power of a Lagrange resolvent $\sum_{\nu=0}^{f-1} \omega^\nu \eta_\nu$ (where $\omega^f = 1$) **.

It is directly to Lagrange that the researches of his compatriot Ruffini are linked, contemporaneous with the *Disquisitiones* ; taking up the question at the point where it had been left by Lagrange, they had as their aim the proof of the impossibility of the solution « by radicals » of the « general » *** equation of the 5th degree. Ruffini's proof, prolix and obscure, remains incomplete, even though revised several times ; but it is already very close to the proof (correct in principle) which Abel obtains later ****. Its chief interest resides in the introduction of a calculus of substitutions and the first notions of the theory of groups, which

* Gauss states explicitly that he has a proof of the fact that this is the only case in which one can construct by ruler and compasses a regular polygon having an odd prime number of sides ([13], vol. I, p. 462) ; but this proof was never published and has not been found among his papers.

** In fact, if we only want to prove that the equation is soluble by radicals we can limit ourselves to the case $e = 1$ and argue by induction on n.

*** By this the 19th century mathematicians understood in essence an equation whose coefficients are *indeterminates* over the field of rationals. But the modern notion of indeterminate had hardly emerged before the closing years of the 19th century ; until then one always understands by « polynomial » or « rational fraction » a *function* of complex variables. A « general » algebraic equation is conceived as an equation whose coefficients are independent complex variables and whose roots are « algebraic functions » of these variables — a notion which in truth is totally devoid of a precise sense if one gives the word « function » its present-day sense. Of course the reasoning with these « algebraic functions » is in general intrinsically correct as may be ascertained by translating it into modern algebraic language.

**** See P. Ruffini, *Opere Matematiche,* 3 vol. Roma (Ed. Cremonese), 1953-1954, as well as H. Burkhardt, *Zeitschr. für Math. und Phys.* vol. XXXVII (1892), Suppl. p. 121-159.

Ruffini develops to show that there exists no function of the 5 roots of the equation taking more than 2 and less than 5 values as the roots are permuted arbitrarily.

We have already said (Hist. Note to Chap. I) how this first outline of the theory of groups of permutations was developed and systematized by Cauchy some years later. But if the notions necessary for the development of Lagrange's ideas concerning substitutions are thus gradually clarified, it was still necessary to state the first principles of the theory of fields in equally clear fashion. It is this which Ruffini lacked, and which Abel and Galois were to undertake, in the last phase of the problem of the solution of algebraic equations.

Throughout his short life Abel never ceased to be preoccupied by this problem. Still almost a child, he believed to have found a formula for the solution by radicals of the general equation of the 5th degree. Later, when he recognizes his error, he does not rest until he succeeds in proving that such a formula does not exist ([14], vol. I, p. 166). But he does not stop there. While his competitor Jacobi develops the theory of elliptic functions as an analyst, it is the algebraic point of view which dominates Abel's work on this question, centred on the theory of equations of the division of elliptic functions ([14], vol. I, p. 265, 377 *et passim*). He thus obtains new types of equations soluble by radicals, by a method modelled on that of Gauss for the equations of the division of the circle ([14], vol. I, p. 310 and 358) * ; a result from which he rises to the concept of « abelian » equations, of which he proves in a celebrated memoir the solubility by radicals ([14], vol. I, p. 478) ; it is on this occasion that he defines in a precise fashion the notion of an irreducible polynomial over a given field (the field generated by the coefficients of the equation he is studying) **. Finally death strikes him down in 1829, while he is attacking the general problem of the characterization of all the equations soluble by radicals, and has just communicated to Crelle and Legendre results that are already very close to those of Galois ([14], vol. II, p. 219-243, 269-270 and 279).

It is for the latter that it was reserved, three years later, to crown the edifice [15]. Like Abel, but in even more precise fashion, he begins by defining (except for terminology) the belonging of a quantity to a field generated by given quantities, the notion of adjunction and irreducible polynomials over a given

* Gauss had already indicated in his *Disquisitiones* the possibility of generalizing his methods to the equations of the division of the lemniscate ([13], vol. I, p. 413), and developed, in notes published only in recent years, the particular case of the division by 5 ([13], vol. X, p. 161-162 and 517). As so many others of the brief and enigmatic indications Gauss was pleased to scatter in his writings, the phrase of the *Disquisitiones* vividly struck the minds of his contemporaries ; we know that it contributed no little to encourage Abel and Jacobi in their researches on this question.

** The very notion of a field (as, more generally, that of a set) is practically foreign to mathematical thought before Cantor and Dedekind. Abel and Galois define the *elements* of their « base field » as being those which may be expressed rationally as a function of known quantities, without thinking of considering explicitly the set formed by these elements.

field. Given an equation $F(x) = 0$ without multiple roots, with coefficients in a given field K, he shows successively how « *it is always possible to form a function* V *of the roots, such that none of the values obtained by permuting the roots in all possible ways in this function is equal to another* », that this function « *enjoys the property that all the roots of the given equation may be expressed rationally as a function of* V » and that, if V, V′, V″, ... are the roots of the irreducible equation satisfied by V, « *if* $a = f(V)$ *is a root of the proposed equation, then so is* $f(V')$ » ([15], p. 36-37) ; in modern terms he thus proves that V, as well as any one of its conjugates over K generates the field N of the roots of F. He then defines the group Γ of F as the set of permutations of the roots x_i which are obtained in substituting for V in the rational expression of each of the x_i as function of V, any one of the conjugates of V ; and he obtains at once the fundamental characterization of the elements of K by the property of being invariant under every permutation of Γ ([15], p. 38-39). He then proves that if N contains the root field L of another polynomial, then the group of N over L is a normal subgroup of Γ (a notion which he introduces on this occasion) ([15], p. 41 and 25-26). From this he finally deduces the criterion for the solubility of an equation by radicals, by means of reasoning which essentially runs as follows : the base field K being assumed to contain all the roots of unity, there exists by hypothesis an ascending sequence $(K_i)_{0 \leqslant i \leqslant m}$ of intermediate fields between K and N with $K_0 = K$, $K_m = N$, the field K_{i+1} being obtained by adjunction to K_i of all the roots of a binomial equation $x^{n_i} - a_i = 0$ (with $a_i \in K_i$). There exists thus in Γ a descending sequence (Γ_i) of subgroups such that $\Gamma_0 = \Gamma$, $\Gamma_m = \{\varepsilon\}$ (neutral element), Γ_{i+1} being normal in Γ_i and the quotient group Γ_i/Γ_{i+1} being cyclic (a case in which the group Γ is said to be *solvable*). Conversely when this is so, the use of a Lagrange resolvent shows that K_{i+1} is obtained from K_i by adjunction of all the roots of a binomial equation and therefore the equation $F(x) = 0$ is soluble by radicals *. The impossibility of the solution by radicals of the « general » equation of degree $n > 4$ is thus a consequence of the fact that the group Γ of this equation, isomorphic to the symmetric group $\mathfrak{S}_n$ (V, p. 59, Example 5), is not soluble (I, p. 142, Ex. 10 and p. 143, Ex. 16).

*

* *

* If K does not contain all roots of unity and if E is the field obtained by adjoining to K all these roots, then E ∩ N is an abelian extension of K ; whence it is easily deduced (using the structure of finite abelian groups) that for the group of N over K to be soluble it is necessary and sufficient for this to be true of the group of E(N) over E. Bearing in mind the fact that the roots of unity are expressible « by radicals », we see that Galois' criterion is independent of any hypothesis on the field K (and is valid more generally for any field of characteristic 0). In actual fact Galois makes no simplifying hypothesis about K and argues by induction on the order of the radicals successively adjoined to K ([15], p. 43).

From the middle of the 19th century the algebraists, as we have already remarked (cf. Hist. Note of Chap. I), enlarged considerably the domain of their investigations, until then almost entirely confined to the study of equations. In the light of Galois' discoveries it was realized that the problem of solution « by radicals » is only a particular somewhat artificial case of the general problem of classifying irrationals. It is the latter which throughout the closing years of the 19th century is attacked from various sides and numerous disparate results accumulate gradually, preparing the way for the synthesis by Steinitz.

In the first place, as far as the algebraic irrationals were concerned, a fundamental principle of classification was provided by Galois theory, which reduced the study of an algebraic equation to that of its group. For this reason it is above all the theory of permutation groups, of which it is not our task to speak here (cf. Hist. Note to Chap. I), which in pure Algebra formed the principal object of research of that period. The remaining progress in the theory of algebraic fields was due to the development, at the same time, of the Theory of Numbers and Algebraic Geometry. Actually this progress concerned mainly the exposition of the theory, and was mostly due to Dedekind [18], who introduced the notions of field and ring *, and (in connexion with his researches on hypercomplex systems) develops systematically the linear aspect of the theory of extensions ([18], vol. 3, p. 33 *et seq.*). It is he also who considers the Galois group as consisting of automorphisms of the extension being considered, and not merely as a group of permutations of the roots of an equation, and he proves (for number fields) his fundamental theorem on the linear independence of automorphisms ([18], vol. 3, p. 29) as well as the existence of normal bases of a Galois extension ([18], vol. 2, p. 433). Finally he tackles the problem of algebraic extensions of infinite degree and notes that Galois theory cannot be applied as it stands (an arbitrary subgroup of a Galois group need not always be identical with the group of the extension with respect to a subextension), and by a daring intuition he already tries to contemplate the Galois group as a topological group ** — an idea which reaches maturity only with the theory of Galois extensions of infinite degree, developed by Krull in 1928 [24].

Parallel to this evolution the notion of a transcendental element over a field is made precise. The existence of transcendental numbers is proved for the first time by Liouville in 1844, by an explicit process of construction, based on the theory of diophantine approximations [17] ; Cantor in 1874 gave another « non-constructive » proof, using simple reflexions on the power of sets (cf. V, p. 147), Ex. 3) ; finally Hermite proved in 1873 the transcendence of e and Lindemann in 1882 that

* The word « field » comes from Dedekind himself ; « ring » was introduced by Hilbert (Dedekind called rings « orders »).

** « ... *The aggregate of all these permutations forms in a certain sense a continuous manifold, a question into which we will enter no further here* » ([18], vol. 2, p. 288).

of π by a method analogous to Hermite's, thus finally closing the discussion on the antique problem of the quadrature of the circle *.

Regarding the role of transcendental numbers in algebraic calculations, Kronecker observes in 1882 that if x is transcendental over a field K, then the field $K(x)$ is isomorphic to the field of rational fractions $K(X)$ ([19], *a*), p. 7). He actually makes the adjunction of indeterminates to a field the corner stone of his exposition of the theory of algebraic numbers ([19], *a*)). On the other hand, Dedekind and Weber show in the same year ([18], vol. I, p. 238) how arithmetic methods may be used to set up a theory of algebraic curves. The analogies between Arithmetic and Algebraic Geometry, which will turn out to be so fruitful for the one and the other are thus seen to appear from several directions.

In all these researches the fields which occur consist of « concrete » elements in the sense of classical mathematics — (complex) numbers or functions of complex variables **. But already Kronecker in 1882 was perfectly aware of the fact (only obscurely glimpsed by Gauss and Galois) that the « indeterminates » play in his theory only the role of basis elements of an algebra and not that of variables in the sense of Analysis ([19], *a*), p. 93-95) ; and in 1887 he develops this idea, combined with a vast programme which aims at nothing less than to re-lay the foundations for all of mathematics by rejecting everything that cannot be reduced to algebraic operations on integers (cf. Hist. Note to Book I, Chap. IV). It is on this occasion that, taking up an idea of Cauchy [16] who had defined the field **C** of complex numbers as the residue class field $\mathbf{R}[X]/(X^2+1)$, Kronecker shows how the theory of algebraic numbers is entirely independent of the « fundamental theorem of algebra » and even of the theory of real numbers, since every algebraic number field (of finite degree) is isomorphic to a residue class field $\mathbf{Q}[X]/(f)$ (where f is an irreducible polynomial over **Q**) [19], *b*). As H. Weber [20] remarks some years later in developing a first sketch of an axiomatic theory of fields, this method of Kronecker's applies in fact to any base field K. Weber indicates in particular that a field $\mathbf{Z}/(p)$ (p prime) may be taken for K, allowing thus the calculus of congruences « modulo p » to be encompassed by the theory of fields ; the former had taken shape in the second half of the 18th century at the hands of Euler, Lagrange, Legendre and Gauss, and the analogy it presented with the theory of

* Simple proofs of these theorems may be found for example in D. HILBERT, *Gesammelte Abhandlungen*, vol. I, p. 1, Berlin (Springer), 1932.

** Like their predecessors neither Kronecker nor Dedekind and Weber actually define the notion of « algebraic function » of one or more complex variables. In effect it is impossible to define correctly an « algebraic function » of a complex variable (in the sense of Analysis) until the corresponding Riemann surface has been defined, and it is precisely the definition of this Riemann surface (by purely algebraic means) that Dedekind and Weber are trying to achieve. This apparent vicious circle of course disappears when a field of « algebraic functions » is defined as an abstract algebraic extension of a field of rational fractions : in fact it is solely this definition that is used by Dedekind and Weber, which clearly legitimizes their results.

algebraic equations had been readily observed ; in developing that analogy, Galois (in the light of his researches in group theory) had not hesitated to introduce « ideal roots » of an irreducible congruence modulo p *, and had indicated its chief properties ([15], p. 15-23) **. When Kronecker's method is applied to $\mathbf{Z}/(p)$, the result is (except for terminology) the presentation which Serret and Dedekind ([18], vol. I, p. 40) had already given of these « Galois imaginaries ».

To all these examples of « abstract fields » there is added, at the turn of the century, the field of formal power series introduced by Veronese [21] and above all the p-adic fields of Hensel [22]. It is the discovery of the latter which leads Steinitz (as he says explicitly) to isolate the abstract notions common to all these theories, in a basic paper [23] which may be considered as having given rise to the current conception of Algebra. Developing systematically the consequences of the axioms of commutative fields, he thus introduces the notions of prime field, of separable (algebraic) elements, of perfect field, he defines the transcendence degree of an extension and finally proves the existence of algebraically closed extensions of an arbitrary field.

More recently the theory of Steinitz has been completed in several important respects. On the one hand the work of Artin has put in evidence the linear character of Galois theory [25]. On the other hand, the general notion of derivation (modelled on the formal properties of classical Differential Calculus) sensed by Dedekind ([18], vol. 2, p. 412), introduced by Steinitz in the particular case of a field of rational fractions ([23], p. 209-212), has been used with success in the study (essential for modern Algebraic Geometry) of transcendental extensions, and notably in the generalization of the notion of separability to the latter [26].

* In a manuscript dating probably from 1799 but published only after his death, Gauss already had the idea of introducing such « imaginaries » and he obtains a substantial part of Galois' results ([13], vol. II, p. 212-240 in particular p. 217).

** Galois is fully aware of the formal character of these algebraic calculations and, for example, does not hesitate to take the derivative of the left-hand side of a congruence to show that the latter has no multiple « imaginary » roots ([15], p. 18). He stresses in particular that the theorem of the primitive element is valid just as much for a finite field as for a number field ([15], p. 17, note [2]), without however giving a proof.

Bibliography

1. O. NEUGEBAUER, *Vorlesungen über Geschichte der antiken Mathematik,* Bd. I : Vorgriechische Mathematik, Berlin (Springer), 1934.
2. *Euclidis Elementa,* 5 vol., ed. J. L. Heiberg, Lipsiae (Teubner), 1883-88.

2 *bis*. T. L. HEATH, *The thirteen books of Euclid's Elements...*, 3 vol., Cambridge, 1908.

3. H. CARDANO, *Opera*, Lyon, 1663.
4. R. BOMBELLI, *L'Algebra,* Bologne (G. Rossi), 1572.
5. FRANCISCI VIETAE, *Opera mathematica...*, Lugduni Batavorum (Elzevir), 1646.
6. A. GIRARD, *Invention nouvelle en Algèbre*, Amsterdam, 1629.
7. R. DESCARTES, *Geometria*, Latin transl. by Fr. van Schooten, 2 ed., 2 vol., Amsterdam (Elzevir), 1659-61.
8. G. W. LEIBNIZ, *Mathematische Schriften*, ed. C. I. Gerhardt, 7 vol., Berlin-Halle (Ascher-Schmidt), 1849-63.
9. *Der Briefwechsel von Gottfried Wilhelm Leibniz mit Mathematikern,* herausg. von C. I. Gerhardt, vol. I, Berlin (Mayer und Müller), 1899.
10. L. EULER, *Opera Omnia* (1), vol. VI, Berlin-Leipzig (Teubner), 1921 : *a*) De Formis Radicum Aequationum...., p. 1-19 ; *b*) De Resolutione Aequationum cujusvis gradus, p. 170-196.
11. J.-L. LAGRANGE, *Œuvres*, t. III, Paris (Gauthier-Villars), 1869 : *a*) Réflexions sur la résolution algébrique des équations, p. 205-421 ; *b*) Sur la forme des racines imaginaires des équations, p. 479.
12. A. VANDERMONDE, Mémoire sur la résolution des équations, *Hist. de l'Acad. royale des sciences*, année 1771, Paris (1774), p. 365-416.
13. C. F. GAUSS, *Werke*, vol. I-X, Göttingen, 1863-1923.
14. N. H. ABEL, *Œuvres*, 2 vol., ed. Sylow et Lie, Christiania, 1881.
15. E. GALOIS, *Œuvres mathématiques*, Paris (Gauthier-Villars), 1897.
16. A.-L. CAUCHY, *Œuvres complètes* (1), t. X, Paris (Gauthier-Villars), 1897, p. 312 et 351.
17. J. LIOUVILLE, Sur des classes très étendues de quantités dont la valeur n'est ni algébrique, ni même réductible à des irrationnelles algébriques, *Journ. de Math.* (1), t. XVI (1851), p. 133.
18. R. DEDEKIND, *Gesammelte mathematische Werke*, 3 vol., Braunschweig (Vieweg), 1932.
19. L. KRONECKER : *a*) Grundzüge einer arithmetischen Theorie der algebraischen Grössen, *Crelle's J.*, XCII (1882), p. 1-122 (= *Werke*, vol. II, Leipzig (Teubner), 1897, p. 245-387) ; *b*) Ein Fundamentalsatz der allgemeinen Arithmetik, *Crelle's J.*, C (1887), p. 490-510 (= *Werke*, vol. III_1, Leipzig (Teubner), 1899, p. 211-240).
20. H. WEBER, Untersuchungen über die allgemeinen Grundlagen der Galoisschen Gleichungstheorie, *Math. Ann.*, vol. XLIII (1893), p. 521-544.
21. G. VERONESE, *Fondamenti di geometria*, Padova, 1891.
22. K. HENSEL, *Theorie der algebraischen Zahlen*, Leipzig-Berlin (Teubner), 1908.
23. E. STEINITZ, Algebraische Theorie der Körper, *Crelle's J.*, vol. CXXXVII (1910), p. 167-309.
24. W. KRULL, Galoissche Theorie der unendlichen algebraischen Erweiterungen, *Math. Ann.*, vol. C (1928), p. 687.
25. E. ARTIN, *Galois Theory...*, Ann Arbor, 1946.
26. A. WEIL, *Foundations of algebraic geometry*, Amer. Math. Soc. Coll. Public., vol. XXIX, New York, 1946.

CHAPTER VI

Ordered groups and fields

§ 1. ORDERED GROUPS. DIVISIBILITY

The notions and results presented in this section concern the study of order relations in commutative monoids (I, p. 12, Def. 2), the most important case being that of *abelian groups*. Unless explicitly stated otherwise, we will use *additive* notation for the composition law in all groups and monoids under study. On the other hand, as we go along we will present certain important algebraic applications of the theory of ordered groups and monoids, and we will accordingly translate certain of our results into the *multiplicative* notation which is appropriate for these applications.

1. Definition of ordered monoids and groups

DEFINITION 1. — *A commutative monoid structure* (written additively) *and an ordering* (written $\leqslant$) *on a set* M *are said to be compatible if they satisfy the following axiom :*

(OM) *For each* $z \in \mathrm{M}$, *the relation* $x \leqslant y$ *implies* $x + z \leqslant y + z$.

A set M *equipped with a commutative monoid structure and an ordering which are compatible is called an ordered monoid ; if its commutative monoid structure is a group structure, then it is called an ordered group.*

In an analogous fashion, we can define the notion of a noncommutative ordered monoid (VI, p. 30, Ex. 1).

If an ordering is compatible with a given monoid structure, then so is the *opposite* ordering.

Examples. — 1) The additive group of the rational integers, and that of the rational numbers, are ordered groups under the ordering defined in I, pp. 21 and 117.

* The same is true for the additive group of the real numbers (*Gen. Top.*, IV, p. 3). *

2) * The additive group of *finite numerical functions* defined on a set E is an ordered group under the ordering « $f(x) \leqslant g(x)$ for all $x \in \mathrm{E}$ », written $f \leqslant g$. This relation says that the graph of the function f lies below that of the function g. The reader may find it convenient to think in terms of this graphical interpretation occasionally. *

According to general definitions (*Set Theory*, IV, p. 264), a bijective map f from an ordered monoid M onto an ordered monoid M′ is called an *isomorphism* of M onto M′ if the structure of M′ is obtained from that of M by transporting it along f. This is equivalent to saying that f is a mapping *onto* M′ such that

$$f(x+y) = f(x) + f(y)$$

(that is to say a homomorphism of the monoid M onto the monoid M′), and that the relations $x \leqslant y$ and $f(x) \leqslant f(y)$ are equivalent (whence in particular $f(x) = f(y)$ implies $x = y$, that is f is injective).

PROPOSITION 1 (« addition of inequalities »). — *In an ordered monoid* M, *let* (x_i) *and* (y_i) $(1 \leqslant i \leqslant n)$ *be two sequences of n elements such that* $x_i \leqslant y_i$ *for all i ; then one has*

$$x_1 + \cdots + x_n \leqslant y_1 + \cdots + y_n .$$

If moreover all the elements x_i, y_i *are cancellable* (I, p. 15, Def. 5) (in particular if M is a *group*), *and if* $x_i < y_i$ *for some i, then*

$$x_1 + \cdots + x_n < y_1 + \cdots + y_n .$$

The general case reduces by induction to the case $n = 2$, using the fact that a sum of cancellable elements is cancellable for the second assertion (I, p. 15, Prop. 2). The first assertion follows from the relations

$$x_1 + x_2 \leqslant x_1 + y_2 \quad \text{and} \quad x_1 + y_2 \leqslant y_1 + y_2 ,$$

which are consequences of the hypotheses and of (OM). This being the case, the relation

$$x_1 + x_2 = y_1 + y_2$$

would imply

$$x_1 + x_2 = x_1 + y_2 = y_1 + y_2 ,$$

whence $x_2 = y_2$ and $x_1 = y_1$ if x_1 and y_2 are cancellable, which proves the second assertion.

PROPOSITION 2. — *In an ordered group* G *the relations* $x \leqslant y$ *and* $x + z \leqslant y + z$ *are equivalent.*

Indeed one can obtain each from the other by adding z or $(-z)$ to both sides.

This fact can be expressed by saying that, in an ordered group G, the ordering is *translation-invariant.* In other words a translation is an *automorphism* of the *ordering* of an ordered group.

COROLLARY. — *In an ordered group* G, *the relations* $x \leqslant y$, $0 \leqslant y - x$, $x - y \leqslant 0$ *and* $-y \leqslant -x$ *are equivalent.*

In fact, we apply Prop. 2, successively taking $z = -x$, $z = -y$ and $z = -(x + y)$.

In particular we deduce from this corollary that if G is an ordered group, the map $x \mapsto -x$ from G to itself takes the ordering of G to the *opposite* ordering.

2. Pre-ordered monoids and groups

Recall that if a relation $x \preccurlyeq y$ between elements of a set E is reflexive and transitive, it is called a *pre-order* relation (*Set Theory*, III, p. 133). The relation « $x \preccurlyeq y$ and $y \preccurlyeq x$ » is an equivalence relation S on E, compatible with the relation $x \preccurlyeq y$; on passing to the quotient, the relation $\preccurlyeq$ induces an order relation on E/S, called the order relation associated to $\preccurlyeq$.

DEFINITION 2. — *A pre-order relation* (written $\preccurlyeq$) *and a commutative monoid structure* (written additively) *on a set* M *are said to be compatible if they satisfy the following axiom :*

(POM) *For each* $z \in M$, *the relation* $x \preccurlyeq y$ *implies* $x + z \preccurlyeq y + z$.

A set M *equipped with a commutative monoid structure and a pre-order relation which are compatible is called a pre-ordered monoid.*

Let M be a pre-ordered monoid, and S the equivalence relation « $x \preccurlyeq y$ and $y \preccurlyeq x$ ». By virtue of (POM) the relation $x \equiv x'$ (mod S) implies $x + y \preccurlyeq x' + y$ and $x' + y \preccurlyeq x + y$ for all $y \in M$, that is $x + y \equiv x' + y$ (mod S). In other words the equivalence relation S is *compatible* with addition in M (I, p. 11). Thus the quotient by S of the addition law on M, together with the ordering associated to $\preccurlyeq$, gives M/S the structure of an ordered monoid. In the case where M is a pre-ordered *group*, the group M/S is the quotient of M by the subgroup consisting of all elements x such that $x \preccurlyeq 0$ and $0 \preccurlyeq x$.

3. Positive elements

Let G be a pre-ordered group with pre-order relation $\preccurlyeq$; if $0 \preccurlyeq x$ and $0 \preccurlyeq y$ then we deduce that $y \preccurlyeq x + y$ by (POM), and so $0 \preccurlyeq x + y$ by transitivity ; this says that the set G_+ of elements $x \in G$ such that $0 \preccurlyeq x$ is closed under addition ; moreover, the relation $x \preccurlyeq y$ is equivalent to $0 \preccurlyeq y - x$, that is to $y - x \in G_+$. Conversely :

PROPOSITION 3. — *If* P *is a subset of an abelian group* G, *containing* 0 *and such that* $P + P \subset P$, *then the relation* $y - x \in P$ *is a pre-order relation compatible with the group structure of* G. *For this relation to make* G *an* ordered *group it is necessary and sufficient to have* $P \cap (-P) = \{0\}$; *for* G *to be a totally ordered group under this ordering it is necessary and sufficient to have in addition* $P \cup (-P) = G$.

It is immediate that the relation $y - x \in \mathrm{P}$ is reflexive and transitive, and (if it is written $x \preccurlyeq y$) satisfies the axiom (POM). To prove the second assertion it is enough to remark that $\mathrm{P} \cap (-\mathrm{P})$ is the subgroup G′ consisting of elements x such that $0 \preccurlyeq x$ and $x \preccurlyeq 0$. Finally, to say that G is totally ordered means that, for each pair of elements x, y of G, at least one of the elements $x - y$, $y - x$ belongs to P, which completes the proof.

DEFINITION 3. — *In an ordered group* G, *any element* x *such that* $0 \leqslant x$ *(resp.* $x \leqslant 0$*) is called a positive* (resp. *negative*) *element.*

Note that 0 is the only element which is *both positive and negative* ; any element x such that $0 < x$ (resp. $x < 0$) is called *strictly positive* (resp. *strictly negative*).

Example. — In the additive group $\mathbf{Z} \times \mathbf{Z}$, let P be the set of elements (x, y) satisfying two inequalities $ax + by \geqslant 0$ and $cx + dy \geqslant 0$, where a, b, c, d are integers (* or real numbers *) such that $ad - bc \neq 0$; the « cone » P satisfies the first two conditions of Prop. 3. In this way various orderings compatible with the group structure of $\mathbf{Z} \times \mathbf{Z}$ can be defined ; the group is not totally ordered under any of these orderings.

Remark. — By virtue of the condition $\mathrm{P} + \mathrm{P} \subset \mathrm{P}$, the relation $x \geqslant 0$ in an ordered group implies that $nx \geqslant 0$ for every natural number n. If in addition the positive element x of G has finite order n, then $-x = (n-1)x$ is positive ; since

$$\mathrm{P} \cap (-\mathrm{P}) = \{0\} ,$$

this implies that $x = 0$. In particular, if every element of G has finite order, then $\mathrm{P} = \{0\}$; the relation $x \leqslant y$ is then equivalent to $x = y$ (the *discrete* ordering).

4. Filtered groups

Recall (*Set Theory*, III, p. 145) that an ordered set G is *right* (resp. *left*) *filtered*[1] if for each pair (x, y) of elements of G there exists $z \in \mathrm{G}$ such that $x \leqslant z$ and $y \leqslant z$ (resp. $z \leqslant x$ and $z \leqslant y$). Every right filtered ordered group G is also left filtered and conversely : indeed, since there exists $z \in \mathrm{G}$ such that $-x \leqslant z$ and $-y \leqslant z$, we have $-z \leqslant x$ and $-z \leqslant y$ (VI, p. 3, Cor.). We will therefore speak simply of filtered groups.

PROPOSITION 4. — *For an ordered group* G *to be filtered it is necessary and sufficient that it be generated by its positive elements, that is that every element be the difference of two positive elements.*

Indeed if G is filtered, then for each $x \in \mathrm{G}$ there exists a positive element z such that $x \leqslant z$, and x is the difference of the two positive elements z and $z - x$. Conversely, if $x = u - v$ and $y = w - t$ with u, v, w, t positive, then the element $u + w$ is greater than x and greater than y.

[1] The terminology used in *loc. cit.* is *directed.*

PROPOSITION 5. — *If* (x_i) *is a finite family of elements of a filtered group* G, *then there exists* $z \in G$ *such that* $x_i + z$ *is positive for each* i.

If $x_i = u_i - v_i$, with u_i and v_i positive, it is enough to take z to be the sum of the family (v_i).

5. Divisibility relations in a field

Here we will define certain ordered groups which play an important part in algebra. The notation usually used for these groups is multiplicative ; the application of results previously obtained in additive notation to these groups thus presupposes the translation to multiplicative notation — which should present no difficulty to the reader. *Throughout this section, the letter* A *will denote an integral domain and* K *its field of fractions* (I, p. 116).

In the multiplicative group K^* of nonzero elements of K, the set P of nonzero elements of A is closed under multiplication, since A is a ring. Hence it defines a pre-order relation on K^* by $x^{-1}y \in P$, that is « there exists $z \in P$ such that $y = zx$ », which makes K^* a *pre-ordered group* (written multiplicatively) (VI, p. 3, Prop. 3). Extending the terminology relating to elements of A (I, p. 97) to K^*, the relation $x^{-1}y \in P$ can also be expressed as : x *divides* y, or x is a *divisor of* y, or y is a *multiple of* x (relative to the ring A) ; and we will call the relation $x^{-1}y \in P$ the *divisibility relation* in K^* relative to the ring A. The relation « x divides y » is denoted $x \mid y$, and its negation $x \nmid y$. The elements of P are none other than the *multiples of* 1.

Remarks. — 1) The divisibility relation in K^* depends in an essential way on the particular ring A. If $A = K$ we obtain the « trivial » relation, under which $x \mid y$ for every pair (x, y) of elements of K^*. Let p and q be prime numbers ; the rational numbers r/s whose denominators are not multiples of p (resp. q) form a subring $\mathbf{Z}_{(p)}$ (resp. $\mathbf{Z}_{(q)}$ of $\mathbf{Q}$; the divisibility relations in $\mathbf{Q}^*$ relative to these two rings are distinct if $p \neq q$, the number p/q being a multiple of 1 under one relation but not under the other.

2) We will sometimes extend the definition of the relation $x \mid y$ to pairs of elements of K (rather than only K^*), this relation being taken to mean « there exists $z \in A$ such that $y = zx$ » ; hence we will have $x \mid 0$ for all $x \in K$. This allows us to state the following results without restrictions : if $x \mid y$ and $x \mid z$ then $x \mid (y - z)$; if $x \mid y$ and $x \nmid z$ then $x \nmid (y - z)$. In the same way we can extend all the corresponding terminology.

To obtain an *order relation* (No. 2) from the divisibility relation we have to pass to the quotient group of K^* by the subgroup A^* of elements $x \in K^*$ such that $x \mid 1$ and $1 \mid x$; these are the elements of P which are *divisors of* 1, that is to say the *invertible* elements of A ; by abuse of language they are often called the units of the ring A. The quotient group K^*/A^* is then an ordered group. Two elements x and y of K^* which belong to the same coset of A^* are said to be *associate* ; this means that $x \mid y$ and $y \mid x$. If on the other hand x divides y without y dividing x, we say that x *strictly* divides y, or that x is a strict divisor of y, or that y is a strict multiple of x.

Note that K^*/A^* is a *filtered* group, since K is the field of fractions of A (VI, p. 4, Prop. 4).

To say that two elements x and y of K^* are associate amounts to saying that they have *the same multiples* in K, by virtue of the transitivity of the divisibility relation. For all $x \in K$, we will denote by Ax the set of all elements zx with $z \in A$; the set Ax is a submodule of K regarded as an A-module. Extending the terminology relating to the case where $x \in A$, we will call it a *principal fractional ideal* of the field K relative to the ring A. In contrast the ideals of the ring A are said to be *integral*.

Note that if $A \neq K$ then a principal fractional ideal $\neq \{0\}$ *is not* an ideal of K considered as a ring.

The principal fractional ideal Ax is also denoted (x). We will write $x \equiv 0 \pmod{y}$ for $x \in Ay$ and $x \equiv x' \pmod{y}$ for $x - x' \in Ay$; if $x \equiv x' \pmod{y}$ then $zx \equiv zx' \pmod{zy}$ for any $z \in K$.

Note that $x \equiv x' \pmod{y}$ *does not imply* $zx \equiv zx' \pmod{y}$ unless $z \in A$. Thus in $\mathbf{Q}$, relative to $\mathbf{Z}$, we have $4 \equiv 2 \pmod{2}$ but not $2 \equiv 1 \pmod{2}$.

The relation $x \mid y$ is obviously equivalent to $(x) \supset (y)$. The mapping $x \mapsto (x)$ from K^* onto the set $\mathscr{P}^*$ of principal fractional ideals $\neq (0)$ of K thus defines, on passing to the quotient, a *bijective* mapping from K^*/A^* onto $\mathscr{P}^*$; translating the group structure of K^*/A^* to $\mathscr{P}^*$ by means of this mapping, we are led to define the product of two principal fractional ideals (x) and (y) to be the ideal (xy), which depends only on (x) and (y). Equipped with this law and the order relation $(x) \supset (y)$, the set $\mathscr{P}^*$ is an ordered group, isomorphic to K^*/A^*. By convention we will identify $\mathscr{P}^*$ with K^*/A^* *via* the above map.

Note that the relation « x divides y » which, in the case of positive integers, implies that x is *smaller* than y, corresponds to the inclusion $(x) \supset (y)$, in which the ideal (x) is « greater » than the ideal (y). We can keep this « order reversal » in mind by noting that for example 7 has « more multiples » than 91.

If we extend the relation $x \mid y$ to all elements of K, this relation is still equivalent to $(x) \supset (y)$ in the set $\mathscr{P}$ of all principal fractional ideals of K (in which (0) is the smallest element under the relation of inclusion).

As in the previous sections, we will generally be using additive notation in the sequel. However, terminology relating to divisibility will be introduced following the corresponding additive terminology, in paragraphs preceded by the sign (DIV) (in which it is understood that the notation used is that of the present section). In order to make the reader's task easier, certain results will be translated into the language of divisibility, the translation of Prop. 7, for example, being denoted « PROPOSITION 7 (DIV) ».

6. Elementary operations on ordered groups

Let H be a subgroup of an ordered group G ; it is clear that the restriction to H of the ordering of G is compatible with the group structure of H ; we will always take H to be ordered in this way, unless otherwise stated. If P is the set of positive elements of G then the set of positive elements of H is $H \cap P$.

Let (G_α) be a family of ordered groups ; according to the definition of the *product* of ordered sets (*Set Theory*, III, p. 137) the product group $G = \prod_\alpha G_\alpha$ is equipped with an ordering, the relation « $(x_\alpha) \leqslant (y_\alpha)$ » between two elements of G being by definition the same as « $x_\alpha \leqslant y_\alpha$ for all α ». It is immediate that this ordering is compatible with the group structure of G ; this ordering makes G an ordered group which we call the *product of the ordered groups* G_α. The positive elements of G are those elements all of whose components are positive. In the case where all the factors G_α are identical to the same ordered group H, then G is the group H^I of maps from the index set I into H, the relation « $f \leqslant g$ » between two maps from I into H being the same as « $f(\alpha) \leqslant g(\alpha)$ for all $\alpha \in I$ » ; the positive maps are those which take only positive values. The *direct sum* of a family (G_α) of ordered groups is defined as an ordered subgroup of their product (II, p. 202).

Let $(G_\iota)_{\iota \in I}$ be a family of ordered groups whose index set I is *well ordered* ; recall (*Set Theory*, III, p. 157) that an order relation, called the *lexicographic* ordering, is defined on the product set $G = \prod_\iota G_\iota$, the relation « $(x_\iota) < (y_\iota)$ » between two elements of G being by definition the same as « if β is the smallest of the indices ι such that $x_\iota \neq y_\iota$, then $x_\beta < y_\beta$ ». Recall that the product of a well ordered family of *totally ordered* sets is totally ordered under the lexicographic ordering. In the general case, the lexicographic ordering on G is compatible with its group structure, as is immediately verified ; equipped with this ordering, the group G is thus an ordered group, called the *lexicographic product* of the well ordered family of ordered groups (G_ι).

Remarks. — 1) In the commonest cases, the well ordered index set I will be a *finite* interval $[1, n]$ in **N**.

2) The set of positive elements of the lexicographic product G consists of 0 and those nonzero elements whose nonzero component of least index is positive.

7. Increasing homomorphisms of ordered groups

Let G and G′ be two ordered groups ; among the homomorphisms f from the underlying additive group of G into that of G′, it makes sense to consider the *increasing* maps, that is those such that $x \leqslant y$ implies $f(x) \leqslant f(y)$. Because of the relation $f(y - x) = f(y) - f(x)$ it follows that the increasing homomorphisms from G into G′ are characterised by the fact that the image of a positive element of G under such a homomorphism is a positive element of G′ ; if P (resp. P′) denotes the set of positive elements of G (resp. G′), this can be written $f(\mathrm{P}) \subset \mathrm{P}'$. It is clear that the canonical injection of a subgroup G into an ordered group G′, and the projection of a product of ordered groups onto one of its factors, are increasing homomorphisms.

An *isomorphism* (VI, p. 2) f from an ordered group G onto an ordered group G′ is a bijective homomorphism from G onto G′ such that *both* f *and* the inverse homomorphism are increasing, in other words $f(\mathrm{P}) = \mathrm{P}'$.

It can happen that an isomorphism from the underlying group of G onto that of G′ can be increasing without the inverse isomorphism also being increasing. This is the case, for example, if $\mathrm{G} = \mathrm{G}'$, if f is the identity map on G, and if $\mathrm{P} \subset \mathrm{P}'$ but $\mathrm{P} \neq \mathrm{P}'$. Thus in **Z** we can take P′ to be the set of (ordinary) positive integers and P to be the set of even positive integers.

(DIV) Let K be the field of rational functions $\mathbf{F}_2(\mathrm{X})$ over the field $\mathbf{F}_2$ of order 2. The divisibility relations relative to the rings $\mathbf{F}_2[\mathrm{X}] = \mathrm{A}'$ and $\mathbf{F}_2[\mathrm{X}^2, \mathrm{X}^3] = \mathrm{A}$ define two distinct ordered group structures on K*, such that $\mathrm{A} \subset \mathrm{A}'$ (they are ordered group structures since 1 is the only unit in A and the only unit in A′).

8. Suprema and infima in an ordered group

Recall (*Set Theory*, III, p. 141) that if the set of upper bounds of a subset F of an ordered set E (that is to say the set of $z \in \mathrm{E}$ such that $x \leqslant z$ for all $x \in \mathrm{F}$) has a least element a, then this element, which is then unique, is called the *supremum* of F. If F is the set of elements in a family $(x_\iota)_{\iota \in \mathrm{I}}$ of elements of E, its supremum, if it exists, is denoted $\sup_{\iota \in \mathrm{I}} x_\iota$ (or $\sup_\iota x_\iota$ or simply $\sup(x_\iota)$) ; if it is a finite family (x_ι) $(1 \leqslant \iota \leqslant n)$, this supremum is also denoted $\sup(x_1, \ldots, x_n)$. The infimum is defined in an analogous manner and is denoted inf. The operations sup and inf are associative and commutative.

Recall (*Set Theory*, *loc. cit.*) that if F is a subset of an ordered set E, and (x_ι) a family of elements of F, then the existence of $\sup(x_\iota)$ in E (which may be denoted $\sup_{\mathrm{E}}(x_\iota)$) does not imply the existence of a supremum of the x_ι in F (which may be denoted $\sup_{\mathrm{F}}(x_\iota)$ when it does exist) ; if both exist we know only that $\sup_{\mathrm{E}}(x_\iota) \leqslant \sup_{\mathrm{F}}(x_\iota)$; however if $\sup_{\mathrm{E}}(x_\iota)$ exists and belongs to F, then $\sup_{\mathrm{F}}(x_\iota)$ exists and is equal to $\sup_{\mathrm{E}}(x_\iota)$. For example, in the polynomial ring $\mathrm{A} = \mathrm{K}[\mathrm{X}, \mathrm{Y}]$ (K a field), the principal ideals AX and AY have the ideal $\mathrm{AX} + \mathrm{AY}$ as supremum (under the relation $\subset$) in the ordered set of ideals of A, but have the ideal A as supremum in the set of all principal ideals of A.

(DIV) An element d of K* is called a *greatest common divisor*, or *gcd* for short, of a family (x_ι) of elements of K*, if the principal fractional ideal (d) is the *supremum in* $\mathscr{P}^*$ (under the relation $\subset$) of the family of ideals $((x_\iota))$, or in other words if the relation $z \mid d$ for $z \in \mathrm{K}^*$ is equivalent to « $z \mid x_\iota$ for all ι ». In the same way we will say that $m \in \mathrm{K}^*$ is a *least common multiple* or an *lcm* of the family (x_ι) if (m) is the infimum in $\mathscr{P}^*$ of the family of ideals $((x_\iota))$, that is if

$m \mid z$ is equivalent to « $x_\iota \mid z$ for all ι ». It amounts to the same thing to say that $(m) = \bigcap_\iota (x_\iota)$; indeed, the condition $x_\iota \mid z$ for all ι is equivalent to $z \in \mathrm{A}x_\iota$ for all ι, that is to $z \in \bigcap (x_\iota)$, and the condition $m \mid z$ is equivalent to $z \in (m)$ [1].

Note that if a principal fractional ideal (d) satisfies $(d) = \sum_\iota (x_\iota)$ then d is a gcd of the family (x_ι) ; but conversely a gcd of (x_ι) *does not necessarily satisfy* the above condition (cf. VI, p. 33, Ex. 24).

The gcd and lcm, if they exist, are well defined modulo the subgroup U of units of K*, that is two gcd's (or two lcm's) of a given family are associate ; by abuse of language we will often write $\gcd(x_\iota)$ and $\operatorname{lcm}(x_\iota)$ for any of the gcd's or lcm's of the family (x_ι), whenever such elements exist.

(DIV) By abuse of language we sometimes extend the notion of gcd to a family (x_ι) of elements of K, some of which may be zero ; this gcd is again defined to be an element d such that the relation $z \mid d$ is equivalent to « $z \mid x_\iota$ for all ι » ; clearly d is 0 if all the x_ι are zero ; otherwise d is a gcd of the nonzero x_ι. Similarly the lcm of a family, some elements of which are zero, is 0.

In an *ordered group* G, an immediate consequence of the invariance under translation of the ordering (VI, p. 2, Prop. 2) is that

$$\sup(z + x_\iota) = z + \sup(x_\iota) \tag{1}$$

in the sense that, whenever one side is defined, then so is the other and the two are equal. Similarly, it follows from the fact that the map $x \mapsto -x$ takes the ordering of G to the opposite ordering (VI, p. 3, Cor.) that

$$\inf(-x_\iota) = -(\sup(x_\iota)), \tag{2}$$

this relation being understood in the same sense as the previous one.

PROPOSITION 6. — *Let* $(x_\alpha)_{\alpha \in \mathrm{A}}$, $(y_\beta)_{\beta \in \mathrm{B}}$ *be two families of elements of an ordered group* G, *each having a supremum. Then the family* $(x_\alpha + y_\beta)_{(\alpha,\beta) \in \mathrm{A} \times \mathrm{B}}$ *has a supremum, and* $\sup_{(\alpha,\beta) \in \mathrm{A}\times\mathrm{B}} (x_\alpha + y_\beta) = \sup_{\alpha \in \mathrm{A}} x_\alpha + \sup_{\beta \in \mathrm{B}} y_\beta$.

Indeed, from $x_\alpha + y_\beta \leqslant z$ for all α and β, we deduce $\sup(x_\alpha) + y_\beta \leqslant z$ for all β, and hence $\sup(x_\alpha) + \sup(y_\beta) \leqslant z$.

[1] When A is the ring of integers (resp. the polynomial ring in one indeterminate with coefficients in a field), these definitions coincide with those of I, p. 112 (resp. IV, p. 12, Def. 1).

9. Lattice ordered groups

Recall that an ordered set in which every *non empty* finite subset has a supremum and an infimum is called a *lattice* (E, III, p. 13). It is clear that a product of lattice ordered groups, and in particular a product of totally ordered groups, is a lattice ordered group. In contrast a subgroup of a lattice ordered group is not necessarily lattice ordered.

Thus, in the product ordered group $\mathbf{Z} \times \mathbf{Z}$, the « antidiagonal » (set of pairs (n, n') such that $n + n' = 0$) is ordered with the discrete ordering, and hence is not a lattice ordered group. * The additive group of polynomials in one real variable (VI, p. 1, example 2) is a filtered group (since both $p(x)$ and $q(x)$ are less than $(p(x))^2 + (q(x))^2 + 1$) which one can show is not lattice ordered. *

PROPOSITION 7. — *If x and y are two elements of an ordered group* G, *and if one of the elements* $\inf(x, y)$, $\sup(x, y)$ *exists, then so does the other, and* $x + y = \inf(x, y) + \sup(x, y)$.

Indeed, according to the relations (1) and (2) (VI, p. 9) we have

$$\sup(a - x, a - y) = a + \sup(-x, -y) = a - \inf(x, y),$$

and it suffices to take $a = x + y$.

PROPOSITION 7 (DIV). — *If $a, b \in \mathrm{K}^*$, and if d is a* gcd *of a and b, and m an* lcm *of a and b, then the product dm is an associate of ab.*

PROPOSITION 8. — *Let* P *be the set of positive elements of an ordered group* G. *For* G *to be lattice ordered, it is necessary and sufficient that* $\mathrm{G} = \mathrm{P} - \mathrm{P}$, *and that in addition* P, *equipped with the induced ordering, satisfies one or other of the following conditions :*

a) Each pair of elements of P *has a supremum in* P.

b) Each pair of elements of P *has an infimum in* P.

The necessity of these conditions is obvious : indeed the relation $\mathrm{G} = \mathrm{P} - \mathrm{P}$ says that G is filtered (VI, p. 4, Prop. 4) ; on the other hand the supremum and infimum *in* G of two elements of P are positive, so are also their supremum and infimum in P.

Conversely, note first that under hypothesis *a*) (resp. *b*)), every pair of elements x, y of P has a supremum (resp. infimum) *in* G equal to its supremum a (resp. infimum b) *in* P. This is obvious for a, since any upper bound for x and y is positive ; for b, let $z \in \mathrm{G}$ be a lower bound for x and y ; then there exists $u \in \mathrm{P}$ such that $z + u \in \mathrm{P}$, since $\mathrm{G} = \mathrm{P} - \mathrm{P}$; now $\inf_{\mathrm{P}}(x + u, y + u)$ is greater than $b + u$, and so is of the form $b + c + u$ $(c \geqslant 0)$; since $b + c$ is less than x and less than y, we have $c = 0$; hence $\inf_{\mathrm{P}}(x + u, y + u) = b + u$, which implies $z + u \leqslant b + u$, thus $z \leqslant b$ and b is certainly the infimum of x and y in G. Now if x and y are arbitrary elements of G we translate them into P : let $v \in \mathrm{P}$ be such that $x + v$ and $y + v$ are positive (VI, p. 5, Prop. 5) ; under hypothesis *a*) (resp. *b*)) there exists a supremum (resp. infimum) for $x + v$ and $y + v$ in P, and hence also in G by what we have just seen ; by translation x and y have a supremum (resp. infimum) in G ; the existence of one of these implies the other, by Prop. 7, and this shows that the conditions are sufficient.

10. The decomposition theorem

THEOREM 1 (decomposition theorem). — *Let $(x_i)_{1 \leq i \leq p}$ and $(y_j)_{1 \leq j \leq q}$ be two finite sequences of positive elements of a lattice ordered group* G *such that* $\sum_{i=1}^{p} x_i = \sum_{j=1}^{q} y_j$; *then there exists a double sequence $(z_{ij})_{1 \leq i \leq p, 1 \leq j \leq q}$ of positive elements of* G *such that $x_i = \sum_{j=1}^{q} z_{ij}$ for all i, and $y_j = \sum_{i=1}^{p} z_{ij}$ for all j.*

1) Let us prove the theorem first for the case $p = q = 2$. Let x, x', y, y' be positive elements of G such that $x + x' = y + y'$, and put $a = \sup(0, x - y')$. Since

$$x - y' = y - x'$$

is less than x and less than y, it follows that $b = x - a$ and $c = y - a$ are positive, as is $d = a - (x - y')$. Also we have

$$x = a + b, x' = c + d, y = a + c \quad \text{and} \quad y' = b + d .$$

2) Let us now show that if the theorem is true for $p < m$ and $q = n$ $(m > 2,\ n \geq 2)$ then it is true for $p = m$ and $q = n$. By hypothesis we have $x_m + \sum_{i=1}^{m-1} x_i = \sum_{j=1}^{n} y_j$. The theorem being true for $p = 2$ and $q = n$, there exist two finite sequences (z_j'), (z_j'') of n positive terms such that $\sum_{i=1}^{m-1} x_i = \sum_{j=1}^{n} z_j'$, $x_m = \sum_{j=1}^{n} z_j''$, and $y_j = z_j' + z_j''$ for $1 \leq j \leq n$. On the other hand, since the theorem is true for $p = m - 1$ and $q = n$, there exists a double sequence $(u_{ij})_{1 \leq i \leq m-1, 1 \leq j \leq n}$ such that $x_i = \sum_{j=1}^{n} u_{ij}$ for $1 \leq i \leq m - 1$, and $z_j' = \sum_{i=1}^{m-1} u_{ij}$ for $1 \leq j \leq n$. Putting

$$z_{ij} = u_{ij} \quad \text{for} \quad 1 \leq i \leq m - 1\,, \quad \text{and} \quad z_{mj} = z_j'' \quad (1 \leq j \leq n)\,,$$

we certainly obtain a double sequence satisfying the conditions of the theorem.

3) By interchanging the x_i and the y_j we see in the same way that, if the theorem is true for $p = m$ and $q < n$ $(m \geq 2,\ n > 2)$, then it is true for $p = m$ and $q = n$. The theorem is thus proved by double induction starting from the case $p = q = 2$, for it is trivially true whenever $p \leq 1$ or $q \leq 1$.

COROLLARY. — *Let $y, x_1, x_2, \ldots, x_n$ be $n+1$ positive elements of* G *such that* $y \leqslant \sum_{i=1}^{n} x_i$; *then there exist n positive elements y_i $(1 \leqslant i \leqslant n)$ such that $y_i \leqslant x_i$ and* $y = \sum_{i=1}^{n} y_i$.

It is sufficient to apply theorem 1 to the sequence (x_i) and the sequence consisting of the two elements y and $z = \left(\sum_{i=1}^{n} x_i\right) - y$.

11. Positive and negative parts

DEFINITION 4. — *In a lattice ordered group* G *the positive part* (resp. *negative part, absolute value*) *of an element $x \in$ G is the element* $\sup(x, 0)$ (resp. $\sup(-x, 0)$, $\sup(x, -x)$), *which is denoted* x^+ (resp. x^-, $|x|$).

Despite its name, the negative part x^- is a *positive* element.

Clearly $x^- = (-x)^+$ and $|-x| = |x|$. Let us also note the following formulae, the first of which is an immediate consequence of the definitions and of the invariance of the ordering under translation, and the second of which follows from the first by Prop. 7 of VI, p. 10 :

$$\text{(3)} \qquad \begin{cases} \sup(x, y) = x + (y - x)^+ , \\ \inf(x, y) = y - (y - x)^+ . \end{cases}$$

PROPOSITION 9. — *a) For each element x of a lattice ordered group* G *we have* $x = x^+ - x^-$ *and* $\inf(x^+, x^-) = 0$.

b) For every expression of x as the difference of two positive elements, say $x = u - v$, *we have* $u = x^+ + w$ *and* $v = x^- + w$ *with* $w = \inf(u, v)$. *In particular if* $\inf(u, v) = 0$ *then* $u = x^+$ *and* $v = x^-$.

c) The relation « $x \leqslant y$ » is equivalent to « $x^+ \leqslant y^+$ and $x^- \geqslant y^-$ ».

d) We have $|x| = x^+ + x^- \geqslant 0$.

e) For any x and y in G, *we have the inequality* $|x + y| \leqslant |x| + |y|$, *and more generally* $\left|\sum_{i=1}^{n} x_i\right| \leqslant \sum_{i=1}^{n} |x_i|$ *for any finite family (x_i) of elements of* G.

f) For any x and y in G *we have* $||x| - |y|| \leqslant |x - y|$.

We will prove *a*) and *b*) simultaneously. If $x = u - v$ with u and v positive, then $u \geqslant x$, so $u \geqslant \sup(x, 0) = x^+$, and $w = u - x^+$ is positive. On the other hand we have

$$x^+ - x = \sup(x, 0) - x = \sup(x - x, -x) = x^-$$

from which it follows that $x = x^+ - x^-$, and $v - x^- = w$. If $z \leqslant x^-$ then $z \leqslant x^+ - x$, and so $x \leqslant x^+ - z$; if also $z \leqslant x^+$, then $x^+ - z$ is positive, and so

$x^+ \leqslant x^+ - z$ by definition of x^+. Hence we have $z \leqslant 0$, which implies $\inf(x^+, x^-) = 0$, whence by translation $\inf(u, v) = w$.

c) The relation $x \leqslant y$ implies $\sup(y, 0) \geqslant x$ and $\sup(y, 0) \geqslant 0$, hence $x^+ \leqslant y^+$; similarly if $-y \leqslant -x$ we deduce $x^- \geqslant y^-$. The converse implication follows immediately from $x = x^+ - x^-$ and $y = y^+ - y^-$.

d) Since $x \leqslant x^+$ and $-x \leqslant x^-$, it is clear that

$$|x| = \sup(x, -x) \leqslant x^+ + x^- .$$

Conversely, if $a \geqslant x$ and $a \geqslant -x$ then we deduce from *c)* that $a^+ \geqslant x^+$, $a^+ \geqslant x^-$, $a^- \leqslant x^-$ and $a^- \leqslant x^+$; since a^- is positive and $\inf(x^+, x^-) = 0$, the last two inequalities imply $a^- = 0$ and $a = a^+$; the first two inequalities then give $a \geqslant \sup(x^+, x^-)$, and $\sup(x^+, x^-)$ is equal to $x^+ + x^-$ by *a)* and by Prop. 7 of VI, p. 10.

e) Since $x \leqslant |x|$ and $y \leqslant |y|$, we have $x + y \leqslant |x| + |y|$; since $-x \leqslant |x|$ and $-y \leqslant |y|$, we have $-x - y \leqslant |x| + |y|$; whence the first inequality. The second follows by induction on n.

f) Replacing x and y by y and $x - y$ in *e)*, we obtain

$$|x| - |y| \leqslant |x - y| ;$$

similarly we have $|y| - |x| \leqslant |y - x| = |x - y|$; whence the stated result.

Remark. — We deduce from *d)* that $|x| = 0$ implies $x = 0$ (for x^+ and x^- are positive); thus $x \neq 0$ implies $|x| > 0$.

PROPOSITION 9 (DIV). — *If the group $\mathscr{P}^*$ of principal fractional ideals of* K *is lattice ordered, then every element x of* K* *can be written in the form $x = uv^{-1}$, where u and v are elements of* A *such that* $1 = \gcd(u, v)$; *for any other expression $x = u'v'^{-1}$ of x as the quotient of two elements of* A, *we have $u' = uw$ and $v' = vw$, where w is a* gcd *of u', v'; in particular if* $1 = \gcd(u', v')$ *then u' and v' are associates of u and v respectively.*

Such an expression uv^{-1} for an element x of K* is often called a *reduced fraction.*

12. Coprime elements

DEFINITION 5. — *In an ordered group two elements x and y are called coprime if* $\inf(x, y) = 0$.

In some cases it is natural to define two elements to be *coprime* if $\inf(|x|, |y|) = 0$ (cf. INT, II, § 1) or to introduce the corresponding terminology in divisibility theory. We shall not do so here.

Two coprime elements are necessarily *positive.* The positive and negative parts x^+ and x^- of x are coprime (VI, p. 12, Prop. 9, *a)*). The elements x_ι of a family

$(x_\iota)_{\iota \in I}$ are said to be *setwise coprime* if $\inf_{\iota \in I} x_\iota = 0$; if the x_ι are all $\geqslant 0$ then it is sufficient for there to exist a finite subset J of I such that the corresponding elements are setwise coprime. The elements of a family (x_ι) are said to be *pairwise coprime* if $\inf(x_\iota, x_\kappa) = 0$ for every pair (ι, κ) of distinct indices.

The x_ι can be setwise coprime without being pairwise coprime.

If x and y are coprime, we also say that x is coprime to y, or that y is coprime to x.

(DIV) Two elements x and y of K are said to be *coprime* if the principal ideals (x) and (y) are nonzero and coprime in $\mathcal{P}^*$; this amounts to saying that 1 is a gcd of x and y, and implies that x and y *belong* to A. For example the numerator and denominator of a reduced fraction are coprime. The notions of pairwise and setwise coprime elements are defined similarly.

(DIV) When x and y are coprime, they are often said to be « prime to one another » ; it is convenient to avoid this terminology, which can lead to confusion with the notion of prime numbers (I, p. 50, Def. 16).

PROPOSITION 10. — *Let x, y and z be three elements of an ordered group ; for $x - z$ and $y - z$ to be coprime, it is necessary and sufficient that $z = \inf(x, y)$.*

Indeed the relations $z = \inf(x, y)$ and $0 = \inf(x - z, y - z)$ are equivalent.

PROPOSITION 10 (DIV). — *Let a, b and c be three elements of* K *with $c \neq 0$; for the quotients ac^{-1} and bc^{-1} to be coprime, it is necessary and sufficient that c be a* gcd *of a and b.*

PROPOSITION 11. — *If (x_i), (y_j) are two finite families of positive elements of a lattice ordered group, then*

$$\inf\left(\sum_i x_i, \sum_j y_j\right) \leqslant \sum_{i,j} \inf(x_i, y_j) .$$

Arguing by induction on the number of elements in the families (x_i) and (y_j), it is enough to prove that, if x, y and z are positive elements, then

$$\inf(x, y + z) \leqslant \inf(x, y) + \inf(x, z) .$$

Indeed, put $t = \inf(x, y + z)$; by VI, p. 12, Cor., we can write $t = t_1 + t_2$ with $0 \leqslant t_1 \leqslant y$ and $0 \leqslant t_2 \leqslant z$; since t_1 and t_2 are positive, we also have $t_1 \leqslant x$ and $t_2 \leqslant x$, whence $t_1 \leqslant \inf(x, y)$ and $t_2 \leqslant \inf(x, z)$.

COROLLARY 1. — *If x and y are two coprime elements, and z a positive element, of a lattice ordered group, then $\inf(x, z) = \inf(x, y + z)$.*

Indeed $\inf(x, y + z) \leqslant \inf(x, z)$ by Prop. 11, and $\inf(x, z) \leqslant \inf(x, y + z)$ since $y \geqslant 0$, whence the corollary.

COROLLARY 2. — *In a lattice ordered group, if x and y are coprime and if $z \geqslant 0$ and $x \leqslant y + z$, then $x \leqslant z$.*

COROLLARY 3. — *In a lattice ordered group, if x is coprime to y and to z, then it is also coprime to $y + z$.*

COROLLARY 4. — *If $(x_i)_{1 \leqslant i \leqslant n}$ and $(y_j)_{1 \leqslant j \leqslant m}$ are two finite families of elements of a lattice ordered group G, such that each x_i is coprime to each y_j, then $x_1 + \cdots + x_n$ is coprime to $y_1 + \cdots + y_m$.*

This follows from Cor. 3 by induction on m and n.

COROLLARY 5. — *For any integer $n \geqslant 0$ we have $(nx)^+ = nx^+$ and $(nx)^- = nx^-$; for all $n \in \mathbf{Z}$ we have $|nx| = |n| \cdot |x|$.*

Indeed $nx = nx^+ - nx^-$; since x^+ and x^- are coprime, so are nx^+ and nx^- if $n \geqslant 0$ (Cor. 4); the first assertion follows by Prop. 9 *b*) of VI, p. 12. The second follows from the first by Prop. 9 *d*) in the case $n \geqslant 0$; the case $n < 0$ follows from this as a result of the relation $|-x| = |x|$.

PROPOSITION 11 (DIV). — *Suppose the set $\mathcal{P}^*$ is a lattice, and let (a_i), (b_j) be two finite families of elements of* A. *Then every* gcd *of $\prod_i a_i$ and $\prod_j b_j$ divides the product $\prod_{i,j} \gcd(a_i, b_j)$.*

COROLLARY 1 (DIV). — *If a, b, c are three elements of* A *such that a is coprime to b, then every* gcd *of a and c is also a* gcd *of a and bc.*

COROLLARY 2 (DIV) (Euclid's lemma). — *Let a, b, c be three elements of* A. *If a is coprime to b and divides bc, then it divides c.*

COROLLARY 3 (DIV). — *If x is coprime to y and to z, then it is coprime to yz.*

COROLLARY 4 (DIV). — *If (x_i) and (y_j) are two finite families of elements of* A *such that each x_i is coprime to each y_j, then the product of the x_i is coprime to the product of the y_j.*

COROLLARY 5 (DIV). — *If d is a* gcd *of x and y, then d^n is a* gcd *of x^n and y^n for each positive integer n.*

Indeed xd^{-1} and yd^{-1} are coprime (Prop. 10 (DIV)), and hence so are $x^n d^{-n}$ and $y^n d^{-n}$ (Cor. 4).

PROPOSITION 12. — *Let x_i $(1 \leqslant i \leqslant n)$ be n pairwise coprime elements in a lattice ordered group. Then*

$$\sup(x_1, \ldots, x_n) = x_1 + \cdots + x_n .$$

This follows from the formula $u + v = \sup(u, v) + \inf(u, v)$ (VI, p. 10, Prop. 7) by induction on n, using the fact that x_i is coprime to $x_1 + \cdots + x_{i-1}$ for $2 \leqslant i \leqslant n$ (Cor. 4 to Prop. 11).

Remark. — Prop. 7 of VI, p. 10 also shows that a necessary and sufficient condition for x and y to be coprime is that $x + y = \sup(x, y)$.

PROPOSITION 12 (DIV). — *Let a_i be a finite number n of pairwise coprime elements of* A ; *then the product $a_1 \ldots a_n$ is an* lcm *of $a_1, \ldots, a_n$.*

PROPOSITION 13. — *In a lattice ordered group* G, *let (x_α) be a family having an infimum* (resp. *supremum*), *and let z be an arbitrary element of* G ; *then the family* $(\sup(z, x_\alpha))$ (resp. $(\inf(z, x_\alpha))$) *has an infimum* (resp. *supremum*) *and*

$$(4)\qquad \begin{cases} \inf\limits_\alpha(\sup(z, x_\alpha)) = \sup\left(z, \inf\limits_\alpha x_\alpha\right) \\ \sup\limits_\alpha(\inf(z, x_\alpha)) = \inf\left(z, \sup\limits_\alpha x_\alpha\right) \end{cases}$$

respectively.

Suppose that the family (x_α) has an infimum y. We will show that $\sup(z, y)$ is an infimum of the family $(\sup(z, x_\alpha))$.

Indeed $\sup(z, x_\alpha) = z + (x_\alpha - z)^+$ and by translating we can reduce to the case $z = 0$, that is we must show that (x_α^+) has an infimum equal to y^+. Since $y \leqslant x_\alpha$ we have $y^+ \leqslant x_\alpha^+$ for all α (VI, p. 12, Prop. 9, *c*)). Conversely, if $a \leqslant x_\alpha^+$ for all α, then it follows that $a \leqslant x_\alpha + x_\alpha^-$ (Prop. 9, *a*)) ; now $y^- \geqslant x_\alpha^-$ since $y \leqslant x_\alpha$; hence $a \leqslant x_\alpha + y^-$ for all α, that is $a \leqslant y + y^- = y^+$.

The other formula follows by changing to the opposite order relation.

COROLLARY. — *If an element z of a lattice ordered group* G *is coprime to each element x_α of a family having an infimum y, then z is coprime to y.*

This is an immediate consequence of the second formula (4).

Remark. — Applying the formulae of Prop. 13 to a family of two elements (x, y), we obtain the following formulae, which express the fact that each of the laws of composition sup, inf in a lattice ordered group is *distributive* with respect to the other :

$$\sup(z, \inf(x, y)) = \inf(\sup(z, x), \sup(z, y))$$
$$\inf(z, \sup(x, y)) = \sup(\inf(z, x), \inf(z, y)) .$$

This distributivity property is peculiar to lattice ordered *groups*, and does not extend to arbitrary lattices, or even to lattice ordered monoids (cf. VI, p. 33, Ex. 24).

13. Irreducible elements

DEFINITION 6. — *An element x of an ordered group G is said to be irreducible if it is a minimal element in the set of strictly positive elements of G.*

Let x be an irreducible element of the ordered group G ; if y is a positive element of G then the element $\inf(x, y)$, if it exists, can only be equal to x or to 0. Thus in a lattice ordered group G, every positive y is either greater than or coprime to the irreducible element x ; in particular, two distinct irreducible elements are coprime.

(DIV) An element p of A is called *irreducible* if the ideal (p) is an irreducible element of the ordered group $\mathscr{P}^*$; this says that p is neither zero nor invertible, and that any element of A which divides p is associate either to p or to 1. If $\mathscr{P}^*$ is lattice ordered, then every $a \in A$ is either coprime to p or a multiple of p.

Examples (DIV). — 1) An integer $p > 0$ is irreducible in **Z** if and only if it is *prime* (I, p. 50).

2) A polynomial in one indeterminate over a field K is irreducible if and only if it is irreducible in the usual sense (IV, p. 13).

PROPOSITION 14. — *For an element $x > 0$ of an ordered group G to be irreducible it is sufficient that it satisfy the following property :*

(P) *The relations $x \leqslant y + z$, $y \geqslant 0$, $z \geqslant 0$ imply $x \leqslant y$ or $x \leqslant z$.*

This condition is necessary when G is lattice ordered.

If G is lattice ordered and x is irreducible, we have just seen that y is either greater than x or coprime to x ; in the latter case Cor. 2 of VI, p. 15 shows that z is greater than x. Conversely, suppose the condition is satisfied : if $0 \leqslant y \leqslant x$ then it follows, by putting $x = y + z$ $(z \geqslant 0)$, that either $x \leqslant y$ or $x \leqslant z$; in the first case we have $x = y$; in the second we have $x \leqslant x - y$, so $y \leqslant 0$ and so $y = 0$; this shows that x is indeed irreducible.

PROPOSITION 14 (DIV). — *For a nonzero element p of A to be irreducible it is sufficient that it not be a unit, and that it cannot divide a product of two elements of A without dividing one or other of them. This condition is necessary is $\mathscr{P}^*$ is lattice ordered.*

Remark. — Proposition 14 (DIV) can also be expressed as follows : if p is a non zero element of A such that the ideal (p) is *prime* (I, p. 117, Def. 3) then p is irreducible ; conversely, if $\mathscr{P}^*$ is lattice ordered and p is irreducible then the ideal (p) is prime.

PROPOSITION 15. — *In an ordered group G, let $(p_\iota)_{\iota \in I}$ be a family of pairwise distinct positive elements of G satisfying condition (P) (and hence irreducible). Then the map*

$$(n_\iota)_{\iota \in I} \mapsto \sum_{\iota \in I} n_\iota p_\iota$$

is an isomorphism of the ordered group $\mathbf{Z}^{(I)}$, *the direct sum of the ordered groups* $\mathbf{Z}$ (VI, p. 7) *onto the ordered subgroup of* G *generated by the* p_ι.

It is enough to show that the relation $\sum_{\iota \in I} n_\iota p_\iota \geqslant 0$ is *equivalent* to $n_\iota \geqslant 0$ for all ι, for in particular the relation $\sum_{\iota \in I} n_\iota p_\iota = 0$ will imply $n_\iota = 0$ for all ι, hence this will show that the family (p_ι) is linearly independent. Now let I′ (resp. I″) be the finite subset of I consisting of those ι such that $n_\iota > 0$ (resp. $n_\iota < 0$) ; we have

$$\sum_{\iota \in I'} n_\iota p_\iota \geqslant \sum_{\iota \in I''} (-n_\iota) p_\iota .$$

In particular, for $\lambda \in I''$, this implies that $p_\lambda \leqslant \sum_{\iota \in I'} n_\iota p_\iota$, and it follows by induction from property (P) that we must have $p_\lambda \leqslant p_\iota$ for some $\iota \in I'$; since p_ι is irreducible, this would imply $p_\lambda = p_\iota$, which is absurd. Hence I″ is empty, which proves the proposition.

THEOREM 2. — *Let* G *be a filtered group. Then the following properties are equivalent :*

a) G *is isomorphic to an ordered group of the form* $\mathbf{Z}^{(I)}$.

b) G *is lattice ordered and satisfies the following condition :*

(MIN) *Every nonempty set of positive elements of* G *has a minimal element.*

c) G *satisfies condition* (MIN) *and every irreducible element of* G *has property* (P).

d) G *is generated by its irreducible elements, and every irreducible element of* G *has property* (P).

Let us show first that *a*) implies *b*). The group $\mathbf{Z}^{(I)}$ is lattice ordered, as the direct sum of totally ordered groups. On the other hand let E be a nonempty set of positive elements of $\mathbf{Z}^{(I)}$ and let $x = \sum_\iota n_\iota e_\iota$ be an element of E (where (e_ι) denotes the natural basis of $\mathbf{Z}^{(I)}$) ; there are a finite number $\prod_\iota (n_\iota + 1)$ of elements y of $\mathbf{Z}^{(I)}$ such that $0 \leqslant y \leqslant x$, so the set F of elements of E which are less than or equal to x is *a fortiori* finite ; since it is nonempty, it contains a minimal element (*Set Theory*, III, p. 170, Cor. 2), which is clearly a minimal element of E.

It is clear that *b*) implies *c*), by Prop. 14. Let us show that *c*) implies *d*). Since G is filtered, it is enough (VI, p. 4, Prop. 4) to check that the set F of positive elements of G which are sums of irreducible elements is equal to $G_+ - \{0\}$. If this were not true, it would follow from (MIN) that the complement of F in $G_+ - \{0\}$ would have a minimal element a ; by definition a is not irreducible, so is the sum of two strictly positive elements x and y ; since $x < a$ and $y < a$, these elements belong to F, and so are sums of irreducible elements, and it follows that so is a, which is a contradiction. Finally, *d*) implies *a*) by Prop. 15.

We will apply Th. 2 to the theory of divisibility in principal ideal domains (VII, p. 4) and in unique factorisation domains (AC, VII, § 3), as well as to the study of ideals in a Dedekind ring (AC, VII, § 2).

§ 2. ORDERED FIELDS

1. Ordered rings

DEFINITION 1. — *Given a commutative ring A, we say that an ordering on A is compatible with the ring structure of A if it is compatible with the additive group structure of A, and if it satisfies the following axiom:*

(OR) *The relations $x \geqslant 0$ and $y \geqslant 0$ imply $xy \geqslant 0$.*

The ring A, together with such an ordering, is called an *ordered ring.*

Examples. — 1) The rings $\mathbf{Q}$ and $\mathbf{Z}$, with the usual orderings, are ordered rings.

2) A product of ordered rings, equipped with the product ordering, is an ordered ring. In particular the ring A^E of mappings from a set E to an ordered ring A is an ordered ring.

3) A subring of an ordered ring, with the induced ordering, is an ordered ring.

In an ordered ring, the relations $x \geqslant y$ and $z \geqslant 0$ imply $xz \geqslant yz$. Indeed these inequalities are equivalent to $x - y \geqslant 0$, $z \geqslant 0$ and $(x - y) z \geqslant 0$ respectively.

Analogously we can show that the relations $x \leqslant 0$ and $y \geqslant 0$ (resp. $y \leqslant 0$) imply $xy \leqslant 0$ (resp. $xy \geqslant 0$). These results are often invoked under the name of *sign rules* (two elements are said to *have the same sign* if they are both $\geqslant 0$ or both $\leqslant 0$). They imply that, if A is a *totally ordered* ring, then every square is positive, and in particular that every idempotent (for example the unit element) is positive.

Example. — There is *only one* totally ordered ring structure on $\mathbf{Z}$: indeed $1 > 0$, whence $n > 0$ for every natural number $n \neq 0$, by induction. In contrast there exist ordered ring structures on $\mathbf{Z}$ which are not totally ordered (see below).

Let P be the set of positive elements of an ordered ring A. It is known (VI, p. 3, Prop. 3) that P determines the ordering on A. To say that A is an ordered ring is equivalent to saying that P satisfies the following properties:

$$(\mathrm{AP_I}) \quad \mathrm{P} + \mathrm{P} \subset \mathrm{P}$$

$$(\mathrm{AP_{II}}) \quad \mathrm{PP} \subset \mathrm{P}$$

$$(\mathrm{AP_{III}}) \quad \mathrm{P} \cap (-\mathrm{P}) = \{0\}\,.$$

Indeed $(\mathrm{AP_I})$ and $(\mathrm{AP_{III}})$ state that the additive group of A is an ordered group (VI, p. 3, Prop. 3), while $(\mathrm{AP_{II}})$ is a translation of (OR).

Recall that the following condition is necessary and sufficient for the order relation on A to be *total*:

$$(\mathrm{AP_{IV}}) \quad \mathrm{P} \cup (-\mathrm{P}) = \mathrm{A}\,.$$

Example. — In **Z**, if we take P to be the set of positive (in the usual sense) even integers, we get a ring which is *not* totally ordered.

Recall also that, in a totally ordered abelian group, the relation $n \,.\, x = 0$ (for a natural number $n \neq 0$) implies $x = 0$ (VI, p. 4) ; this gives us the following result.

PROPOSITION 1. — *A totally ordered ring is torsion free as a* **Z**-*module* (II, p. 313).

2. Ordered fields

DEFINITION 2. — *A* commutative *field, equipped with a* total *ordering, is called an ordered field if its ordering and its ring structure are compatible.*

We restrict ourselves to *total* order relations on fields because the others are very « pathological » (VI, p. 38, Ex. 6).

Examples. — 1) The field **Q** of rational numbers is an ordered field.

2) A subfield of an ordered field, with the induced ordering, is an ordered field.

3) * The field of real numbers is an ordered field. *

Let K be an ordered field. For all $x \in K$ we put

$$\begin{aligned} \operatorname{sgn}(x) &= 1 && \text{if } x > 0\,, \\ \operatorname{sgn}(x) &= -1 && \text{if } x < 0\,, \\ \operatorname{sgn}(x) &= 0 && \text{if } x = 0\,. \end{aligned}$$

Then we have $\operatorname{sgn}(xy) = \operatorname{sgn}(x)\operatorname{sgn}(y)$; we call $\operatorname{sgn}(x)$ the *sign* of x. The map $x \mapsto \operatorname{sgn}(x)$ from K^* to the multiplicative group $\{-1, +1\}$ is a surjective homomorphism whose kernel, the set of strictly positive elements of K, is a subgroup of K^* of index 2.

Conversely, if K is a commutative field and $s : K^* \to \{-1, +1\}$ is a surjective homomorphism whose kernel is closed under addition, then s is the sign map for a unique ordered field structure, where the set of strictly positive elements is the kernel of s.

For all x and y in K we have $x = \operatorname{sgn}(x)|x|$ and $|xy| = |x|\,|y|$.

On the other hand every ordered field is of characteristic zero (Prop. 1).

PROPOSITION 2. — *Let* A *be a totally ordered integral domain, and let* K *be its field of fractions. Then there exists one and only one ordering on* K *which restricts to the given ordering on* A *and makes* K *an ordered field.*

Every $x \in K$ can be expressed in the form $x = ab^{-1}$, with a and b in A and $b \neq 0$. If x is positive, then a and b have the same sign, and conversely. Thus we see that, if there exists an ordering on K satisfying the prescribed conditions, then it is unique, and the set P of positive elements is identical to the set of ab^{-1}, where a and b are elements of A of the same sign, and $b \neq 0$. It remains to show that P satisfies conditions (AP_I), (AP_{II}), (AP_{III}) and (AP_{IV}). This is obvious for (AP_{II}) and (AP_{IV}). For (AP_I), consider $ab^{-1} + cd^{-1}$, where we may assume that a, b, c and d are positive ; this sum is $(ad + bc)(bd)^{-1}$, and $ad + bc$ and bd are positive.

To show (AP_{III}), consider an identity of the form $ab^{-1} = -cd^{-1}$, so that $ad + bc = 0$. If we assume that a and b have the same sign and that c and d have the same sign, then the sign rules show that ad and bc have the same sign ; whence $ad = bc = 0$, so $a = c = 0$; hence P does indeed satisfy (AP_{III}).

Example. — Since **Z** admits only one totally ordered ring structure (VI, p. 19, *example*), the field **Q** admits only one ordering which makes it an ordered field : this is the usual ordering.

3. Extensions of ordered fields

DEFINITION 3. — *Let* K *be an ordered field. An ordered extension of* K *is a pair* (E, u), *where* E *is an ordered field and* u *is an increasing homomorphism from* K *to* E.

Let K be a field, let E be an ordered field and let $u : K \to E$ be a homomorphism. The relation

$$x \leqslant y \quad \text{if} \quad u(x) \leqslant u(y)$$

is a total order relation on K which gives it an ordered field structure, said to be *induced* by that of E. If K and E are ordered fields, then a homomorphism $u : K \to E$ is increasing if and only if the ordered field structure of K is induced by that of E. We will usually identify K with its image in E under u.

Examples. — 1) Every ordered field K is an ordered extension of **Q**. Indeed K is an extension of **Q**, since it is of characteristic zero, and on the other hand **Q** can only be ordered in one way, as we have just seen.

2) Let K be an ordered field, and let K(X) be the field of rational functions in one indeterminate over K. Let us define an ordering on the polynomial ring K[X] by taking the positive elements to be 0 and those polynomials whose leading coefficient is positive. In this way we obtain a totally ordered ring whose ordering extends that of K. By applying Prop. 2 we give K(X) the structure of an ordered extension of K. * For K = **R** it can be shown that the order relation defined on K(X) in this way is that of growth near $+\infty$ (cf. VI, p. 24, Prop. 4). *

THEOREM 1. — *For an extension* E *of* K *to admit the structure of an ordered extension of* K, *the following condition is necessary and sufficient :*

(OE) *The relation* $p_1x_1^2 + \cdots + p_nx_n^2 = 0$ *implies*

$$p_1x_1 = \cdots = p_nx_n = 0$$

for any finite sequence (x_i, p_i) *of pairs of elements* x_i *of* E *and positive elements* p_i *of* K.

Condition (OE) is clearly equivalent to :

(OE') *The element* -1 *is not a sum of elements of the form* px^2 ($x \in E$, $p \in K$, $p \geqslant 0$).

Condition (OE) is necessary : if E is an ordered extension of K then the elements $p_i x_i^2$ are positive in E, so zero if their sum is zero. On the other hand $p_i x_i^2 = 0$ is equivalent to $p_i x_i = 0$.

Conversely, suppose condition (OE) is satisfied, then we will define an ordering on E by constructing a subset P of E which satisfies conditions (AP_I), (AP_{II}), (AP_{III}) and (AP_{IV}), and which contains the set K_+ of positive elements of K. Such a subset P will certainly make E an ordered extension of K, for we will have $K \cap P = K_+$; indeed, if P were to contain an element $-a < 0$ of K, then a would belong to $P \cap (-P)$, contradicting (AP_{III}).

To define P, let us consider the set $\mathfrak{M}$ of subsets of E which satisfy (AP_I), (AP_{II}) and (AP_{III}), and which contain the union of K_+ and the set C of squares of elements of E. This set $\mathfrak{M}$ is nonempty, for it contains the set P_0 of elements of the form $\sum_i p_i x_i^2$ (that P_0 satisfies (AP_{III}) follows immediately from (OE)).

Moreover $\mathfrak{M}$ is inductive (*Set Theory*, III, p. 154, Def. 3). Thus there exists, by Th. 2 of *Set Theory*, III, p. 154, a maximal element in $\mathfrak{M}$, which it remains for us to prove satisfies (AP_{IV}) ; now this follows from the following lemma :

Lemma. — *Let* $P \in \mathfrak{M}$ *and* $x \notin P$; *then there exists* $P' \in \mathfrak{M}$ *such that* $P \subset P'$ *and* $-x \in P'$.

Take $P' = P - xP$, and check that P' has the required properties. Since $0 \in C \subset P$, we have $P \subset P'$. Whence $C \subset P'$ and $K_+ \subset P'$. Since $1 \in C \subset P$ we have $-x \in P'$. We have

$$P' + P' = P - xP + P - xP = P + P - x\,(P + P) \subset P - xP = P' ,$$

whence (AP_I). We have

$$P'P' = (P - xP)(P - xP) \subset \subset PP + x^2PP - x(PP + PP) \subset P + CP - xP \subset P - xP = P' ,$$

whence (AP_{II}.) Finally, let us check (AP_{III}) : suppose given an identity of the form $p - xq = -(r - xs)$ where p, q, r, s belong to P ; we deduce from this the relation $x(s + q) = p + r$; if $s + q \neq 0$ we have

$$x = (s + q)^{-2}(s + q)(p + r) \in CPP \subset P ,$$

contrary to the hypothesis ; hence $s + q = 0$, whence $p + r = 0$; since P satisfies (AP_{III}) we deduce that $s = q = r = p = 0$, which completes the proof.

COROLLARY 1 (« Artin-Schreier Theorem »). — *A necessary and sufficient condition for there to exist an ordering on a commutative field* E *which makes it an ordered field, is that the relation* $x_1^2 + \cdots + x_n^2 = 0$ *imply* $x_1 = \cdots = x_n = 0$.

The necessity is obvious. Conversely, the stated condition implies that E is of characteristic zero, hence an extension of $\mathbf{Q}$; then condition (OE) is satisfied, and Th. 1 shows that there exists on E the structure of an ordered extension of $\mathbf{Q}$, that is an ordered field structure.

There does *not* exist any ordered field structure on a field E in which -1 is a square, in particular on an algebraically closed field.

COROLLARY 2. — *Let* E *be an extension of* K *admitting the structure of an ordered extension of* K. *For an element $x \in$ E to be positive under every such structure on* E, *it is necessary and sufficient that x be of the form $\sum_i p_i x_i^2$, where $x_i \in$ E and the p_i are positive elements of* K.

The condition is obviously sufficient ; it is also necessary, for (in the notation of the proof of Th. 1), if $x \notin P_0$ there exists a maximal element P of $\mathfrak{M}$ such that $x \notin$ P ; then $-x \in$ P by the Lemma, and x is not positive under the ordering defined by P, since $x \neq 0$.

4. Algebraic extensions of ordered fields

Let K be an ordered field, and f a polynomial in K[X]. We will say that f *changes sign in* K if there exist two elements a and b in K such that $f(a)\,f(b) < 0$; then we say that f *changes sign between a and b.*

PROPOSITION 3. — *Let* K *be an ordered field and f an irreducible polynomial over* K *which changes sign between a and b in* K. *Then the extension* $\mathrm{E} = \mathrm{K}[\mathrm{X}]/(f)$ *of* K *admits the structure of an ordered extension.*

We will argue by induction on the degree n of f. For $n = 1$ the proof is trivial. Suppose the result holds for degrees $\leqslant n-1$, and let us prove it for degree n by contradiction ; by Th. 1 we are thus assuming a relation of the form

$$1 + \sum_i p_i f_i^2(\mathrm{X}) \equiv 0 \pmod{f(\mathrm{X})}, \quad \text{where} \quad f_i \in \mathrm{K}[\mathrm{X}], \quad p_i \in \mathrm{K} \quad \text{and} \quad p_i \geqslant 0 .$$

Without loss of generality we may suppose that the f_i have degrees $\leqslant n-1$ (IV, p. 11, Cor). Then

$$1 + \sum_i p_i f_i^2(\mathrm{X}) = h(\mathrm{X})\, f(\mathrm{X})$$

where $h \neq 0$ has degree at most $n-2$. Replacing X by a and b in the above inequality, we see that $h(a)\,f(a) > 0$ and $h(b)\,f(b) > 0$. Since f changes sign between a and b by hypothesis, we conclude that $h(a)\,h(b) < 0$. Then we have a similar inequality for one of the irreducible factors $g(X)$ of $h(X)$: that is $g(a)\, g(b) < 0$. But $1 + \sum_i p_i f_i^2(X) \equiv 0 \pmod{g(X)}$, which shows that the field $K[X]/(g)$ cannot be an ordered extension of K (Th. 1), contrary to the induction hypothesis.

Remark. — There exist irreducible polynomials f over an ordered field K which do not change sign in K, but such that $K[X]/(f)$ admits the structure of an ordered extension of K (cf. VI, p. 43, Ex. 26, *c*)).

In order to apply the previous proposition we will need the following result :

PROPOSITION 4. — *Let* K *be an ordered field and let* $f \in K[X]$. *There exists an interval in* K, *in the complement of which f takes the same sign as its highest degree term.*

We can immediately reduce ourselves to the case of a monic polynomial ; then one can write $f(x) = x^n(1 + a_1 x^{-1} + \cdots + a_n x^{-n})$ for $x \neq 0$. Let

$$M = \sup(1, |a_1| + \cdots + |a_n|) .$$

For $|x| > M$ we have $1 + a_1 x^{-1} + \cdots + a_n x^{-n} > 0$, which completes the proof of the proposition.

COROLLARY 1. — *Every extension of an ordered field of odd finite degree admits the structure of an ordered extension.*

Such an extension, being monogenous (V, p. 40, Th. 1), is isomorphic to $K[X]/(f)$, where f is an irreducible polynomial of odd degree. Then it is enough to show that f changes sign in K (Prop. 3), which follows immediately from Prop. 4.

COROLLARY 2. — *If a is a positive element of an ordered field* K, *then every splitting field* E *of the polynomial* $X^2 - a$ *admits the structure of an ordered extension of* K.

The result is trivial if a is a square in K. Otherwise the polynomial $f(X) = X^2 - a$ is irreducible and changes sign, since $f(0) < 0$ and $f(x)$ has the same sign as x^2, so positive, for x in the complement of some interval of K. We can now complete the proof by applying Prop. 3.

Remark. — When the ordered field K contains the « square roots » of a positive element a of K (roots of the polynomial $X^2 - a$) then the notation $\sqrt{a}$ is generally reserved for the *positive* square root. If K does *not* contain the square roots b and $-b$ of a in the field E, then the latter can be made an ordered extension of K in *two* ways, each induced from the other *via* the K-automorphism which sends b to $-b$; the choice of one of these orderings determines $\sqrt{a}$: it is whichever of the elements b and $-b$ is positive.

If a and a' are two positive elements of K, whose square roots are in K, then $\sqrt{aa'} = \sqrt{a}\,\sqrt{a'}$, which follows from the definition of $\sqrt{a}$ and the sign rule.

5. Maximal ordered fields

DEFINITION 4. — *An ordered field* K *is maximal if every ordered* algebraic *extension of* K *is trivial.*

Example. — * We will see later (*Gen. Top.*, VIII, p. 1) that the field **R** of real numbers is a maximal ordered field. *

The existence of maximal ordered fields is a consequence of the following theorem :

THEOREM 2. — *Every ordered field* K *admits an ordered algebraic extension which is a maximal ordered field.*

One can show that this ordered extension is unique up to K-isomorphism (VI, p. 40, Ex. 15).

Let Ω be an algebraic closure of K, and let $\mathfrak{N}$ be the set of pairs (A, ω), where A is a sub-K-extension of Ω, and ω is an ordering on A making A an ordered extension of K. Order $\mathfrak{N}$ by the relation « L is an ordered extension of M » between M and L. Equipped with this ordering , $\mathfrak{N}$ is an *inductive* ordered set : indeed if (L_ι) is a totally ordered family of elements of $\mathfrak{N}$, then the field $L = \bigcup_\iota L_\iota$, ordered by taking $L_+ = \bigcup_\iota (L_\iota)_+$ is an upper bound for the L_ι. Then $\mathfrak{N}$ has a maximal element, by *Set Theory*, III, p. 154, Th. 2, which completes the proof.

PROPOSITION 5. — *Let* K *be a maximal ordered field, and let f be a polynomial in* K[X] *which changes sign between two elements a and b of* K *(with $a < b$). Then f has a root x in* K *such that $a < x < b$.*

At least one of the irreducible factors of f, say h, changes sign between a and b. Then the field $K[X]/(h)$ admits the structure of an ordered extension of K (VI, p. 23, Prop. 3), and h has degree 1 (Def. 4). Since $h(a)\,h(b) < 0$, the unique root x of h is such that $a < x < b$, since a polynomial function of degree 1 is monotonic.

PROPOSITION 6. — *Every positive element of a maximal ordered field* K *has a square root in* K. *Every polynomial of odd degree in* K[X] *has at least one root in* K.

This follows immediately from Cor. 2 and 1 to Prop. 4 of VI, p. 24.

COROLLARY. — *On a maximal ordered field* K *there exists only one ordering compatible with the field structure.*

Indeed the positive elements of K are determined by its algebraic structure : they are the squares.

6. Characterisation of maximal ordered fields. Euler-Lagrange Theorem

The property expressed by Prop. 6 of VI, p. 25 characterises maximal ordered fields. More precisely :

THEOREM 3 (Euler-Lagrange). — *Let* K *be an ordered field. Then the following three properties are equivalent :*

a) The field K(i) *is algebraically closed (where i denotes a square root of* -1).

b) The ordered field K *is maximal.*

c) Every positive element of K *is a square, and every polynomial of odd degree in* K[X] *has a root in* K.

It is clear that *a*) implies *b*) : indeed K has only two algebraic extensions up to isomorphism, the field K itself and K(i), which cannot be ordered since -1 is a square.

The fact that *b*) implies *c*) is nothing other than Prop. 6 of VI, p. 25.

It remains for us to prove that *c*) implies *a*). That will follow from the next two propositions.

PROPOSITION 7. — *Let* K *be an ordered field in which every positive element is a square. Then every element of* K(i) *is a square, and every polynomial of degree* 2 *over* K(i) *has a root in* K(i).

Let us show first that the second assertion reduces to the first. One can put the second degree polynomial $aX^2 + bX + c$ $(a \neq 0)$ in the following form, often called the *canonical trinomial form :*

$$a((X + (b/2a))^2 - (b^2 - 4ac)/4a^2)\,.$$

If d is a square root of $(b^2 - 4ac)/4a^2$, then $d - (b/2a)$ is a root of the quadratic polynomial under consideration.

Now we show that every element $a + bi$ $(a \in K,\ b \in K)$ is a square ; we are looking for an element $x + yi$ such that

$$(x + yi)^2 = a + bi\ ;$$

this translates into $x^2 - y^2 = a$ and $2xy = b$. From this we deduce that

$$(x^2 + y^2)^2 = a^2 + b^2\,.$$

Let c denote the positive square root of $a^2 + b^2$; then $c \geqslant |a|$, $c \geqslant |b|$ and $x^2 + y^2 = c$. Whence $x^2 = (c + a)/2$ and $y^2 = (c - a)/2$. Since $c \geqslant |a|$ these equations are soluble in K, and if x_0 and y_0 are two solutions then $x_0^2 - y_0^2 = a$ and $2x_0y_0 = \pm b$. We obtain the desired square root by taking $x = x_0$ and $y = b/2x_0$.

PROPOSITION 8. — *Let* K *be a commutative field* (of arbitrary characteristic) *and let* K' *be a splitting field for the polynomial* $X^2 + 1 \in K[X]$ (V, p. 21). *Suppose :*

a) every polynomial in K[X] *of odd degree has a root in* K' ;

b) every polynomial in $K'[X]$ *of degree* 2 *has a root in* K'.
Then K' *is algebraically closed.*

Note first that it is enough to prove that every non-constant polynomial in $K[X]$ has a root in K': this is indeed clear if $K' = K$; if $K' \neq K$ then $[K' : K] = 2$; let $a \mapsto \bar{a}$ denote the unique K-automorphism of K' distinct from the identity map; if $f \in K'[X]$ and if $\bar{f}$ denotes the polynomial obtained by applying $a \mapsto \bar{a}$ to the coefficients of f, then $f\bar{f} \in K[X]$; if $a \in K'$ is a root of $f\bar{f}$ then a is either a root of f or of $\bar{f}$; thus either a or $\bar{a}$ is a root of f.

Thus let f be a polynomial over K of degree $2^n p$, p odd. We will proceed by induction on n, the property being true for $n = 0$ by hypothesis a). Let E be an extension of K in which f splits into linear factors:

$$f(X) = \prod_i (X - a_i).$$

Let $b \in K$; put $y_{ij} = a_i + a_j + ba_ia_j \in E$ and

$$h(X) = \prod_{i<j} (X - y_{ij}) \in E[X].$$

The coefficients of this polynomial are symmetric functions in the a_i, with coefficients in K; it therefore belongs to $K[X]$ (IV, p. 62, Th. 1); since it has degree $2^n p(2^n p - 1)/2 = 2^{n-1}p'$ (p' odd), it has a root y_{ij} in K' by inductive hypothesis. If we note that this holds for all $b \in K$, and that K is an infinite field (indeed a finite field, which has monogenous extensions of arbitrarily large odd degree (V, p. 94, Prop. 3), cannot satisfy a)), then we can deduce the existence of at least one pair (i, j) such that

$$a_i + a_j + ba_ia_j \in K' \quad \text{and} \quad a_i + a_j + b'a_ia_j \in K',$$

with $b \neq b'$. Then $a_i + a_j$ and a_ia_j are elements of K', hence so are a_i and a_j, since they are the roots of the quadratic equation

$$x^2 - (a_i + a_j)x + a_ia_j = 0. \qquad \text{Q.E.D.}$$

For a generalisation and an alternative proof of Prop. 8, based on Galois theory, see VI, p. 46, Ex. 33.

Let K be an ordered field and let $K' = K(i)$; for every element $z = a + bi$ of K', the *norm* $z\bar{z} = a^2 + b^2$ of z relative to K (III, p. 544, example 1) is a positive element of K, which vanishes only for $z = 0$. If every positive element in K is a square (in particular if K is a maximal ordered field), then the positive square root of the norm $z\bar{z}$ is called the *absolute value* of z, and is written $|z|$. Since $|zz'|^2 = |z|^2 |z'|^2$ we have $|zz'| = |z| \cdot |z'|$.

Moreover, the *triangle inequality*

$$|z + z'| \leq |z| + |z'|$$

holds for every pair of elements z, z' of K'. Indeed, if $z = a + bi$ and $z' = a' + b'i$, then this inequality is equivalent to

$$(a + a')^2 + (b + b')^2 \leqslant a^2 + b^2 + a'^2 + b'^2 + 2\sqrt{(a^2 + b^2)(a'^2 + b'^2)}$$

and hence also to

$$(aa' + bb')^2 \leqslant (a^2 + b^2)(a'^2 + b'^2)$$

which can be written $(ab' - ba')^2 \geqslant 0$.

Th. 3 enables us to determine all the irreducible polynomials over a maximal ordered field :

PROPOSITION 9. — *If* K *is a maximal ordered field, then the only irreducible polynomials in* K[X] *are the first degree polynomials, and the second degree polynomials* $aX^2 + bX + c$ *such that* $b^2 - 4ac < 0$.

Since K(i) is algebraically closed, every algebraic extension of K, and hence also every irreducible polynomial over K, has degree 1 or 2. To see which second degree polynomials are irreducible, it is enough to consider the canonical form $a((X + (b/2a))^2 - (b^2 - 4ac)/4a^2)$ (cf. VI, p. 26, Prop. 7).

Remark. — Translating into the canonical trinomial form yields this stronger result : a necessary and sufficient condition for the polynomial $aX^2 + bX + c$ over a given ordered field K to have constant sign in K is that $b^2 - 4ac < 0$, and then the sign of the polynomial is that of a.

7. Vector spaces over an ordered field

Let K be an ordered field, and let E be a vector space over K. The relation « there exists $\lambda > 0$ in K such that $y = \lambda x$ » between two elements x and y in the set E – $\{0\}$ is an *equivalence relation*. The equivalence classes under this relation are called *open half-lines with origin* 0 ; the union of an open half-line and $\{0\}$ is called a *closed half-line* (or sometimes simply a *half-line*) with origin 0. Every vector $a \neq 0$ contained in an open (resp. closed) half-line Δ is called a *direction vector* of Δ, and Δ is the set of vectors λa for all scalars $\lambda > 0$ (resp. $\lambda \geqslant 0$). Every line D through 0 contains exactly two open (resp. closed) half-lines with origin 0 ; if Δ is one of these, then $-\Delta$ is the other (called the *opposite* of Δ).

Now if F is an *affine space* over K, and E the space of translations of F, then any subset of F of the form $\Delta = a + \Delta_0$, where Δ_0 is an open (resp. closed) half-line of E, is called an *open* (resp. *closed*) *half-line with origin* $a \in$ F. The half-line Δ_0 is completely determined by Δ (for it is the half-line with direction vector $b - a$, for any $b \neq a$ in Δ), and is called the *direction* of Δ ; a direction vector of Δ_0 is also called a *direction vector* of Δ.

Suppose now that E has *finite* dimension n over K ; then it is known (III, p. 518, Cor. 1) that the *n-th exterior power* $\bigwedge^n$ E is a vector space of dimension 1 over K, and so the union of two opposite closed half-lines of origin 0. These half-lines are called *orientations of* E ; the space E together with a given one of these half-lines Δ is said to be *oriented ;* an n-vector z is then called *positive* (resp. *negative*) under this orientation if it belongs to Δ (resp. to $-\Delta$) ; it is negative (resp. positive) under the opposite orientation.

An orientation of an *affine* space F over K is by definition an orientation of the space of translations of F ; the space F, together with such an orientation, is called an *oriented affine space.*

Let E be an oriented vector space over K, of dimension n ; an ordered basis $(a_i)_{1 \leqslant i \leqslant n}$ of E is called *positive* or *direct* (resp. *negative* or *inverse*) if the n-vector $a_1 \wedge a_2 \wedge \dots \wedge a_n$ is positive (resp. negative). If u is an automorphism of the vector space E then $(\bigwedge^n u)(z) = \det(u) \,.\, z$ for all $z \in \bigwedge^n \mathrm{E}$, so a necessary and sufficient condition for $\bigwedge^n u$ to leave invariant the orientation of E (or, as we also say, to *preserve the orientation*) is that $\det(u) > 0$; the automorphisms having this property are precisely the automorphisms of E as an *oriented vector space ;* they form a normal subgroup $\mathbf{GL}^+(\mathrm{E})$ of the linear group $\mathbf{GL}(\mathrm{E})$, which has index 2 whenever $\mathrm{E} \neq 0$.

When $\mathrm{E} = 0$ then $\mathbf{GL}(\mathrm{E}) = \mathrm{End}(\mathrm{E})$ contains only the identity map 1_{E} and by definition $\det(1_{\mathrm{E}}) = 1$. Note that $\bigwedge^n \mathrm{E} = \bigwedge^0 \mathrm{E} = \mathrm{K}$ by definition in this case ; the half-line of K formed by the positive scalars is called the *canonical* orientation of the zero space.

Let M and N be two complementary subspaces of dimensions p and $n-p$ respectively in the vector space E of dimension n ; if z' (resp. z'') is a nonzero vector in $\bigwedge^p \mathrm{M}$ (resp. $\bigwedge^{n-p} \mathrm{N}$), then $z' \wedge z''$ is a nonzero vector in $\bigwedge^n \mathrm{E}$. Given an orientation on M and an orientation on N, the vectors $z' \wedge z''$ for positive z' and z'' form an orientation of E, called the *product orientation of the orientation of* M *by the orientation of* N (which depends on the order of the factors when $p(n-p)$ is odd). Conversely, given orientations on E and on M, there exists a unique orientation on N such that the given orientation on E is the product of the given orientation on M and this orientation on N (in that order) ; this orientation is said to be *complementary* to the orientation of M with respect to that of E. If N′ is a second complementary subspace to M, the canonical projection $\mathrm{N} \to \mathrm{N}'$ parallel to M takes the complementary orientation of N onto that of N′. The image of the complementary orientation of N under the canonical map $\mathrm{N} \to \mathrm{E}/\mathrm{M}$ is thus independent of the choice of complement N ; it is called the *quotient* orientation on E/M of the orientation of E by that of M.

Exercises

§ 1

1) A not necessarily commutative ordered monoid M is a (multiplicatively written) monoid equipped with an ordering such that the relation $x \leqslant y$ implies $zx \leqslant zy$ and $xz \leqslant yz$ for all $z \in$ M. Let G be a not necessarily commutative ordered group. Show that :
a) If P denotes the set of elements greater than the identity e of G, then $\mathrm{P} \cdot \mathrm{P} = \mathrm{P}$, $\mathrm{P} \cap \mathrm{P}^{-1} = \{e\}$ and $a\mathrm{P}a^{-1} = \mathrm{P}$ for all $a \in \mathrm{G}$. Prove the converse. Find a condition for G to be totally ordered.
b) If one of the elements $\sup(x, y)$, $\inf(x, y)$ exists, then so does the other, and

$$\sup(x, y) = x(\inf(x, y))^{-1}y = y(\inf(x, y))^{-1}x .$$

c) The elements $\sup(x, e)$ and $\sup(x^{-1}, e)$ commute. Two coprime elements commute.
d) The subgroup G′ generated by the irreducible elements of G is commutative. Deduce that Th. 2 of VI, p. 18 is still valid.

2) Let E be a lattice with a composition law $(x, y) \mapsto xy$ (not necessarily associative), such that, for all $a \in \mathrm{E}$, the maps $x \mapsto ax$ and $x \mapsto xa$ are order-automorphisms of E. Let x_a (resp. ${}_ax$) denote the element of E defined by $(x_a)\,a = x$ (resp. $a({}_ax) = x$) and suppose that for each $a \in \mathrm{E}$ the maps $x \mapsto a_x$ and $x \mapsto {}_xa$ are isomorphisms from the ordering of E to the opposite ordering. Show that, for arbitrary x, y, z,

$$(z_{\inf(x, y)}) \cdot x = (z_y) \cdot \sup(x, y) .$$

3) Let G be an ordered group such that the set P of positive elements is not 0 ; show that G is an infinite group, and cannot have a greatest (or a least) element.

4) Let G be an ordered group, let P be the set of positive elements of G, and let f be the natural map from G onto a *quotient group* G/H. For $f(\mathrm{P})$ to make G/H an ordered group it is necessary and sufficient that $0 \leqslant y \leqslant x$ and $x \in \mathrm{H}$ imply $y \in \mathrm{H}$; then H is called a *convex* subgroup of G, and G/H is considered as an ordered group in this way. If G is lattice ordered, then a necessary and sufficient condition for G/H to be lattice ordered and for $f(\sup(x, y)) = \sup(f(x), f(y))$ is that the relations $|y| \leqslant |x|$ and $x \in \mathrm{H}$ imply $y \in \mathrm{H}$; show that this says that H is a filtered convex subgroup. Show that a lattice ordered group with no filtered convex subgroups other than itself and $\{0\}$ is totally ordered (consider filtered convex subgroups generated by two positive elements of G) ; * deduce (*Gen. Top.*, V, p. 25, Ex. 1) that G is then isomorphic to an additive subgroup of the real numbers. *

5) Give an example of an ordered group whose ordering is non-discrete, having nonzero elements of finite order (take the quotient of a suitable ordered group by a subgroup H such that $\mathrm{P} \cap \mathrm{H} = \{0\}$).

6) Let G be an ordered group, let P be its set of positive elements. Show that $P - P$ is the largest filtered subgroup of G, and that it is convex. What is the order relation on the quotient ?

7) In the group **Z**, if P is taken to be the set consisting of 0 and the integers $\geqslant 2$, then the resulting ordered group is filtered but not lattice ordered (show that the set of elements x such that $x \geqslant 0$ and $x \geqslant 1$ has two distinct minimal elements).

8) Let x be an element of an ordered group such that $y = \inf(x, 0)$ is defined ; if $n > 0$ is an integer, then $nx \geqslant 0$ implies $x \geqslant 0$ (we have

$$ny = \inf(nx, (n-1)x, \ldots, 0) \geqslant \inf((n-1)x, \ldots, 0) = (n-1)y) ;$$

and hence $nx = 0$ implies $x = 0$.

9) In a lattice ordered group G, show that the sum of a family (H_α) of filtered convex subgroups is a filtered convex subgroup (use VI, p. 12, Cor.).

10) Show that, for every finite sequence (x_i) $(1 \leqslant i \leqslant n)$ of elements of a lattice ordered group G,

$$\sup(x_i) = \sum_i x_i - \sum_{i<j} \inf(x_i, x_j) + \cdots + (-1)^{p+1} \sum_{i_1 < i_2 < \ldots < i_p} \inf(x_{i_1}, \ldots, x_{i_p}) + \\ + \cdots + (-1)^{n+1} \inf(x_1, \ldots, x_n) .$$

(Argue by induction based on Prop. 7 (VI, p. 10), using the distributivity of sup over inf.)

11) Let (x_i) be a family of n elements of a lattice ordered group G ; for each integer k $(0 \leqslant k \leqslant n)$, let d_k (resp. m_k) denote the inf (resp. sup) of the $\binom{n}{k}$ sums of k terms x_i with distinct indices. Show that $d_k + m_{n-k} = x_1 + x_2 + \cdots + x_n$.

12) Given a lattice ordered group G, a subgroup H of G is called a *colattice* if for each pair x, y of elements of H, $\sup_G(x, y) \in H$ (and so equals $\sup_H(x, y)$).
a) If $G = \mathbf{Q} \times \mathbf{Q} \times \mathbf{Q}$ (where **Q** has the usual ordering), then the subgroup H of (x, y, z) such that $z = x + y$ is a lattice but not a colattice.
b) Every filtered convex (Ex. 4) subgroup H of a lattice ordered group G is a colattice.
c) Let G be a lattice ordered group and H an arbitrary subgroup of G. Define H′ to be the set of infima of finite subsets of H, and H″ to be the set of suprema of finite subsets of H′. Show that H″ is the smallest colattice of G containing H (use the remark on VI, p. 16).

13) *a*) A monoid M is said to be *lower semi-lattice ordered* (or simply *semi-lattice ordered*) if it is an ordered monoid, if $\inf(x, y)$ exists for all $x, y \in M$, and if

$$\inf(x + z, y + z) = \inf(x, y) + z$$

for all x, y, $z \in M$. Prove the identities :

$$\inf(x, z) + \inf(y, z) = \inf(x + y, z + \inf(x, y, z))$$

$$\inf(x, y, z) + \inf(x + y, y + z, z + x) = \inf(x, y) + \inf(y, z) + \inf(z, x) .$$

From the first of these relations, deduce that $x \leqslant z$ and $y \leqslant z$ imply $x + y \leqslant z + \inf(x, y)$.

Show that Prop. 11 and Cor. (VI, pp. 14-15) are valid in a semi-lattice ordered monoid.

b) Let M be a semi-lattice ordered monoid and N a submonoid such that $\inf_M(x, y) = \inf_N(x, y)$ for $x, y \in N$ and such that the elements of N are *invertible* in M. Show that the subgroup G of M consisting of elements $x - y$ for $x, y \in N$, is lattice ordered.

14) Show that, in a semi-lattice ordered monoid M,

$$\inf(x_i + y_i) \geqslant \inf(x_i) + \inf(y_i)$$

for all finite sequences of n terms (x_i), (y_i). Deduce the inequalities

$$(x + y)^+ \leqslant x^+ + y^+ , \quad |x^+ - y^+| \leqslant |x - y|$$

in any lattice ordered group.

15) Show that

$$|x^+ - y^+| + |x^- - y^-| = |x - y|$$

in a lattice ordered group (notice that $x - y \leqslant |x - y|$ and $|x| - |y| \leqslant |x - y|$).

16) Show that

$$n \,.\, \inf(x, y) + \inf(nx, ny) = 2n \,.\, \inf(x, y)$$

in a semi-lattice ordered monoid (cf. Ex. 8). Deduce that $\inf(nx, ny) = n \,.\, \inf(x, y)$ if $\inf(x, y)$ is a regular element.

17) Let x, y, z, t be elements of a semi-lattice ordered monoid M such that $z \geqslant 0$ and $t \geqslant 0$. Prove the inequality

$$\inf(x + z, y + t) + \inf(x, y) \geqslant \inf(x + z, y) + \inf(x, y + t) .$$

18) Let $(G_\iota)_{\iota \in I}$ be a family of totally ordered groups indexed by a well-ordered set I ; define an ordering on the *direct sum* G' of the G_ι by taking the strictly positive elements to be the nonzero (x_ι) such that $x_\iota > 0$ for the first index ι such that $x_\iota \neq 0$. Show that G' is totally ordered.

19) Let G be an additive group and (P_α) a nonempty family of subsets of G such that $P_\alpha + P_\alpha \subset P_\alpha$ and $P_\alpha \cap (-P_\alpha) = \{0\}$; let G_α denote the ordered group obtained from G by taking P_α as the set of positive elements. Show that $P + P \subset P$ and $P \cap (-P) = \{0\}$, where $P = \bigcap_\alpha P_\alpha$; if H is the ordered group obtained from G by taking P as the set of positive elements, show that H is isomorphic to the diagonal subgroup of the product of the G_α.

¶ 20) Let G be an additive group and let P be a subset of G satisfying the following conditions :

1. $P + P = P$; 2. $P \cap (-P) = \{0\}$; 3. for every integer $n > 0$, $nx \in P$ implies $x \in P$ (conditions (C)).

a) If a is an element of G such that $a \notin P$, show that there exists a subset P′ of G, satisfying conditions (C), such that $P \subset P'$ and $-a \in P'$ (take P′ to be the set of $x \in G$ such that there exist integers $m > 0$ and $n \geqslant 0$ and an element $y \in P$ with $mx = -na + y$.

b) Deduce from *a*) that P is the intersection of those subsets T of G such that $T + T = T$, $T \cap (-T) = \{0\}$, $T \cup (-T) = G$ (that is, making G a *totally ordered* group) and $P \subset T$ (use Zorn's Lemma).

c) In particular if G is an additive group in which every nonzero element has infinite order, then the intersection of all the subsets T of G such that $T + T = T$, $T \cap (-T) = \{0\}$ and $T \cup (-T) = G$ is $\{0\}$.

¶ 21) An ordered group is called lattice orderable if it is isomorphic to a subgroup of a lattice ordered group. Show that it is necessary and sufficient for an ordered group to be lattice orderable that, for every integer $n > 0$, the relation $nx \geqslant 0$ imply $x \geqslant 0$ (to show that the condition is sufficient, use Ex. 20, *b*) and 19). Show that every lattice orderable group is isomorphic to a subgroup of a product of totally ordered groups.

22) Let G be a lattice orderable group (considered as a **Z**-module) and E the vector space $G_{(\mathbf{Q})}$ (II, p. 277); show that there is a unique ordering on E compatible with the additive group structure of E, inducing the given ordering on G, and such that E is lattice orderable.

23) Let G be the lattice ordered group $\mathbf{Z} \times \mathbf{Z}$ (where **Z** has the usual ordering) and H the convex subgroup of G generated by $(2, -3)$; show that the ordered group G/H is not lattice orderable (cf. VI, p. 30, Ex. 4).

24) Let A be a commutative ring and I the set of all *ideals* of A. Then $\mathfrak{a}(\mathfrak{b} + \mathfrak{c}) = \mathfrak{ab} + \mathfrak{ac}$ for all $\mathfrak{a}$, $\mathfrak{b}$, $\mathfrak{c}$ in I, by I, p. 107; in other words I, with the order relation $\mathfrak{a} \supset \mathfrak{b}$ and the law of composition $(\mathfrak{a}, \mathfrak{b}) \mapsto \mathfrak{ab}$, is a semi-lattice ordered monoid (VI, p. 31, Ex. 13), the sup of two elements $\mathfrak{a}$, $\mathfrak{b}$ of I being $\mathfrak{a} \cap \mathfrak{b}$, and their inf $\mathfrak{a} + \mathfrak{b}$.

a) Let K be a field and $A = K[X, Y]$ the ring of polynomials in two indeterminates over K. Consider the principal ideals $\mathfrak{a} = (X)$, $\mathfrak{b} = (Y)$, $\mathfrak{c} = (X + Y)$ in A. Show that

$$(\mathfrak{a} \cap \mathfrak{b}) + \mathfrak{c} \neq (\mathfrak{a} + \mathfrak{c}) \cap (\mathfrak{b} + \mathfrak{c}),$$
$$(\mathfrak{a} + \mathfrak{b}) \cap \mathfrak{b} \neq (\mathfrak{a} \cap \mathfrak{c}) + (\mathfrak{b} \cap \mathfrak{c}),$$
$$(\mathfrak{a} \cap \mathfrak{b})(\mathfrak{a} + \mathfrak{b}) \neq (\mathfrak{a}(\mathfrak{a} + \mathfrak{b})) \cap (\mathfrak{b}(\mathfrak{a} + \mathfrak{b})).$$

b) In the ring A, 1 is a gcd of X and Y, but $(X) + (Y) \neq A$.

¶ 25) Let I be the semilattice ordered monoid of ideals of a commutative ring A (Ex. 24). For a finite system of congruences $x \equiv a_i (\mathfrak{a}_i)$ to admit a solution whenever any two of them admit a common solution (that is to say whenever

$$a_i \equiv a_j (\mathfrak{a}_i + \mathfrak{a}_j)$$

for every pair of indices i, j), it is necessary and sufficient that each of the laws

$$(\mathfrak{a}, \mathfrak{b}) \mapsto \mathfrak{a} \cap \mathfrak{b}, \ (\mathfrak{a}, \mathfrak{b}) \mapsto \mathfrak{a} + \mathfrak{b},$$

in I be distributive over the other (« *Chinese remainder theorem* »). One may proceed as follows:

a) If $(\mathfrak{a}_1 \cap \mathfrak{a}_2) + (\mathfrak{a}_1 \cap \mathfrak{a}_3) = \mathfrak{a}_1 \cap (\mathfrak{a}_2 + \mathfrak{a}_3)$ and if any two of the three congruences $x \equiv a_i (\mathfrak{a}_i)$ $(i = 1, 2, 3)$ admit a common solution, then the three congruences have a

common solution (let x_{12} be a common solution of $x \equiv a_1(\mathfrak{a}_1)$ and $x \equiv a_2(\mathfrak{a}_2)$ and let x_{13} be a common solution of $x \equiv a_1(\mathfrak{a}_i)$ and $x \equiv a_3(\mathfrak{a}_3)$; show that the congruences $x \equiv x_{12}(\mathfrak{a}_1 \cap \mathfrak{a}_2)$ and $x \equiv x_{13}(\mathfrak{a}_1 \cap \mathfrak{a}_3)$ have a common solution).
b) If any system of three congruences, any two of which have a common solution, admits a common solution, show that the following distributive laws hold

$$\mathfrak{a} + (\mathfrak{b} \cap \mathfrak{c}) = (\mathfrak{a} + \mathfrak{b}) \cap (\mathfrak{a} + \mathfrak{c}), \quad \text{and} \quad \mathfrak{a} \cap (\mathfrak{b} + \mathfrak{c}) = (\mathfrak{a} \cap \mathfrak{b}) + (\mathfrak{a} \cap \mathfrak{c}).$$

(For the first, note that for all $x \in (\mathfrak{a} + \mathfrak{b}) \cap (\mathfrak{a} + \mathfrak{c})$, there exists $y \in \mathfrak{b} \cap \mathfrak{c}$ such that $y \equiv x(\mathfrak{a})$; for the second, note that for all $x \in \mathfrak{a} \cap (\mathfrak{b} + \mathfrak{c})$, there exists $y \in \mathfrak{a} \cap \mathfrak{b}$ such that $y \equiv x(\mathfrak{c})$.) Note that by *a*) and *b*), the second distributive law implies the first (cf. *Set Theory*, III, p. 217, Ex. 16).
c) Prove the « Chinese remainder theorem » by induction on the number of congruences under consideration, by an analogous argument to that of *a*).
d) * Translation in the case of principal ideal domains. *

26) Show that the ideal (X) in the monoid of ideals of the polynomial ring K[X, Y] (where K is a commutative field) satisfies condition (P) of Prop. 14 of VI, p. 17, but is not maximal.

27) Let A be the quadratic **Z**-algebra with basis $(1, e)$, where $e^2 = -5$. Show that A is an integral domain ; in A we have $9 = 3 . 3 = (2 + e)(2 - e)$; show that 3, $(2 + e)$ and $(2 - e)$ are irreducible elements of A, but do not satisfy the condition in Prop. 14 (DIV) of VI, p. 17.

¶ 28) Let G be a lattice ordered group with positive elements P. Two elements x and y of P are said to be *equivalent* if any element coprime to one is also coprime to the other ; the equivalence classes under this relation are called the *threads* of P ; let $\bar{x}$ denote the thread containing x.
a) Let $\bar{a}$ and $\bar{b}$ be threads ; let x, x_1 be elements of $\bar{a}$ and y, y_1 be elements of $\bar{b}$; show that if every element coprime to x is also coprime to y, then every element coprime to x_1 is also coprime to y_1. Define a relation in this way between $\bar{a}$ and $\bar{b}$, written $\bar{a} \geqslant \bar{b}$; show that this is an ordering of the set F of threads.
b) Show that F is a lattice under this ordering, and that

$$\inf(\bar{a}, \bar{b}) = \overline{\inf(a, b)}, \quad \sup(\bar{a}, \bar{b}) = \overline{\sup(a, b)} = \overline{a + b}$$

(use Cor. 3 of VI, p. 15). Show that F has a smallest element, namely the thread $\bar{0} = \{0\}$.
c) Call two threads $\bar{a}$, $\bar{b}$ coprime if $\inf(\bar{a}, \bar{b}) = \bar{0}$. Show that if $\bar{a}$ and $\bar{b}$ are two threads $\neq \bar{0}$, and if $\bar{a} < \bar{b}$, then there exists a thread $\bar{c}$, coprime to $\bar{a}$, such that $\bar{c} < \bar{b}$.
d) Call a thread $\bar{m} \neq \bar{0}$ irreducible if it is a minimal element of the set of threads $\neq \bar{0}$. Show that an irreducible thread is totally ordered (if a and b are two elements of $\bar{m}$, consider the threads containing $a - \inf(a, b)$ and $b - \inf(a, b)$ and use *b*)). Show also that the union of $\bar{m}$, $-\bar{m}$ and $\{0\}$ is a totally ordered subgroup $H(\bar{m})$ of G.
e) Given a family $(\bar{m}_\iota)$ of distinct irreducible threads of G, show that the subgroup H generated by the union of the $\bar{m}_\iota$ is isomorphic to the direct sum of the groups $H(\bar{m}_\iota)$ (reduce to the case of a finite family, and argue by induction).

f) Show that in a product (resp. direct sum) of nonzero totally ordered groups, the elements of irreducible threads are those in which all the coordinates are zero except for one (which is positive). Deduce that an ordered group can be a product (resp. direct sum) of nonzero totally ordered groups in at most one way (in other words the factors are uniquely determined).

29) An ordered group G is *completely lattice ordered* if it is filtered and if every nonempty subset of G which is bounded above has a supremum in G. A lower semi-lattice ordered monoid is said to be *completely semi-lattice ordered* if $\inf_\alpha(x_\alpha)$ exists for every nonempty family of elements (x_α) of M which is bounded below, and if

$$\inf_{\alpha,\beta}(x_\alpha + y_\beta) = \inf_\alpha(x_\alpha) + \inf_\beta(y_\beta)$$

for any two nonempty families (x_α), (y_β) which are bounded below. A completely lattice ordered group is a completely semi-lattice ordered monoid.
a) Show that a necessary and sufficient condition for a filtered group G to be completely lattice ordered is that every nonempty subset of the set P of positive elements of G admit an infimum in P.
b) Every product of completely lattice ordered groups is completely lattice ordered.
c) Every filtered convex subgroup of a completely lattice ordered group is completely lattice ordered.

¶ 30) Let G be an ordered group. For every subset A of G let $m(\mathrm{A})$ (resp. $\mathrm{M}(\mathrm{A})$) denote the set of lower (resp. upper) bounds for A in G ; then $m(\mathrm{A}) = -\mathrm{M}(-\mathrm{A})$.
a) The relation $\mathrm{A} \subset \mathrm{B}$ implies $m(\mathrm{B}) \subset m(\mathrm{A})$ and $\mathrm{M}(m(\mathrm{A})) \subset \mathrm{M}(m(\mathrm{B}))$.
b) We have $\mathrm{A} \subset \mathrm{M}(m(\mathrm{A}))$ and $\mathrm{M}(m(\mathrm{M}(\mathrm{A}))) = \mathrm{M}(\mathrm{A})$.
c) If (A_ι) is an arbitrary family of subsets of G then $m\left(\bigcup_\iota \mathrm{A}_\iota\right) = \bigcap_\iota m(\mathrm{A}_\iota)$.

d) Every set $\mathrm{M}(\mathrm{B})$, where B is a nonempty subset bounded above in G, is called a *major* set in G. For any nonempty subset A bounded below in G, the set $\mathrm{M}(m(\mathrm{A}))$ is the smallest major set containing A ; denote it $\langle \mathrm{A} \rangle$; if $\mathrm{A} \subset \mathrm{B}$ then $\langle \mathrm{A} \rangle \subset \langle \mathrm{B} \rangle$; if A has an infimum a in G, then $\langle \mathrm{A} \rangle = \mathrm{M}(a)$, which we also write $\langle a \rangle$.
e) If A and B are two nonempty subsets bounded below in G, then $\langle \mathrm{A} + \mathrm{B} \rangle = \langle \langle \mathrm{A} \rangle + \mathrm{B} \rangle$. Deduce that, in the set $\mathfrak{M}(\mathrm{G})$ of major sets in G, the map $(\mathrm{A}, \mathrm{B}) \mapsto \langle \mathrm{A} + \mathrm{B} \rangle$ is an associative and commutative composition law, for which $\langle 0 \rangle = \mathrm{P}$ is an identity, that the order relation $\mathrm{A} \supset \mathrm{B}$ is compatible with this law, and that $\mathfrak{M}(\mathrm{G})$ is a completely semi-lattice ordered monoid (Ex. 29) under this structure if G is filtered. Moreover, the map $x \mapsto \langle x \rangle$ from G to $\mathfrak{M}(\mathrm{G})$ is an isomorphism from the ordered group G onto a subgroup of the monoid $\mathfrak{M}(\mathrm{G})$.
f) If an element A of $\mathfrak{M}(\mathrm{G})$ is invertible under the composition law of this monoid, then its inverse is $\mathrm{M}(-\mathrm{A}) = -m(\mathrm{A})$ (notice that if B is the inverse of A, then the relations $\mathrm{B} \subset \mathrm{M}(-\mathrm{A})$ and $\mathrm{A} + \mathrm{M}(-\mathrm{A}) \subset \langle 0 \rangle$ hold, whence $\langle \mathrm{A} + \mathrm{B} \rangle = \langle \mathrm{A} + \mathrm{M}(-\mathrm{A}) \rangle$).
g) For an element A of $\mathfrak{M}(\mathrm{G})$ to be invertible, it is necessary and sufficient that $x + \mathrm{A} \subset \mathrm{A}$ imply $x \geqslant 0$ (write down an expression for $0 \in \langle \mathrm{A} + \mathrm{M}(-\mathrm{A}) \rangle$).

¶ 31) An ordered group G is called *archimedean* if the only elements $x \in \mathrm{G}$ such that the set of nx (n a positive integer) is bounded below are the positive elements of G.

a) The monoid $\mathfrak{M}(G)$ of major sets in an ordered group G is a group if and only if G is archimedean (use Ex. 30, *g*) and *d*)). If in addition G is filtered, then $\mathfrak{M}(G)$ is a completely lattice ordered group.
b) Deduce that a filtered group G is isomorphic to a subgroup of a completely lattice ordered group if and only if G is archimedean.

¶ 32) *a*) Let G be a completely lattice ordered group, and H a subgroup of G. For every major subset A *of the group* H, let x_A denote the infimum of A *in* G ; show that the map $A \mapsto x_A$ from $\mathfrak{M}(H)$ to G is injective (show that A is the set of $y \in H$ such that $y \geqslant x_A$).
b) For every subset B of H, bounded below in H, let $\langle B \rangle$ denote the major set generated by B *in* H (an element of $\mathfrak{M}(H)$). If for any subset B of H, bounded below in H, the equation $\inf B = \inf \langle B \rangle$ holds (infima taken in G), show that the map $A \mapsto x_A$ is an isomorphism of the ordered group $\mathfrak{M}(H)$ onto a subgroup of G (cf. Ex. 30, *e*)).
c) If every element of G is the infimum of some subset of H, show that for any subset B of H, bounded below in H, we have $\inf B = \inf \langle B \rangle$ (infima taken in G). (Notice that this holds if $\inf B \in H$; in general, consider an element $x \in G$ such that $x + \inf B \in H$ and use the fact that x is the infimum of some subset of G, as well as Ex. 30, *e*).)
d) * Let G be the completely lattice ordered group $\mathbf{Z} \times \mathbf{Z} \times \mathbf{R}$. Let $\theta > 0$ be an irrational number ; let H be the subgroup of (x, y, z) such that $\theta(z - x) + y = 0$. Show that there exists *no* isomorphism of the ordered group $\mathfrak{M}(H)$ onto any subgroup of G which restricts to the identity map on H (notice that $\mathfrak{M}(H)$ is isomorphic to $K = \mathbf{Z} \times \mathbf{R}$; consider the subgroup of K consisting of elements u such that, for every integer $n > 0$, there exists $v \in K$ with $nv = u$, and the analogous subgroup of G). *

33) *a*) A totally ordered group G is archimedean if and only if, for any pair $x > 0$, $y > 0$ of elements of G, there exists an integer $n > 0$ such that $y \leqslant nx$.
* *b*) Any totally ordered, archimedean, completely lattice ordered group G is isomorphic to $\{0\}$, $\mathbf{Z}$ or $\mathbf{R}$ (ruling out the first two cases, fix $a > 0$ in G, and make each $x \in G$ correspond to the infimum of the rational numbers p/q such that $qx \leqslant pa$).
c) Deduce that every archimedean totally ordered group G is isomorphic to a subgroup of $\mathbf{R}$ (notice that $\mathfrak{M}(G)$ is totally ordered). *
d) The lexicographic product $\mathbf{Z} \times \mathbf{Z}$ is totally ordered and not archimedean.

34) Let $G = \mathbf{Z}^{\mathbf{N}}$ be the product of a countably infinite family of totally ordered groups $\mathbf{Z}$, and let $H = \mathbf{Z}^{(\mathbf{N})}$ be the filtered convex subgroup, the direct sum of the factors of G. Show that the lattice ordered group G/H (VI, p. 30, Ex. 4) is not archimedean, although G and H are completely lattice ordered.

¶ 35) A major set A in an ordered group G is said to be *finitely generated* if there exists a finite set F such that $A = \langle F \rangle$. We say that G is *semi-archimedean* if every finitely generated major set is invertible in the monoid $\mathfrak{M}(G)$. Every lattice ordered (resp. archimedean) group is semi-archimedean (Ex. 31, *a*)).
a) Show that every filtered semi-archimedean group is lattice ordered (if $nx \geqslant 0$, consider the major set $\langle F \rangle$, where $F = \{0, x, \ldots, (n-1)x\}$, and use Ex. 30, *g*) and the fact that it is invertible).
b) * Let K be the lexicographic product $\mathbf{R} \times \mathbf{R}$, and G the (usual) product group $K \times \mathbf{R}$. Let θ be an irrational number such that $0 < \theta < 1$, and let H be the subgroup generated by $((1, 0), 0)$, $((\theta, 0), \theta)$ and $((0, x), 0)$ where x runs through $\mathbf{R}$. Show that H,

a subgroup of a product of two totally ordered groups, is not semi-archimedean (consider the major set generated by the two elements $((1, 0), 0)$ and $((\theta, 0), \theta)$ of H, and show that it is not invertible). $_*$

c) Let G be a semi-archimedean filtered group ; show that the submonoid of $\mathfrak{M}(G)$ generated by the finitely generated major sets and their inverses is a *lattice ordered group* (cf. VI, p. 31, Ex. 13, *b*)).

d) * Let $\theta > 0$ be an irrational number, and let G be the group $\mathbf{Z} \times \mathbf{Z}$ in which the set P of positive elements is taken to be the set of (x, y) such that $x \geqslant 0$ and $y \geqslant \theta x$. Show that G is an archimedean group, but that the inverse in $\mathfrak{M}(G)$ of a finitely generated major set, not of the form $\langle a \rangle$, is not finitely generated. $_*$

36) Let G be a filtered group with positive elements P. Then G is isomorphic to a group of the form $\mathbf{Z}^{(I)}$ if and only if there exists a map $x \mapsto d(x)$ from P into N such that, for all b, either $b \geqslant a$ or there exists an element c in the major set $\langle a, b \rangle$ such that $d(c) < d(a)$. (Show that this condition is satisfied in $\mathbf{Z}^{(I)}$ by the map $d(x_\iota) = \sum_\iota x_\iota$ (when $x_\iota \geqslant 0$ for all ι). Conversely, if the condition is satisfied, show that every major set $A \subset P$ has the form $\langle a \rangle$, by taking an element a of A such that $d(a) \leqslant d(x)$ for all $x \in A$. Apply Th. 2 of VI, p. 18.)

§ 2

All rings under consideration are assumed to be *commutative* unless explicitly stated otherwise.

1) Let A be a totally ordered ring and B an additive subgroup of A which is closed under multiplication.

a) An element $x \in A$ is said to be *infinitely large with respect to* B if $|y| < |x|$ for all $y \in B$. Show that the set F(B) of elements of A which are not infinitely large with respect to B is a subring of A containing B.

b) An element $x \in A$ is said to be infinitely small with respect to B if, for all $y > 0$ in B, we have $|x| < y$. If, for all $y > 0$ in B, there exists $z \in B$ such that $0 < z < y$, then the set of elements of A which are infinitely small with respect to B is a sub-pseudo-ring I(B) of A. If in addition, for every pair of elements y, z of B with $0 < y < z$, there exists $x \in B$ with $0 < xz < y$, then I(B) is an *ideal* in the ring F(B).

2) Let A be a totally ordered ring and $\mathfrak{n}$ the set of nilpotent elements of A (which is an ideal of A). Every element of A not belonging to $\mathfrak{n}$ is infinitely large with respect to $\mathfrak{n}$; the quotient ring $A/\mathfrak{n}$ is totally ordered (VI, p. 30, Ex. 4) and is an integral domain.

3) In the field Q, let P be the set consisting of 0 and the rational numbers $\geqslant 1$ (in the usual ordering). Show that P satisfies axioms (AP_I), (AP_{II}) and (AP_{III}).

¶ 4) *a*) Let K be a field, and P a subset of K satisfying conditions (AP_I), (AP_{II}) and (AP_{III}), and such that $K^2 \subset P$ (where K^2 denotes the set of squares of elements of K) ; show that if $x > 0$ in the ordering defined by P, then $x^{-1} > 0$; deduce that the set K_+^* of elements > 0 in K is a subgroup of the multiplicative group of K.

b) In the field $\mathbf{Q}$, the only set P which satisfies the conditions of *a*) is the set of rational numbers $\geqslant 0$ (in the usual ordering).
c) * Let P be a subset of K satisfying (AP_I), (AP_{II}) and (AP_{III}), such that $1 \in P$ and such that $x > 0$ implies $x^{-1} > 0$ in the ordering defined by P. Show that $y^2 \geqslant z^2$ implies $y \geqslant z$, for positive y and z. Deduce that, for arbitrary $y > 0$, we have

$$\frac{1}{2}(y + y^{-1}) \geqslant 1 - n^{-1}$$

for every integer $n > 0$ (notice that $(y + y^{-1})^{2m} \geqslant \binom{2m}{m}$ for every integer $m > 0$). Deduce that, if the ordered additive group structure defined by P is archimedean (VI, p. 35, Ex. 31), then $(y - z)^2 \geqslant 0$ for every pair of elements y, z of P ; if K′ is the subfield of K generated by P then $x^2 \geqslant 0$ for all $x \in K'$. *
d) * Let $K = \mathbf{R}(X)$ be the field of rational functions in one indeterminate over $\mathbf{R}$; let P be the set consisting of 0 and the rational functions $u \in K$ such that $u(t)$ is defined and > 0 for every real number t. Show that P satisfies the conditions of *c*) and generates K, but that there exist elements $v \in K$ such that $v^2 \notin P$. *

¶ 5) Let A be a *lattice ordered* ring. A *convex* ideal (VI, p. 30, Ex. 4) of A is called *irreducible* if it is not the intersection of two convex ideals distinct from itself.
a) Show that the intersection of the irreducible convex ideals of A is 0 (if $a \in A$ is nonzero, consider a convex ideal of A maximal among those which do not contain a). Deduce that A is isomorphic to a subring A′ of a product $\prod_\iota A_\iota$, such that each A_ι is lattice ordered, $pr_\iota(A') = A_\iota$ for all ι and the ideal $\{0\}$ is irreducible in A_ι (as a convex ideal) for all ι.
b) Show that the following conditions are equivalent in A :
α) $|xy| = |x| \cdot |y|$ for all x, y ;
β) $x \cdot \sup(y, z) = \sup(xy, xz)$ for all $x \geqslant 0$, y, z ;
γ) $x \cdot \inf(y, z) = \inf(xy, xz)$ for all $x \geqslant 0$, y, z ;
δ) $\inf(y, z) = 0$ implies $\inf(xy, z) = 0$ for all $x \geqslant 0$.
(To see that γ) implies δ), observe that $xy \leqslant y \cdot \sup(x, 1)$).
When these conditions are satisfied, then A is said to be *strongly lattice ordered*.
c) Show that a ring A is strongly lattice ordered if and only if it is isomorphic to a subring of a product $\prod_\iota A_\iota$ of totally ordered rings. (Using *a*), reduce to showing that if $\{0\}$ is irreducible in a strongly lattice ordered ring A, then A is totally ordered ; for this, note that if B is a strongly lattice ordered ring then for all $b \in B$ the set $\mathfrak{m}$ of $x \in B$ such that $\inf(|x|, |b|) = 0$ and the set $\mathfrak{n}$ of $y \in B$ such that $|y| \leqslant |zb|$ for some $z \in B$ are two convex ideals of B such that $\mathfrak{m} \cap \mathfrak{n} = \{0\}$.)
d) Show that in a strongly lattice ordered ring A the relation $\inf(x, y) = 0$ implies $xy = 0$; for all $z \in A$ we have $z^2 \geqslant 0$; for all $x, y \in A$ we have $xy \leqslant \sup(x^2, y^2)$.
e) Let A be a lattice ordered ring with no nilpotent element > 0 ; show that if $\inf(x, y) = 0$ implies $xy = 0$ then A is strongly lattice ordered (check condition δ) of *b*)).

¶ 6) *a*) Let K be a lattice ordered field. Show that the following conditions are equivalent :
α) $x^2 \geqslant 0$ for all $x \in K$;
β) $x > 0$ implies $x^{-1} > 0$;

γ) K is strongly lattice ordered (Ex. 5) ;
δ) K is totally ordered.
(To see that β) implies δ), notice that β implies $xy \leqslant x^2 + y^2$ for $x, y \in K$).
b) Let K be the field $\mathbf{Q}(\sqrt{2})$ obtained by adjoining $\sqrt{2}$ to the field of rational numbers ; as an additive group K is identified with $\mathbf{Q} \times \mathbf{Q}$ *via* the bijection $x + y\sqrt{2} \mapsto (x, y)$; show that if the *product* ordering on $\mathbf{Q} \times \mathbf{Q}$ (where $\mathbf{Q}$ has the usual ordering) is carried over to K *via* this bijection, then K is lattice ordered but not strongly lattice ordered.
c) The field $K = \mathbf{Q}(X)$ of rational functions in one indeterminate X over $\mathbf{Q}$ is generated by its set of squares K^2. Show that the set P of sums of squares of elements of K defines a non-lattice ordering on K compatible with its ring structure, and such that $u > 0$ implies $u^{-1} > 0$.

7) Every element of a commutative field K of characteristic $\neq 2$ is a sum of squares if and only if -1 is a sum of squares (notice that every element of K is a difference of two squares).

¶ 8) Let A be a nonempty subset of a commutative field K of characteristic $\neq 2$; then K is said to be A-*orderable* if no element of A is a sum of squares in K. We say that K is *orderable* if there exists a total ordering of K which is compatible with its ring structure.
a) Show that if K is A-orderable then it is orderable (Ex. 7), and hence of characteristic 0.
b) Show that pure extensions and algebraic extensions of odd degree of an A-orderable field K are A-orderable (argue as in Prop. 3 of VI, p. 23).
c) Let K be an A-orderable field and let b be an element of K not of the form $ca - d$, where $a \in A$ and c and d are sums of squares in K. Show that the field $K(\sqrt{b})$ is A-orderable.
d) An A-orderable field K is called *maximal* if no proper algebraic extension of K is A-orderable. Show that a maximal A-orderable field K has the following properties : 1. K is *pythagorean,* in other words every sum of squares is a square ; 2. no element of A is a square ; 3. every element of K which is not a square has the form $c^2a - d^2$, where $a \in A$; 4. every odd degree polynomial in K[X] has at least one root in K (use *b*) and *c*)).
e) Show that if K satisfies the conditions of *d*) then every element algebraic over K has degree 2^q over K for some q (argue as in Prop. 8 of VI, p. 26, by induction on the exponent of 2 in the degree of the element under consideration). Show that on the other hand no quadratic extension of K is A-orderable. Deduce that K is a maximal A-orderable field (use Galois theory and Prop. 12 of I, p. 77).
f) Let K be an A-orderable field and Ω an algebraically closed extension of K. Show that there exists a maximal A-orderable field contained in Ω and containing K.

9) Let $q > 0$ be a non-square natural number ; let A denote the set $\{-1, \sqrt{q}\}$ in the field $K = \mathbf{Q}(\sqrt{q})$; show that K is A-orderable. Show that there exists an extension E of K such that E is pythagorean, orderable, and that every odd degree polynomial over E has a root in E, but that E does not admit the structure of a maximal ordered field (consider a maximal A-orderable extension of K).

¶ 10) *a*) Let K be an orderable field and E a Galois extension of K. Show that either E is orderable or there exists an orderable algebraic extension F of K, contained in E, such that E is a quadratic extension of F (use Th. 6 of V, p. 70).

b) Show that the polynomial $X^4 + 2$ is irreducible over **Q**, and that the extension K of **Q** of degree 4, obtained by adjoining a root of this polynomial to **Q**, contains no orderable subfield other than **Q** (find all the subfields of K by Galois theory).

11) Let K be an ordered field and G a subfield of K. *a*) Show that $x \neq 0$ is infinitely large with respect to G if and only if x^{-1} is infinitely small with respect to G. We say that K is *comparable* to G if there is no element of K which is infinitely large with respect to G (nor, consequently, any nonzero element of K which is infinitely small with respect to G). A necessary and sufficient condition for K to be comparable to its prime subfield **Q** is that K be *archimedean* (VI, p. 35, Ex. 31), * and hence that K be isomorphic to a subfield of **R** (VI, p. 36, Ex. 33*c*)). *
b) Show that, in the ring F(G) of elements of K which are not infinitely large with respect to G, the set I(G) of elements infinitely small with respect to G is a maximal ideal ; moreover the ordering on the quotient field K(G) = F(G)/I(G) induced from that of F(G) is compatible with the ring structure and is total.
c) Show that a class modulo I(G) can contain at most one element of G ; deduce that the natural map from G to K(G) is an isomorphism from the ordered field G onto a subfield G′ of K(G), and that K(G) is *comparable* to G′.

12) Let K be a maximal ordered field, let *f* be a polynomial over K and let *a* and *b* be two roots of F such that $a < b$ and such that F has no roots between *a* and *b*. Show that if *g* is a rational function over K whose denominator does not vanish for $a \leqslant x \leqslant b$ then the equation $f(x)\,g(x) + f'(x) = 0$ has an odd number of solutions in $[a, b]$ (where each solution is counted with its multiplicity ; use Prop. 5 of VI, p. 25). Deduce that if *h* is a rational function over K having *a* and *b* as roots and whose denominator does not vanish in $]a, b[$, then the equation $h'(x) = 0$ has at least one root in $]a, b[$.

13) Let K be a maximal ordered field, let *h* be a rational function over K and let $[a, b]$ be a closed interval in which *h* is defined. Show that there exists $c \in \,]a, b[$ such that $h(b) - h(a) = (b - a)\,h'(c)$ (use Ex. 12). Deduce that *h* is an increasing function in $[a, b]$ if and only if $h'(x) \geqslant 0$ in this interval (to see that this condition is necessary, split the interval by the roots of $h'(x) = 0$).

¶ 14) *a*) Let K be a maximal ordered field, let E be a subfield of K and let *f* be a polynomial over E ; show that all the roots of *f* in K belong to F(E) (Ex. 11, *b*)) (use Prop. 4 of VI, p. 24). Deduce that if G is a subfield of K and $E \subset K$ is an extension comparable to G (Ex. 11) then the set of elements of K algebraic over E is a maximal ordered field comparable to G.
b) Deduce from *a*) that, under the hypotheses of *a*), the field K(G) (Ex. 11, *b*)) is a maximal ordered field.
c) Let *f* be a polynomial of degree $\geqslant 1$ over G. Show that $f(t)$ is infinitely large with respect to G if and only if *t* is infinitely large with respect to G ; the element $f(t)$ is infinitely small with respect to G if and only if *t* is congruent modulo I(G) to a root of *f* in K (split K into intervals using the roots of *f* and f' in K, and apply Ex. 13 ; notice that if $x < t$ and if $x \in K$ is not congruent modulo I(G) to *t*, then there exists $y \in K$ such that $x < y < t$ and $y - x \in G$).

¶ 15) Let E be an ordered field, let K be a subfield of E such that E is algebraic over K, and let R be a maximal ordered extension of K.

a) Among the elements of $E - K$, let x_0 be one whose minimal polynomial f over K has the least possible degree. Show that $f'(T)$ is a product of factors in $R[T]$ of the form $(T - a_i)^2 + c_i^2$ (a_i, $c_i \in R$), and first order factors $T - b_j$ with the b_j pairwise distinct elements of K. Deduce that there exists an element y_0 in R such that $f(y_0) = 0$ and such that $x_0 - z$ and $y_0 - z$ *have the same sign*, in E and R respectively, for all $z \in K$ (cf. VI, p. 43, Ex. 26, *c*)).

b) Deduce from *a*) that there exists a K-isomorphism from the ordered field E onto a subfield of R. (First show that there exists such an isomorphism for the subfield $K(x_0)$ of E ; for any polynomial $g \in K[T]$ such that $\deg(g) < \deg(f)$, apply the same argument used for f' in *a*) to evaluate the signs of $g(x_0)$ and $g(y_0)$; then apply Zorn's Lemma).

c) Show that if R and R′ are two maximal ordered algebraic extensions of K then there exists precisely one K-isomorphism from R onto R′ (apply *b*)).

16) Let K be a maximal ordered field and let

$$g(T) = a_0 + a_1 T + \cdots + a_n T^n$$

be a nonzero polynomial in $K[T]$ whose roots all lie in K. Show that for any polynomial $f \in K[T]$, the number of roots of the polynomial

$$a_0 f(T) + a_1 f'(T) + a_2 f''(T) + \cdots + a_n f^{(n)}(T)$$

which do not belong to K, counted with multiplicity, is at most equal to the number of roots of f, counted with multiplicity, which do not belong to K (use Ex. 12 for $n = 1$, then proceed by induction). Deduce that the polynomial

$$a_0 + \frac{a_1}{1!} T + \frac{a_2}{2!} T^2 + \cdots + \frac{a_n}{n!} T^n$$

has all its roots in K.

17) Let K be a maximal ordered field, let g be a nonzero polynomial in $K[T]$ whose roots are all in K, and do not belong to the interval $[0, n]$ of K. Show that for every polynomial $f(T) = a_0 + a_1 T + \cdots + a_n T^n$ of $K[T]$, the number of roots of the polynomial

$$a_0 g(0) + a_1 g(1) T + a_2 g(2) T^2 + \cdots a_n g(n) T^n$$

which do not belong to K, counted with multiplicity, is at most equal to the number of roots of f, counted with multiplicity, which do not belong to K (same method as Ex. 16).

18) Let K be a maximal ordered field and let f be a polynomial of degree n in $K[T]$, all of whose roots are in K. Show that for all $c \neq 0$ in K the polynomial $f^2 + cf'$ has at least $n - 1$ and at most $n + 1$ roots in K, counted with multiplicity (use Ex. 13).

19) Let K be a maximal ordered field and f a nonzero polynomial in $K[T]$. A necessary and sufficient condition that, for every polynomial $g \in K[T]$, the number of roots of $fg + g'$, counted with multiplicity, not belonging to K, be at most equal to the number of roots of g, counted with multiplicity, not belonging to K, is that $f(T) = a - bT$ with

$b \geqslant 0$ in K. (To see that the condition is sufficient use Ex. 12 ; to show that it is necessary, take $g = 1$ and $g = f$).

20) Let K be a maximal ordered field and

$$f(\mathrm{T}) = a_0 + \binom{n}{1} a_1\mathrm{T} + \binom{n}{2} a_2\mathrm{T}^2 + \cdots + a_n\mathrm{T}^n$$

a polynomial in K[T]. For all integers p, q such that $0 \leqslant p < p + q \leqslant n$, the number of roots of the polynomial

$$a_p + \binom{q}{1} a_{p+1}\mathrm{T} + \binom{q}{2} a_{p+2}\mathrm{T}^2 + \cdots + a_{p+q}\mathrm{T}^q$$

counted with multiplicity, which do not belong to K is at most equal to the number of roots of f, counted with multiplicity, which do not belong to K (use Ex. 12).

Deduce that if $b_1, \ldots, b_n$ are n distinct elements > 0 in K, and if

$$(\mathrm{T} + b_1) \ldots (\mathrm{T} + b_n) = \mathrm{T}^n + \binom{n}{1} m_1\mathrm{T}^{n-1} + \cdots + m_n$$

then

$$m_k^{1/k} > m_{k+1}^{1/(k+1)} \quad \text{for} \quad 1 \leqslant k \leqslant n - 1 .$$

21) Let $(a_i)_{1 \leqslant i \leqslant n}$ be a finite sequence of n elements of an ordered field K, not all zero ; let $(a_{i_k})_{1 \leqslant k \leqslant p}$ be the subsequence of (a_i) consisting of the nonzero a_i $(i_1 < i_2 < \cdots < i_p)$; then the number of indices $k \leqslant p - 1$ such that a_{i_k} and $a_{i_{k+1}}$ have opposite signs is called the *number of variations* of the sequence (a_i).

Let K be a maximal ordered field and let f be a nonzero polynomial in K[T] of degree $n > 0$; for all $a \in \mathrm{K}$ let $w(a)$ denote the number of variations of the sequence $(f^{(i)}(a))_{0 \leqslant i \leqslant n}$. If v is the number of roots of f, counted with multiplicity, in the interval $]a, b]$ $(a < b)$, show that $v \leqslant w(a) - w(b)$ and that $w(a) - w(b) - v$ is even (« Budan-Fourier rule » ; split the interval $]a, b]$ by the roots of f, and evaluate the amount of variation in $w(x)$ as x passes one of these roots).

Deduce that if $f(\mathrm{T}) = a_0 + a_1\mathrm{T} + \cdots + a_n T^n$, then the number of roots > 0 of f, counted with multiplicity, is at most equal to the number of variations of the sequence $(a_i)_{0 \leqslant i \leqslant n}$, and that the difference between these two numbers is even (« Descartes' rule »).

22) Let K be a maximal ordered field and let

$$f(\mathrm{T}) = a_0 + a_1\mathrm{T} + \cdots + a_n\mathrm{T}^n$$

be a polynomial in K[T] such that $a_0 \neq 0 \neq a_n$ and

$$a_p = a_{p+1} = \cdots = a_{p+2m-1} = 0 \quad (1 \leqslant p < p + 2m - 1 < n) .$$

Show that f has at most $n - 2m$ roots in K, counted with multiplicity (apply Descartes' rule).

Deduce that if $g(\mathrm{T}) = 1 + c_1\mathrm{T} + \cdots + c_n\mathrm{T}^n$ has all its roots in K, and if $h(\mathrm{T}) = 1 + b_1\mathrm{T} + \cdots + b_{2m}\mathrm{T}^{2m}$ is the polynomial consisting of the first $2m + 1$ terms of the formal power series $1/g \in \mathrm{K}[[\mathrm{T}]]$, then h has no roots in K.

23) Let K be a maximal ordered field and let $a_1, \dots, a_n, b_1, \dots, b_n$ be $2n$ elements of K such that $\sum a_i \neq 0$ and $b_1 < b_2 < \cdots < b_n$. Show that for any integer $m \geqslant 1$ the number v of roots in K, counted with multiplicity, of the polynomial

$$f(T) = a_1(T - b_1)^m + a_2(T - b_2)^m + \cdots + a_n(T - b_n)^m$$

is at most equal to the number of variations w of the sequence

$$(a_1, a_2, \dots, a_n, (-1)^m a_1),$$

and that the difference $w - v$ is even. (Argue by induction on n, applying Ex. 13 (VI, p. 40) to a rational function of the form $f(T)/(T - c)^m$.)

¶ 24) Let K be a maximal ordered field ; consider an ordering on the field K(X) of rational functions which makes it an ordered extension of K. Show that this ordering is determined by the set A of $x \in K$ such that $x < X$ (show that the sign of every polynomial f over K is determined by A, using Prop. 9 of VI, p. 28). Conversely, show that to every set $A \subset K$ such that $y \in A$ whenever $y \leqslant x$ and $x \in A$, there corresponds an ordering of K(X) making it an ordered extension of K and such that A is the set of $x \in K$ such that $x < X$ (same method). When A has a greatest element, or $\complement A$ has a least element, or when $A = K$ or $A = \varnothing$, then the corresponding ordering of K(X) is such that K(X) is not comparable to K. On the other hand, if neither A nor $\complement A$ is empty, and there is no greatest element in A nor least element in $\complement A$, then K(X) is comparable to K.

25) Use Ex. 24 to find an example of a field E admitting several ordered field structures, none of which is induced from another *via* an automorphism of E. Hence find an example of a field, equipped with an ordering compatible with its ring structure but not a lattice ordering (consider the diagonal in $E \times E$ and use VI, p. 38, Ex. 5).

¶ 26) *a*) If K is an archimedean ordered field (VI, p. 40, Ex. 11, *a*)), and x and y are two elements of K with $x < y$, show that there exists $r \in \mathbf{Q}$ such that $x < r < y$. * Deduce that there is only one ordered subfield of $\mathbf{R}$ isomorphic to K. *

b) Consider the ordering of the field $K = \mathbf{Q}(X)$ in which X is infinitely large with respect to $\mathbf{Q}$ (Ex. 24). Show that K is comparable to the subfield $\mathbf{Q}(X^2)$, and give an example of two elements x, y of K such that $x < y$ and no element of $\mathbf{Q}(X^2)$ lies between x and y.

c) Let K be the ordered field defined in *b*) ; show that the polynomial

$$f(Y) = (Y^2 - X)(Y^2 - 4X) - 1$$

over K is irreducible, that it has roots in every maximal ordered extension of K, and that $f(a) > 0$ for all $a \in K$.

¶ 27) Let K be an ordered field, let E be a purely transcendental extension of K and let $(X_\iota)_{\iota \in I}$ be a transcendence basis for E (V, p. 109).

a) * If K is archimedean then E has a structure of an ordered extension of K for which E is comparable to K if and only if the cardinality of I is at most equal to that of a transcendence basis for $\mathbf{R}$ over K (where K is considered as an ordered subfield of $\mathbf{R}$, *cf.* Ex. 26, *a*)) ; the set of such structures is then bijective with the set of injective maps from I into $\mathbf{R}$ such that $f(I)$ is algebraically independent over K. *

b) If K is not archimedean then there always exists (at least) one structure on E of an ordered extension of K such that E is comparable to K (when I contains only one element, use Ex. 24 and notice that **Q** does not have a supremum in K ; pass to the general case by Zorn's Lemma).

* 28) *a*) Let K be a subfield of **R** and let θ be a real number, algebraic over K. Show that the number of structures on $K(\theta)$ of an ordered extension of K is equal to the number of real conjugates of θ (use Ex. 14, *a*) (VI, p. 40) and 26, *a*)).
b) Let K be a subfield of **R**, equipped with n distinct structures of an ordered extension of **Q**, all archimedean ; let P_i $(1 \leqslant i \leqslant n)$ be the sets of positive elements under these orderings. Show that there exists a family of n elements b_i of K $(1 \leqslant i \leqslant n)$ such that $b_i \in P_i$ for all i and $b_i \notin P_k$ for all $k \neq i$. (For each pair of distinct indices i, k, let $a_{ik} \in P_k$ be such that $a_{ik} \notin P_i$; consider the elements

$$c_i = \sum_{k \neq i} \left(\frac{1 + a_{ik}}{1 - a_{ik}} \right)^{2r}$$

where r is a sufficiently large integer.)
c) Let $K = \mathbf{Q}(\theta)$ be an algebraic extension of **Q** contained in **R**, and let n be the number of real conjugates of θ. Let C be the set of sums of squares of elements of K ; show that $C^* = C \cap K^*$ is a subgroup of K^* and that $(K^* : C^*) = 2^n$. (Use *b*) and Cor. 2 of VI, p. 23.) *

¶ 29) Let K be a maximal ordered field and let G be a subfield of K. Show that the set of extensions of G contained in K, and which are comparable to G, is inductively ordered ; if E_0 is a maximal element of this set, show that E_0 is isomorphic to the field $K(G)$ defined in Ex. 11, *b*) of VI, p. 40 (prove that the natural map from $F(G)$ onto $K(G)$ maps E_0 onto $K(G)$, by first showing that E_0 is a maximal ordered field, using Ex. 14, *a*) of VI, p. 40, and then that no element of $K(G)$ is transcendental over the image of E_0, using Ex. 24).

30) *a*) Let K be a maximal ordered field and let m and M be two elements of K with $m < M$. Show that every polynomial $f \in K[X]$ which is positive in the interval (m, M) is a sum of polynomials of the form $(\alpha X + \beta) g^2$, where $g \in K[X]$ and $\alpha X + \beta$ is positive in (m, M). (Reduce to the case of first or second degree polynomials, and use the formulae

$$(X - a)(X - b) = (X - b)^2 + (b - a)(X - b)$$
$$(X - a)(b - X) = ((X - a)(b - X)^2 + (b - X)(X - a)^2)/(b - a)$$

for $a < b$.)
b) Show that the result of *a*) is no longer necessarily true when K is an arbitrary ordered field (notice that a polynomial can be > 0 in K, but not in a maximal ordered extension of K ; cf. Ex. 26, *c*)).

¶ 31) *a*) Let K be a field whose algebraic closure E is an extension of K of prime degree q. Show that K is perfect (V, p. 7).
b) Show that q is distinct from the characteristic of K (V, p. 162, Ex. 9).

c) Show that K contains the q-th roots of unity, and that E is the splitting field of an irreducible polynomial of the form $X^q - a$ over K; deduce that $q = 2$ (otherwise, use Ex. 5, V, p. 161 to deduce that $X^{q^2} - a$ would be irreducible). Show also that $-a$ is a square in K (V, *loc. cit.*), that -1 is not a square in K, and that $E = K(i)$ $(i^2 = -1)$.
d) Now suppose that K is such that its algebraic closure E has arbitrary finite degree $\neq 1$ over K. Show that $i \notin K$ and that $E = K(i)$ (if $E \neq K(i)$, show by Galois theory that there would exist a field F such that $K(i) \subset F \subset E$, and that E has prime degree over F; then apply *c*)). Deduce that K is a maximal ordered field (under some suitable ordering) (show by induction that any sum of n squares in K is a square in K; to prove that $a^2 + b^2$ is a square, consider a square root $x + iy$ of $a + ib$ in $K(i)$).

¶ 32) Let A be an algebraically closed field of characteristic 0.
a) The only elements of finite order in the group Aut(A) of automorphisms of A are the identity element and elements of order 2 (*involutions* of A); these latter correspond bijectively to the maximal ordered subfields E of A such that $[A : E] = 2$, that is $A = E(i)$ (cf. Ex. 31). If σ is an involution of A and E is the fixed subfield of σ, then the group Aut(E) of automorphisms of E is isomorphic to the quotient $Z(\sigma)/\{e, \sigma\}$, where $Z(\sigma)$ is the centraliser of σ in Aut(A).
b) If A is algebraic over its prime field **Q** then any two involutions of A are conjugate in Aut(A) and the centraliser of an involution σ is $\{e, \sigma\}$; the group Aut(A) and the set of involutions of A have the cardinality of the continuum, and the centre of Aut(A) is $\{e\}$.
c) If A has finite transcendence degree over **Q**, show that the centraliser $Z(\sigma)$ of any involution σ of A is countable (observe that the fixed subfield E of σ is countable, and any automorphism of E which fixes the elements of some transcendence basis for E over **Q** is the identity). If $\deg.\mathrm{tr}_{\mathbf{Q}} A = n$, show that there exist at least $n + 1$ conjugacy classes of involutions of A (if $E \subset A$ is maximal orderable with $[A : E] = 2$, consider the transcendence degree over **Q** of the subfield of E consisting of elements fixed by Aut(E), and use Ex. 24).
d) If A has infinite transcendence degree $\mathfrak{m}$ over **Q**, show that there exist involutions σ of A such that $\mathrm{Card}(Z(\sigma)) = \mathrm{Card}(\mathrm{Aut}(A)) = 2^{\mathfrak{m}}$ (use Ex. 24 to show that there exist maximal orderable fields E such that $\deg.\mathrm{tr}_{\mathbf{Q}} E = \mathfrak{m}$, such that for any subset M of any transcendence basis B for E over **Q**, there exists an automorphism u_M of E such that $u_M(x) \neq x$ for $x \in M$ and $u_M(x) = x$ for $x \in B - M$).
e) If $\deg.\mathrm{tr}_{\mathbf{Q}} A = 1$, show that there is only one conjugacy class of involutions of A whose centralisers are infinite. (If E is a maximal orderable extension of **Q** of transcendence degree 1 such that Aut(E) is nontrivial, use Ex. 24 to show that E cannot be comparable to **Q**; if $t \in E$ is infinitely large with respect to **Q**, show that so is $u(t)$ for any automorphism u of E, and that conversely if t' is infinitely large with respect to **Q** then there exists an automorphism u of E such that $u(t) = t'$, using Ex. 15 of VI, p. 40.)
f) Let u be an automorphism of A which commutes with all the involutions of A. Show that if $a \in A$ is transcendental over **Q** then a and $u(a)$ are algebraically dependent over **Q**. (Consider two elements b, c of A such that $b^2 = a$, $c^2 = -u(a)$ and observe that there exists a maximal orderable extension $E \subset A$ such that $[A : E] = 2$ and b, c are in E; obtain a contradiction by noticing that $u(E) = E$.)
g) Deduce from *f*) that $u(a) = a$ for every element $a \in A$ transcendental over **Q** (consider a maximal ordered extension E of $\mathbf{Q}(a)$ contained in A, algebraic over $\mathbf{Q}(a)$ and

comparable to **Q**, using Ex. 24). Conclude that u is necessarily the identity (observe that if a is algebraic over **Q** and b is transcendental over **Q** then ab is transcendental over **Q**).

33) Let K be a commutative field (of arbitrary characteristic q), and let l be a prime number. Let K′ be an algebraic extension of K with the following two properties :

(i) every polynomial in K[X] whose degree is not divisible by l has a root in K′ ;

(ii) every polynomial in K′[X] of degree equal to l has a root in K′.

a) Show that the perfect closure K_1 of K is contained in K′ and satisfies the condition analogous to (i).

b) Every algebraic extension of K whose degree is not divisible by l is isomorphic to a subextension of K′ (use the primitive element theorem).

c) Let L be a subfield of K′ containing K, and let M be a Galois extension of L, whose degree is a power of l. Show that M is isomorphic to a sub-extension of K′ (use the fact that an l-group is nilpotent).

d) Deduce that K′ is an algebraic closure of K (if F is a (separable) extension of K of finite degree, consider a Galois extension $F_1 \supset F$ of finite degree, and the fixed subfield of F_1 under a Sylow l-subgroup of the Galois group of F_1 over K).

34) Let E be a totally ordered set. A (Dedekind) *cut* is a partition (S, D) of E into two nonempty subsets such that D is of the form M(A), the set of upper bounds for some nonempty subset A of E ; a necessary and sufficient condition for D to have a least element x is that $D = [x, \rightarrow[$, so $S =]\leftarrow, x[$.

In a totally ordered abelian group G, a cut (S, D) is called *proper* if for all $e > 0$ in G there exist $x' \in S$ and $x'' \in D$ such that $x'' - x' < e$. Show that (S, D) is proper if and only if D is *invertible* in the monoid $\mathfrak{M}(G)$ of major subsets of G (VI, p. 35, Ex. 30) ; all cuts are proper if and only if G is archimedean (VI, p. 35, Ex. 31) ; and hence * isomorphic to a subgroup of **R** (VI, p. 36, Ex. 33). *

35) A subfield K of an ordered field E is *dense* in E if for every nonempty open interval $]a, b[$ in E there exists $x \in K$ such that $a < x < b$. Show that E is then *comparable* to K (Ex. 11, *a*)). If E is taken to be R(X), where R is a maximal ordered field and X is infinitely large with respect to R (VI, p. 43, Ex. 24), show that E is comparable to the subfield $K = R(X^2)$, but that K is not dense in E.

For K to be dense in E, it is necessary and sufficient that for all $x \in E$ the partition of K consisting of $K \cap]\leftarrow, x[$ and $K \cap [x, \rightarrow[$ be a proper cut (Ex. 34).

36) Let K be an ordered field and let $\tilde{K}$ be the set of major sets D such that the cut (S, D) is proper ; show that $\tilde{K}$ is an ordered field, taking addition to be that of the monoid $\mathfrak{M}(K)$, and the product of two elements D_1 and D_2 of $\tilde{K}$ contained in K_+ to be the set $\langle D_1D_2\rangle$ (VI, p. 35, Ex. 30), where D_1D_2 denotes the set of x_1x_2 for $x_1 \in D_1$ and $x_2 \in D_2$. Show that for any increasing homomorphism f from K onto a dense subfield F of an ordered field E, there exists an increasing homomorphism g from E into $\tilde{K}$ such that $g \circ f$ is the natural injection of K into $\tilde{K}$.

37) Let K be an ordered field and let $n > 0$ be an integer such that every element $x > 0$ of K is equal to a power y^n of some element $y > 0$. Show that the field $\tilde{K}$ has the same property.

38) Suppose that the field K is maximal ordered. Show that the field $\tilde{K}$ is also maximal ordered. (One can argue by contradiction, by considering an ordered algebraic extension E of $\tilde{K}$ which is maximal ordered, and supposing that $E \neq \tilde{K}$. Then there exists among the elements of $E \cap \complement\tilde{K}$ an element w whose minimal polynomial $f \in \tilde{K}[X]$ has least degree $n > 1$. Show that there exists an interval $]a, b[$ in E such that $a, b \in K$ and $a < w < b$, and an element $e > 0$ of E such that $f'(x) > e$ for all $x \in [a, b]$, or $f'(x) < -e$ for all $x \in [a, b]$. Show that there exists $r > 0$ in K such that $f(a) \leqslant -r$ and $r \leqslant f(b)$, or $f(a) \geqslant r$ and $f(b) \leqslant -r$. Approximate the coefficients of f by elements of K, and show that we obtain in this way a polynomial $g \in K[X]$ such that the function $x \mapsto g(x)$ is monotone in $[a, b]$ and vanishes at some point of K in this interval, arbitrarily close to w.)

39) Let K be an ordered field and let E be an ordered algebraic extension which is maximal ordered. Show that the only K-automorphism of the field E is the identity. (First show that every K-homomorphism from E to itself is surjective, and is an isomorphism of *ordered* fields ; if $x \in E$ has minimal polynomial $f \in K[X]$ then an automorphism of E necessarily permutes the roots of f belonging to E. Deduce that it fixes x.)

40) Let K be an ordered field. Make the field of formal power series $E = K((X))$ an ordered field by taking the elements > 0 to be the formal power series $\sum_{n=-h}^{\infty} r_n X^n$ whose nonzero coefficient of least degree is > 0.

a) Show that if $L = K(X)$ is the subfield of E consisting of rational functions in one indeterminate, then $E = \tilde{L}$ (Ex. 36) up to isomorphism of ordered fields (observe that for every element $a > 0$ of L there exists an integer $n > 0$ such that $0 < X^n < a$).

b) Suppose that K is maximal ordered. Show that the union F of the fields $K((X^{1/n}))$ (ordered in the same way as E) is an ordered algebraic extension of E which is maximal ordered (use Ex. 2 of V, p. 150).

c) Show that the field $\tilde{F}$ (Ex. 36) is the field of formal series $\sum_r \alpha_r X^r$, where r runs through the set **Q** of rational numbers, and where $\alpha_r \in K$ and the set of r such that $\alpha_r \neq 0$ is an increasing sequence which is either finite or tends to $+\infty$; order this field in the same manner as E. In particular $\tilde{F} \neq F$.

41) Let K be an ordered field, and E a vector space of dimension 2 over K.

a) Show that the set of triples of pairwise distinct lines (D_1, D_2, D_3) passing through 0 has two orbits under the action of the group $\mathbf{GL}^+(E)$ (consider the subgroup fixing two lines).

b) The set of triples of pairwise distinct half-lines $(\Delta_1, \Delta_2, \Delta_3)$ with origin 0 has 7 orbits under the action of $\mathbf{GL}(E)$, and 14 under the action of $\mathbf{GL}^+(E)$ (same method).

CHAPTER VII

Modules over principal ideal domains

§ 1. PRINCIPAL IDEAL DOMAINS

1. Definition of a principal ideal domain

Recall (I, p. 104) that an ideal of a commutative ring A is said to be *principal* if it has the form $(a) = Aa$ for some $a \in A$.

DEFINITION 1. — *A principal ideal domain is an integral domain* (I, p. 116) *in which every ideal is principal.*

Examples. — The ring **Z** of rational integers is a principal ideal domain (I, p. 111). If K is a commutative field, then the polynomial ring K[X] in one indeterminate over K is a principal ideal domain (IV, p. 11, Prop. 11) ; the same is true of the ring of formal power series K[[X]], for every ideal of this ring has the form (X^n) (IV, p. 38, Prop. 12). * The ring of integers in a $\mathfrak{p}$-adic field is a principal ideal domain. *

If $\mathbf{Q}(i)$ denotes the field obtained from the field **Q** of rational numbers by adjoining a root i of the irreducible polynomial $X^2 + 1$, then the elements $a + bi$ of $\mathbf{Q}(i)$, where a and b are rational integers, form a subring A of $\mathbf{Q}(i)$, called « the *Gaussian integers* », which is a principal ideal domain (VII, p. 50, Ex. 7). By contrast, in the field $\mathbf{Q}(\rho)$, where ρ is a root of $X^2 + 5$, the subring B consisting of elements $a + b\rho$ (a and b rational integers) is not a principal ideal domain (VII, p. 51, Ex. 12).

The polynomial ring K[X, Y] in two indeterminates over a field K is not a principal ideal domain. Indeed the nonzero constants are the only elements which divide both X and Y, and none of them generates the ideal generated by X and Y.

2. Divisibility in principal ideal domains

Let A be a principal ideal domain and let K be its field of fractions (I, p. 116) ; we will see that the ordered group $\mathscr{P}^*$ of principal fractional ideals (VI, p. 6) of K is lattice ordered ; more precisely :

PROPOSITION 1. — *Let* K *be the field of fractions of a principal ideal domain* A, *and let* $(x_\iota)_{\iota \in I}$ *be a family of elements of* K *having a common denominator* $b \in K^*$ (in other words $bx_\iota \in A$ for all ι). *Then* :

a) The family (x_ι) *has a gcd in* K.

b) Every gcd of (x_ι) *can be expressed in the form* $d = \sum_\iota a_\iota x_\iota$, *where the* a_ι *are elements of* A, *all but finitely many of which are zero.*

Indeed the ideal $\sum_\iota \mathrm{A}bx_\iota$ of A is principal, and so of the form $\mathrm{A}d'$. Put $d' = bd$ ($d \in \mathrm{K}$). From the relation $d' = \sum_\iota a_\iota bx_\iota$, we deduce $d = \sum_\iota a_\iota x_\iota$, where $a_\iota \in \mathrm{A}$. Hence every common divisor of the x_ι divides d. On the other hand, since bd is a common divisor of the bx_ι by construction, it follows that d is a common divisor of the x_ι.

Remark. — Prop. 1 applies with no restrictions to an arbitrary family (x_ι) of elements *of* A (take $b = 1$), and also to any *finite* family (x_i) of elements of K (if $x_i = c_i b_i^{-1}$ with $c_i \in \mathrm{A}$ and $b_i \in \mathrm{A}$, then take b to be the product of the b_i).

COROLLARY. — *Let* (x_ι) *be an arbitrary family of elements of a principal ideal domain* A *contained as a subring in an integral domain* B, *and let d be a gcd of the family* (x_ι) *in* A. *Then the family* (x_ι) *has gcd's in* B, *and d is one of them.*

Indeed d is a common divisor of the x_ι in B. On the other hand the relation $d = \sum_\iota a_\iota x_\iota$ shows that every common divisor of the x_ι in B divides d.

An important application of this corollary is when $\mathrm{A} = \mathrm{K}[\mathrm{X}]$ and $\mathrm{B} = \mathrm{E}[\mathrm{X}]$, where K is a field and E is an extension of K (IV, p. 12, Cor. 1).

The first assertion of Prop. 1 shows that the ordered group $\mathscr{P}^*$ is *lattice ordered* (VI, p. 10). In particular every finite family of elements of K admits an lcm. We can thus apply the results denoted (DIV) in VI, pp. 10 to 17 to principal ideal domains.

The following result is a consequence of the second assertion of Prop. 1 :

THEOREM 1 (« Bezout's identity »). — *Elements* x_ι $(\iota \in \mathrm{I})$ *of a principal ideal domain* A *are setwise coprime if and only if there exist elements* a_ι $(\iota \in \mathrm{I})$ *of* A, *all but finitely many of them zero, such that* $\sum_\iota a_\iota x_\iota = 1$.

The necessity of the condition is just Prop. 1. Conversely, if $\sum_\iota a_\iota x_\iota = 1$ then every common divisor of the x_ι divides 1, and so 1 is a gcd of the x_ι.

PROPOSITION 2. — *Let a, b, d, m and p be elements of the field of fractions* K *of a principal ideal domain* A.

a) « *d is a gcd of a and b* » *is equivalent to* « $(d) = (a) + (b)$ ».

b) « *m is an lcm of a and b* » *is equivalent to* « $m = (a) \cap (b)$ ».

c) « *p is an irreducible element of* A » *is equivalent to* « (p) *is a nonzero maximal ideal of* A » *and to* « (p) *is a nonzero prime ideal of* A ».

We have already proved *a*) (Prop. 1). Since the common multiples of a and b are the elements of $(a) \cap (b)$, and since, by hypothesis, the ideal $(a) \cap (b)$ is principal, say $(a) \cap (b) = (m)$, it follows that m is an lcm of a and b, which proves *b*). Finally, to say that $p \neq 0$ is an irreducible element of A means by definition (VI, p. 17) that (p) is a maximal element of the family of principal ideals $\neq$ A of A, ordered by inclusion ; since A has no ideals other than the principal ideals, this means that (p) is a maximal ideal of A, whence *c*), by the remark in VI, p. 17.

In a principal ideal domain A, the sum (resp. the intersection) of a finite number of ideals is also called the gcd (resp. lcm) of these ideals.

PROPOSITION 3. — *Let a, b, c be elements of the field of fractions of a principal ideal domain* A, *and let d be a gcd of a and c ; then the congruence $ax \equiv b$ (mod c) has a solution $x_0 \in$ A if and only if d divides b ; in this case the elements x of* A *which satisfy $ax \equiv b$ (mod c) are precisely those which satisfy $x \equiv x_0$ (mod cd^{-1}).*

If $ax \equiv b$ (mod c), with $x \in$ A, then there exists $y \in$ A such that $b = ax + cy$, so d divides b. Conversely, if d divides b then $b = ax_0 + cy_0$ with $x_0, y_0 \in$ A (Prop. 1), so $ax_0 \equiv b$ (mod c) ; moreover, the relation $ax \equiv b$ (mod c) is then equivalent to $a(x - x_0) \equiv 0$ (mod c) ; putting $a = da'$ and $c = dc'$, the latter becomes $a'(x - x_0) \equiv 0$ (mod c'). But this is equivalent (for $x \in$ A) to $x - x_0 \equiv 0$ (mod c'), since a' and c' are coprime (VI, p. 14, Prop. 10 (DIV) and VI, p. 15, Cor. 2 to Prop. 11 (DIV)).

PROPOSITION 4. — *Let $(a_i)_{1 \leqslant i \leqslant n}$ be a finite family of pairwise coprime elements of the principal ideal domain* A. *Then the canonical homomorphism* (I, p. 110) *from* $A/\left(\prod_{i=1}^{n} a_i\right)$ *into the product* $\prod_{i=1}^{n} A/(a_i)$ *is an isomorphism of* A*-algebras.*

This follows from I, p. 110, Prop. 9 and from Prop. 2 *a*).

The conclusion of Prop. 4 is not valid if it is not assumed that the a_i are pairwise coprime (cf. VII, p. 24, Prop. 9).

3. Decomposition into irreducible factors in principal ideal domains

We will now apply the results of VI, p. 18, relating to decomposition into irreducible elements, to principal ideal domains. By Prop. 2, an element $p \neq 0$ of A is irreducible if and only if the ring $A/(p)$ is a field (I, p. 115, Cor. 1), that is if and only if the congruence $ax \equiv b$ (mod p) admits a solution in A for each $b \in$ A and for each $a \in$ A which is not a multiple of p.

DEFINITION 2. — *Let* A *be an integral domain. A system of representatives of irreducible elements of* A *is a family (p_α) of irreducible elements of* A *such that every irreducible element of* A *is an associate of precisely one p_α.*

THEOREM 2. — *Let A be a principal ideal domain and let (p_α) be a system of representatives of irreducible elements of A. Then every nonzero element x of the field of fractions of A can be uniquely expressed in the form*

$$x = u \prod_\alpha p_\alpha^{n_\alpha}, \tag{1}$$

where u is an invertible element of A, and where the n_α are integers, all but finitely many of which are zero. For x to belong to A it is necessary and sufficient that all the n_α be positive.

We will use the theorem about decomposition as a sum of irreducible elements (VI, p. 18, Th. 2), of which the above statement is only a translation. Since $\mathscr{P}^*$ is a lattice ordered group it will be sufficient for us to show that every non empty set of principal ideals of A contains a maximal element, in order to check that the hypotheses of this theorem are indeed satisfied ; now this follows from the next Lemma :

Lemma 1. — *Let A be a ring such that every left ideal of A is finitely generated. Then every nonempty set Φ of left ideals of A, ordered by inclusion, has a maximal element.*

By Zorn's Lemma (*Set Theory*, III, p. 154, Th. 2) it is enough to prove that Φ is inductive. Now if $(\mathfrak{a}_\lambda)$ is a totally ordered family of elements of Φ then the union $\mathfrak{a}$ of the ideals $\mathfrak{a}_\lambda$ is a left ideal of A, and so admits a finite system of generators $(a_i)_{1 \leqslant i \leqslant n}$. Since each a_i belongs to an ideal $\mathfrak{a}_{\lambda_i}$, and since the family $(\mathfrak{a}_\lambda)$ is totally ordered, the a_i all belong to the largest of the ideals $\mathfrak{a}_{\lambda_i}$, say $\mathfrak{a}_\mu$. Then $\mathfrak{a} = \mathfrak{a}_\mu$ belongs to Φ, which is thus indeed an inductive set.

Later we will study those rings B, called *noetherian rings,* such that every nonempty set of ideals of B contains a maximal element.

Remark. — The family $(u, (n_\alpha))$ is called the decomposition of x into irreducible factors ; by abuse of language, we also say that the formula (1) is the decomposition of x into irreducible factors. If $x = u \prod_\alpha p_\alpha^{n_\alpha}$ and $y = v \prod_\alpha p_\alpha^{m_\alpha}$ are the decompositions of x and y into irreducible factors, then a necessary and sufficient condition for x to divide y is that $n_\alpha \leqslant m_\alpha$ for all α ; from this we deduce the formulae

$$\gcd(x, y) = \prod_\alpha p_\alpha^{\inf(n_\alpha, m_\alpha)} \tag{2}$$

$$\operatorname{lcm}(x, y) = \prod_\alpha p_\alpha^{\sup(n_\alpha, m_\alpha)}. \tag{3}$$

The property expressed by Th. 2 is true for a more general class of rings than principal ideal domains ; we will study them later as *unique factorisation domains* ; and we will see that polynomial rings and formal power series rings in arbitrarily many indeterminates are unique factorisation domains (*Comm. Alg.*, VII, § 3).

4. Divisibility of rational integers

As was pointed out in section 1, the ring **Z** of rational integers is a principal ideal domain ; its field of fractions is **Q**. The multiplicative group U of invertible elements of **Z** has two elements 1 and -1. The group $\mathbf{Q}_+^*$ of rational numbers > 0 contains precisely one element from each class of associate elements of **Q** ; it is thus isomorphic to the multiplicative group $\mathcal{P}^* = \mathbf{Q}^*/\mathrm{U}$ of principal fractional ideals of **Q**, with which it is usually identified. In particular, whenever gcd or lcm is used in the field **Q** (with respect to the ring **Z**), it is understood that these are elements $\geqslant 0$; this convention allows us to speak of *the* gcd and *the* lcm of a family of rational numbers.

The irreducible integers > 0 in **Z** are precisely those we have called *prime numbers* (I, p. 50) (they are sometimes called *rational prime numbers*) ; every irreducible element of **Z** is thus of the form p or $-p$, where p is a prime number, and the set P of prime numbers is a system of representatives of irreducible elements of **Z**.

PROPOSITION 5. — *The set of prime numbers is infinite.*

Indeed, given an arbitrary finite family (p_i) $(1 \leqslant i \leqslant n)$ of distinct prime numbers, any prime divisor q of the number $\left(\prod_{i=1}^{n} p_i\right) + 1$ (which is > 1) is distinct from all the p_i, for otherwise it would divide 1.

5. Divisibility of polynomials in one indeterminate over a field

The polynomial ring K[X] in one indeterminate over a commutative field K is a principal ideal domain (IV, p. 11, Prop. 11). Its field of fractions is the field K(X) of rational functions in X with coefficients from K. The ring K[X] contains the subring of polynomials of degree 0, that is the field of constants, which is identified with K ; the elements of K* are invertible in K, and hence in K[X] ; conversely the formula $\deg(uv) = \deg(u) + \deg(v)$ shows that every invertible polynomial has degree 0 ; the group U of invertible elements of K[X] is thus precisely K*. Thus two associate polynomials differ only by a nonzero constant factor ; in particular every class of associate polynomials contains a unique *monic* polynomial. The subgroup of the multiplicative group K(X)* generated by the monic polynomials thus contains a unique element from each class of associate rational functions, and consequently is isomorphic to the group

$$\mathcal{P}^* = \mathrm{K}(\mathrm{X})^*/\mathrm{U}$$

of principal fractional ideals of K(X). In particular, whenever gcd or lcm in the field K(X) (with respect to the ring K[X]) is mentioned, it will usually be understood that these are quotients of monic polynomials (or 0) ; this convention allows us to speak of *the* gcd or *the* lcm of a family of rational functions.

The irreducible elements of K[X] are precisely the *irreducible polynomials* in the usual sense (IV, p. 13, Def. 2), and the set of monic irreducible polynomials is a system of representatives of irreducible elements of K[X].

A first degree polynomial is always irreducible. If K is an *algebraically closed* field then the converse is true (V, p. 19, Prop. 1) ; thus in this case every polynomial $p(X)$ of degree n in K[X] can be written uniquely (up to the order of the factors) in the form

$$p(X) = c(X - a_1)(X - a_2) \dots (X - a_n)$$

where c and the a_i are elements of K.

PROPOSITION 6. — *For any field* K, *the set of monic irreducible polynomials in* K[X] *is infinite.*

Indeed, given an arbitrary nonempty finite family (p_i) $(1 \leqslant i \leqslant n)$ of distinct monic irreducible polynomials, the polynomial $\left(\prod_{i=1}^{n} p_i\right) + 1$ is not invertible, and any monic irreducible factor q of this polynomial is necessarily distinct from all the p_i, otherwise it would divide 1.

§ 2. TORSION MODULES OVER A PRINCIPAL IDEAL DOMAIN

1. Modules over a product of rings

Let A be a ring and $(\mathfrak{b}_i)_{i \in I}$ a direct decomposition of A, that is (I, p. 110, Def. 7) a finite family of two-sided ideals of A such that the natural homomorphism from A into the product of the $A/\mathfrak{b}_i$ is bijective. By *loc. cit.*, Prop. 10, there exists a family $(e_i)_{i \in I}$ of central idempotents of A such that $\mathfrak{b}_i = A(1 - e_i)$, $\sum_{i \in I} e_i = 1$ and $e_i e_j = 0$ for $i \neq j$.

For every left A-module M, let M_i denote the set of $m \in M$ such that $\mathfrak{b}_i m = 0$; since $\mathfrak{b}_i$ is a two-sided ideal, this is a submodule of M ; moreover, if $a, b \in A$ and $a - b \in \mathfrak{b}_i$, then the homotheties a_{M_i} and b_{M_i} coincide ; there is thus a unique $(A/\mathfrak{b}_i)$-module structure on M_i such that the A-module structure of M_i is induced *via* the homomorphism $A \to A/\mathfrak{b}_i$.

PROPOSITION 1. — *The* A*-module* M *is the direct sum of its submodules* M_i.

Let $p_i : M \to M$ denote the homothety $m \mapsto e_i m$; the map p_i is A-linear since e_i is central ; since $e_i^2 = e_i$, $\sum_{i \in I} e_i = 1$ and $e_i e_j = 0$ for $i \neq j$, we have

$$p_i \circ p_i = p_i , \quad \sum_{i \in I} p_i = 1_M , \quad p_i \circ p_j = 0 \quad \text{for} \quad i \neq j ,$$

and the p_i form an orthogonal family of projectors whose sum is the identity (II, p. 209, Def. 7). By *loc. cit.*, Prop. 12, the module M is the direct sum of the submodules $p_i(M) = e_iM$. In addition e_iM is annihilated by $\mathfrak{b}_i = A(1-e_i)$; if $i \neq j$ and $m \in M$ then $(1-e_i)e_jm = e_jm$, whence no nonzero element of e_jM is annihilated by $1-e_i$, and so *a fortiori* by $\mathfrak{b}_i$. It follows that $e_iM = M_i$, and the proposition is proved.

Remarks. — 1) Conversely, let M_i' be an $(A/\mathfrak{b}_i)$-module for each i, and consider the A-module M, the direct sum of the A-modules M_i' ; then the submodules M_i constructed above coincide with the M_i' (it is enough to note that if $i \neq j$ then no nonzero element of M_j' is annihilated by $\mathfrak{b}_i$ because $\mathfrak{b}_i + \mathfrak{b}_j = A$). Thus, roughly speaking, it amounts to the same thing to consider an A-module M or a family (M_i) of modules over the rings $A/\mathfrak{b}_i = A_i$.

2) By the above proof, the projectors of M onto the components M_i are homotheties.

3) The A-module M is cyclic if and only if each M_i is cyclic : if $M = Am$, then $M_i = A_ie_im$; conversely if $M_i = A_im_i$ and $m = \sum_{i \in I} m_i$, then $M = Am$; indeed, if $n \in M$ projects onto a_im_i for each i, and if $a \in A$ is congruent to a_i mod $\mathfrak{b}_i$ for each i, then am and n have the same image in each M_i, so coincide.

4) Let M and N be two A-modules, with components (M_i) and (N_i). Let $u \in \operatorname{Hom}_A(M, N)$ be an A-linear map from M to N ; then for all i and for all $m \in M_i$ we have $u(m) \in N_i$, so u induces an A_i-linear map $u_i \in \operatorname{Hom}_{A_i}(M_i, N_i)$. It is easy to check that the map $u \mapsto (u_i)$ is an isomorphism of **Z**-modules (resp. of A-modules when A is commutative)

$$\operatorname{Hom}_A(M, N) \to \prod_{i \in I} \operatorname{Hom}_{A_i}(M_i, N_i) .$$

2. Canonical decomposition of a torsion module over a principal ideal domain

Let M be a module over a commutative ring A. For each $\alpha \in A$ let $M(\alpha)$ denote the kernel of the endomorphism $x \mapsto \alpha x$ of M. If α and β are two elements of A such that α divides β, then clearly $M(\alpha) \subset M(\beta)$. In particular, as n runs through the set of rational integers $\geqslant 1$, the submodules $M(\alpha^n)$ form an increasing sequence ; the union M_α of the $M(\alpha^n)$ is thus a submodule of M, consisting of those elements of M which are annihilated by some power of α. For each submodule N of M, it is clear that $N_\alpha = N \cap M_\alpha$.

DEFINITION 1. — *Let π be an irreducible element of a principal ideal domain* A ; *an* A-*module* M *is called π-primary if, for all $x \in$* M, *there exists an integer $n \geqslant 1$ such that $\pi^n x = 0$* (in other words, if M is equal to the submodule M_π).

Clearly every cyclic module of the form $A/(\pi^s)$ is π-primary. For an arbitrary A-module M, the submodule M_π is π-primary.

Lemma 1. — *Let* M *be a module over a principal ideal domain* A ; *for all* $\alpha \in$ A *such that* $\alpha \neq 0$, *let* $\alpha = \varepsilon \sum_{i=1}^{r} \pi_i^{n(i)}$ *be a decomposition of* α *into irreducible factors* (VII, p. 4). *The submodule* N = M(α) *of elements of* M *annihilated by* α *is the direct sum of the submodules* $\mathrm{M}(\pi_i^{n(i)})$, *and the map which sends each* $x \in \mathrm{M}(\alpha)$ *to its component in* $\mathrm{M}(\pi_i^{n(i)})$ *has the form* $x \mapsto \gamma_i x$ ($\gamma_i \in$ A). *Moreover*

$$\mathrm{M}(\pi_i^{n(i)}) = \mathrm{N} \cap \mathrm{M}_{\pi_i} = \mathrm{N}_{\pi_i}\,.$$

Note first that N is annihilated by α, so has a natural A/(α)-module structure. By Prop. 4 of VII, p. 3, the canonical homomorphism from A/(α) into the product of the rings $\mathrm{A}/(\pi_i^{n(i)})$ is a ring isomorphism ; now applying Prop. 1 of VII, p. 6, we deduce that N is the direct sum of the $\mathrm{M}(\pi_i^{n(i)})$; the projectors of this decomposition are homotheties, by VII, p. 7, Remark 2. The inclusion $\mathrm{M}(\pi_i^{n(i)}) \subset \mathrm{M}(\alpha) \cap \mathrm{M}_{\pi_i}$ is obvious ; conversely, let $x \in \mathrm{M}(\alpha) \cap \mathrm{M}_{\pi_i}$. Then there exists a power π_i^s of π_i which annihilates x ; we may assume $s \geqslant n(i)$; by Bezout's identity, there exist λ, $\mu \in$ A such that $\pi_i^{n(i)} = \lambda \pi_i^s + \mu\alpha$, so $\pi_i^{n(i)} x = 0$ and so finally $x \in \mathrm{M}(\pi_i^{n(i)})$.

Lemma 2. — *Let* M *be a torsion module* (II, p. 313) *over an integral domain* A. *For every finite family* $(x_i)_{1 \leqslant i \leqslant n}$ *of elements of* M, *there exists an element* $\gamma \neq 0$ *in* A *such that the* x_i *all belong to* M(γ).

Indeed, for each index i there exists an element $\alpha_i \neq 0$ in A which annihilates x_i, and the element $\gamma = \prod_{i=1}^{n} \alpha_i$ will fit the bill.

THEOREM 1. — *Let* M *be a torsion module over a principal ideal domain* A ; *for each irreducible element* π *of* A, *let* M_π *be the submodule of* M *consisting of elements annihilated by some power of* π. *If* P *is a system of representatives of irreducible elements of* A, *then* M *is the direct sum of its submodules* M_π *for* $\pi \in$ P.

Every element $x \in$ M belongs to the submodule M(α) for some $\alpha \neq 0$, so by Lemma 1 is a sum of a finite number of elements, each of which belongs to some submodule M_π. On the other hand, if $\sum_{\pi \in \mathrm{P}} x_\pi = \sum_{\pi \in \mathrm{P}} y_\pi$, where x_π, $y_\pi \in \mathrm{M}_\pi$ for all $\pi \in$ P, and where all but finitely many of the x_π and y_π are zero, then Lemma 2 shows that there exists $\gamma \neq 0$ in A such that all the x_π and y_π belong to the same submodule M(γ) ; applying Lemma 1 to M(γ) shows that $x_\pi = y_\pi$ for all $\pi \in$ P, which completes the proof.

Clearly, if π and π' are two associate irreducible elements, then $\mathrm{M}_\pi = \mathrm{M}_{\pi'}$; thus, for a given module M, the submodule M_π depends only on the ideal (π) of A ; it is called the π-*primary component* of the module M, and the decomposition of M as a direct sum of the M_π is called the *canonical decomposition* of M as a direct sum of its π-primary components.

COROLLARY 1. — *Every submodule* N *of a torsion module* M *is the direct sum of its submodules* $N \cap M_\pi$.

This follows from the fact that $N \cap M_\pi$ is the π-primary component N_π of N.

COROLLARY 2. — *The submodule* N *of the torsion* A-*module* M *is a direct factor if and only if* N_π *is a direct factor of* M_π *for every irreducible element* π *of* A.

Indeed, if N and N′ are two submodules of M, then $M = N \oplus N'$ if and only if $M_\pi = N_\pi \oplus N'_\pi$ for every irreducible element π of A (Cor. 1).

COROLLARY 3. — *Let* N *be a submodule of the torsion* A-*module* M. *If, for every irreducible element* π *of* A, *either* $N_\pi = 0$ *or* $(M/N)_\pi = 0$, *then* N *is a direct factor of* M.

Indeed, the condition $(M/N)_\pi = 0$ implies $N_\pi = M_\pi$, and Cor. 2 applies.

An A-module M is called *semi-simple* if every submodule of M is a direct factor (cf. A, VIII, § 3).

COROLLARY 4. — *Let* A *be a principal ideal domain which is not a field, and let* M *be an* A-*module. Then* M *is semi-simple if and only if* M *is a torsion module and* $M_\pi = M(\pi)$ *for every irreducible element* π *of* A.

First suppose that M is semi-simple ; let $x \in M$ and let π be an irreducible element of A. If N is a complement of $A\pi x$ in M, then we can write $x = \alpha\pi x + y$, with $\alpha \in A$ and $y \in N$; but that implies $y = (1 - \alpha\pi)x$, so

$$\pi(1 - \alpha\pi)x \in A\pi x \cap N = 0 .$$

It follows first of all that M is a torsion module ; if moreover $x \in M_\pi$, then $\pi(1 - \alpha\pi)x = 0$, thus $\pi x = \alpha\pi^2 x = \alpha^2\pi^3 x = \cdots = \alpha^n\pi^{n+1}x$ is zero and $M_\pi = M(\pi)$.

Conversely, by Cor. 2 it is enough to prove that an A-module M annihilated by an irreducible element π is semi-simple ; but that is clear, since M then has a natural structure of a vector space over the field $A/(\pi)$, and the submodules of M are precisely the vector subspaces under this structure.

Remark 1. — Clearly the annihilator of every element $\neq 0$ of a π-primary module has the form $A\pi^k$ ($k > 0$ an integer), since it is a principal ideal containing a power of π. Let x be an element of M ; for each $\pi \in P$, let x_π be the component of x in M_π ; the annihilator of x is the lcm of the annihilators of the nonzero x_π, but by the above it is equal in this case to the *product* of the annihilators of the nonzero x_π (VI, p. 16, Prop. 12 (DIV)).

PROPOSITION 2. — *If* M *is a finitely generated torsion module over a principal ideal domain* A, *then the* π-*primary components of* M *are zero except for a finite number of them, and the projectors of* M *onto these components* M_π *are homotheties.*

This follows immediately from Lemma 1, for by Lemma 2 there exists $\alpha \neq 0$ in A such that $M = M(\alpha)$.

Remark 2. — By VII, p. 7, Remark 3, a finitely generated torsion A-module is cyclic if and only if each of its π-primary components is cyclic.

An important special case where Th. 1 and Prop. 2 apply is when $A = \mathbf{Z}$; a $\mathbf{Z}$-module is just an *abelian group*. An abelian group is called a *torsion group* if it is a torsion $\mathbf{Z}$-module, that is if all its elements have *finite order*. Then P is taken to be the set of prime numbers > 0 ; for each prime number $p > 0$, an (abelian) group is said to be p-torsion if all its elements have orders which are powers of p. In this terminology, Th. 1 shows that *every torsion abelian group is a direct sum of p-torsion groups*. In the case of finite groups, this also follows from I, p. 80, Th. 4.

3. Applications : I. Canonical decompositions of rational numbers and of rational functions in one indeterminate

THEOREM 2. — *Let* A *be a principal ideal domain, let* K *be its field of fractions and let* P *be a system of representatives of irreducible elements of* A. *Given an element* $x \in K$, *there exist a finite subset* H *of* P, *elements* $a_0 \in A$ *and* $a_p \in A$ *not divisible by* p *in* A ($p \in H$), *and integers* $s(p) > 0$ ($p \in H$) *such that*

$$x = a_0 + \sum_{p \in H} a_p p^{-s(p)}, \tag{1}$$

where H *and the* $s(p)$ *are uniquely determined by these conditions.*

Moreover, if R_p *denotes a subset of* A *containing precisely one element of each residue class* mod p *in* A ($p \in P$), *then each* $x \in K$ *can be uniquely expressed in the form*

$$x = a + \sum_{p \in P} \left(\sum_{h=1}^{\infty} r_{ph} p^{-h} \right) \tag{2}$$

where $a \in A$ *and* $r_{ph} \in R_p$ *for all* h *and* p, *and all but finitely many of the* r_{ph} *are zero.*

Consider K as an A-module ; then A is the submodule generated by 1. The quotient module K/A is the quotient of K by the equivalence relation $x' - x \in A$, which is also written, in the notation of VI, p. 6, as $x \equiv x' \pmod 1$; let f denote the natural homomorphism from K onto $M = K/A$.

The quotient module M is a *torsion module*, for every element of M has the form $f(a/b)$ ($a \in A$, $b \in A$, $b \neq 0$), whence $bf(a/b) = f(a) = 0$. Hence Th. 1 of VII, p. 8 applies. Let M_p ($p \in P$) be the submodule of elements of M annihilated by powers of p ; then $f^{-1}(M_p)$ is the subring A_p of elements of K of the form ap^{-n} where $a \in A$ and $n \geqslant 0$ is an integer. The module M is the direct sum of the M_p, so every $x \in K$ is congruent mod 1 to an element in the sum of the A_p ; in other words, formula (1) holds, with the $s(p)$ integers > 0, and the a_p finitely many elements of A such that a_p is not a multiple of p.

We now show that these conditions on the $s(p)$ and the a_p completely determine H and the $s(p)$. Indeed H is then the set of $p \in P$ such that the component of $f(x)$ in M_p is $\neq 0$. On the other hand if s and s' are two integers > 0 such that

$s \geqslant s'$ and if a and a' are elements of A not divisible by p such that $ap^{-s} \equiv a'p^{-s'}$ (mod 1), then we deduce that $a \equiv a'p^{s-s'}$ (mod p^s) ; if $s > s'$ then $a \equiv 0$ (mod p), contradicting the hypotheses. This argument also shows that each a_p is well defined mod $p^{s(p)}$.

To complete the proof, we note first of all that in each residue class mod p^s in A there exists a unique element of the form $\sum_{h=0}^{s-1} r_h p^h$ with $r_h \in \mathrm{R}_p$ for $0 \leqslant h \leqslant s-1$. In fact, we proceed by induction on s (the property follows from the definition of R_p for $s = 1$) : let $x \in \mathrm{A}$; by hypothesis there is a unique element of the form $\sum_{h=0}^{s-2} r_h p^h$ ($r_h \in \mathrm{R}_p$) in the residue class of x mod p^{s-1} ; then $x - \sum_{h=0}^{s-2} r_h p^h$ is a multiple ap^{s-1} of p^{s-1} ; now there exists a unique element r_{s-1} of R_p such that $a \equiv r_{s-1}$ (mod p) ; whence $x \equiv \sum_{h=0}^{s-1} r_h p^h$ (mod p^s). To obtain formula (2) it is now enough to apply this fact to each a_p in formula (1). The uniqueness is clear in view of the above.

The following are the most important applications of Th. 2 :

I. *The ring* A *is the ring* **Z** *of rational integers, and* K = **Q**. Let P be the set of prime numbers > 0, and for each $p \in \mathrm{P}$ let R_p be the interval $(0, p-1)$ in **Z**. Thus we have the canonical decomposition

$$x = a + \sum_{p \in \mathrm{P}} \left(\sum_{h=1}^{\infty} e_{ph} p^{-h} \right)$$

with $a \in \mathbf{Z}$, $e_{ph} \in \mathbf{Z}$ and $0 \leqslant e_{ph} \leqslant p - 1$.

II. *The ring* A *is the ring* E[X] *of polynomials in one indeterminate over a commutative field* E, *and* K = E(X). Let P be the set of monic irreducible polynomials in E[X] (VII, p. 5). For $p \in \mathrm{P}$ we can, by virtue of euclidean division of polynomials (IV, p. 10), take R_p to be the set of polynomials of degree strictly less than that of p. Thus we have the decomposition (called canonical) of a rational function $r(\mathrm{X}) \in \mathrm{E}(\mathrm{X})$:

$$r(\mathrm{X}) = a(\mathrm{X}) + \sum_{p \in \mathrm{P}} \left(\sum_{h=1}^{\infty} v_{ph}(\mathrm{X}) \, . \, p(\mathrm{X})^{-h} \right)$$

where $a(\mathrm{X})$ is a polynomial and $v_{ph}(\mathrm{X})$ is a polynomial of degree strictly less than that of $p(\mathrm{X})$, for all p and h. In particular, if the field E is *algebraically closed*, then the $p(\mathrm{X})$ have the form $\mathrm{X} - \alpha$ with $\alpha \in \mathrm{E}$, and the $v_{ph}(\mathrm{X})$ are thus constants.

Hence we can say that the vector space E(X) over the field E has a basis consisting of the monomials X^n ($n \geqslant 0$ an integer) and the rational functions of the form $\mathrm{X}^m/(p(\mathrm{X}))^h$, where $p \in \mathrm{P}$ and h and m are integers with $h \geqslant 1$ and $0 \leqslant m < \deg(p)$.

4. Applications : II. The multiplicative group of units of the integers modulo a

Let a be a rational integer > 1, and let $(\mathbf{Z}/a\mathbf{Z})^*$ be the multiplicative group of invertible elements of the ring $\mathbf{Z}/a\mathbf{Z}$. If $a = \prod_i p_i^{n(i)}$ is the decomposition of a into prime factors, then the ring $\mathbf{Z}/a\mathbf{Z}$ is isomorphic to the product of the rings $\mathbf{Z}/p_i^{n(i)}\mathbf{Z}$ (VII, p. 3, Prop. 4) and the group $(\mathbf{Z}/a\mathbf{Z})^*$ is isomorphic to the product of the groups $(\mathbf{Z}/p_i^{n(i)}\mathbf{Z})^*$. We are thus reduced to the study of the groups $(\mathbf{Z}/p^n\mathbf{Z})^*$, where p is a prime number ; recall (V, p. 80) that the order $\varphi(p^n)$ of $(\mathbf{Z}/p^n\mathbf{Z})^*$ is $p^n - p^{n-1} = p^{n-1}(p-1)$.

Suppose first of all that $p > 2$; the natural homomorphism $\mathbf{Z}/p^n\mathbf{Z} \to \mathbf{Z}/p\mathbf{Z}$ restricts to a homomorphism of groups from $(\mathbf{Z}/p^n\mathbf{Z})^*$ onto $(\mathbf{Z}/p\mathbf{Z})^*$, whose kernel we denote $\mathrm{U}(p^n)$; by VII, p. 3, Prop. 3 the residue class mod p^n of an integer m is invertible if and only if m is coprime to p, that is if and only if the residue class of m mod p is invertible. It follows that $\mathrm{U}(p^n)$ consists of all the residue classes mod p^n of integers congruent to 1 mod p, so has p^{n-1} elements, and that there is an exact sequence

(3) $$\{1\} \to \mathrm{U}(p^n) \to (\mathbf{Z}/p^n\mathbf{Z})^* \to (\mathbf{Z}/p\mathbf{Z})^* \to \{1\} .$$

Similarly, for $n \geqslant 2$ let $\mathrm{U}(2^n)$ denote the kernel of the natural homomorphism from $(\mathbf{Z}/2^n\mathbf{Z})^*$ to $(\mathbf{Z}/4\mathbf{Z})^*$; this is a group of order 2^{n-2}, consisting of all the residue classes mod 2^n of integers congruent to 1 mod 4, and there is an exact sequence

(4) $$\{1\} \to \mathrm{U}(2^n) \to (\mathbf{Z}/2^n\mathbf{Z})^* \to (\mathbf{Z}/4\mathbf{Z})^* \to \{1\} .$$

Lemma 3. — *Let x, y, k be integers with $k \geqslant 0$, and let $p > 2$ be a prime number. If $x \equiv 1 + py \bmod p^2$ then $x^{p^k} \equiv 1 + p^{k+1}y \bmod p^{k+2}$. If $x \equiv 1 + 4y \bmod 8$ then $x^{2^k} \equiv 1 + 2^{k+2}y \bmod 2^{k+3}$.*

To prove the first assertion, it is enough to show that, if $k \geqslant 1$ and $x \equiv 1 + p^k y \bmod p^{k+1}$, then $x^p \equiv 1 + p^{k+1}y \bmod p^{k+2}$, and then to argue by induction on the integer k. For all $a \in \mathbf{Z}$ and $k \geqslant 1$, it is immediate that

$$(1 + p^k a)^p \equiv 1 + p^{k+1}a \mod p^{k+2} ,$$

hence

$$(1 + p^k y + p^{k+1}z)^p = (1 + p^k(y + pz))^p \equiv$$
$$\equiv 1 + p^{k+1}(y + pz) \equiv 1 + p^{k+1}y \mod p^{k+2} .$$

Similarly, for $k \geqslant 1$ we have

$$(1 + 2^{k+1}a)^2 \equiv 1 + 2^{k+2}a \mod 2^{k+3} ,$$

so

$$(1 + 2^{k+1}y + 2^{k+2}z)^2 \equiv 1 + 2^{k+2}y \mod 2^{k+3} ,$$

whence the second assertion by induction on k.

PROPOSITION 3. — *Let $p > 2$ be a prime number and let $n > 0$ be an integer ; then the group* $U(p^n)$ *is cyclic of order* p^{n-1}; *if* $n \geq 2$ *then the residue class* mod p^n *of an integer x congruent to* 1 mod p *is a generator of* $U(p^n)$ *if and only if x is not congruent to* 1 mod p^2. *Let $m > 1$ be an integer ; then the group* $U(2^m)$ *is cyclic of order* 2^{m-2} ; *if* $m \geq 3$ *then the residue class* mod 2^m *of an integer x congruent to* 1 mod 4 *is a generator of* $U(2^m)$ *if and only if x is not congruent to* 1 mod 8.

Since $U(p^n)$ has order p^{n-1}, the order of every element u of $U(p^n)$ is a power of p, and u is a generator of $U(p^n)$ if and only if $u^{p^{n-2}} \neq 1$. Now if u is the class of $x = 1 + py$, then $u^{p^{n-2}}$ is the class of $1 + p^{n-1}y$, by Lemma 3, whence u generates $U(p^n)$ if and only if $y \not\equiv 0 \bmod p$, in other words $x \not\equiv 1 \bmod p^2$. For example, the class $1 + p$ generates $U(p^n)$. Similarly, the class u of x mod 2^n generates $U(2^n)$ if and only if $u^{2^{n-3}} \neq 1$, which means that x is not congruent to 1 mod 8, by Lemma 3 ; this is satisfied by $x = 5$.

Lemma 4. — *Let A be a principal ideal domain and let* $0 \to N \to M \to P \to 0$ *be an exact sequence of A-modules. Suppose that there exist coprime elements a, $b \in A$ such that* $aN = 0$ *and* $bP = 0$. *Then the exact sequence splits. If in addition* N *and* P *are both cyclic, then* M *is cyclic.*

The module M is torsion, since $ab\mathrm{M} = 0$. The first assertion follows from Cor. 3 of VII, p. 9. If N and P are cyclic, then they are finitely generated, and hence so is M (II, p. 17, Cor. 5) ; since each p-primary component of M is isomorphic to a p-primary component either of N or of P, it follows from Remark 2 of VII, p. 10, that M is cyclic.

THEOREM 3. — *If* $a = \prod_i p_i^{n(i)}$ *is the prime decomposition of the integer* $a > 1$, *then the group* $(\mathbf{Z}/a\mathbf{Z})^*$ *of invertible elements of the ring* $\mathbf{Z}/a\mathbf{Z}$ *is isomorphic to the product of the groups* $(\mathbf{Z}/p_i^{n(i)}\mathbf{Z})^*$. *If* $p > 2$ *is a prime number and* $n \geq 1$ *an integer, then the group* $(\mathbf{Z}/p^n\mathbf{Z})^*$ *is cyclic of order* $p^{n-1}(p-1)$. *The group* $(\mathbf{Z}/2\mathbf{Z})^*$ *is trivial ; for* $n \geq 2$ *the group* $(\mathbf{Z}/2^n\mathbf{Z})^*$ *is the direct product of the cyclic group of order* 2^{n-2} *generated by the residue class of* 5 mod 2^n *and the cyclic group of order* 2 *consisting of the residue classes of* 1 *and* -1 mod 2^n.

The orders p^{n-1} of $U(p^n)$ and $p - 1$ of $(\mathbf{Z}/p\mathbf{Z})^*$ are coprime ; since $U(p^n)$ and $(\mathbf{Z}/p\mathbf{Z})^*$ are cyclic (Prop. 3 and V, p. 78, Lemma 1), the group $(\mathbf{Z}/p^n\mathbf{Z})^*$ is cyclic (apply Lemma 4 to the exact sequence (3)). If $n \geq 2$ then the restriction of the homomorphism $v : (\mathbf{Z}/2^n\mathbf{Z})^* \to (\mathbf{Z}/4\mathbf{Z})^*$ to the subgroup $\{1, -1\}$ is bijective ; the group $(\mathbf{Z}/2^n\mathbf{Z})^*$ is thus the direct product of this subgroup and the kernel $U(2^n)$ of v ; the result follows from Prop. 3.

Remark. — Let $p > 2$ be a prime number and let x be an integer congruent to 1 mod p and not congruent to 1 mod p^2 ; there is an exact sequence

(5) $$\{0\} \to \mathbf{Z}/p^{n-1}\mathbf{Z} \xrightarrow{u} (\mathbf{Z}/p^n\mathbf{Z})^* \xrightarrow{v} (\mathbf{Z}/p\mathbf{Z})^* \to \{1\}$$

of groups, where v is the natural projection and where u is induced on the quotients by the map $r \mapsto x^r$. Let $\mathbf{Z}_p$ be the ring of p-adic integers (V, p. 96), and let x be an element of $\mathbf{Z}_p$ such that $x - 1 \in p\mathbf{Z}_p$ and $x - 1 \notin p^2\mathbf{Z}_p$; on passing to inverse limits, the exact sequences (5) induce an exact sequence

$$\{0\} \to \mathbf{Z}_p \xrightarrow{u} \mathbf{Z}_p^* \xrightarrow{v} (\mathbf{Z}/p\mathbf{Z})^* \to \{1\}$$

where v is the natural map, and the continuous map u extends the map $n \mapsto x^n$ $(n \in \mathbf{Z})$. We often put $x^n = u(x)$ for $n \in \mathbf{Z}_p$.

Similarly, if $x \in \mathbf{Z}_2$ with $x - 1 \in 4\mathbf{Z}_2$ and $x - 1 \notin 8\mathbf{Z}_2$, then there is a split exact sequence

$$\{0\} \to \mathbf{Z}_2 \xrightarrow{u} \mathbf{Z}_2^* \xrightarrow{v} (\mathbf{Z}/4\mathbf{Z})^* \to \{1\} ,$$

where u is a continuous extension of the map $n \mapsto x^n$.

§ 3. FREE MODULES OVER A PRINCIPAL IDEAL DOMAIN

THEOREM 1. — *Let* A *be a ring such that every left ideal of* A *is projective* (II, p. 231, Def. 1) *as an* A*-module. Then every submodule* M *of a free left* A*-module* L *is a direct sum of modules isomorphic to ideals of* A.

Let $(e_\iota)_{\iota \in I}$ be a basis for L, and let p_ι be the coordinate functions corresponding to this basis. Choose a *well-ordering* (*Set Theory*, III, p. 153) of I and let L_ι denote the submodule generated by the e_λ for $\lambda \leq \iota$; put $M_\iota = M \cap L_\iota$. The coordinate function p_ι maps M_ι onto an ideal $\mathfrak{a}_\iota$ of A ; since $\mathfrak{a}_\iota$ is a projective A-module, there exists (II, p. 231, Prop. 4) a submodule N_ι of M_ι such that the map $x \mapsto p_\iota(x)$ from N_ι into $\mathfrak{a}_\iota$ is bijective. Let M'_ι be the submodule of L generated by the N_λ for $\lambda \leq \iota$; we will show that $M'_\iota = M_\iota$ for all ι, which will imply that M is generated by the family $(N_\iota)_{\iota \in I}$. In fact, suppose $M'_\lambda = M_\lambda$ for all $\lambda < \iota$; then $p_\iota(x) \in \mathfrak{a}_\iota$ for all $x \in M_\iota$; thus there exists $y \in N_\iota$ such that $x - y$ is a linear combination of finitely many elements e_λ with $\lambda < \iota$; in other words $x - y$ is an element of M_λ for some $\lambda < \iota$; the inductive hypothesis shows that $x - y \in M'_\lambda \subset M'_\iota$, that is that $x \in M'_\iota$, and so $M'_\iota = M_\iota$. It remains to show that the sum of the N_ι is direct ; now suppose there exists a linear relation $\sum_\iota a_\iota = 0$, with $a_\iota \in N_\iota$, where the a_ι (all but finitely many of which are zero) are not all zero. Let μ be the greatest index ι such that $a_\iota \neq 0$; since $p_\mu(a_\lambda) = 0$ for $\lambda < \mu$, we have $p_\mu(a_\mu) = p_\mu\left(\sum_\iota a_\iota\right) = 0$, so $a_\mu = 0$, contradicting the choice of μ.

COROLLARY 1. — *If every left ideal of* A *is projective, then every submodule of a projective left* A*-module is projective.*

Indeed every projective A-module is a submodule of a free A-module (II, p. 231, Prop. 4), and Th. 1 applies.

COROLLARY 2. — *Every submodule of a free module over a principal ideal domain is free.*

This follows immediately from Th. 1, since every ideal of a principal ideal domain is free.

COROLLARY 3. — *Every projective module over a principal ideal domain is free.*

Remark. — The proof of Th. 1 shows that every submodule of $A^{(I)}$ is isomorphic to a direct sum $\bigoplus_{\iota \in I} \mathfrak{a}_\iota$, where each $\mathfrak{a}_\iota$ is an ideal of A.

PROPOSITION 1. — *If* L *is a free module of finite rank* n *over a principal ideal domain* A, *then every submodule* M *of* L *is a free module of rank* $\leqslant n$.

Indeed M is a free module by Cor. 2 to Th. 1, and it has rank $\leqslant n$ by the previous remark, or by the following lemma :

Lemma 1. — *Let* L *be a module over a commutative ring* A, *generated by* n *elements, and let* M *be a free submodule of* L ; *then* M *has rank* $\leqslant n$.

Suppose first that L is free. Let i denote the canonical injection of M into L. By III, p. 520, Cor., the homomorphism $\bigwedge^{n+1} i : \bigwedge^{n+1} M \to \bigwedge^{n+1} L$ is injective ; by III, p. 511, Prop. 6, $\bigwedge^{n+1} L = \{0\}$, so $\bigwedge^{n+1} M = \{0\}$; it follows that M has rank $\leqslant n$ (III, p. 518, Cor. 1). Now consider the general case ; the module L is a quotient of a free module L′ of rank n. There exists a submodule M′ of L isomorphic to M (II, p. 218, Prop. 21). By the first part of the argument M′ has rank $\leqslant n$, and the result follows.

COROLLARY. — *Let* E *be a module over a principal ideal domain* A, *generated by* n *elements. Then any submodule* F *of* E *can be generated by at most* n *elements.*

Indeed there exists a homomorphism f from A^n onto E (II, p. 216, Cor. 3), and $f^{-1}(F)$, which is a free module of rank $m \leqslant n$, is generated by n elements ; the images of these elements under f generate F.

§ 4. FINITELY GENERATED MODULES OVER A PRINCIPAL IDEAL DOMAIN

1. Direct sums of cyclic modules

Let A be a commutative ring. Recall (II, p. 220, Prop. 22) that a cyclic A-module is isomorphic to a *quotient module* $A/\mathfrak{a}$, where $\mathfrak{a}$ is an ideal of A. We will see later (Sect. 4) that every finitely generated module over a principal ideal domain is a direct sum of finitely many cyclic modules.

PROPOSITION 1. — *Let* E *be a module over a commutative ring* A ; *suppose* E *is a direct sum of* n *cyclic modules* $A/\mathfrak{a}_k$ $(1 \leqslant k \leqslant n)$, *where the* $\mathfrak{a}_k$ *are ideals of* A ;

then, for each integer $p > 0$, *the* A-*module* $\bigwedge^p \mathrm{E}$ *is isomorphic to the direct sum of the modules* $\mathrm{A}/\mathfrak{a}_{\mathrm{H}}$, *where for each* p-*element subset* $\mathrm{H} = \{k_1, \ldots, k_p\}$ *of* $[1, n]$, *the ideal* $\mathfrak{a}_{\mathrm{H}}$ *is* $\sum_{j=1}^{p} \mathfrak{a}_{k_j}$.

Let x_k be the canonical generator of $\mathrm{A}/\mathfrak{a}_k$, that is the image of the unit element of A, so that E is the direct sum of the $\mathrm{A}x_i$ $(1 \leqslant i \leqslant n)$. Then we know (III, p. 515, Prop. 10) that the exterior algebra $\bigwedge \mathrm{E}$ is isomorphic as an A-module to the tensor product $\bigotimes_{i=1}^{n} (\bigwedge(\mathrm{A}x_i))$. Now $\bigwedge(\mathrm{A}x_i)$ is just the direct sum $\mathrm{A} \oplus \mathrm{A}x_i$, since the exterior product of any two elements of $\mathrm{A}x_i$ is zero, and so $\bigwedge^p \mathrm{E}$ is the direct sum of the modules $\mathrm{M}_{\mathrm{H}} = (\mathrm{A}x_{k_1}) \otimes \ldots \otimes (\mathrm{A}x_{k_p})$ as $\mathrm{H} = \{k_1, \ldots, k_p\}$ runs over the set of p-element subsets of $[1, n]$ (with $k_1 < \ldots < k_p$) ; now M_{H} is known to be isomorphic to $\mathrm{A}/\mathfrak{a}_{\mathrm{H}}$ (II, p. 257, Cor. 4), which completes the proof.

We now see that, in the notation of Prop. 1, if the ideals $\mathfrak{a}_k$ form an *increasing* sequence, they are completely determined by the module E. More precisely :

PROPOSITION 2. — *Let* A *be a commutative ring, and let* E *be a direct sum of* n *cyclic modules* $\mathrm{A}/\mathfrak{a}_k$, *where the* $\mathfrak{a}_k$ *satisfy* $\mathfrak{a}_1 \subset \mathfrak{a}_2 \subset \ldots \subset \mathfrak{a}_n$. *Then, for* $1 \leqslant p \leqslant n$, *the ideal* $\mathfrak{a}_p$ *is the annihilator of* $\bigwedge^p \mathrm{E}$; *if* $\mathfrak{a}_n \neq \mathrm{A}$ *then* $\bigwedge^p \mathrm{E} \neq 0$ *for* $1 \leqslant p \leqslant n$ *and* $\bigwedge^m \mathrm{E} = 0$ *for* $m > n$.

Indeed, in the notation of Prop. 1, we have $\mathfrak{a}_{\mathrm{H}} = \mathfrak{a}_{s(\mathrm{H})}$, where $s(\mathrm{H})$ denotes the greatest element of the subset H. Since $s(\mathrm{H}) \geqslant p$ for every p-element subset H, and since $s(\mathrm{H}) = p$ for $\mathrm{H} = \{1, \ldots, p\}$, it follows that $\mathfrak{a}_p$ is the intersection of $\mathfrak{a}_{\mathrm{H}}$ as H varies over the set of p-element subsets of $[1, n]$; the ideal $\mathfrak{a}_p$ is thus indeed the annihilator of $\bigwedge^p \mathrm{E}$, by Prop. 1.

COROLLARY. — *In the notation of Prop.* 2, *if* $\mathfrak{a}_n \neq \mathrm{A}$, *and if* E *is also isomorphic to the direct sum of* m *cyclic modules* $\mathrm{A}/\mathfrak{a}'_j$ *with* $\mathfrak{a}'_1 \subset \mathfrak{a}'_2 \subset \ldots \subset \mathfrak{a}'_m \neq \mathrm{A}$, *then* $m = n$ *and* $\mathfrak{a}_k = \mathfrak{a}'_k$ *for* $1 \leqslant k \leqslant n$ (« uniqueness of the $\mathfrak{a}_k$ »).

2. Content of an element of a free module

Let A be a principal ideal domain, let L be a free A-module, and let x be an element of L. As f runs through the set L^* of linear forms on L, the elements $f(x)$ form an ideal $\mathfrak{c}_{\mathrm{L}}(x)$ of A, called the *content* of x in L. An element c of A is called a *content* of x in L if it generates the ideal $\mathfrak{c}_{\mathrm{L}}(x)$; this amounts to saying that there exists a linear form f on L such that $f(x) = c$ and that c divides $g(x)$ for every linear form g on L. Let $(e_i)_{i \in \mathrm{I}}$ be a basis of L ; put $x = \sum a_i e_i$, $a_i \in \mathrm{A}$; then the ideal $\mathfrak{c}_{\mathrm{L}}(x)$ consists of sums $\sum a_i b_i$, as (b_i) runs through the set A^{I} ; it follows immediately that an element c of A is a content of x in L if and only if it is a gcd of the family (a_i) of coordinates of x.

We say that x is *indivisible* if $\mathfrak{c}_L(x) = A$, that is if the coordinates of x with respect to a basis of L are setwise coprime.

Lemma 1. — *Let* L *be a free module over a principal ideal domain* A *and let x be an element of* L. *Then the following conditions are equivalent :*

(i) *x is indivisible ;*

(ii) *there exists a linear form f on* L *such that* $f(x) = 1$;

(iii) *x is nonzero and the submodule* A*x is a direct factor of* L ;

(iv) *x is an element of some basis of* L.

(i) $\Rightarrow$ (ii) : this follows from the definition.

(ii) $\Rightarrow$ (iii) : let f be a linear form on L such that $f(x) = 1$; then $x \neq 0$ and the map $y \mapsto f(y)x$ is a projector of L, with image Ax.

(iii) $\Rightarrow$ (iv) : let L′ be a complement of Ax in L, and let B′ be a basis of L′ (VII, p. 15, Cor. 2) ; then $B' \cup \{x\}$ is a basis of L.

(iv) $\Rightarrow$ (i) : trivial.

Remarks. — 1) If x is a nonzero element of L and c a content of x, then there exists a unique element y of L such that $x = cy$; denote this element x/c ; then x/c is an indivisible element of L.

2) The content $\mathfrak{c}_L(x)$ is the annihilator of the torsion module of L/Ax.

Let L be a free module over a principal ideal domain A and let M be a submodule of L ; by VII, p. 4, Lemma 1, the family $\mathfrak{c}_L(x)$, $x \in M$, has a maximal element ; if $M \neq \{0\}$ then such a maximal element is nonzero.

PROPOSITION 3. — *Let* L *be a free module over a principal ideal domain* A *and let* M *be a nonzero submodule of* L. *Let x be an element of* M *such that* $\mathfrak{c}_L(x)$ *is maximal among the contents of elements of* M, *let c be a content of x in* L, *and let f be a linear form on* L *such that* $f(x) = c$.

a) L *is the direct sum of* $A(x/c)$ *and the kernel* K *of f.*

b) M *is the direct sum of* A*x and* $K \cap M$.

c) $g(M) \subset Ac$ *for every linear form g on* L.

Put $y = x/c$; clearly $Ay \cap K = \{0\}$, since $f(y) = 1$. Furthermore, for each $u \in L$ we have

$$u = f(u)\,y + (u - f(u)\,y)\,,$$

with $f(u)\,y \in Ay$ and $u - f(u)\,y \in K$; this proves *a*). Now note that for $u \in M$ we have $f(u) \in Ac$: indeed, let $u \in M$, and let d be a gcd of $f(u)$ and c ; then there exist $\lambda, \mu \in A$ such that $d = \lambda f(u) + \mu c = f(\lambda u + \mu x)$; hence the content of the element $\lambda u + \mu x$ of M divides d ; by the maximality of c, this implies that d is an associate of c, so $f(u) \in Ac$. Hence for all u in M, we can write

$$u = (f(u)/c)\,x + (u - (f(u)/c)\,x) \in Ax + (K \cap M)\,,$$

which shows *b*). Finally, let g be a linear form on L ; by *a*) there exists a scalar $\alpha \in A$ and a linear form h on K such that $g(u) = \alpha f(u) + h(u - f(u)\,y)$; thus by *b*) we have $g(\mathrm{M}) \subset \mathrm{A}c + h(\mathrm{K} \cap \mathrm{M})$. To prove *c*), it is therefore sufficient to prove that $h(\mathrm{K} \cap \mathrm{M}) \subset \mathrm{A}c$ for every linear form h on K, or equivalently for every linear form h on L such that $h(x) = 0$; now, if $u \in \mathrm{K} \cap \mathrm{M}$ and d is a gcd of $h(u)$ and c, then there exist λ, $\mu \in \mathrm{A}$ with $d = \lambda h(u) + \mu c$; then $(f + h)(\lambda u + \mu x) = d$, which implies as above that $h(u) \in \mathrm{A}c$, and *c*) follows.

3. Invariant factors of a submodule

THEOREM 1. — *Let* L *be a free module over a principal ideal domain* A, *and let* M *be a submodule of* L *of finite rank n. Then there exist a basis* B *of* L, *n elements* e_i *of* B, *and n nonzero elements* α_i *of* A $(1 \leqslant i \leqslant n)$ *such that :*

a) *the* $\alpha_i e_i$ *form a basis of* M ;

b) α_i *divides* α_{i+1} *for* $1 \leqslant i \leqslant n - 1$.

Moreover the module M′ *generated by the* (e_i) *and the principal ideals* $\mathrm{A}\alpha_i$ *are uniquely determined by the above conditions ; the module* M′/M *is the torsion submodule of* L/M, *and is isomorphic to the direct sum of the* A-*modules* $\mathrm{A}/\mathrm{A}\alpha_i$; *finally* L/M *is the direct sum of* M′/M *and a free module isomorphic to* L/M′.

1) *Existence of the* e_i *and the* α_i.

If $\mathrm{M} = \{0\}$ then the theorem is trivial. If $\mathrm{M} \neq \{0\}$ then it follows from Prop. 3 that there exist an element e_1 of L, a nonzero element α_1 of A, and a submodule L_1 of L, such that L is the direct sum of $\mathrm{A}e_1$ and L_1, such that M is the direct sum of $\mathrm{A}\alpha_1 e_1$ and the submodule $\mathrm{M}_1 = \mathrm{M} \cap \mathrm{L}_1$ of L, and such that $g(\mathrm{M}) \subset \mathrm{A}\alpha_1$ for every linear form g on L.

Now we can proceed by induction on the rank n of M. Since L_1 is a free module (VII, p. 15, Cor. 2) and M_1 has rank $n - 1$, there exist a basis B_1 of L_1, $(n - 1)$ elements $e_2, \dots, e_n$ of B_1, and nonzero elements $\alpha_2, \dots, \alpha_n$ of A such that $(\alpha_2 e_2, \dots, \alpha_n e_n)$ is a basis of M_1, and α_i divides α_{i+1} for $2 \leqslant i \leqslant n - 1$. If L′ is the submodule of L_1 generated by the elements of B_1 distinct from $e_2, \dots, e_n$, then L is the direct sum of L′ and the module M′ generated by $e_1, \dots, e_n$; now $(e_1, \dots, e_n)$ is a basis for M′, and $(\alpha_1 e_1, \dots, \alpha_n e_n)$ is a basis for M. It now remains only to show that α_1 divides α_2 ; but $\mathrm{A}\alpha_2$ has the form $g(\mathrm{M}_1)$, where g is the linear form on L defined by $g(e_2) = 1$, $g(e_i) = 0$ for $i \neq 2$, and $g(\mathrm{L}') = \{0\}$; and we have seen above that $g(\mathrm{M}_1) \subset \mathrm{A}\alpha_1$.

2) *Uniqueness properties.*

Since the α_i are nonzero, the module M′ is the set of $x \in \mathrm{L}$ such that $\beta x \in \mathrm{M}$ for some $\beta \neq 0$ in A ; in other words M′/M is the torsion submodule of L/M. This uniquely determines M′.

It is clear that M′/M is isomorphic to the direct sum of the n cyclic modules $\mathrm{A}/\mathrm{A}\alpha_i$ (II, p. 204, formula (26)). Let r be the number of the ideals $\mathrm{A}\alpha_i$ which are distinct from A : so that the first $n - r$ ideals $\mathrm{A}\alpha_i$ are equal to A, and the last r are distinct from A. Then M′/M is also isomorphic to the direct sum of the modules $\mathrm{A}/\mathrm{A}\alpha_n, \dots, \mathrm{A}/\mathrm{A}\alpha_{n-r+1}$, where $\mathrm{A}\alpha_n \subset \mathrm{A}\alpha_{n-1} \subset \dots \subset \mathrm{A}\alpha_{n-r+1} \neq \mathrm{A}$.

Thus the conditions of the Cor. to Prop. 2 (VII, p. 16) are satisfied : the ideals $A\alpha_i$ $(1 \leqslant i \leqslant n)$ are thus uniquely determined by the module M′/M.

Since L is the direct sum of M′ and L′, it follows that L/M is the sum of M′/M and (L′ + M)/M, and this sum is direct since (L′ + M) ∩ M′ = M ; on the other hand (L′ + M)/M is isomorphic to L′/(M ∩ L′) (I, p. 41, Th. 4, *c*)), that is to L′, which shows that (L′ + M)/M is a free module isomorphic to L/M′.

COROLLARY. — *A submodule* M *of finite rank in a free module* L *over a principal ideal domain* A *has a complement in* L *if and only if* L/M *is torsion free.*

In the notation of Th. 1, if L/M is torsion free, then M = M′, and M′ has a complement L′ in L. Conversely if M has a complement L′ in L, then L/M is isomorphic to L′, which is free (VII, p. 15, Cor. 2), and *a fortiori* torsion free.

Remark. — It can happen that a submodule M of *infinite* rank in a free module L can be such that L/M is torsion free, but M has no complement in L (VII, p. 60, Ex. 6, *b*)).

DEFINITION 1. — *In the notation and hypotheses of Th. 1, the ideals* $A\alpha_i$ *of* A *are called the invariant factors of the submodule* M *with respect to the module* L.

In the case where A is either the ring **Z** of rational integers or the polynomial ring K[X] in one indeterminate over a field K, there is a canonical way of choosing a generator for each ideal of A : a positive integer in the case of **Z**, or a monic polynomial in the case of K[X] (VII, p. 5). In each of these cases, the canonical generator of the invariant factor $A\alpha_i$ is also called an *invariant factor* of M with respect to L, by abuse of language.

4. Structure of finitely generated modules

THEOREM 2. — *Every finitely generated module* E *over a principal ideal domain* A *is isomorphic to a direct sum of a finite number m of cyclic modules* $A/\mathfrak{a}_k$, *where the* $\mathfrak{a}_k$ *are ideals of* A *(some of which may be zero) such that* $\mathfrak{a}_1 \subset \mathfrak{a}_2 \subset \ldots \subset \mathfrak{a}_m \neq A$, *and which are uniquely determined by these conditions.*

If E can be generated by q elements then it is isomorphic to a quotient module L/M, where $L = A^q$ (II, p. 218). Since M has finite rank $n \leqslant q$ (VII, p. 15, Prop. 1), the conditions of Th. 1 (VII, p. 18) are satisfied. Then in the notation of Th. 1 L/M is isomorphic to a direct sum of a complement L′ of M′ in L and the torsion module M′/M. The module L′ is free of finite rank $p = q - n$, so isomorphic to A^p. If r is the smallest index such that $A\alpha_r \neq A$, then M′/M is isomorphic to the direct sum of modules $A/A\alpha_i$ for $r \leqslant i \leqslant n$. The stated conditions will then be satisfied if we take $m = p + (n - r + 1)$, $\mathfrak{a}_k = (0)$ for $1 \leqslant k \leqslant p$, and $\mathfrak{a}_{p+j} = A\alpha_{n-j+1}$ for $1 \leqslant j \leqslant n - r + 1$. Uniqueness follows from the Cor. of VII, p. 16.

COROLLARY 1. — *Every finitely generated module* E *over a principal ideal domain is the direct sum of the torsion submodule of* E *and a free module.*

The torsion submodule of E has in general several distinct complements. For example, if $E = \mathbf{Z} \times (\mathbf{Z}/(2))$ then the torsion submodule of E is $\{0\} \times (\mathbf{Z}/(2))$; it has as a complement the submodule $\mathbf{Z} \times \{0\}$, and also the submodule consisting of all elements $(n, \bar{n})$, where n runs through $\mathbf{Z}$ and $\bar{n}$ is the residue class of n mod 2.

COROLLARY 2. — *Every torsion free finitely generated module over a principal ideal domain is free of finite rank.*

This follows immediately from Cor. 1.

The condition that the module be finitely generated is essential. For example the additive group of the field of fractions K of A is torsion free as an A-module ; however it is not a free module if $A \neq K$, for on the one hand any two elements of K have a common divisor, and on the other hand K is not a cyclic A-module, for otherwise $K = ab^{-1}A$ $(a, b \in A)$, whence $b^{-2} = acb^{-1}$ $(c \in A)$, so $b^{-1} = ac \in A$, and $K = A$.

DEFINITION 2. — *In the notation and hypotheses of Th.* 2, *the ideals* $\mathfrak{a}_k$ *are called the invariant factors of the module* E.

As in Def. 1 (VII, p. 19), when $A = \mathbf{Z}$ or $A = K[X]$, the canonical generator of the ideal $\mathfrak{a}_k$ (a positive integer or a monic polynomial) is also called, by abuse of language, an *invariant factor* of the finitely generated module E.

We must be careful not to confuse the invariant factors of a module E with those of a submodule M of a free module L *with respect to the module* L (Def. 1).

5. Calculation of invariant factors

PROPOSITION 4. — *Let* A *be a principal ideal domain, let* L *be a free* A-*module with finite basis* (u_j) $(1 \leqslant j \leqslant k)$, *let* M *be a submodule of* L, *let* (x_ι) *be a system of generators of* M *and let* $A\alpha_i$ $(1 \leqslant i \leqslant n)$ *be the invariant factors of* M *with respect to* L. *Then for* $1 \leqslant m \leqslant n$ *the product* $\delta_m = \alpha_1 \ldots \alpha_m$ *is a gcd of the m-th order minors of the matrix whose columns are the coordinate vectors of the* x_ι *with respect to the basis* (u_j).

By Th. 1 it is clear that $M \subset \alpha_1 L$; hence the coordinates of any element of M are multiples of α_1. On the other hand, there exists an element x of M for which α_1 is a content in L. Expressing x as a linear combination of the x_ι, it follows that α_1 is an element of the ideal generated by the coordinates of the x_ι. As these are all multiples of α_1, it follows that α_1 is indeed their gcd, and our assertion is proved in the case $m = 1$.

For general m, consider the m-th *exterior power* $\bigwedge^m M$ of M (III, p. 507). In the notation of Th. 1 there exists a basis (a_i) for M where $a_i = \alpha_i e_i$ $(1 \leqslant i \leqslant n)$; thus $\bigwedge^m M$ has a basis consisting of the elements $a_{i_1} \wedge \ldots \wedge a_{i_m}$, as $(i_1, \ldots, i_m)$ runs through the set of strictly increasing sequences of m elements of $[1, n]$. Now the elements $e_{i_1} \wedge \dots \wedge e_{i_m}$ belong to a basis B_m of $\bigwedge^m L$. Hence the canonical map from $\bigwedge^m M$ into $\bigwedge^m L$ is an isomorphism from $\bigwedge^m M$ onto the submodule of $\bigwedge^m L$ having as a basis the elements $(\alpha_{i_1} \dots \alpha_{i_m}) e_{i_1} \wedge \dots \wedge e_{i_m}$, and we identify this submodule with $\bigwedge^m M$. Since α_j is a multiple of α_k for $j \geqslant k$, the elements $\alpha_{i_1} \dots \alpha_{i_m}$ are all multiples of $\delta_m = \alpha_1 \dots \alpha_m$, and one of them is equal to δ_m; thus δ_m is a gcd of the coordinates with respect to B_m of the elements of a system of generators of $\bigwedge^m M$. The first part of the argument shows that then δ_m is a gcd of the set of coordinates of any system of generators of $\bigwedge^m M$, with respect to any basis of $\bigwedge^m L$. Taking the basis for $\bigwedge^m L$ induced from the basis (u_j) of L, and the system of generators for $\bigwedge^m M$ consisting of exterior products of the (x_ι), the expression for the coordinates of these products in terms of determinants (III, p. 528, Prop. 9) gives the stated result.

6. Linear mappings of free modules, and matrices over a principal ideal domain

Let A be a principal ideal domain. Consider a linear mapping f from a free A-module L of rank m into a free A-module L′ of rank n. The preceding results allow us, by choosing suitable bases for L and L′, to put the matrix for f in a particularly simple form, called the *canonical form* of the matrix.

PROPOSITION 5. — *Let A be a principal ideal domain, and f a linear map of rank r from a free A-module L of rank m into a free A-module L′ of rank n. Then there exist bases (e_i) $(1 \leqslant i \leqslant m)$ of L and (e'_j) $(1 \leqslant j \leqslant n)$ of L′ such that $f(e_i) = \alpha_i e'_i$ for $1 \leqslant i \leqslant r$ and $f(e_i) = 0$ for $i > r$, where the α_i are nonzero elements of A, each of which divides the next; the ideals $A\alpha_i$ are the invariant factors of $f(L)$ in L′, and are thus uniquely determined.*

Let $L_0 = f^{-1}(0)$ be the kernel of f; the quotient L/L_0 is isomorphic to the module $f(L)$, which is a submodule of L′ and so free (VII, p. 15, Cor. 2); thus L_0 has a complement L_1 in L (II, p. 218, Prop. 21), and the restriction of f to L_1 is an isomorphism from L_1 onto $f(L) = M'$. If the ideals $A\alpha_i$ $(1 \leqslant i \leqslant r)$ are the invariant factors of M′ in L′, then Th. 1 of VII, p. 18 shows that there exists a basis (e'_j) $(1 \leqslant j \leqslant n)$ of L′ such that $(\alpha_i e'_i)$ $(1 \leqslant i \leqslant r)$ is a basis of M′. Since the restriction of f to L_1 is an isomorphism from L_1 onto M′, there exists a basis (e_i) $(1 \leqslant i \leqslant r)$ of L_1 such that $f(e_i) = \alpha_i e'_i$. This basis extends to a basis (e_k) $(1 \leqslant k \leqslant m)$ of L by taking (e_s) $(r+1 \leqslant s \leqslant m)$ to be a basis of the kernel L_0.

COROLLARY 1. — *Let X be a matrix of rank r, with n rows and m columns, over a principal ideal domain* A ; *then there exists a matrix X_0 equivalent to X* (II, p. 354) *of the form*

$$\begin{pmatrix} \alpha_1 & 0 & \dots & 0 & 0 & \dots & 0 \\ 0 & \alpha_2 & \dots & 0 & 0 & \dots & 0 \\ & & & & & & \\ 0 & 0 & \dots & \alpha_r & 0 & \dots & 0 \\ 0 & 0 & \dots & 0 & 0 & \dots & 0 \\ & & & & & & \\ 0 & 0 & \dots & 0 & 0 & \dots & 0 \end{pmatrix}$$

where the α_i are nonzero elements of A, *each of which divides the next. Under these conditions the α_i are uniquely determined up to multiplication by invertible elements.*

Given that two matrices X and X' are equivalent if there exist invertible square matrices P and Q, of orders n and m, over A, such that $X' = PXQ$, Cor. 1 is just Prop. 5 expressed in terms of matrices.

In the notation of Prop. 5 and Cor. 1, the (nonzero) ideals $A\alpha_i$ are called the *invariant factors* of the linear map f, or of the matrix X. It then follows immediately from Cor. 1 that :

COROLLARY 2. — *Two matrices X and X' with n rows and m columns over a principal ideal domain* A *are equivalent if and only if they have the same invariant factors.*

Note that when A is a *field* we can take the α_i to be equal to 1, and then we recover Prop. 13 of II, p. 360.

If X is the matrix of the linear mapping f with respect to an arbitrary basis of L and an arbitrary basis of L′, then the columns of X are the coordinate vectors, with respect to the basis of L′, of elements of L′ which form a system of generators for $f(\mathrm{L})$. The following result is thus an immediate consequence of Prop. 4.

PROPOSITION 6. — *Let X be a matrix of rank r over a principal ideal domain* A, *and let $A\alpha_i$ $(1 \leqslant i \leqslant r)$ be its sequence of invariant factors. Then α_1 is a gcd of the elements of X ; and the product $\alpha_1 \dots \alpha_q$ is a gcd of the q-th order minors of X for $q \leqslant r$.*

7. Finitely generated abelian groups

In the case A = **Z**, the results of section 4 can be expressed :

THEOREM 3. — *Every finitely generated abelian group* G *is a direct sum of its torsion subgroup* F (subgroup of elements of finite order in G) *and a free abelian group of finite rank p* (isomorphic to $\mathbf{Z}^p$). *The group* F *is a direct sum of a finite number of cyclic groups of orders* $n_1, n_2, \ldots, n_q$, *where the* n_i *are integers* > 1, *each of which divides the previous one ; moreover, the integers* p, q *and* n_i $(1 \leqslant i \leqslant q)$ *are uniquely determined by* G.

Remark. — While the *orders* $n_1, \ldots, n_q$ of the cyclic groups of which F is the direct sum are well defined by the divisibility condition of Th. 3, it is *not* the same for the groups themselves : for example, in the product G of $\mathbf{Z}/(p)$ with itself (for p prime), the subgroups are precisely the $\mathbf{F}_p$-vector subspaces, and G is the direct sum of two 1-dimensional subspaces in $p(p+1)$ different ways.

COROLLARY 1. — *In a finite abelian group* G, *there exists an element whose order is the lcm of all the orders of elements of* G ; *this order* n *is the first invariant factor of* G.

COROLLARY 2. — *Any finite abelian group* G *whose order is not divisible by the square of any integer* > 1 *is cyclic.*

Let us keep to the notation of Th. 3. Then $p = 0$ since G is finite, and $q \leqslant 1$, for otherwise the order of G would be divisible by n_q^2. Hence G is cyclic.

COROLLARY 3. — *Let* L *and* M *be two free* $\mathbf{Z}$*-modules of rank* n, *let* (e_i) *be a basis for* L *and* (f_i) *a basis for* M $(1 \leqslant i \leqslant n)$, *let* u *be a homomorphism from* L *into* M, *and let* U *be its matrix with respect to the bases* (e_i) *and* (f_i). *Then* $\operatorname{Coker} u = \mathrm{M}/u(\mathrm{L})$ *is finite if and only if* $\det(U) \neq 0$, *and then* $\operatorname{Card}(\operatorname{Coker}(u)) = |\det(U)|$.

By changing bases in L and M if necessary, we may assume that U has the form described in VII, p. 22, Cor. 1 to Prop. 5 (where the α_i are in this case integers) ; the corollary then becomes obvious, since the order of a direct sum of $\mathbf{Z}$-modules $\mathbf{Z}/\alpha_i\mathbf{Z}$ $(1 \leqslant i \leqslant n)$ is infinite if one of the α_i is zero, and is equal to $|\alpha_1\alpha_2 \ldots \alpha_n|$ otherwise.

8. Indecomposable modules. Elementary divisors

DEFINITION 3. — *A left module* M *over a ring* A *is said to be decomposable if it is the direct sum of a family of proper nonzero submodules. Otherwise it is said to be indecomposable.*

The zero module is thus *decomposable,* being the direct sum of the empty family of submodules.

Let $\mathfrak{a}$ be a left ideal of the ring A ; the submodules of $\mathrm{A}/\mathfrak{a}$ are just the quotients $\mathfrak{b}/\mathfrak{a}$, where $\mathfrak{b}$ is an ideal of A containing $\mathfrak{a}$ (I, p. 41, Th. 4) ; if $\mathfrak{b}$ and $\mathfrak{c}$ are two ideals of A containing $\mathfrak{a}$, then the module $\mathrm{A}/\mathfrak{a}$ is the direct sum of the submodules $\mathfrak{b}/\mathfrak{a}$ and $\mathfrak{c}/\mathfrak{a}$ if and only if $\mathrm{A} = \mathfrak{b} + \mathfrak{c}$ and $\mathfrak{b} \cap \mathfrak{c} = \mathfrak{a}$. As a result :

Lemma 2. — *The module* $\mathrm{A}/\mathfrak{a}$ *is indecomposable if and only if* $\mathfrak{a} \neq \mathrm{A}$ *and there is no pair* $(\mathfrak{b}, \mathfrak{c})$ *of ideals of* A, *distinct from* A *and* $\mathfrak{a}$, *such that* $\mathrm{A} = \mathfrak{b} + \mathfrak{c}$ *and* $\mathfrak{b} \cap \mathfrak{c} = \mathfrak{a}$.

PROPOSITION 7. — *Let* A *be a commutative ring, let* $\mathfrak{p}$ *be a prime ideal of* A (I, p. 117, Def. 3), *and let* $\mathfrak{q}$ *be an ideal of* A *contained in* $\mathfrak{p}$. *Suppose that for every* $x \in \mathfrak{p}$ *there exists an integer* $n > 0$ *such that* $x^n \in \mathfrak{q}$. *Then the* A*-module* $A/\mathfrak{q}$ *is indecomposable.*

Let $\mathfrak{b}$ and $\mathfrak{c}$ be two ideals of A, such that $A = \mathfrak{b} + \mathfrak{c}$ and $\mathfrak{b} \cap \mathfrak{c} = \mathfrak{q}$. Then $\mathfrak{b}\mathfrak{c} \subset \mathfrak{b} \cap \mathfrak{c} = \mathfrak{q} \subset \mathfrak{p}$; if $x \notin \mathfrak{p}$ and $x \in \mathfrak{c}$, then $x\mathfrak{b} \subset \mathfrak{p}$, so $\mathfrak{b} \subset \mathfrak{p}$ (I, p. 116, Prop. 4) ; hence either $\mathfrak{b} \subset \mathfrak{p}$ or $\mathfrak{c} \subset \mathfrak{p}$. Suppose for example that $\mathfrak{c} \subset \mathfrak{p}$, so $\mathfrak{b} + \mathfrak{p} = A$; then there exist $x \in \mathfrak{b}$ and $y \in \mathfrak{p}$ such that $1 = x + y$; let $n \in \mathbf{N}$ be such that $y^n \in \mathfrak{q}$; then $1 = (x+y)^n$, so $1 \in xA + y^nA \subset \mathfrak{b} + \mathfrak{q} \subset \mathfrak{b}$, so $\mathfrak{b} = A$. Lemma 2 now shows that $A/\mathfrak{q}$ is indecomposable.

Now suppose that A is a principal ideal domain ; by VII, p. 2, Prop. 2, the prime ideals of A are the ideals (p), where p is an irreducible element of A, and the ideal 0 ; by the previous proposition, the modules A and $A/(p^n)$, for p irreducible and $n > 0$, are indecomposable. Since every cyclic module is a direct sum of modules of this type (VII, p. 3, Prop. 4) and since every finitely generated A-module is a direct sum of cyclic modules (VII, p. 19, Th. 2), we deduce :

PROPOSITION 8. — *Let* A *be a principal ideal domain and let* M *be a finitely generated* A*-module.*

a) M *is indecomposable if and only if it is isomorphic to* A *or to a module of the form* $A/(p^n)$, *where* p *is an irreducible element of* A *and* $n > 0$ *is an integer.*

b) M *is a direct sum of a finite family of indecomposable submodules.*

Part *b*) of the above proposition can be made more precise as follows :

PROPOSITION 9. — *Let* A *be a principal ideal domain, let* P *be a system of representatives of irreducible elements of* A *and let* M *be a finitely generated* A*-module. Then there exist positive integers* $m(0)$ *and* $m(p^n)$ ($p \in P$, $n > 0$), *uniquely determined by* M *and zero except for finitely many of them, such that* M *is isomorphic to the direct sum of* $A^{m(0)}$ *and the* $(A/(p^n))^{m(p^n)}$ ($p \in P$, $n > 0$).

The existence of the integers $m(0)$ and $m(p^n)$ ($p \in P$, $n > 0$) follows from Prop. 8. The integer $m(0)$ is uniquely determined : it is the rank of the free module which is the quotient of M by its torsion submodule. Finally, the p-primary component of M is isomorphic to the direct sum of the $(A/(p^n))^{m(p^n)}$; since the family of ideals (p^n) $(n \geqslant 1)$ is totally ordered by inclusion, the uniqueness of the $m(p^n)$ follows from the Cor. to Prop. 2 of VII, p. 16.

DEFINITION 4. — *In the notation of Prop.* 9, *those ideals* (p^n) ($p \in P$, $n \geqslant 1$ *an integer*) *such that* $m(p^n) > 0$ *are called the elementary divisors of the module* M, *and the integers* $m(p^n)$ *are called their multiplicities ; if the integer* $m(0)$ *is* > 0, *it is called the multiplicity of the elementary divisor* 0.

As for the invariant factors (VII, p. 19, Def. 1), when $A = \mathbf{Z}$ or $A = K[X]$ (K a commutative field), then the canonical generator of the ideal (p^n) (a positive integer or a monic polynomial) is also called, by abuse of language, an *elementary divisor* of the finitely generated module M.

Remarks. — 1) If M is a finite abelian group, then its structure can be described by writing down its elementary divisors, each repeated as often as its multiplicity. We will say, for example, that M is « of type (2, 2, 4, 27, 27, 25) » (or that it is « a group (2, 2, 4, 27, 27, 25) ») if it is isomorphic to the product of two groups $\mathbf{Z}/(2)$, one group $\mathbf{Z}/(2^2)$, two groups $\mathbf{Z}/(3^3)$ and one group $\mathbf{Z}/(5^2)$.

2) If a finitely generated torsion module M over a principal ideal domain A is given as a direct sum of cyclic modules isomorphic to $A/(a_i)$ (in particular if the invariant factors of M are known), then the elementary divisors of M, and their multiplicities, can be determined by noticing that $A/(a)$ is isomorphic to the product of the $A/(p^{n(p)})$, where $a = \varepsilon \prod_{p \in P} p^{n(p)}$ is the decomposition of a into irreducible factors (VII, p. 3). Let us study for example the multiplicative group G(464 600), where $G(n)$ denotes the multiplicative group $(\mathbf{Z}/n\mathbf{Z})^*$ (VII, p. 12). Since $464\,600 = 2^3 . 5^2 . 23 . 101$, this group is isomorphic to the product of the groups $G(2^3)$, $G(5^2)$, G(23) and G(101) (VII, p. 13, Th. 3) ; now the last three groups are cyclic of orders 20, 22 and 100, and $G(2^3)$ is the product of two cyclic groups of order 2 (*loc. cit.*) ; since $20 = 2^2 . 5$, $22 = 2 . 11$ and $100 = 2^2 . 5^2$, the group G(464 600) is of type $(2, 2, 2, 2^2, 2^2, 5, 5^2, 11)$.

3) To calculate the invariant factors of a torsion module whose elementary divisors are known, we again lean on the fact that, if the a_i are pairwise coprime elements of A, then the product $\prod_i A/(a_i)$ is a cyclic module isomorphic to $A/(a_1a_2...a_n)$ (VII, p. 3, Prop. 4). Let us illustrate the method by looking at the example of the group G(464 600) = M : write the elementary divisors p^n of M which are powers of the same irreducible p on the same line, beginning with those of greatest exponent ; extend these lines to lines of equal length by putting in 1's where necessary :

$$\begin{matrix} 2^2, & 2^2, & 2, & 2, & 2 \\ 5^2, & 5, & 1, & 1, & 1 \\ 11, & 1, & 1, & 1, & 1 . \end{matrix}$$

Then the invariant factors are the products of elements in the same column : 1100, 20, 2, 2, 2. Indeed M is isomorphic to a product of cyclic groups of orders 1100, 20, 2, 2, 2 by Prop. 4 of VII, p. 3 ; since each of these orders is a multiple of the next, these are the invariant factors of M (VII, p. 22, Th. 3).

4) An A-module is called *simple* (I, p. 37) if it is nonzero and has no submodules other than itself and 0 ; it is then necessarily cyclic, so finitely generated, and indecomposable ; since the modules $A/(p^n)$ are not simple for $n \neq 1$, while the modules $A/(p)$ are, and since A is simple only if the ring A is a field, we deduce that the simple A-modules are :

a) free modules of rank 1, when A is a field ;

b) modules isomorphic to quotients $A/(p)$, where p is an irreducible element of A, when A is not a field.

9. Duality in modules of finite length over a principal ideal domain

In this section A denotes a principal ideal domain which is not a field (and hence has at least one irreducible element), and K the field of fractions of A. For every A-module M, put

$$D(M) = \mathrm{Hom}_A(M, K/A) ;$$

we know that $D(M)$ is equipped with the structure of an A-module in a natural way, namely for every homomorphism $u : M \to K/A$ and every $\alpha \in A$, the homomorphism αu maps x to $\alpha u(x) = u(\alpha x)$. To every homomorphism $f : M \to N$ of A-modules is associated the homomorphism $D(f) : D(N) \to D(M)$, where $D(f)(v) = v \circ f$ (II, p. 196). For $x \in M$ and $x' \in D(M)$, we put $\langle x, x' \rangle = x'(x) \in K/A$; then $(x, x') \mapsto \langle x, x' \rangle$ is an A-*bilinear* map from $M \times D(M)$ into K/A, called *canonical*.

If M and N are two A-modules, then to every A-bilinear map $\varphi : M \times N \to K/A$ are associated the A-linear maps $d_\varphi : N \to D(M)$ and $s_\varphi : M \to D(N)$, where $d_\varphi(y)(x) = \varphi(x, y) = s_\varphi(x)(y)$ (II, p. 268, Cor. to Prop. 1). In particular the canonical A-bilinear map $M \times D(M) \to K/A$ defines an A-linear map (also called *canonical*)

$$c_M : M \to D(D(M))$$

such that $\langle x', c_M(x) \rangle = \langle x, x' \rangle$ for $x \in M$ and $x' \in D(M)$.

PROPOSITION 10. — *If* M *is an* A-*module of finite length, then* $D(M)$ *is* (in general non-naturally) *isomorphic to* M, *and the canonical map* $c_M : M \to D(D(M))$ *is an isomorphism.*

Using VII, p. 19, Th. 2 and II, p. 203, Cor. 1, we reduce to the case where M is cyclic. Thus we may suppose that $M = A/tA$ with $t \neq 0$. Note that any homomorphism $u : A/tA \to K/A$ is completely determined by the image $\xi \in K/A$ under u of the class ε of 1 mod tA, and this element must satisfy the relation $t\xi = 0$; conversely, for any $\xi \in K/A$ such that $t\xi = 0$, there exists a homomorphism $u : A/tA \to K/A$ such that $u(\varepsilon) = \xi$. It follows that $D(M)$ is isomorphic to $t^{-1}A/A$, and since the homothety by t is a bijection on K, we also have $D(M)$ isomorphic to A/tA, which proves the first assertion. This proves that M and $D(D(M))$ are isomorphic, so have the same length; on the other hand c_M is injective, for if $y \in A$ is such that the relation $tz \in A$ (for $z \in K$) implies $yz \in A$, then taking $z = t^{-1}$ we have $y \in tA$. It follows that the image $c_M(M)$ is necessarily equal to $D(D(M))$.

COROLLARY. — *Let* M, N *be two* A-*modules of finite length, and let* φ *be an* A-*bilinear map from* $M \times N$ *into* K/A, *such that*: 1) *the relation* $\varphi(x, y) = 0$ *for all* $y \in N$ *implies* $x = 0$; *and* 2) *the relation* $\varphi(x, y) = 0$ *for all* $x \in M$ *implies* $y = 0$. *Then the* A-*linear maps* $s_\varphi : M \to D(N)$ *and* $d_\varphi : N \to D(M)$ *associated to* φ *are isomorphisms.*

Indeed the hypotheses on φ mean that s_φ and d_φ are *injective* and since $\operatorname{long}(D(N)) = \operatorname{long}(N)$ and $\operatorname{long}(D(M)) = \operatorname{long}(M)$ by Prop. 10, this implies that $\operatorname{long}(M) = \operatorname{long}(N)$, and consequently s_φ and d_φ are bijective.

PROPOSITION 11. — *If* $M' \xrightarrow{u} M \xrightarrow{v} M''$ *is an exact sequence of* A-*modules of finite length, then the sequence* $D(M'') \xrightarrow{D(v)} D(M) \xrightarrow{D(u)} D(M')$ *is exact* [1].

[1] We will see later (A, X, p. 18) that the A-module K/A is injective. It follows that Prop. 11 remains valid for arbitrary A-modules M, M′ and M″.

Let us show first of all that given an exact sequence

$$0 \to M' \to M \to M'' \to 0 \tag{1}$$

the corresponding sequence

$$0 \to D(M'') \to D(M) \to D(M') \to 0$$

is exact ; indeed we know that the sequence

$$0 \to D(M'') \to D(M) \to D(M')$$

is exact (II, p. 227, Th. 1) ; on the other hand, it follows from (1) that

$$\operatorname{long}(M) = \operatorname{long}(M') + \operatorname{long}(M'')$$

(II, p. 212, Prop. 16) ; by Prop. 10 we thus have

$$\operatorname{long}(D(M)) = \operatorname{long}(D(M')) + \operatorname{long}(D(M'')),$$

in other words $\operatorname{long}(D(M')) = \operatorname{long}(D(M)/D(M''))$. Since $D(M)/D(M'')$ is naturally identified with a submodule of $D(M')$, it must be equal to $D(M')$, which proves our assertion.

This implies immediately that if $u : M' \to M$ is injective, then $D(u) : D(M) \to D(M')$ is surjective ; the conclusion then follows from II, p. 199, remark 4.

For every A-module M, let $\mathfrak{S}(M)$ denote the set of submodules of M. For every submodule N of M (resp. every submodule N' of $D(M)$), let N^0 (resp. N'^0) denote the submodule of $D(M)$ (resp. M) consisting of those $x' \in D(M)$ (resp. $x \in M$) such that $\langle y, x' \rangle = 0$ for all $y \in N$ (resp. $\langle x, y' \rangle = 0$ for all $y' \in N'$).

PROPOSITION 12. — *Let* M *be an* A*-module of finite length. Then the map which sends every submodule* N *of* M *to* N^0 *is a bijection from* $\mathfrak{S}(M)$ *onto* $\mathfrak{S}(D(M))$, *and the inverse bijection sends every submodule* N' *of* $D(M)$ *to the submodule* N'^0 *of* M ; *the module* $D(N)$ *is naturally identified with* $D(M)/N^0$ *and* $D(M/N)$ *with* N^0. *Moreover, we have*

$$(N_1 + N_2)^0 = N_1^0 \cap N_2^0, \quad (N_1 \cap N_2)^0 = N_1^0 + N_2^0 \tag{2}$$

for all submodules N_1, N_2 *of* M.

For each submodule N of M, there is an exact sequence

$$0 \to N \to M \to M/N \to 0$$

and hence (Prop. 11) an exact sequence

$$0 \to D(M/N) \to D(M) \to D(N) \to 0$$

and since the image of $D(M/N)$ in $D(M)$ is obviously N^0, we see (Prop. 10) that $\operatorname{long}(N^0) = \operatorname{long}(M) - \operatorname{long}(N)$; since M is identified with $D(D(M))$ by Prop. 10, we have similarly

$$\operatorname{long}(N^{00}) = \operatorname{long}(M) - \operatorname{long}(N^0) = \operatorname{long}(N) ;$$

in addition it is clear that $N \subset N^{00}$, so $N^{00} = N$. Furthermore the first relation in (2) is obvious, and by applying it to the submodules N_1^0 and N_2^0 of $D(M)$, we have $(N_1^0 + N_2^0)^0 = N_1 \cap N_2$, so $N_1^0 + N_2^0 = (N_1^0 + N_2^0)^{00} = (N_1 \cap N_2)^0$. This completes the proof of the proposition.

Examples. — 1) For $A = \mathbf{Z}$, the $\mathbf{Z}$-modules of finite length are precisely the *finite abelian groups ;* then $K = \mathbf{Q}$, so $K/A = \mathbf{Q}/\mathbf{Z}$. Then to define $D(M)$, we sometimes take, instead of $\mathbf{Q}/\mathbf{Z}$, a $\mathbf{Z}$-module isomorphic to it, such as (V, p. 79, Prop. 2) the group R of roots of unity (under multiplication) in an algebraically closed field of characteristic 0 ; we then put $D(M) = \operatorname{Hom}_{\mathbf{Z}}(M, R)$. We leave the reader to rewrite the preceeding results for this special case in the corresponding notation.

2) Let a be a nonzero element of A. The map $x \mapsto x/a$ from A into K induces an isomorphism on quotient modules from $A/(a)$ onto the submodule $(K/A)(a)$ of K/A consisting of elements annihilated by a. If M is an A-module annihilated by a, or equivalently an $A/(a)$-module, then the A-module $D(M)$ is identified with $\operatorname{Hom}_{A/(a)}(M, A/(a))$. We leave the reader to rewrite the preceding results for this special case in the corresponding notation (cf. V, p. 86).

§ 5. ENDOMORPHISMS OF VECTOR SPACES

Notation. — Given a module M, an element x of M, and two endomorphisms u and v of M, we will write $u \,.\, x$, $uv \,.\, x$ and uv in place of $u(x)$, $(u \circ v)(x)$ and $u \circ v$ respectively ; we will denote the identity map from M to itself by 1 when no confusion can arise.

1. The module associated to an endomorphism

Let A be a commutative ring, let M be an A-module, and let u be an A-endomorphism of M. Recall (III, p. 538) that the map $(p(X), x) \mapsto p(u) \,.\, x$ from $A[X] \times M$ into M makes M an $A[X]$-module, written M_u. Recall also (III, pp. 538 and 539) that if $M[X]$ denotes the $A[X]$-module obtained from M by extension of scalars from A to A[X], and if $\bar{u}$ denotes the A[X]-endomorphism of M[X] induced by u, then there is an exact sequence of A[X]-modules [1]

(1) $$0 \to M[X] \overset{\psi}{\to} M[X] \overset{\varphi}{\to} M_u \to 0,$$

where $\varphi(p(X) \otimes x) = p(u) . x$ and $\psi = X - \bar{u}$.

An endomorphism u of an A-module M and an endomorphism u' of an A-module M′ are said to be *similar* if there exists an isomorphism g from M onto M′ such that $u' \circ g = g \circ u$, that is (III, p. 540, Prop. 19) an isomorphism g from M_u onto $M'_{u'}$. If M (resp. M′) is free on the finite basis B (resp. B′), and if $M(u)$ (resp. $M(u')$) is the matrix of u (resp. u') with respect to B (resp. B′), then u and u' are similar if and only if $M(u)$ and $M(u')$ are similar matrices (II, p. 356, Def. 6). The *characteristic polynomials* (III, p. 541, Def. 3) of two similar endomorphisms of finitely generated free modules are equal (III, p. 540, Prop. 19).

Let K be a commutative field ; then any pair (E, u) consisting of a vector space E over K and an endomorphism u of E corresponds to a K[X]-module E_u. Since the ring K[X] is a *principal ideal domain* (IV, p. 11, Prop. 11), the results of the preceding sections can be applied to E_u.

Let us first show how to translate certain notions from the language of modules to that of endomorphisms of vector spaces :

« V is a submodule of E_u » means : « V is a vector subspace of E closed under u ».

« V is a cyclic submodule of E_u » means : « there exists $x \in V$ such that the vector subspace V is spanned by the elements $u^i . x$ ($i \in \mathbf{N}$) ». Then V is said to be *cyclic* (with respect to u) and x is called a generator.

« V is an indecomposable submodule of E_u » means : « V is nonzero and is not the direct sum of two nonzero subspaces each closed under u ».

« $\mathfrak{a}$ is the annihilator of the submodule V » means : « $\mathfrak{a}$ is the ideal consisting of those polynomials $p(X) \in K[X]$ such that $p(u) . x = 0$ for all $x \in V$ ».

The monic polynomial g such that $\mathfrak{a}$ is equal to the principal ideal (g) is then called the *minimal polynomial* of the restriction of u to V.

« E_u is cyclic with annihilator $\mathfrak{a} = (g)$ »

$$(\text{with } g(X) = X^n + \alpha_{n-1}X^{n-1} + \cdots + \alpha_0)$$

[1] The injectivity of ψ, which is not stated in Prop. 18 of III, p. 539, is proved as follows : in the notation of *loc. cit.*, we have

$$\psi(\textstyle\sum (X^k \otimes x_k)) = \sum X^k \otimes (x_{k-1} - u(x_k)).$$

If $\sum X^k \otimes x_k$ belongs to the kernel of ψ, then it follows that $x_{k-1} = u(x_k)$ for all k, and the x_k are all zero, since the family (x_k) has finite support.

means : « there exists $x \in \mathrm{E}$ such that $(u^i . x)$ $(0 \leqslant i \leqslant n-1)$ is a basis of the vector space E, and $g(u) . x = 0$ ». In other words, we can find a basis of E such that the matrix U of u with respect to this basis is

$$U = \begin{pmatrix} 0 & 0 & 0 \ldots 0 & -\alpha_0 \\ 1 & 0 & 0 \ldots 0 & -\alpha_1 \\ 0 & 1 & 0 \ldots 0 & -\alpha_2 \\ \cdots & \cdots & \cdots & \cdots \\ 0 & 0 & 0 \ldots 0 & -\alpha_{n-2} \\ 0 & 0 & 0 \ldots 1 & -\alpha_{n-1} \end{pmatrix} . \tag{2}$$

« E_u is a torsion module » means, by the characterisation of cyclic torsion modules given above : « every cyclic submodule of E_u is finite dimensional over K ». In particular :

« E_u is a finitely generated torsion module » means : « E is finite dimensional over K ».

2. Eigenvalues and eigenvectors

DEFINITION 1. — *Let* E *be a vector space over a commutative field* K, *and* u *an endomorphism of* E. *An element* x *of* E *is said to be an eigenvector of* u *if there exists* $\alpha \in \mathrm{K}$ *such that* $u . x = \alpha x$; *if* $x \neq 0$ *then the scalar* α *is called the eigenvalue of* u *corresponding to the eigenvector* x. *For every scalar* α, *the vector subspace* V_α *consisting of* $x \in \mathrm{E}$ *such that* $u . x = \alpha x$ *is called the eigenspace of* E *corresponding to* α.

The *geometric multiplicity* of the eigenvalue α is the cardinal $\dim \mathrm{V}_\alpha$.

Suppose E is finite dimensional. The eigenvalues of u are those elements α of K such that the endomorphism $\alpha . 1 - u$ of E is not injective, in other words (III, p. 524, Prop. 3) such that $\det(\alpha . 1 - u) = 0$. But, by the definition of the characteristic polynomial χ_u of u (III, p. 541, Def. 3), we have $\det(\alpha . 1 - u) = \chi_u(\alpha)$. Consequently :

PROPOSITION 1. — *Suppose* E *is finite dimensional. Then an element* α *of* K *is an eigenvalue of the endomorphism* u *if and only if it is a root of the characteristic polynomial of* u.

If L is an extension of the field K, then the roots of χ_u in L are eigenvalues of the endomorphism $1_{\mathrm{L}} \otimes u$ of the L-vector space $\mathrm{L} \otimes_{\mathrm{K}} \mathrm{E}$. They are often referred to as *eigenvalues of u in* L. By abuse of language, we say that all the eigenvalues of u belong to L if this holds for all the eigenvalues of u in an algebraically closed extension of L ; this means that χ_u decomposes into linear factors in L[X].

Let U be a square matrix of order n with coefficients in K. Then by definition the characteristic polynomial of U is

$$\chi_U(\mathrm{X}) = \det(\mathrm{X} . I_n - U) ;$$

the *eigenvalues* of U (in an extension L of K) are the roots (in L) of the polynomial χ_U ; these are also the scalars α (in L) such that there exists a nonzero solution to the system of linear equations $UX = \alpha X$, where X is a column matrix of order n ; a column matrix X satisfying this equation is called an eigenvector of U corresponding to α.

If U is the matrix of an endomorphism u of an n-dimensional vector space with respect to a basis B, then $\chi_U = \chi_u$, the eigenvalues of U are the eigenvalues of u, and the eigenvectors of U are the matrices of the eigenvectors of u with respect to the basis B.

PROPOSITION 2. — *Let u be an endomorphism of a vector space* E *over a commutative field* K ; *for each scalar α, let V_α be the eigenspace corresponding to α. Then the subspaces V_α are closed under u and the sum of the V_α is direct.*

The first assertion is clear. By definition the subspace V_α is annihilated by the element $X - \alpha$ of $K[X]$; the $X - \alpha$, $\alpha \in K$, are irreducible and pairwise non-associate ; the second assertion thus follows from VII, p. 8, Th. 1.

3. Similarity invariants of an endomorphism

If we translate the decomposition of a finitely generated torsion module in VII, p. 8, Th. 1 and p. 9, Prop. 2, then we obtain :

PROPOSITION 3. — *Let* E *be a vector space of finite dimension n over a commutative field* K, *and let u be an endomorphism of* E ; *for every monic irreducible polynomial $p(X)$, let M_p be the vector subspace consisting of elements x of* E *such that $(p(u))^k . x = 0$ for some integer k. Then M_p is closed under u, the vector space* E *is the direct sum of the M_p, and there exist polynomials s_p such that, for all $x \in E$, the component of x in M_p is equal to $s_p(u) . x$.*

Remark 1. — Clearly the minimal polynomial of the restriction of u to M_p is the greatest power of p which divides the minimal polynomial of u. Moreover we have $s_p(u) . x = x$ for $x \in M_p$, from which it follows immediately that, if $M_p \neq 0$, then s_p is coprime to p.

Similarly, by Th. 2 of VII, p. 19, the module E_u is isomorphic to a direct sum of cyclic modules $F_j = K[X]/\mathfrak{a}_j$ $(1 \leqslant j \leqslant r)$, where the ideals $\mathfrak{a}_j$ are distinct from $K[X]$ and $\mathfrak{a}_j \subset \mathfrak{a}_{j+1}$; and the $\mathfrak{a}_j$ are determined by these conditions. Moreover, since E_u is a torsion module, we have $\mathfrak{a}_1 \neq (0)$; since E has dimension n, we have $r \leqslant n$. Put $\mathfrak{a}_j = (h_j)$ $(1 \leqslant j \leqslant r)$, with h_j a monic polynomial, and consider the sequence (q_i) $(1 \leqslant i \leqslant n)$ of polynomials defined by :

$$(3) \qquad \begin{cases} q_i(X) = 1 & \text{if } \ i \leqslant n - r \\ q_i(X) = h_{n-i+1}(X) & \text{if } \ n - r < i \leqslant n . \end{cases}$$

It is clear that the polynomials q_i determine the polynomials h_j and conversely, and that E_u is isomorphic to the direct sum of the n modules $K[X]/(q_i)$, the first $n - r$ of which are 0.

In other words :

PROPOSITION 4. — *Let* E *be a vector space of finite dimension n over a commutative field* K, *and let u be an endomorphism of* E. *Then there exist n monic polynomials* $q_i(X) \in K[X]$ $(1 \leq i \leq n)$ *such that* q_i *divides* q_{i+1} *for* $1 \leq i \leq n-1$, *and* E *is the direct sum of n subspaces* V_i $(1 \leq i \leq n)$ *closed under u, cyclic (with respect to u), and such that the minimal polynomial of the restriction of u to* V_i *is equal to* q_i $(1 \leq i \leq n)$. *The polynomials* q_i *are uniquely determined by these conditions, and* q_n *is the minimal polynomial q of u.*

Remark 2. — By the above proposition, there exists a basis of E with respect to which the matrix U of u has the form

$$\begin{pmatrix} A_{n-r+1} & 0 & \dots & 0 & 0 \\ 0 & A_{n-r+2} & \dots & 0 & 0 \\ \dots & \dots & \dots & \dots & \dots \\ 0 & 0 & \dots & A_{n-1} & 0 \\ 0 & 0 & & 0 & A_n \end{pmatrix}$$

where each matrix A_i has the form (2) (taking $g(X) = q_i(X)$) (cf. VII, pp. 29-30).

DEFINITION 2. — *In the notation of Prop.* 4, *the n monic polynomials* $q_i(X)$ $(1 \leq i \leq n)$ *are called the similarity invariants of the endomorphism u.*

Thus the n-th similarity invariant q_n is the minimal polynomial of u (Prop. 4) ; in other words, for a polynomial $p(X) \in K[X]$ to satisfy $p(u) = 0$, it is necessary and sufficient that p be a multiple of q_n.

COROLLARY 1. — *Let* K *be a field, let* E *and* E' *be two finite dimensional vector spaces over* K, *and let u* (*resp. u'*) *be an endomorphism of* E (*resp.* E'). *Then u and u' are similar* (VII, p. 29) *if and only if they have the same similarity invariants.*

Indeed u and u' are similar if and only if the K[X]-modules E_u and $E'_{u'}$ are isomorphic.

COROLLARY 2. — *Let u be an endomorphism of a finite dimensional vector space* E *over a field* K, *let* $(q_1, \dots, q_n)$ *be the family of similarity invariants of u, let* L *be an extension of* K, *let* $E_{(L)} = L \otimes_K E$ *be the* L*-vector space induced from* E *by extension of scalars and let* $u_{(L)} = 1_L \otimes u$ *be the endomorphism of* $E_{(L)}$ *induced by u. Then the similarity invariants of* $u_{(L)}$ *are the images* $\bar{q}_1, \dots, \bar{q}_n$ *of* $q_1, \dots, q_n$ *in* L[X].

This follows immediately from Prop. 4 and the fact that the L[X]-modules $E_{(L)u_{(L)}}$ and $(K[X]/(q_i))_{(L)}$ are isomorphic to $L[X] \otimes_{K[X]} E_u$ and $L[X]/(\bar{q}_i)$ respectively.

Let U be a square matrix of order n with coefficients in a commutative field K. Then the similarity invariants of the endomorphism of K^n defined by U are called the *similarity invariants of U*. It then follows from Cor. 1 above that two square matrices are similar if and only if they have the same similarity invariants, and that if u is an endomorphism of a finite dimensional vector space E over K, and U is the matrix of u with respect to some basis B of E, then the similarity invariants of U and u coincide. By Cor. 1 and 2 above, we have :

COROLLARY 3. — *Let U and V be two square matrices of order n with coefficients in a commutative field* K. *If there exists an invertible square matrix P over some extension* K' *of* K *such that* $V = P^{-1}UP$, *then there exists an invertible square matrix Q over K such that* $V = Q^{-1}UQ$.

Let E be a finite dimensional vector space over a commutative field K, let $(e_i)_{1 \leqslant i \leqslant n}$ be a basis of E and let u be an endomorphism of E. Then by the exact sequence (1) of VII, p. 29, the K[X]-module E_u associated to u is isomorphic to the quotient of the *free* K[X]-module E[X], with basis $(1 \otimes e_i)$, by the image of E[X] under the K[X]-linear map $X - \bar{u}$. The similarity invariants $q_i(X)$ of u (VII, p. 32, Def. 2) are thus the *invariant factors* of $X - \bar{u}$ (VII, p. 22). Thus Prop. 6 of VII, p. 22, implies :

PROPOSITION 5. — *Let* E *be a vector space of finite dimension n over a commutative field* K, *let u be an endomorphism of* E, *and let U be its matrix with respect to some basis of* E. *Then for each integer m with* $1 \leqslant m \leqslant n$, *the product*

$$d_m(X) = q_1(X)\, q_2(X) \ldots q_m(X)$$

of the first m similarity invariants of u is equal to the gcd of the m-th order minors of the matrix $XI_n - U$.

COROLLARY 1. — *Let u be an endomorphism of a vector space of finite dimension n over a commutative field* K, *with characteristic polynomial* $\chi_u(X)$ *and similarity invariants* $q_i(X)$ $(1 \leqslant i \leqslant n)$. *Then*

$$\chi_u(X) = q_1(X)\, q_2(X) \ldots q_n(X)\,.$$

COROLLARY 2. — *In the notation of Cor.* 1, *let* $q(X)$ *be the minimal polynomial of u ; then* $q(X)$ *divides* $\chi_u(X)$ *and* $\chi_u(X)$ *divides* $q(X)^n$. *In particular the minimal polynomial and the characteristic polynomial of u have the same roots, and these are the eigenvalues of u.*

Since $q(X) = q_n(X)$, it is clear that $q(X)$ divides $\chi_u(X)$. On the other hand, since each q_i divides q, their product χ_u divides q^n.

COROLLARY 3. — *An endomorphism u is nilpotent if and only if its characteristic polynomial has the form* X^n.

This follows immediately from Cor. 2.

Let us now rewrite Prop. 9 of VII, p. 24, which gives the decomposition of a module as a direct sum of indecomposable submodules.

PROPOSITION 6. — *Let* E *be a vector space of finite dimension* n *over a commutative field* K, *and let* u *be an endomorphism of* E. *Then* E *is the direct sum of subspaces* E_k, *closed under* u *and cyclic with respect to* u, *such that the minimal polynomial of the restriction of* u *to* E_k *has the form* $p_k^{n(k)}$, *where* p_k *is an irreducible polynomial, and* E_k *cannot be expressed as a direct sum of two nonzero subspaces each closed under* u. *For every monic irreducible polynomial* $p \in K[X]$ *and every integer* $n \geqslant 1$, *the number* $m(p^n)$ *of subspaces* E_k *in any such decomposition, such that* p^n *is the minimal polynomial of the restriction of* u *to* E_k, *is uniquely determined.*

The $p_k^{n(k)}$ determine the similarity invariants of u and *vice versa* ; we can get from one to the other by the procedure explained in VII, p. 25, Remarks 2 and 3. Furthermore, we can immediately get from the decomposition considered in Prop. 6 to those considered in Prop. 3 and 4.

Note that the monic irreducible polynomials $p \in K[X]$ such that $m(p^n) > 0$ for some integer $n \geqslant 1$ are precisely the monic irreducible factors of the minimal polynomial of u. Thus, in contrast to the similarity invariants, these polynomials depend in general on the field K in which we are working.

4. Triangularisable endomorphisms

In this section we will be interested in the case where the minimal polynomial $p(X)$ of u splits into a product of linear factors in $K[X]$, in other words (VII, p. 33, Cor. 2) in the case where all the eigenvalues of u belong to K. This will hold in particular when K is *algebraically closed.* Prop. 3 of VII, p. 31 gives immediately :

PROPOSITION 7. — *Let* E *be a finite dimensional vector space over a commutative field* K, *and let* u *be an endomorphism of* E *whose eigenvalues all belong to* K. *For each eigenvalue* α *of* u, *let* M_α *be the vector subspace of E consisting of those elements* x *for which there exists an integer* $k \geqslant 1$ *such that* $(u - \alpha)^k \cdot x = 0$. *Then* M_α *is closed under* u, *the vector space* E *is the sum of the* M_α, *and there exist polynomials* $s_\alpha \in K[X]$ *such that, for all* $x \in E$, *the component of* x *in* M_α *is equal to* $s_\alpha(u) \cdot x$.

The submodule M_α, being finitely generated as a $K[X]$-module, then has an annihilator of the form $(X - \alpha)^r$; in other words, there exists an integer $r \geqslant 1$ such that

$$(u - \alpha)^r \cdot x = 0$$

for *all* $x \in M_\alpha$; the restriction of $u - \alpha$ to M_α is a *nilpotent* endomorphism.

Still assuming that the eigenvalues of u belong to K, we now apply Prop. 6 of VII, p. 34 to u. The polynomials p_k are nothing other than the $X - \alpha$ (as α runs through the set of eigenvalues of u), and we see that E is the direct sum of subspaces E_i closed under u, cyclic (with respect to u), and such that the minimal polynomial of the restriction of u to E_i has the form $(X - \alpha)^m$. Let E_i' be the K[X]-module associated to E_i ; then E_i' is isomorphic to one of the modules $K[X]/((X - \alpha)^m)$. Now the residue classes $\bmod (X - \alpha)^m$ of the elements $(X - \alpha)^k$ $(0 \leqslant k \leqslant m - 1)$ form a K-basis of $K[X]/((X - \alpha)^m)$ (IV, p. 11, Cor.), and

$$X(X - \alpha)^k = \alpha(X - \alpha)^k + (X - \alpha)^{k+1}$$

for $0 \leqslant k \leqslant m - 1$; it follows that E_i has dimension m, and if α is the unique eigenvalue of the restriction u_i of u to E_i, then there exists a basis of E_i with respect to which the matrix of u_i is the $m \times m$ matrix

$$U_{m,\alpha} = \begin{pmatrix} \alpha & 0 & 0 & \dots & 0 & 0 \\ 1 & \alpha & 0 & \dots & 0 & 0 \\ 0 & 1 & \alpha & \dots & 0 & 0 \\ \dots & \dots & \dots & \dots & \dots & \dots \\ 0 & 0 & 0 & \dots & \alpha & 0 \\ 0 & 0 & 0 & \dots & 1 & \alpha \end{pmatrix} \tag{4}$$

DEFINITION 3. — *For every field* K, *every integer* $m \geqslant 1$, *and every* $\alpha \in K$, *the matrix* $U_{m,\alpha}$ *is called the Jordan matrix of order* m *and eigenvalue* α.

PROPOSITION 8. — *Let* E *be a finite dimensional vector space over a commutative field* K, *and let* u *be an endomorphism of* E. *Then the following conditions are equivalent :*

(i) *the eigenvalues of* u *(in some algebraically closed extension of* K) *belong to* K ;

(ii) *there exists a basis of* E *with respect to which the matrix of* u *is lower* (resp. *upper*) *triangular ;*

(iii) *there exists a basis of* E *with respect to which the matrix of* u *is a diagonal block of Jordan matrices.*

We have (i) ⇒ (iii) by Prop. 7 and the above remarks, and the assertions (iii) ⇒ (ii) and (ii) ⇒ (i) are trivial.

DEFINITION 4. — *An endomorphism satisfying conditions* (i), (ii) *and* (iii) *of Prop. 8 is called triangularisable.*

In particular if K is algebraically closed, then every endomorphism of a K-vector space is triangularisable.

For matrices, Prop. 8 implies :

COROLLARY. — *Let* U *be a square matrix over a commutative field* K *such that all the eigenvalues of* U *are in* K ; *then there exists a matrix similar to* U *which is a diagonal block of Jordan matrices.*

Remarks. — 1) It follows from Prop. 6 of VII, p. 34, that, if U is similar to a diagonal block of Jordan matrices (J_k), then the number of J_k of the form $U_{m,\alpha}$ (for given m and α) is uniquely determined by U.

2) More generally, if U is similar to a diagonal block of Jordan matrices U_{m_i,α_i}, then the similarity invariants of U can be readily calculated by a method modelled on that presented in VII, p. 25, Remark 3 : write all the $(X-\alpha_i)^{m_i}$ with the same α on the same line, in decreasing order of exponents, and complete with 1's to have lines whose lengths are equal to the order of U ; this done, the similarity invariants of U are obtained, in decreasing order of indices, by forming the products of terms in the same column. For example, for the matrix

$$\begin{pmatrix} 2 & 0 & 0 \\ 0 & 3 & 0 \\ 0 & 1 & 3 \end{pmatrix}$$

we write

$$\begin{matrix} (X-2), 1, 1 \\ (X-3)^2, 1, 1 \end{matrix}$$

and the similarity invariants are 1, 1 and $(X-2)\,(X-3)^2$.

By noticing that the minimal polynomial of the Jordan matrix $U_{m,\alpha}$ is $(X-\alpha)^m$, and that it is equal to its characteristic polynomial, we obtain the following result :

PROPOSITION 9. — *If the square matrix U is similar to a diagonal block of Jordan matrices (U_{m_i,α_i}), then the minimal polynomial of U is the lcm of the $(X-\alpha_i)^{m_i}$, and the characteristic polynomial is the product of the $(X-\alpha_i)^{m_i}$.*

COROLLARY. — *In the notation of Prop. 7, the dimension of the subspace M_α is the multiplicity of the eigenvalue α as a root of the characteristic polynomial of u.*

5. Properties of the characteristic polynomial : trace and determinant

Let E be a vector space of finite dimension n over a commutative field K, and let u be an endomorphism of E. By III, p. 541, the characteristic polynomial of u has the form :

$$\chi_u(X) = X^n - \mathrm{Tr}(u)\,X^{n-1} + \cdots + (-1)^n \det(u)\,. \tag{5}$$

PROPOSITION 10. — *Let E be a vector space of finite dimension n over a commutative field K, let u be an endomorphism of E, and let $\chi_u(X) = \prod_{i=1}^{n}(X-\alpha_i)$ be a decomposition into linear factors of its characteristic polynomial (in a suitable extension of K,* cf. V, p. 21). *If q is a polynomial with coefficients in K, then the characteristic polynomial of $q(u)$ is given by*

$$\chi_{q(u)}(X) = \prod_{i=1}^{n}(X-q(\alpha_i))\,, \tag{6}$$

and its trace and determinant are given by

(7) $$\operatorname{Tr}(q(u)) = \sum_{i=1}^{n} q(\alpha_i),$$

(8) $$\det(q(u)) = \prod_{i=1}^{n} q(\alpha_i).$$

It is clear that (7) and (8) follow from (6) by virtue of (5). To prove the formula (6), we may assume that K is algebraically closed. We then take a basis for E with respect to which the matrix U of u is lower triangular (VII, p. 35, Cor. to Prop. 8) ; we will make use of the following easy lemma :

Lemma 1. — *If B and C are lower triangular matrices of order n with diagonals (β_i) and (γ_i), then the matrices $B + C$ and BC are lower triangular with diagonals $(\beta_i + \gamma_i)$ and $(\beta_i \gamma_i)$.*

Since the matrix U of u is lower triangular with diagonal (α_i) say, it follows from Lemma 1 that $q(U)$ is a lower triangular matrix with diagonal $(q(\alpha_i))$. Then $X . I_n - q(U)$ is a lower triangular matrix with diagonal $(X - q(\alpha_i))$, which proves (6).

COROLLARY 1. — *For $q(u)$ to be invertible, it is necessary and sufficient that q be coprime to χ_u.*

Indeed, to say that q and χ_u are coprime is equivalent to saying that they have no common root in an algebraically closed extension of K, in other words (8) that $\det(q(u)) \neq 0$.

Remark 1. — A polynomial is coprime to χ_u if and only if it is coprime to the minimal polynomial of u (VII, p. 33, Cor. 2).

COROLLARY 2. — *Let $r \in K(X)$ be a rational function over K. Then u is substitutable* (IV, p. 21) *in r if and only if each of its eigenvalues is. In this case the following formulae hold :*

$$\chi_{r(u)}(X) = \prod_{i=1}^{n} (X - r(\alpha_i)), \quad \operatorname{Tr}(r(u)) = \sum_{i=1}^{n} r(\alpha_i), \quad \det(r(u)) = \prod_{i=1}^{n} r(\alpha_i).$$

Write $r = p/q$ where p and q are coprime polynomials. Then u is substitutable in r if and only if $\det(q(u)) \neq 0$, so the first assertion follows from (8). By Cor. 1, we may suppose q coprime to χ_u, so by the Bezout identity there exist polynomials g and h such that $qg + h\chi_u = 1$. Then $q(\alpha_i)\, g(\alpha_i) = 1$ and $q(u)\, g(u) = 1$ by the Cayley-Hamilton theorem (III, p. 541). The stated formulae can then be obtained by applying formulae (6), (7) and (8) to $p(u)\, g(u) = r(u)$.

COROLLARY 3. — *For each integer $s \geqslant 0$ we have $\operatorname{Tr}(u^s) = \sum_{i=1}^{n} \alpha_i^s$. This formula is also valid for $s < 0$ provided u is invertible.*

This is a special case of the previous corollary.

COROLLARY 4. — *Suppose the field* K *is of characteristic zero; then the endomorphism* u *is nilpotent if and only if* $\operatorname{Tr}(u^s) = 0$ *for* $1 \leqslant s \leqslant n$.

If u is nilpotent then the α_i are all zero, and $\operatorname{Tr}(u^s) = 0$ for all $s > 0$ (Cor. 3). Conversely, if $\operatorname{Tr}(u^s) = 0$ for $1 \leqslant s \leqslant n$, then the α_i are all zero since K is of characteristic zero (IV, p. 72, Cor.), and u is nilpotent (VII, p. 33).

COROLLARY 5. — *Let* Y *be an indeterminate and let* $\tilde{u}$ *denote the endomorphism of the* K(Y)*-vector space* $K(Y) \otimes_K E$ *induced from* u *by extension of scalars from* K *to the field* K(Y) *of rational functions in* Y *with coefficients from* K. *Then the endomorphism* $Y.1 - \tilde{u}$ *is invertible. Moreover, if* χ'_u *denotes the derivative of the polynomial* χ_u, *then*

$$\operatorname{Tr}((Y.1 - \tilde{u})^{-1}) = \chi'_u(Y)/\chi_u(Y).$$

The endomorphism $Y.1 - \tilde{u}$ is invertible because its determinant is the nonzero element $\chi_u(Y)$ of $K(Y)$. It follows that $\tilde{u}$ is substitutable in the rational function $r(X) = (Y - X)^{-1}$ in $K(Y)(X)$. The second assertion now follows from Cor. 2, by the relation

$$\chi'_u(Y)/\chi_u(Y) = \sum_i (Y - \alpha_i)^{-1} = \sum_i r(\alpha_i).$$

COROLLARY 6. — *Suppose the field* K *is of characteristic zero. Then, in the ring* K[[T]] *of formal power series, we have*

$$-T\frac{d}{dT}\log\det(1 - Tu) = \sum_{m \geqslant 1} \operatorname{Tr}(u^m)\,T^m.$$

Let us first of all work in the field K(T) of rational functions, and put $P(T) = \det(I_n - T.U)$, where U is the matrix of u with respect to some basis of E. Then

$$P(T) = \det(T(T^{-1}.I_n - U)) = T^n\chi_U(T^{-1}),$$

so $P'(T)/P(T) = n/T - \chi'_u(T^{-1})/T^2\chi_u(T^{-1})$. Moreover, by Cor. 5 we have

$$\chi'_U(T^{-1})/T\chi_U(T^{-1}) = \operatorname{Tr}((T^{-1}.I_n - U)^{-1})/T = \operatorname{Tr}((I_n - T.U)^{-1}).$$

It follows that $-TP'(T)/P(T) = -n + \operatorname{Tr}((I_n - TU)^{-1})$. The corollary is now obtained by expanding each side of this equation as a formal power series.

Remark 2. — By IV, p. 80, Cor. 1 and formula (8), we have, for each polynomial $q \in K[X]$, that

$$\det q(u) = \operatorname{res}(\chi_u, q), \tag{9}$$

where $\operatorname{res}(\chi_u, q)$ is the resultant of the polynomials χ_u and q. In particular if we take $q = \chi_u'$ we obtain

$$\det \chi_u'(u) = (-1)^{n(n-1)/2} \operatorname{dis}(\chi_u) , \tag{10}$$

where $\operatorname{dis}(\chi_u)$ is the discriminant of the polynomial χ_u (IV, p. 82, formula (47)). Furthermore :

COROLLARY 7. — We have $\det(\operatorname{Tr}(u^{i+j})_{0 \leqslant i,j \leqslant n}) = \operatorname{dis}(\chi_u)$.

Let D be the Vandermonde matrix $(\alpha_j^{i-1})_{1 \leqslant i,j \leqslant n}$. Then (III, p. 532, formula (29)) :

$$\det(D)^2 = \prod_{i<j} (\alpha_i - \alpha_j)^2 = \operatorname{dis}(\chi_u) .$$

Moreover, the (i, j)-th entry of $D \,.\, {}^tD$ is $\sum_k \alpha_k^{i+j-2} = \operatorname{Tr}(u^{i+j-2})$, and the corollary follows.

6. Characteristic polynomial of the tensor product of two endomorphisms

PROPOSITION 11. — *Let* E (resp. E′) *be a finite dimensional vector space over a commutative field* K *and let* u (resp. u') *be an endomorphism of* E (resp. E′). *Let*

$$\chi_u(X) = \prod_i (X - \alpha_i) , \quad \chi_{u'}(X) = \prod_j (X - \beta_j)$$

be decompositions into linear factors of the characteristic polynomials of u and u' in some suitable extension of K. *Then the characteristic polynomial of the endomorphism $u \otimes u'$ of the vector space* $E \otimes_K E'$ *is given by the formula*

$$\chi_{u \otimes u'}(X) = \prod_{i,j} (X - \alpha_i \beta_j) .$$

Arguing as in the proof of Prop. 10 of VII, p. 36, we see that it is sufficient to prove the following lemma :

Lemma 2. — *Let B and C be two lower triangular matrices of orders m and n respectively, with diagonals $(\beta_i)_{1 \leqslant i \leqslant m}$ and $(\gamma_j)_{1 \leqslant j \leqslant n}$. Let us identify the lexicographic product of the ordered sets $\{1, 2, \ldots, m\}$ and $\{1, 2, \ldots, n\}$ with the interval $\{1, 2, \ldots, mn\}$. Then the tensor product matrix* (II, p. 357) *$B \otimes C$ is lower triangular with diagonal $(\beta_i \gamma_j)$.*

This follows immediately from the definition of the tensor product of two matrices (*loc. cit.*) and of the lexicographic product (*Set Theory*, III, p. 157).

7. Diagonalisable endomorphisms

DEFINITION 5. — *Let* E *be a finite dimensional vector space over a commutative field* K *and let* $\mathfrak{F}$ *be a set of endomorphisms of* E. *Then* $\mathfrak{F}$ *is said to be diagonal with respect to a basis* (e_i) *of* E *if the matrix of each* $u \in \mathfrak{F}$ *with respect to* (e_i) *is diagonal. The set* $\mathfrak{F}$ *is said to be diagonalisable if there exists a basis of* E *with respect to which* $\mathfrak{F}$ *is diagonal.*

This definition applies in particular to the case when $\mathfrak{F}$ contains only one element u ; we then say that u is diagonal (diagonalisable). Note also that $\mathfrak{F}$ is diagonal with respect to a basis (e_i) if and only if the (e_i) are common eigenvectors of all the elements of $\mathfrak{F}$; it follows that $\mathfrak{F}$ is diagonalisable if and only if E is generated by eigenvectors common to all the elements of $\mathfrak{F}$.

Let A be a subalgebra of $\mathrm{End}_K(E)$ containing Id_E. Then A is diagonalisable if and only if it is isomorphic to an algebra K^r (in other words is diagonalisable in the sense of V, p. 28, Def. 1) ; indeed, if A is isomorphic to K^r, then A is diagonalisable by V, p. 29, Prop. 1 ; conversely, if A is diagonalisable, then it is isomorphic to a subalgebra of the algebra of diagonal matrices, which is isomorphic as an algebra to K^n, $n = \dim(E)$, hence A is isomorphic to some algebra K^r (V, p. 30, Prop. 3).

PROPOSITION 12. — *Let* E *be a finite dimensional vector space over a commutative field* K, *and let* u *be an endomorphism of* E. *Then the following conditions are equivalent :*

(i) *u is diagonalisable.*

(ii) E *is the direct sum of the eigenspaces of* u.

(iii) *All the roots of the minimal polynomial of* u *are in* K, *and these roots are all simple.*

Moreover, if these conditions are satisfied, then every subspace of E *closed under* u *is the direct sum of its intersections with the eigenspaces of* u.

The equivalence of (i) and (ii) follows from the preceding remarks and VII, p. 31, Prop. 2. Suppose u is diagonalisable, and let (α_i) be its family of eigenvalues, and (V_i) the corresponding family of eigenspaces ; since the restriction of u to V_i is the homothety defined by α_i, it annihilates the polynomial $X - \alpha_i$; it follows that u annihilates the polynomial $\prod_i (X - \alpha_i)$, which is therefore a multiple of the minimal polynomial of u, and hence coincides with it, which proves (iii). Conversely, if (iii) is satisfied then there exists a basis of E with respect to which the matrix U of u is a diagonal block of Jordan matrices $U_{m,\alpha}$ (VII, p. 35, Prop. 8) ; then by Prop. 9 the integers m are all equal to 1 and so U is diagonal. Finally, the last assertion follows from the fact that if u is diagonalisable then its eigenspaces are the primary components of E_u, and from VII, p. 9, Cor. 1.

COROLLARY. — *If all the roots of the characteristic polynomial of* u *are in* K, *and they are all simple, then* u *is diagonalisable.*

Indeed the minimal polynomial divides the characteristic polynomial.

PROPOSITION 13. — *Let* E *be a finite dimensional vector space over a commutative field* K, *let* $\mathfrak{F}$ *be a set of endomorphisms of* E, *and let* A *be the subalgebra of* $\mathrm{End}_K(E)$ *generated by* $\mathfrak{F}$ *and* Id_E. *Then the following conditions are equivalent :*

(i) $\mathfrak{F}$ *is diagonalisable.*

(ii) *The* K-*algebra* A *is diagonalisable.*

(iii) *The elements of* $\mathfrak{F}$ *are diagonalisable and commute with one another.*

If (e_i) is a basis of E with respect to which $\mathfrak{F}$ is diagonal, then A is contained in the algebra of endomorphisms which are diagonal with respect to this basis, so is also diagonalisable ; if A is diagonalisable, then the same argument shows that $\mathfrak{F}$ is diagonalisable. This shows the equivalence of (i) and (ii). Since any two diagonal matrices commute, we have (i) $\Rightarrow$ (iii), and it remains to prove the converse. Suppose then that the elements of $\mathfrak{F}$ are diagonalisable and commute with each other. We will make use of the following lemma :

Lemma 3. — *Let g and h be two commuting endomorphisms of a vector space* E. *Then each eigenspace of g is closed under h.*

Indeed, if W_λ is the eigenspace of g corresponding to the eigenvalue λ, then for all $x \in W_\lambda$ we have

$$gh \,.\, x = hg \,.\, x = h \,.\, \lambda x = \lambda h \,.\, x\,,$$

which says that $h \,.\, x \in W_\lambda$.

Let us now return to the proof of Prop. 13. Among all decompositions of E as a direct sum of nonzero subspaces each closed under all the elements of $\mathfrak{F}$, choose one with the greatest number of components (the dimension of E is an upper bound for this number), say $E = \sum_{i \in I} E_i$. Let $u \in \mathfrak{F}$ and let $E = \sum_\alpha V_\alpha$ be the decomposition of E as the direct sum of the eigenspaces of u. By Lemma 3, each V_α is closed under $\mathfrak{F}$, and hence so is each $V_\alpha \cap E_i$; by Prop. 12 each E_i is the direct sum of the $V_\alpha \cap E_i$. The choice of the E_i thus forces each E_i to be contained in one of the V_α ; thus the restriction of u to each E_i is a homothety. Since this is true for all the elements of $\mathfrak{F}$, it follows that $\mathfrak{F}$ is diagonalisable.

COROLLARY. — *The sum and the composite of two commuting diagonalisable endomorphisms of* E *are diagonalisable.*

8. Semi-simple and absolutely semi-simple endomorphisms

DEFINITION 6. — *Let* E *be a finite dimensional vector space over a commutative field* K. *Then an endomorphism u of* E *is said to be semi-simple if every subspace of* E *which is closed under u has a complement which is closed under u.*

This means that every submodule of the K[X]-module E_u is a direct factor, in other words the K[X]-module E_u is semi-simple (VII, p. 9).

PROPOSITION 14. — *An endomorphism u of a finite dimensional vector space over a commutative field is semi-simple if and only if the minimal polynomial of u has no multiple factors.*

This follows immediately from VII, p. 9, Cor. 4 and p. 31, Remark 1.

Let E be a vector space over a commutative field K, let L be an extension of K, and u an endomorphism of E ; let $u_{(L)}$ denote the L-endomorphism $1_L \otimes u$ of the L-vector space $E_{(L)} = L \otimes_K E$ induced from E by extension of scalars. In the same way, if $\mathfrak{F}$ is a set of endomorphisms of E, let $\mathfrak{F}_{(L)}$ denote the set of $u_{(L)}$ for u in $\mathfrak{F}$.

COROLLARY. — *Let u be an endomorphism of a finite dimensional vector space over a commutative field* K, *and let* L *be an extension of* K. *If $u_{(L)}$ is semi-simple then u is semi-simple. If u is semi-simple and* L *is separable over* K, *then $u_{(L)}$ is semi-simple.*

This follows immediately from Prop. 14 and from V, p. 120, Cor. 1 (note that the minimal polynomials of u and $u_{(L)}$ coincide).

PROPOSITION 15. — *Let* E *be a finite dimensional vector space over a commutative field* K, *let u be an endomorphism of* E *and let $q(X)$ be its minimal polynomial. Then the following conditions are equivalent :*

(i) *For every extension* L *of* K, *the endomorphism $u_{(L)}$ is semi-simple.*

(ii) *There exists an extension* L *of* K *such that the endomorphism $u_{(L)}$ is diagonalisable.*

(iii) *The polynomial $q(X)$ is separable over* K.

Indeed, condition (i) means that the polynomial $1 \otimes q(X)$ in L[X] has no multiple factors, for any extension L of K (Prop. 14), condition (ii) means that there exists an extension L of K such that all the roots of $q(X)$ belong to L and that these roots are all simple (VII, p. 40, Prop. 12), and these conditions are each equivalent to (iii) by definition (V, p. 38).

DEFINITION 7. — *An endomorphism u satisfying conditions* (i), (ii) *and* (iii) *of Prop.* 15 *is said to be absolutely semi-simple.*

COROLLARY. — *A necessary and sufficient condition for u to be absolutely semi-simple is that there exist an extension* L *of* K *such that* L *is perfect and $u_{(L)}$ is semi-simple.*

The condition in the corollary means that there exists an extension L of K such that L is perfect and $q(X)$ has no multiple factors in L[X] (Prop. 14) ; this condition is equivalent to (iii) by V, p. 38, Cor. 2.

PROPOSITION 16. — *Let* E *be a finite dimensional vector space over a commutative field* K, *let $\mathfrak{F}$ be a set of endomorphisms of* E *and let* A *be the subalgebra of* $\mathrm{End}_K(E)$ *generated by $\mathfrak{F}$ and* Id_E. *Then the following conditions are equivalent :*

(i) *There exists an extension* L *of* K *such that $\mathfrak{F}_{(L)}$ is diagonalisable.*

(ii) *The* K-*algebra* A *is étale* (V, p. 28, Def. 1).

(iii) *The elements of* $\mathfrak{F}$ *are absolutely semi-simple and commute with one another.*

Note first that, for every extension L of K, the L-algebra generated by $\mathfrak{F}_{(L)}$ and $\mathrm{Id}_{E_{(L)}}$ coincides with $L \otimes_K A$; hence by Prop. 13 $\mathfrak{F}_{(L)}$ is diagonalisable if and only if the L-algebra $L \otimes_K A$ is diagonalisable. The equivalence of conditions (i) and (ii) thus follows from V, p. 28, Def. 1. On the other hand, it is immediate that (i) $\Rightarrow$ (iii). Finally suppose (iii) holds, and let L be an algebraic closure of K ; then the elements of $\mathfrak{F}_{(L)}$ are diagonalisable (VII, p. 40, Prop. 12) and commute with one another ; hence $\mathfrak{F}_{(L)}$ is diagonalisable by VII, p. 41, Prop. 13.

COROLLARY. — *The sum and product of two commuting, absolutely semi-simple endomorphisms are absolutely semi-simple.*

Remark. — Suppose the conditions of Prop. 16 are satisfied and let L be an extension of K. By Prop. 13 the set $\mathfrak{F}_{(L)}$ is diagonalisable if and only if the algebra $L \otimes_K A$ is diagonalisable. It follows from V, p. 30, Prop. 2 that there exists a finite extension L of K such that $\mathfrak{F}_{(L)}$ is diagonalisable. In fact L may be taken to be *Galois* ; indeed, taking a finite subset $\mathfrak{F}'$ of $\mathfrak{F}$ which generates A, we may take L to be a splitting field for the minimal polynomials of the elements of $\mathfrak{F}'$ (Prop. 12 and 13).

9. Jordan decomposition

DEFINITION 8. — *Let* E *be a finite dimensional vector space over a commutative field and let* u *be an endomorphism of* E. *Then a Jordan decomposition of* u *is a pair* (u_s, u_n), *where* u_s *is a semi-simple endomorphism of* E *and* u_n *is a nilpotent endomorphism of* E, *such that* $u_s u_n = u_n u_s$ *and* $u = u_s + u_n$.

THEOREM 1. — *Let* E *be a finite dimensional vector space over a commutative field* K *and let* u *be an endomorphism of* E. *Then* u *has a Jordan decomposition* (u_s, u_n) *if and only if the eigenvalues of* u *are separable over* K. *Moreover the decomposition is unique, the characteristic polynomials of* u *and* u_s *coincide, and there exist polynomials* P, $Q \in K[X]$, *with no constant terms, such that* $u_s = P(u)$ *and* $u_n = Q(u)$.

A) Let us first of all prove the following special case :

Lemma 4. — *Let* E *be a finite dimensional vector space over a commutative field* K *and let* u *be a triangularisable endomorphism of* E. *Then there exists a unique diagonalisable endomorphism* v *of* E *which commutes with* u *and is such that* $u - v$ *is nilpotent. Moreover, under these conditions the characteristic polynomials of* u *and* v *coincide, and there exists a polynomial* $P \in K[X]$ *such that* $v = P(u)$.

Let v be a diagonalisable endomorphism of E such that $uv = vu$ and $v - u$ is nilpotent ; let α be an eigenvalue of v and let V_α be the corresponding eigenspace. By Lemma 3 (VII, p. 41), V_α is closed under u, and the restriction of

$u - \alpha$ to V_α is also the restriction of $u - v$, so is nilpotent ; hence V_α is contained in the subspace M_α consisting of those $x \in E$ which are annihilated by some power of $u - \alpha$. Since E is the direct sum of the V_α and also of the M_α (VII, p. 34, Prop. 7), this shows that $V_\alpha = M_\alpha$ for all α. By the corollary to Prop. 9 (VII, p. 36) it follows that $\chi_u = \chi_v$; it also follows that v is uniquely determined by u ; its restriction to each M_α is the homothety defined by α.

Conversely, let us define v by the above condition ; it is clear that v is diagonalisable and that $u - v$ is nilpotent. By Prop. 7 of VII, p. 34 there exist polynomials q_α such that, for all $x \in E$, the component of x in M_α is $q_\alpha(u) \cdot x$. Then $v = \sum_\alpha \alpha q_\alpha(u)$; this implies that u and v commute and completes the proof.

B) Now let us return to the proof of Th. 1.

First of all, suppose that u can be written in the form $s + n$, where s is absolutely semi-simple and n is nilpotent, and where s and n commute. Let Ω be an algebraic closure of K ; then $u_{(\Omega)} = s_{(\Omega)} + n_{(\Omega)}$, where $s_{(\Omega)}$ is diagonalisable and $n_{(\Omega)}$ is nilpotent, and where $s_{(\Omega)}$ and $n_{(\Omega)}$ commute ; by Lemma 4 it follows that $s_{(\Omega)}$, and so also s, is unique, that the polynomials $\chi_{u_{(\Omega)}}$ and $\chi_{s_{(\Omega)}}$ in $\Omega[X]$ coincide, hence also the polynomials χ_u and χ_s, and that s can be expressed as a polynomial in u with coefficients in Ω. This shows first of all that the eigenvalues of u are the same as those of s, so are separable over K (VII, p. 42, Prop. 15) ; in addition, since s is an Ω-linear combination of powers of u, it is also a K-linear combination of these same powers (II, p. 311, Prop. 19), and there exists a polynomial $P \in K[X]$ such that $s = P(u)$, hence $n = Q(u)$ where $Q(X) = X - P(X)$. Now let us show that Q (and hence P) may be chosen with no constant term. If u is invertible then its characteristic polynomial has a nonzero constant term, and the Cayley-Hamilton theorem (III, p. 541, Prop. 20) shows that 1 can be expressed as a polynomial in u with no constant term, so the assertion holds in this case. If u is not invertible, then its kernel W is nonzero and closed under n (VII, p. 41, Lemma 3) ; since the restriction of n to W is nilpotent, there exists a vector $x \neq 0$ in W such that $u(x) = n(x) = 0$, which shows that Q cannot have a constant term.

Conversely, suppose that the eigenvalues of u are separable over K, and let L be a finite Galois extension of K containing these eigenvalues. By Lemma 4 we can write $u_{(L)} = v + w$, where v is diagonalisable and w is nilpotent, and where $vw = wv$. Let B be a basis of E, let B′ be the corresponding basis of $L \otimes_K E$, and let U, V, W be the matrices of $u_{(L)}, v, w$ with respect to B′ ; note that U is also the matrix of u with respect to B, so has entries in K. For every K-automorphism σ of L, and every matrix A with entries in L, let A^σ denote the matrix obtained by applying σ to the entries of A. Let σ be a K-automorphism of L ; then $U = U^\sigma = (V + W)^\sigma = V^\sigma + W^\sigma$, $V^\sigma W^\sigma = (VW)^\sigma = (WV)^\sigma = W^\sigma V^\sigma$; since V^σ is the matrix of a diagonalisable endomorphism and W^σ is nilpotent, it follows from Lemma 4 that $V^\sigma = V$ and $W^\sigma = W$. Since this is valid for all σ, the entries of V and W are in K ; if u_s and u_n are the endomorphisms of E with matrices V and W with respect to B, then $(u_s)_{(L)} = v$ and $(u_n)_{(L)} = w$. It follows that u_s is absolutely semi-simple, that u_n is nilpotent, that u_s and u_n commute, and that $u = u_s + u_n$. This completes the proof.

Whenever an endomorphism f admits a Jordan decomposition, we write it (f_s, f_n), and the endomorphisms f_s and f_n are called the *absolutely semi-simple component* and the *nilpotent component* of f respectively. When K is *perfect*, every endomorphism has a Jordan decomposition ; in this case also there is no distinction between absolutely semi-simple endomorphisms and semi-simple endomorphisms, and we sometimes say « *semi-simple component* » for « absolutely semi-simple component ».

COROLLARY 1. — *Suppose u has a Jordan decomposition, and let* L *be an extension of* K. *Then $u_{(L)}$ has a Jordan decomposition, with $(u_{(L)})_s = (u_s)_{(L)}$ and $(u_{(L)})_n = (u_n)_{(L)}$.*

COROLLARY 2. — *Suppose u has a Jordan decomposition. Then every endomorphism of* E *which commutes with u also commutes with u_s and u_n.*

COROLLARY 3. — *Let u and v be two commuting endomorphisms of* E *which have Jordan decompositions.*

a) The endomorphisms u, v, u_s, v_s, u_n, v_n all commute.

b) The endomorphisms $u + v$ and uv have Jordan decompositions with

$$(u+v)_s = u_s + v_s\,, \quad (u+v)_n = u_n + v_n\,, \quad (uv)_s = u_s v_s\,,$$
$$(uv)_n = u_s v_n + u_n v_s + u_n v_n\,.$$

Part *a*) follows from Cor. 2. To prove part *b*) it is enough to notice that $u_s + v_s$ and $u_s v_s$ are absolutely semi-simple (VII, p. 43, Cor.) and that $u_n + v_n$ and $u_s v_n + u_n v_s + u_n v_n$ are nilpotent (as sums of commuting nilpotent endomorphisms).

COROLLARY 4. — *Suppose u has a Jordan decomposition, and let* R *be a polynomial in* K[X]. *Then the endomorphism* R(u) *has a Jordan decomposition with* R$(u)_s$ = R(u_s).

Remarks. — 1) We have $\det(u_s) = \det(u)$ and $\mathrm{Tr}(u_s) = \mathrm{Tr}(u)$.

2) A necessary and sufficient condition for u to be triangularisable is that u have a Jordan decomposition with u_s diagonalisable. Then there exists a basis of E with respect to which the matrix of u is lower triangular, and that of u_s is diagonal, with the same diagonal as the matrix of u (cf. Lemma 4 and Prop. 19 below).

Note however that if the matrix of u with respect to some basis is triangular, it does not in general follow that the matrix of u_s with respect to the same basis is diagonal.

3) The notion of Jordan decomposition for a square matrix can be defined in an analogous manner. For example, for the Jordan matrix $U_{m,\alpha}$ we have

$$(U_{m,\alpha})_s = \alpha \cdot I_m, \quad (U_{m,\alpha})_n = U_{m,0}.$$

4) If u is semi-simple but not absolutely semi-simple, then it has no Jordan decomposition.

An endomorphism u of a vector space V over a commutative field is said to be *unipotent* if the endomorphism $u - \mathrm{Id}_V$ is nilpotent, that is if there exists an integer r such that $(u - \mathrm{Id}_V)^r = 0$; then u is an *automorphism* of V, since if $u = \mathrm{Id}_V - n$ with $n^r = 0$, then

$$(\mathrm{Id}_V + n + \cdots + n^{r-1})\, u = u\,(\mathrm{Id}_V + n + \cdots + n^{r-1}) = \mathrm{Id}_V .$$

If V has finite dimension m, then u is unipotent if and only if $\chi_u(X) = (X - 1)^m$ (VII, p. 33, Cor. 3 to Prop. 5).

PROPOSITION 17. — *Let* E *be a finite dimensional vector space over a commutative field, and let f be an endomorphism of* E. *Then the following conditions are equivalent:*

(i) *f has a Jordan decomposition and is an automorphism;*

(ii) *f has a Jordan decomposition and f_s is an automorphism;*

(iii) *there exists an absolutely semi-simple automorphism a of E and a unipotent endomorphism u of* E *such that $f = ua = au$.*

Moreover, under these conditions, in the notation of (iii), *we must have $a = f_s$ and $u = 1 + f_s^{-1} f_n$.*

(i) $\Rightarrow$ (ii) : this follows from Remark 1.

(ii) $\Rightarrow$ (iii) : take $a = f_s$ and $u = 1 + f_s^{-1} f_n$; then $f = ua = au$, while a is an absolutely semi-simple automorphism, and u is unipotent.

(iii) $\Rightarrow$ (i) : in the notation of (iii), take $n = a(u - 1) = (u - 1)a$. Then $an = na$ and $f = a + n$, and n is nilpotent. It follows that (a, n) is the Jordan decomposition of f. This implies (i) as well as the relations $a = f_s$ and $u = 1 + f_s^{-1} f_n$.

Put $f_u = f_s^{-1} f = f f_s^{-1} = 1 + f_s^{-1} f_n$, and call this the *unipotent* component of f. The pair (f_s, f_u) is often called the *multiplicative Jordan decomposition* of the automorphism f.

PROPOSITION 18. — *Let* E *be a finite dimensional vector space over a commutative field* K, *let u be an endomorphism of* E *and let* E′ *be a subspace of* E *closed under u. Let u′* (resp. *u″*) *be the endomorphism of* E′ (resp. E/E′) *induced by u. Then* $\chi_u = \chi_{u'} \cdot \chi_{u''}$.

For u to have a Jordan decomposition, it is necessary and sufficient that u′ and u″ have; moreover, if this holds then the absolutely semi-simple (resp. *nilpotent) components of u′ and u″ are the endomorphisms of* E′ *and* E/E′ *induced by the absolutely semi-simple* (resp. *nilpotent) component of u.*

Let B be a basis of E containing a basis B′ of E′, and let B″ be the basis of

$E'' = E/E'$ which is the image of $B - B'$. Let U, U', U'' be the matrices of u, u', u'' with respect to B, B′, B″ respectively. Then U has the form

$$\begin{pmatrix} U' & Z \\ 0 & U'' \end{pmatrix}$$

and $\chi_u = \chi_U = \chi_{U'}\chi_{U''} = \chi_{u'}\chi_{u''}$ (cf. III, p. 533, Ex. 2). We deduce that the set of eigenvalues of u is the union of the sets of eigenvalues of u' and u''. If u' and u'' have Jordan decompositions then the eigenvalues of u' and u'' are separable over K, hence so are the eigenvalues of u, and u has a Jordan decomposition (VII, p. 43, Th. 1). Conversely, if u has a Jordan decomposition (s, n) then s and n leave E′ invariant because they are polynomials in u ; let s', n', s'', n'' denote the endomorphisms of E′, E′, E″, E″ induced by s, n, s, n respectively. Then s' and s'' are absolutely semi-simple, since their minimal polynomials divide that of s (VII, p. 42, Prop. 15) ; also n' and n'' are nilpotent. Finally $u' = s' + n'$, $u'' = s'' + n''$, $s'n' = n's'$, and $s''n'' = n''s''$, which completes the proof.

PROPOSITION 19. — *Let* E *be a finite dimensional vector space over a commutative field* K, *and let* $\mathfrak{F}$ *be a set of commuting triangularisable endomorphisms of* E. *Then there exists a basis of* E *with respect to which the matrix of each element* u *of* $\mathfrak{F}$ *is lower triangular and the matrix of* u_s *is diagonal, with the same diagonal elements as that of* u.

By Cor. 3 of VII, p. 45, the set $\mathfrak{F}_s$ of absolutely semi-simple components of elements of $\mathfrak{F}$ consists of diagonalisable elements which commute with one another, so is diagonalisable (VII, p. 41, Prop. 13), the set $\mathfrak{F}_n$ of nilpotent components of elements of $\mathfrak{F}$ consists of nilpotent elements which commute with one another, and each element of $\mathfrak{F}_n$ commutes with each element of $\mathfrak{F}_s$. Arguing as in the proof of Prop. 13 (VII, p. 41), we see that there exists a decomposition of E as a direct sum of subspaces E_i, which are invariant under $\mathfrak{F}_s$ and $\mathfrak{F}_n$, and such that the restriction of each element of $\mathfrak{F}_s$ to each E_i is a homothety. Replacing E by each E_i in turn, we may assume that the elements of $\mathfrak{F}_s$ are homotheties ; it is enough to prove that there exists a basis of E with respect to which the elements of $\mathfrak{F}_n$ are represented by lower triangular matrices with zero diagonals ; we are thus reduced to the case where $\mathfrak{F}$ consists of nilpotent elements.

Now suppose $E \neq 0$, and let F be a nonzero subspace of E, invariant under $\mathfrak{F}$, of minimum dimension. Then for each $u \in \mathfrak{F}$, the kernel of the restriction of u to F is nonzero and invariant under $\mathfrak{F}$ (VII, p. 41, Lemma 3) ; by the choice of F the restriction of u to F is thus zero for all $u \in \mathfrak{F}$. Let $x \in F$, $x \neq 0$; then $u(x) = 0$ for all $u \in \mathfrak{F}$; arguing by induction on the dimension of E, we may suppose that there exists a basis $(\bar{e}_1, \ldots, \bar{e}_{n-1})$ of the quotient $E' = E/Kx$ such that, for all $u \in \mathfrak{F}$, the endomorphism $\bar{u}$ of E′ induced by u has a matrix with respect to this basis which is lower triangular with zero diagonal ; if $e_i \in E$ projects onto $\bar{e}_i$ for $i = 1, \ldots, n-1$, then the basis $(e_1, \ldots, e_{n-1}, x)$ satisfies the required conditions.

Exercises

§ 1

1) Show that if A is an integral domain then the polynomial ring $A[X_\iota]_{\iota\in I}$ is not a principal ideal domain when one of the following conditions holds : 1) Card(I) $\geqslant 2$; 2) A is not a field (consider the ideals $(X_\alpha) + (X_\beta)$ for $\alpha \neq \beta$, and $(a) + (X_\alpha)$ where $a \neq 0$ is a noninvertible element of A).

2) Show that, in the ring K[[X]] of formal power series over a field K, every irreducible element is an associate of X.

3) Let A be an integral domain in which every nonempty set of ideals has a maximal element and in which every maximal ideal is principal ; show that A is a principal ideal domain. (Notice that if $\mathfrak{a} \neq A$ is an ideal of A then there exists a maximal ideal $(p) \supset \mathfrak{a}$ such that $\frac{1}{p}\mathfrak{a}$, which is an ideal of A, strictly contains $\mathfrak{a}$.)

b) Let K be a field, let $K_1 = K((X))$ be the field of formal power series in one indeterminate X over K (IV, p. 38), and let $B = K_1[[Y]]$ be the ring of formal power series in Y with coefficients in K_1. Let A be the subring of B consisting of those power series $\sum_{n=0}^{\infty} a_n(X)\,Y^n$ such that the power series $a_0(X)$ involves only powers of X with exponent $\geqslant 0$. Show that the principal ideal (X) is the unique maximal ideal of A, but the ideal of A generated by the elements YX^{-n} ($n \geqslant 0$ an arbitrary integer) is not principal.

4) Let A be an integral domain and let S be a multiplicatively closed subset of A not containing 0. Let $S^{-1}A$ denote the ring of elements $s^{-1}a$ ($s \in S$, $a \in A$) in the field of fractions K of A (I, p. 116).

a) Show that if A is a principal ideal domain then so is $S^{-1}A$.

b) Let A be a principal ideal domain and let p be an irreducible element of A ; show that the complement S of the ideal (p) is multiplicatively closed ; in the ring $S^{-1}A$ the ideal $pS^{-1}A$ is the unique maximal ideal, and the quotient field $S^{-1}A/pS^{-1}A$ is isomorphic to $A/(p)$.

5) Let A be a commutative ring in which every ideal is finitely generated.

a) An element $p \neq 0$ of A is said to be *indivisible* if it is not invertible and if the relation $p = xy$ ($x, y \in A$) implies that either x or y belongs to Ap. Show that every element $a \neq 0$ of A can be written (in at least one way) as a product of indivisible elements and of one invertible element (use Lemma 1 of VII, p. 4).

b) An idempotent e in a commutative ring A is said to be *indecomposable* if there does not exist a pair of nonzero idempotents f, g such that $f + g = e$ and $fg = 0$. Two distinct indecomposable idempotents have product zero. Show that every idempotent in the ring A is a sum of indecomposable idempotents (show that if e is an idempotent then so is $1 - e$, and use Lemma 1 of VII, p. 4). Deduce that A is the direct product of a finite family of rings A_i, in each of which every ideal is finitely generated, and the unit element is the only nonzero idempotent. Suppose this family has at least two elements. Then an element $a = \sum_i a_i$ (with $a_i \in A_i$ for all i) is indivisible if and only if there is an index k such that a_k is either indivisible or zero in A_k, and a_i is invertible in A_i for all $i \neq k$; if $a_k = 0$ then A_k is necessarily an integral domain.

c) Show that if 1 is the only nonzero idempotent of A, and if $x \in A$ is such that $x^k \in Ax^{k+1}$ for some integer $k \geqslant 1$, then either x is invertible or $x^k = 0$ (if $x^k = ax^{k+1}$, then consider $a^k x^k$).

d) Let $p \neq 0$ be a zero divisor in A. Show that there exist a finite sequence $(a_k)_{1 \leqslant k \leqslant m}$ of nonzero elements of A such that $0 = pa_1$, $a_k = pa_{k+1}$ for $1 \leqslant k < m$, $a_m \notin A_p$, and $(a_k) \neq (a_{k+1})$ for $1 \leqslant k < m$.

¶ 6) A commutative ring A is called a principal ideal ring if every ideal of A is principal.

a) Show that a quotient ring of a principal ideal ring and a finite product of principal ideal rings are principal ideal rings.

b) Show that every principal ideal ring A is a direct product of a finite family $(A_i)_{1 \leqslant i \leqslant m}$ of principal ideal rings in each of which the unit element is the only nonzero idempotent (use Ex. 5, *b*)).

c) Suppose that A is a principal ideal ring in which the unit element is the only nonzero idempotent, and which contains zero divisors other than 0. Show that there exists an *indivisible nilpotent* element p in A. (Using Ex. 5, *a*), show that there exists an indivisible element $p \neq 0$ in A which is a zero divisor ; apply Ex. 5 *d*), then prove that the principal ideal Ab of $y \in A$ such that $yp^m \in Ap^{m+1}$ is the whole of A ; finally apply Ex. 5, *c*)).

d) With the hypotheses and notation as in *c*), suppose that $p^m = 0$ and $p^{m-1} \neq 0$. Show that the annihilator of p^{m-1} is Ap (otherwise it would be a principal ideal Ac with $p \in Ac$ and $c \notin Ap$; using the fact that p is indivisible, deduce that c would be invertible) ; deduce that, for $1 \leqslant k \leqslant m - 1$, the annihilator of p^{m-k} is Ap^k. Conclude that the ideal Ap is maximal (argue as for the first assertion of *d*), using the fact that $1 - xp$ is invertible in A for all $x \in A$).

e) Deduce from *d*) that for all $x \in A$ there exists an integer $k \geqslant 0$ and an invertible element u such that $x = up^k$, and that the only ideals of A are the Ap^k $(0 \leqslant k \leqslant m)$ (observe that every element of A not belonging to Ap is invertible in A).

f) Let A be a principal ideal ring, let (x_ι) be a family of elements of A, and let Ad be the ideal generated by the family (x_ι) ; show that there exist elements x'_ι of A such that $x_\iota = dx'_\iota$ for all ι, and $\sum_\iota Ax'_\iota = A$ (reduce to the case where 1 is the only nonzero idempotent of A).

g) Give an example of a principal ideal ring and an indivisible element p of A such that the ideal Ap is not maximal (use Ex. 5, *b*)).

7) Given an integral domain A, we call a map w from $A' = A - \{0\}$ into $\mathbf{N}$ a *Euclidean function* if it satisfies the following conditions :

(S_I) $w(xy) \geqslant w(y)$ for each pair x, y of elements of A'.

(S_{II}) If $a \in A'$ and $b \in A'$, then there exist elements q, r of A such that $a = bq + r$, and such that either $r = 0$ or $w(r) < w(b)$.

If there exists a Euclidean function on A then A is called a *Euclidean domain.*
a) Show that $\mathbf{Z}$ and $K[X]$ (where K is a field) are Euclidean domains.
b) Show that every Euclidean domain is a principal ideal domain (if $\mathfrak{a}$ is an ideal of A, choose $a \in \mathfrak{a}$ such that $w(a)$ is as small as possible).
c) Show that the ring A of Gaussian integers (VII, p. 1) is a Euclidean domain with respect to the function $w(a+bi) = a^2 + b^2 = N_{\mathbf{Q}(i)/\mathbf{Q}}(a+bi)$ (notice that, for all $x \in \mathbf{Q}(i)$, there exists $y \in A$ such that $N(x-y) \leqslant 1/2$).
d) Show that the quadratic extension ring B of $\mathbf{Z}$ with basis $(1, e)$ $(e^2 = 2)$, is a Euclidean domain with respect to the function

$$w(a+be) = |a^2 - 2b^2| = |N_{\mathbf{Q}(e)/\mathbf{Q}}(a+be)|$$

(for all $x \in \mathbf{Q}(e)$, there exists $y \in B$ such that $|N(x-y)| \leqslant 1/2$).

8) Let K be a *quadratic field,* that is an extension of $\mathbf{Q}$ of degree 2.
a) Show that $K = \mathbf{Q}(e)$ where e^2 is a rational integer $d \neq 1$ not divisible by a square > 1 (in $\mathbf{Z}$).
b) If $\alpha = a + be$ $(a, b \in \mathbf{Q})$, let $\bar{\alpha}$ denote the conjugate $a - be$ of α. We say that α is an *integer* of the quadratic field K if its norm $\alpha\bar{\alpha}$ and its trace $\alpha + \bar{\alpha}$ are rational integers. Show that the integers of K form a ring A (notice that if α and β are integers of K then the sum and product of the rational numbers $\alpha\beta + \overline{\alpha\beta}$ and $\alpha\bar{\beta} + \bar{\alpha}\beta$ are integers, hence that these rational numbers are the roots of a monic polynomial over $\mathbf{Z}$; deduce that they are rational integers).
c) Show that the integers of $K = \mathbf{Q}(e)$ form a $\mathbf{Z}$-module with basis $(1, e)$ if $d - 1 \not\equiv 0 \pmod 4$, or $(1/2(1+e), 1/2(1-e))$ if $d - 1 \equiv 0 \pmod 4$ (if $a + be$ is an integer of K, check that $2a$ and $2b$ are rational integers such that $(2a)^2 - d(2b)^2 \equiv 0(4)$, and investigate whether they can be odd). Show that the discriminants of these bases (III, p. 549, Def. 3) are $4d$ and d respectively.

9) * (Minkowski's Theorem) If a, b, c, d are real numbers, and $D = \begin{vmatrix} a & b \\ c & d \end{vmatrix}$, then the inequalities $|na + mb| \leqslant A$, $|nc + md| \leqslant B$ have nonzero integer solutions (n, m) if $A > 0$, $B > 0$, $D > 0$ and $AB \geqslant D$, and such solutions are finite in number (consider the discrete subgroup G of $\mathbf{R}^2$ generated by (a, c) and (b, d), and show that the rectangle E defined by the relations $0 \leqslant x \leqslant A$, $0 \leqslant y \leqslant B$, whose area is greater than that of the parallelogram constructed on the vectors (a, c) and (b, d), contains two distinct vectors which are congruent mod G ; argue by contradiction, noticing that otherwise the p^2 rectangles obtained from E by means of the translations $(ha + kb, hc + kd)$, where $0 \leqslant h < p$ and $0 \leqslant k < p$, would be pairwise disjoint). *

¶ 10) Let d be a rational integer not divisible by a square > 1. In the quadratic field $K = \mathbf{Q}(\sqrt{d})$, we shall examine the *units*, that is to say the invertible elements of the ring A of integers of K (Ex. 8, *b*)).
a) If $d < 0$ (« imaginary quadratic fields »), then show that the only units of K are 1 and -1, except for $d = -1$, when i and $-i$ are also units, and for $d = -3$, where the four numbers $1/2(\mp 1 \mp \sqrt{-3})$ are also units.
b) Suppose from now on that $d > 0$ (« real quadratic fields »), and suppose that K is embedded in a maximal ordered extension of $\mathbf{Q}$ (VI, p. 25), * for example in $\mathbf{R}$. * Show

that the positive units form a subgroup U′ of the group U of units of K, and that U is the direct product of U′ and $\{-1, 1\}$.
c) Show that the units of K are the integers with norm 1 or -1.
d) Show that the group U′ is nontrivial (let D denote the discriminant of the integer basis (Ex. 8, *c*)) ; show that the integers α of K such that $N(\alpha) \leqslant D$ are partitioned into a finite number of congruence classes mod U ; choose an integer β_i in each class ; if B is a rational number such that $0 < B < \inf(|\beta_i|)$, show (Ex. 9) that there exists an integer β of K such that $0 < |\beta| \leqslant B$ and $N(\beta) \leqslant D$; deduce that some $\beta\beta_i^{-1}$ is a unit η such that $|\eta| < 1$).
e) Show that U′ is isomorphic to **Z** (show that there exists a unit ε such that $|\varepsilon| < 1$ and $|\varepsilon|$ is as great as possible with this property, using Ex. 9 and the fact that $|\bar{\xi}| = |\xi|^{-1}$ for a unit).
f) Show that in $\mathbf{Q}(\sqrt{2})$, the element $\sqrt{2} - 1$ is a generator of U′, and has norm -1.
g) Show that if d has a prime factor p of the form $4n - 1$, then every generator of U′ has norm 1. (Observe that -1 cannot be a square in the field $\mathbf{F}_p$, using V, p. 93, Prop. 1, *c*)).

¶ 11) We investigate the quadratic fields $\mathbf{Q}(\sqrt{d})$ for which the map $\alpha \mapsto |N(\alpha)|$ is a *Euclidean function* on the integers of the field (Ex. 7).
a) Show that a necessary and sufficient condition for this is that, for all $\alpha \in \mathbf{Q}(\sqrt{d})$, there exist an integer β of the field such that $|N(\alpha - \beta)| < 1$; write down this condition, expressing α and β in terms of an integer basis of the field (Ex. 8, *c*)).
b) In the case of an imaginary quadratic field ($d < 0$), show that the only values of d for which $|N(\alpha)|$ is a Euclidean function are $-1, -2, -3, -7, -11$.
c) For real quadratic fields, we will restrict ourselves to the cases $d \equiv 2$ or 3 (mod 4). Show that there are then only *finitely* many values of d such that $|N(\alpha)|$ is a Euclidean function. (If $|N(\alpha)|$ is a Euclidean function and a is a natural number such that $0 \leqslant a \leqslant d$ and $a \equiv t^2(d)$, then show that either a or $d - a$ has the form $x^2 - dy^2$, where x and y are rational integers, by considering the element $t/\sqrt{d}$ of $\mathbf{Q}(\sqrt{d})$; next show that for large enough d there exists an odd integer z such that $5d < z^2 < 6d$ (if $d \equiv 3$ mod 4) or $2d < z^2 < 3d$ (if $d \equiv 2$ mod 4) ; then take $a = z^2 - d[z^2/d]$ and show that, under these conditions, neither a nor $d - a$ has the form $x^2 - dy^2$, by examining residues mod 8). Show that the only values of d for which no such integer z exists are 3, 7, 11, 19, 35, 47 and 59 (for $d \equiv 3$ mod 4) and 2, 6, 14 and 26 (for $d \equiv 2$ mod 4). Show that there is indeed a Euclidean function for $d = 2$ and $d = 3$.

12) Show that the ring of integers of the quadratic field $\mathbf{Q}(\alpha)$, where $\alpha^2 = -5$, is not a principal ideal domain (see VI, p. 34, Ex. 27). Same question for $\mathbf{Q}(\beta)$, where $\beta^2 = 10$ (we have

$$2.3 = (4 + \beta)(4 - \beta)).$$

13) Let f be a nonconstant polynomial in one indeterminate over **Z**. Show that for each integer $k \geqslant 1$ there exist infinitely many integers $n \in \mathbf{N}$ such that $f(n)$ is nonzero and divisible by at least k distinct primes (one can proceed by induction on k : if $n \in \mathbf{N}$ is such that $f(n)$ is nonzero and divisible by at least k distinct primes, consider the polynomial $f(n + Xf(n)^2)$).

14) The set of prime numbers of the form $4n - 1$ (resp. $6n - 1$) is infinite (same method as Prop. 5 (VII, p. 5) ; notice that every prime number except 2 (resp. except 2 and 3) has the form $4n \pm 1$ (resp. $6n \pm 1$)).

15) A number of the form $2^k + 1$ ($k \geq 1$ an integer) can be a prime only if k is a power of 2. If a prime number p divides $2^{2^n} + 1$, then it cannot divide $2^{2^m} + 1$ for $m > n$ ($2^{2^m} \equiv 1\ (p)$) ; use this to give a new proof of Prop. 5. Show that $2^{2^5} + 1$ is not prime (put $a = 2^7$ and $b = 5$; show that $1 + ab - b^4 = 2^4$ and that $1 + ab$ divides

$$n = 2^4 a^4 + 1 = (1 + ab - b^4) a^4 + 1 .$$

16) A number of the form $a^k - 1$ (a and k integers, $a > 0$, $k > 1$) can be prime only if $a = 2$ and k is prime.

17) Show that if $a^n + 1$ is prime, where $a > 1$ and $n > 1$, then a is even and $n = 2^m$.

18) Let M be the multiplicative monoid generated by a finite number of integers $q_i > 1$ ($1 \leq i \leq m$). Show that, for each integer $k > 0$, there exist k consecutive integers which do not belong to M (if $h_1, \ldots, h_m$ are m integers > 1, and k is the greatest number of consecutive integers none of which is divisible by any of the h_i, then $(k+1)^{-1} \leq h_1^{-1} + \cdots + h_m^{-1}$; now notice that, for all $r > 0$, every sufficiently large number in M is divisible by at least one q_i^r). Obtain a new proof of Prop. 5 of VII, p. 5 from this result.

19) For every integer $n > 0$, show that the exponent of a prime p in the prime factorisation of $n!$ is $\sum_{k=1}^{\infty} [np^{-k}]$ (which has only finitely many nonzero terms).

¶ 20) * Let $\pi(x)$ be the number of primes less than the real number $x > 0$. For each integer $n > 0$, show that

$$n^{\pi(2n) - \pi(n)} \leq \binom{2n}{n} \leq (2n)^{\pi(2n)}$$

(notice that $\binom{2n}{n}$ is a multiple of the product of all the primes q such that $n < q \leq 2n$) ; on the other hand, use Ex. 19 to show that $\binom{2n}{n}$ divides the product $\prod p^{r(p)}$, where p runs through the set of prime numbers $\leq 2n$, and where $r(p)$ is the largest integer such that $p^{r(p)} \leq 2n$. Deduce from this result that there exist two real numbers a and b such that $0 < a < b$ and [1]

$$an(\log n)^{-1} \leq \pi(n) \leq bn(\log n)^{-1}$$

(notice that $2^{2n}(2n+1)^{-1} \leq (2n)!\,(n!)^{-2} \leq 2^{2n}$). *

[1] It has been shown that $\pi(x) \sim \dfrac{x}{\log x}$ as x tends to $+\infty$ (cf. for example A. E. INGHAM, *The distribution of prime numbers* (Cambridge tracts, No. 30), Cambridge University Press, 1932).

21) Show that for $m \geqslant 1$ and $n \geqslant 1$ the rational number $\sum_{k=n}^{n+m} k^{-1}$ is never an integer (if 2^q is the largest power of 2 which divides at least one of the numbers n, $n+1$, ..., $n+m$, then it divides only one of these numbers).

22) Let a and b be two integers > 0, let d be the gcd of a and b, and let n be an integer $\geqslant 1$; put $a = da_1$ and $b = db_1$; show that the denominator in the irreducible expression for the fraction $a(a+b) \dots (a+(n-1)b)/n!$ admits only prime factors which divide b_1 and are $\leqslant n$ (observe that if p is a prime number not dividing b_1, then b_1 is invertible mod p^r for all $r > 0$). Hence obtain a direct proof of the fact that the binomial coefficients $n!/(m!(n-m)!)$ are integers.

23) Let $n > 1$ be a rational integer ; show that the gcd of the $n-1$ binomial coefficients $\binom{n}{k}$ $(1 \leqslant k \leqslant n-1)$ is equal to 1 if n is divisible by two distinct prime numbers, and is equal to p if n is a power of the prime number p.

24) *a*) Let $n = \prod_k p_k^{\nu_k} > 0$ be a rational integer expressed in terms of its prime factors. Show that the sum of the positive divisors of n is given by the formula

$$\sigma(n) = \prod_k \frac{p_k^{\nu_k+1} - 1}{p_k - 1}.$$

b) We say that n is *perfect* if $2n = \sigma(n)$. Show that an even number n is perfect if and only if it has the form $2^h(2^{h+1}-1)$, where $h \geqslant 1$ and $2^{h+1}-1$ is prime. (If n has the form $2^m q$, where $m \geqslant 1$ and q is odd, show that q is divisible by $2^{m+1}-1$, and deduce that, if q were not prime, then we would have $\sigma(n) > 2n$).

25) Let a and b be two coprime integers > 0. If a and b are > 2, then there exist integers x and y such that $ax + by = 1$ and $|x| < 1/2\, b$, $|y| < 1/2\, a$; moreover the integers x and y are uniquely determined by these conditions. What happens in the case where one of a, b is equal to 2 ?

26) $^{\bullet}$Let x, y be two coprime integers ; then every integer z such that $|z| < |xy|$ can be expressed, in either one or two ways, in the form $z = ux + vy$ where u and v are integers with $|u| < |y|$ and $|v| < |x|$. If z can be expressed in two different ways in this form, then neither x nor y can divide z ; the example $x = 5$, $y = 7$, $z = 12$ shows that the converse fails.

27) Let f and g be two coprime polynomials in K[X] ; show that every polynomial h such that $\deg(h) < \deg(fg)$ can be expressed uniquely in the form $h = uf + vg$ where u and v are polynomials such that $\deg(u) < \deg(g)$ and $\deg(v) < \deg(f)$. First give a proof analogous to that of Ex. 26 (based on the Euclidean algorithm), then give a proof in which the problem is considered as a linear problem.

¶ 28) *a*) Let K be an algebraically closed commutative field, let f be a nonconstant monic polynomial in $K[X]$, and let g be a polynomial in $K[X, Y]$; suppose that for each of the roots α_i of f there exists a root β_i of $g(\alpha_i, Y) = 0$ such that $\frac{\partial g}{\partial Y}(\alpha_i, \beta_i) \neq 0$. Then show that there exists a polynomial $h(X) \in K[X]$ such that $g(X, h(X)) \equiv 0 \pmod{f(X)}$. (Using Prop. 4, reduce to the case $f(X) = X^m$ (for some $m \geqslant 1$); notice that the ring $K[X]/(X^m)$ is isomorphic to the quotient ring $K[[X]]/(X^m)$, and use the Cor. of IV, p. 37).
b) In particular show that if m is an integer not divisible by the characteristic of K, then for every pair of coprime monic polynomials f and f_0 in $K[X]$, there exists a polynomial $h(X)$ in $K[X]$ such that $(h(X))^m \equiv f_0(X) \pmod{f(X)}$.

§ 2

1) Let A be a principal ideal domain, let M be a torsion A-module, let x, y be two elements of M, and let $A\alpha$ and $A\beta$ be the annihilators of x, y respectively; if δ is a gcd of α and β, show that the annihilator $A\gamma$ of $x + y$ is such that γ divides the lcm $\alpha\beta/\delta$ of α and β, and is a multiple of $\alpha\beta/\delta^2$. When α and β are *distinct* powers π^μ, π^ν of the same irreducible element π of A, show that $\gamma = \pi^{\sup(\mu, \nu)}$.

2) Let A be a principal ideal domain, let π be an irreducible element of A and let M be a π-primary module; show that for all $x \in M$ and all $\alpha \in A$ not a multiple of π, there exists a unique $y \in M$ such that $x = \alpha y$ (use the Bezout identity); put $y = \alpha^{-1}x$. Deduce that, if $A_{(\pi)}$ denotes the principal ideal domain consisting of elements β/α of the field of fractions of A, such that $\alpha, \beta \in A$ and α is not a multiple of π (VII, p. 48, Ex. 4), then there exists a unique $A_{(\pi)}$-module structure on M such that restriction of the ring of operators to A gives the original A-module structure on M. The $A_{(\pi)}$-module thus defined is said to be *canonically associated* to M; its submodules are identical to the A-submodules of M.

¶ 3) Let A be a principal ideal domain.
a) An A-module M is *injective* (II, p. 389, Ex. 11) if and only if, for all $x \in M$ and all $\alpha \neq 0$ in A, there exists $y \in M$ such that $x = \alpha y$; the A-module M is then said to be *divisible*. If E is an arbitrary A-module, then the sum of all the divisible submodules of E is the greatest divisible submodule $\Delta(E)$ of E. We have $\Delta(F) \subset \Delta(E)$ for every submodule F of E.
b) Let M be a divisible A-module, let π be an irreducible element of A and let x_0 be an element of M whose annihilator has the form $A\pi^n$ ($n > 0$ an integer). Define a sequence (x_s) of elements of M recursively by the conditions $x_s = \pi x_{s+1}$ for each integer $s \geqslant 0$. Show that the submodule of M generated by the sequence (x_s) is isomorphic to the π-primary component U_π of the module $U = K/A$, where K is the field of fractions of A.
c) Show that U_π is a divisible module, and that every submodule of U_π other than U_π and 0 is cyclic and generated by the class mod A of some power of π; deduce that U_π is indecomposable (II, p. 393, Ex. 21).
d) Show that every indecomposable divisible A-module is isomorphic either to K or to one of the modules U_π (use *a*) and *b*)).
e) Show that every divisible A-module is isomorphic to the direct sum of a family of indecomposable divisible submodules (apply Zorn's Lemma to families of indecomposable divisible submodules whose sums are direct).

¶ 4) A family (x_ι) of nonzero elements of a module M over a principal ideal domain A is said to be *pseudofree* if every relation of the form $\sum_\iota \alpha_\iota x_\iota = \beta y$ $(y \in M,\ \alpha_\iota \in A,\ \beta \in A)$ implies the existence of a family of elements $\alpha'_\iota \in A$ such that $\alpha_\iota x_\iota = \beta \alpha'_\iota x_\iota$ for each index ι. The relation $\iota \neq \kappa$ then implies $x_\iota \neq x_\kappa$; the set of elements of a pseudofree family is called a pseudofree subset of M.

a) Let π be an irreducible element of A. Show that, in an A-module with annihilator $A\pi^n$ $(n \geqslant 1)$, any element with annihilator $A\pi^n$ is pseudofree (use Bezout's identity).

b) Let (x_ι) be a pseudofree family in an A-module M, and let N be the submodule of M generated by this family. Show that for every element $\dot{x}$ of M/N there exists an element $x \in \dot{x}$ whose annihilator is equal to the annihilator of $\dot{x}$ in M/N. If also $\dot{x}$ is pseudofree in M/N, show that the family consisting of the (x_ι) and x is pseudofree in M.

c) Deduce from *a*) and *b*) that if the A-module M has nonzero annihilator, then it is a direct sum of cyclic modules (reduce to the case of a π-primary module, and apply Zorn's Lemma to pseudofree families of elements of M) (cf. VII, p. 19, Th. 2).

5) Let M be a finitely generated torsion module over a principal ideal domain A. Show that M is a module of finite length ; consequently every nonempty set of submodules of M has a maximal element and a minimal element.

6) Let M be a torsion module over a principal ideal domain A, which is not finitely generated, but all of whose proper submodules are finitely generated. Show that there exists an irreducible element π of A such that M is isomorphic to the π-primary component U_π of the module U = K/A (Ex. 3, *c*)). (First of all show that M cannot have more than one nonzero π-primary component, and hence is a π-primary module for some irreducible element $\pi \in A$. Then show that $M = \pi M$ by observing that $M/\pi M$ could not be finitely generated if it were nonzero ; but $M/\pi M$ is a vector space over the field $A/\pi A$, and would contain a proper subspace which was not finitely generated.)

¶ 7) Let M be a module over a principal ideal domain A ; a submodule N of M is said to be *pure* if, for all $\alpha \in A$, we have $N \cap (\alpha M) = \alpha N$.

a) The submodule N of M is pure if and only if, for all $\alpha \in A$ and every element $\dot{x} \in M/N$ whose annihilator is $A\alpha$, there exists an element $x \in \dot{x}$ whose annihilator is $A\alpha$.

b) Show that the torsion submodule of M is pure. A family (x_ι) of elements of M is pseudofree (Ex. 4) if and only if the sum of the submodules Ax_ι is direct and is a pure submodule of M. Show that if P and Q are submodules of M such that $P \cap Q$ and $P + Q$ are pure then P and Q are pure. If Q is a pure submodule of M and $P \supset Q$ a submodule of M, then P is pure in M if and only if P/Q is pure in M/Q.

c) Show that every submodule N of M which is a direct factor of M is pure. The same is true for the union of a filtered family of direct factors of M. Conversely, if N is a pure submodule of M, and M/N is a direct sum of cyclic submodules, show that N is a direct factor of M (use *a*)). Give examples of direct factors P and Q of M such that $P \cap Q$ (resp. $P + Q$) is not a pure submodule (for the first example, take $M = (\mathbf{Z}/4\mathbf{Z}) \oplus (\mathbf{Z}/2\mathbf{Z})$, and for the second, take $M = \mathbf{Z}^2$).

d) Let N be a pure submodule of M, whose annihilator $A\alpha$ is nonzero. If f is the natural homomorphism from M onto $M/\alpha M$, show that f restricts to an isomorphism from N onto $f(N)$, and that $f(N)$ is a pure submodule of $M/\alpha M$ (use Bezout's identity). Deduce (with the help of *c*) and Ex. 4, *c*)) that $f(N)$ is a direct factor in $M/\alpha M$, and that N is a direct factor in M. In particular, if the torsion submodule of a module M has nonzero annihilator (which is the case when it is finitely generated), then it is a direct factor in M (cf. VII, p. 20, Cor. 1 to Th. 2).

e) Let p be a prime number; let N denote the direct sum of the cyclic **Z**-modules $\mathbf{Z}/(p^n)$ $(n \geqslant 1)$, and let e_n denote the element of N whose coordinates are all zero except for the n-th, which is equal to the class of 1 mod p^n. Let M be the submodule of the **Z**-module $N \times \mathbf{Q}$ generated by the elements (e_n, p^{-n}). Show that the torsion submodule P of M is not a direct factor in M (notice that every element of M/P is divisible by p^n for all n, and that no element of M has this property except for multiples of $(0, 1)$).

8) Let A be a principal ideal domain, let π be an irreducible element of A and let M be a π-primary module. Then an element x of M is said to have *height n* if $x \in \pi^n M$ and $x \notin \pi^{n+1}M$. If $x \in \pi^n M$ for all $n \geqslant 1$ then x is said to have *infinite height*.

a) Show that M is divisible if and only if all its elements have infinite height (Ex. 3).

b) Let x be an element of M of height n, such that $\pi x = 0$; show that if $x = \pi^n y$ then the cyclic submodule Ay of M is a direct factor of M (notice that Ay is a pure submodule, and apply Ex. 7, *d*)).

c) Show that a nondivisible π-primary module M always contains at least one cyclic nonzero direct factor (if $x \in M$ has finite height, and there exist integers m such that $\pi^m x \neq 0$ has infinite height, then let s be the least such integer; write $\pi^s x = \pi y$, where y has height strictly greater than $\pi^{s-1}x$, and apply *b*) to $y - \pi^{s-1}x$). Deduce that any indecomposable π-primary module is either cyclic or isomorphic to a module U_π (Ex. 3).

d) Show that if every pure submodule of the π-primary module M is a direct factor in M, then M is a direct sum of a divisible module P and a π-primary module Q with nonzero annihilator. (Reduce to the case where M contains no nonzero divisible submodule, and argue by contradiction. Use *b*) to show that there would exist an infinite sequence (x_k) of elements of height 1 in M such that: 1) the annihilator of x_k is $A\pi^{n_k}$, where the sequence (n_k) of integers is strictly increasing; and 2) the sum of the Ax_k is direct, and the sum of any finite number of these submodules is a direct factor in M. Then show that the submodule N of M generated by the elements $x_k - \pi^{n_{k+1}-n_k}x_{k+1}$ is pure, but not a direct factor of M, arguing as in Ex. 7, *e*)).

9) *a*) Let M be a π-primary module, and let M_0 be the submodule of M consisting of elements of infinite height in M. Show that the π-primary module M/M_0 contains no nonzero element of infinite height.

b) Let p be a prime number, and let E be the direct sum of the p-groups $E_n = \mathbf{Z}/(p^n)$ $(n \geqslant 1)$; let e_n denote the residue class of 1 mod p^n in E_n. Let H be the submodule of E generated by the elements $e_1 - p^{n-1}e_n$ for all integers $n \geqslant 2$. Let M be the quotient module E/H. Show that the submodule M_0 of elements of infinite height in M is the submodule $(E_1 + H)/H$, which is isomorphic to E_1, and show that there is no nonzero element in M_0 which has infinite height *in* M_0.

¶ 10) *a*) Let M be a π-primary module with no nonzero elements of infinite height and let N be a pure submodule of M generated by a maximal pseudofree family of elements of M (Ex. 4 and 7, *b*)). Show that M/N is a divisible module (argue by contradiction, applying Ex. 8, *c*), Ex. 7, *c*), Ex. 7, *b*), and finally Ex. 4, *c*)).

b) Consider the metrisable topology $\mathcal{T}$ on M defined by taking as a basis of neighbourhoods of 0 the submodules π^nM for all integers $n \geqslant 0$ (Gen. Top., III, p. 5). Then the quotient M/P of M by a submodule P is divisible if and only if P is everywhere dense in M under the topology $\mathcal{T}$.

c) Let P_1, P_2 be two submodules of M which are pure, everywhere dense in M under the topology $\mathcal{T}$, and direct sums of cyclic submodules. Show that P_1 and P_2 are isomorphic (notice that $P_1/\pi^n P_1$ is isomorphic to $M/\pi^n M$ for all n).

d) Let $\hat{M}$ be the completion of M with respect to the topology $\mathcal{T}$ (Gen. Top., III, p. 24). Show that $x \mapsto \alpha x$ is a strict morphism from $\hat{M}$ onto $\alpha\hat{M}$ for all $\alpha \in A$, that $\pi^n\hat{M}$ is an open and closed submodule, and the closure of π^nM, for all n, and that the quotient modules $\hat{M}/\pi^n\hat{M}$, $M/\pi^n M$ are isomorphic.

e) Let $\bar{M}$ be the torsion submodule of $\hat{M}$; show that $\bar{M}$ is a π-primary module, that $\bar{\bar{M}} = \bar{M}$, and that M is a pure submodule of $\bar{M}$.

f) Suppose the π-primary module M is the direct sum of a family (E_α) of cyclic submodules. Then $\mathcal{T}$ is discrete if and only if the annihilator of M is 0. In this case, let E be the product module $\prod_\alpha E_\alpha$ and let F be the submodule of E consisting of elements all but countably many of whose coordinates are zero; show that $\hat{M}$ can then be identified (as a non-topological module) with the submodule of F consisting of (x_α) such that, for each integer n, there are only *finitely* many indices α for which $x_\alpha \notin \pi^n E_\alpha$. Then show that the four modules M, $\hat{M}$, $\bar{M}$ and F are distinct. Deduce that $\bar{M}$ is not a direct sum of cyclic modules (use *e*)), and that it is not a direct sum of indecomposable modules (notice that $\bar{M}$ contains no elements of infinite height, and apply Ex. 8, *c*)).

¶ 11) Let M be a π-primary module and let N be a submodule of M whose nonzero elements all have height $\leqslant h$ *in* M.

a) Let P be a pure submodule of M such that $P(\pi) \subset N(\pi)$ and $P(\pi) \neq N(\pi)$. Let a be an element of $N(\pi)$ not belonging to $P(\pi)$, and let x_0 be an element of $a + P(\pi)$ whose height k in M is as large as possible; let $x_0 = \pi^k y_0$. Show that the sum $P + Ay_0$ is direct, and is a pure submodule of M.

b) Let B be a pseudofree subset of M such that, if P is the pure submodule generated by B, then $P(\pi) \subset N(\pi)$. Show that there exists a pseudofree subset $C \supset B$ of M such that, if Q is the pure submodule of M generated by C, then $Q(\pi) = N(\pi)$ (apply Zorn's Lemma, using *a*)).

¶ 12) A π-primary module M is a direct sum of cyclic submodules if and only if there exists a pseudofree subset H in M such that, if N is the pure submodule of M generated by H, then $M(\pi) = N(\pi)$.

b) The module M is a direct sum of cyclic submodules if and only if M is the union of an increasing sequence (P_k) of submodules, such that the heights in M of the nonzero elements in each P_k are *bounded*. (To see that the condition is sufficient, use *a*) and Ex. 11, *b*).)

c) Deduce from *b*) that if M is a direct sum of cyclic submodules, then every submodule of M is a direct sum of cyclic submodules.

d) Let M be a π-primary module with no elements of infinite height, having a *countable* system of generators (x_k). Show that M is a direct sum of cyclic submodules (if

P_k is the submodule generated by x_i for $i \leqslant k$, then apply *b*) to the increasing sequence (P_k), using Ex. 5 applied to the decreasing sequence $(P_k \cap \pi^j M)$ (for fixed k)).

e) Let M be a π-primary module having a countable system of generators. Then M is a direct sum of indecomposable submodules if and only if the submodule N of elements of infinite height in M is pure (use *d*) and Ex. 7, *c*)).

¶ 13) Let A be a principal ideal domain, let π be an irreducible element of A, and let M be a π-primary module over A. Let I be the set of ordinals α (*Set Theory*, III, p. 225, Ex. 14) such that the cardinality of the set of ordinals $< \alpha$ is at most equal to that of M. Define submodules $M^{(\alpha)}$ of M for all $\alpha \in I$ by transfinite induction as follows: take $M^{(0)} = M$; take $M^{(\alpha+1)} = \pi M^{(\alpha)}$; and if α is a limit ordinal take $M^{(\alpha)}$ to be the intersection of all the $M^{(\beta)}$ for $\beta < \alpha$.

a) Show that there exists a least ordinal $\tau \in I$ such that $M^{(\tau+1)} = M^{(\tau)}$ (the *terminal* ordinal of M). Show that $M^{(\tau)}$ admits a complement in M, and is the direct sum of submodules isomorphic to U_π (Ex. 3). Say that M is *reduced* if $M^{(\tau)} = 0$.

b) Let M be a reduced π-module. For all $x \in M$, show that there exists a greatest ordinal $\gamma(x) \leqslant \tau$ such that $x \in M^{(\gamma(x))}$ (in particular, $\gamma(0) = \tau$). Generalising the definition in Ex. 8, we say that $\gamma(x)$ is the *height* of x in M. Show that if $\gamma(x) < \gamma(y)$, then $\gamma(x+y) = \gamma(x)$, and that if $\gamma(x) = \gamma(y)$, then $\gamma(x+y) \geqslant \gamma(x)$. We say that x is *proper* with respect to a submodule N of M if for all $y \in N$ *the height* $\gamma(x+y)$ is at most equal to $\gamma(x)$; show that then $\gamma(x+y) = \inf(\gamma(x), \gamma(y))$, for all $y \in N$.

c) Let $\alpha < \tau$ be an ordinal, and let $\tilde{M}^{(\alpha)}$ be the submodule of $M^{(\alpha)}$ consisting of those x such that $\pi x = \pi y$ for some $y \in M^{(\alpha+1)}$. For those $y \in M^{(\alpha+1)}$ having this property, the elements of the form $u = x - y$ form a class mod $M(\pi) \cap M^{(\alpha+1)}$ in $M(\pi) \cap M^{(\alpha)}$; if φ is the canonical homomorphism from $M(\pi) \cap M^{(\alpha)}$ onto the quotient of this module by $M(\pi) \cap M^{(\alpha+1)}$, then show that the map ψ which sends x to $\varphi(u)$ is a homomorphism with kernel $M^{(\alpha+1)}$, and consequently that $\tilde{M}^{(\alpha)}/M^{(\alpha+1)}$ is isomorphic to $(M(\pi) \cap M^{(\alpha)})/(M(\pi) \cap M^{(\alpha-1)})$.

d) Let N be a submodule of M, and let $x \in \tilde{M}^{(\alpha)}$ be an element of height α; show that x is proper with respect to N if and only if the corresponding elements $u = x - y$ of $M(\pi) \cap M^{(\alpha)}$ (cf. *c*)) are proper with respect to N. There exists such an element x if and only if the image of $N \cap \tilde{M}^{(\alpha)}$ under the homomorphism ψ is distinct from $(M(\pi) \cap M^{(\alpha)})/(M(\pi) \cap M^{(\alpha+1)})$.

¶ 14) Let M be a reduced π-primary module. In the notation of Ex. 13, for each ordinal $\alpha < \tau$ we can consider $(M(\pi) \cap M^{(\alpha)})/(M(\pi) \cap M^{(\alpha+1)})$ as a vector space over the field $F_\pi = A/A\pi$; let $c(\alpha)$ denote the dimension of this vector space, and call $c(\alpha)$ the *multiplicity* of the ordinal α in the module M (cf. VII, p. 24).

a) Show that if M and N are two π-primary modules each of which is a direct sum of cyclic submodules (which implies that $\tau \leqslant \omega$, the first infinite ordinal), then M and N are isomorphic if and only if they have the same terminal ordinal τ and the multiplicity of every integer $n < \tau$ is the same in M and in N.

b) Give an example of two π-primary modules with no elements of infinite height, which are not isomorphic, but in which each integer n has the same multiplicity (cf. Ex. 10, *f*)).

c) Let M and N be two reduced π-primary modules, each having a *countably* infinite system of generators. Show that if M and N have the same terminal ordinal τ, and if each ordinal $\alpha < \tau$ has the same multiplicity in M and in N, then M and N are *isomorphic*

(« *Ulm-Zippin Theorem* »). (Given two submodules $S \subset M$ and $T \subset N$, an isomorphism h from S onto T is said to be *indicial* if, for all $x \in S$, the height of x in M is equal to the height of $h(x)$ in N. Define an indicial isomorphism from M onto N by induction, and by alternating the rôles played by M and N, in such a way as to reduce the problem to the following: given two *finitely* generated submodules $S \subset M$ and $T \subset N$, an indicial isomorphism h from S onto T, and an element $x \in M \cap \complement S$, extend h to an indicial isomorphism from $S + Ax$ onto a submodule of N containing T. Show that this can be reduced to the case where $\pi x \notin S$; using Ex. 5, choose an element w_1 in $(S + Ax) \cap \complement S$ with greatest possible height α in M, then choose an element w_2 in $S \cap M^{(\alpha)}$ such that the height of $\pi(w_1 + w_2)$ in M is greatest possible; show that $w = w_1 + w_2$ is proper with respect to S (Ex. 13, *b*)). Let $z = h(\pi w) \in T$; show that, if there exists an element t of N, of height α in N, proper with respect to T, and such that $\pi t = z$, then an extension $\bar{h}$ of h of the desired form can be obtained by putting $\bar{h}(w) = t$. It remains to prove the existence of the element t. If

$$\gamma(z) = \gamma(\pi w) = \alpha + 1 < \tau,$$

show that there exist elements u in $N^{(\alpha)} \cap \complement T$ such that $z = \pi u$, and that t may be taken to be any such element. If on the other hand $\gamma(z) = \gamma(\pi w) > \alpha + 1$ or else if $\gamma(z) = \gamma(\pi w) = \alpha + 1 = \tau$, use Ex. 13, *d*) to prove the existence of the element t, noticing that (in the notation of Ex. 13) the image under ψ of $S \cap \tilde{M}^{(\alpha)}$ in $(M(\pi) \cap M^{(\alpha)})/(M(\pi) \cap M^{(\alpha+1)})$ is finite dimensional over the field F_π.)

* *d*) Obtain from *c*) a new proof of Ex. 12, *d*) (use also Th. 1 of VII, p. 14). *

15) *a*) Let A be a principal ideal domain, let M be a divisible A-module (Ex. 3), and let N be a torsion A-module. Show that $M \otimes_A N = 0$.
b) Let M and N be two π-primary modules with no elements of infinite height; show that $M \otimes_A N$ is a direct sum of cyclic submodules (use *a*) and Ex. 10, *a*)).

§ 3

1) Let A be a commutative ring. Show that, if every submodule of every free A-module is free, then every ideal of A is principal, and A is an integral domain (and consequently a principal ideal domain).

¶ 2) Let M be a module over a principal ideal domain A; suppose M is a direct sum of cyclic submodules. Show that every submodule N of M is a direct sum of cyclic submodules (if T is the torsion submodule of N, show first of all that N/T is a free module, using Th. 1, and deduce that T has a complement in N; then show that T is a direct sum of cyclic submodules, using Ex. 12, *c*) of VII, p. 57).

3) Let (p_n) be a strictly increasing sequence of prime numbers such that p_n does not divide any of the numbers

$$a + b(1 + p_1 + p_1p_2 + \cdots + p_1p_2 \dots p_{n-1}) \quad \text{for} \quad |a| \leqslant n \quad \text{and} \quad |b| \leqslant n.$$

In the rational plane $\mathbf{Q}^2$, consider the $\mathbf{Z}$-module M generated by the sequence (x_n) defined as follows:

$$x_0 = (1,0), \quad x_1 = (0,1), \quad x_{n+1} = p_n^{-1}(x_0 + x_n) \quad \text{for} \quad n \geqslant 1.$$

Show that M has rank 2, and that every rank 1 submodule of M is cyclic (notice that x_0 and x_n form a basis for the submodule of M generated by the x_i with $i \leqslant n$).

4) Let E be a torsion-free module over a principal ideal domain A.
a) Show that, if F is a pure submodule of E (VII, p. 55, Ex. 7), then E/F is torsion-free, and conversely.
b) Let F be a submodule of E generated by a maximal pseudofree family (VII, p. 55, Ex. 4); then F is a pure, free submodule of E. Show that, for all $\dot{x} \in E/F$, there exist a noninvertible element α of A and an element $\dot{y} \in E/F$ such that $\dot{x} = \alpha\dot{y}$, but that E/F is not necessarily a divisible module (cf. Ex. 3).

5) Let A be a principal ideal domain having a unique maximal ideal $A\pi$ (cf. VII, p. 48, Ex. 4). Let E be a torsion-free A-module, and M a free submodule of E, such that E/M is torsion-free, divisible and of rank 1. Let (e_α) be a basis of M, let $\dot{u}$ be a nonzero element of E/M, and u an element of the coset $\dot{u}$; for each integer $n > 0$, let u_n be an element of E such that $u \equiv \pi^n u_n$ (mod M). Put $u = \pi^n u_n + \sum_\alpha \lambda_{n\alpha} e_\alpha$.

a) Show that M is generated by M and the u_n, and that $\lambda_{n\alpha} - \lambda_{n+1,\alpha} \in A\pi^n$ for all α and for every integer $n > 0$ (cf. Ex. 4, *a*)).
b) Suppose in addition that A is *complete* with respect to the topology $\mathcal{T}$ defined by taking the ideals $A\pi^n$ as a basis of open neighbourhoods of 0 in A (Gen. Top., III, p. 5); put $\lambda_\alpha = \lim_{n\to\infty} \lambda_{n\alpha}$. Show that, if μ and μ_α are such that $\mu u - \sum_\alpha \mu_\alpha e_\alpha$ belongs to all the submodules π^nE, then necessarily $\mu_\alpha = \lambda_\alpha \mu$ for all α.

¶ 6) Let A be a principal ideal domain with a unique maximal ideal $A\pi$, and complete with respect to the topology $\mathcal{T}$ defined in Ex. 5.
a) Show that any torsion-free A-module E of *finite* rank, containing no nonzero divisible submodules, is free (consider a submodule M of E generated by a maximal pseudofree family; if M $\neq$ E show, with the help of Ex. 4, *b*) and 5, *b*)), that E contains a nonzero divisible submodule). Generalise this result to the case where E has countable rank (proceed by induction).
b) In the free module $E = A^{(\mathbf{N})}$ (the direct sum of countably infinitely many copies of A), let $(a_n)_{n\geqslant 0}$ be the canonical basis, and let $e_n = a_{n-1} - \pi a_n$ for each integer $n \geqslant 1$. Show that the submodule M of E generated by the e_n is pure, and that the e_n form a maximal pseudofree family, but that E/M is a divisible module, and hence M does not admit a complement in E.

7) Let p be a prime number and let A denote the ring of rational numbers of the form r/s, where s is coprime to p (VII, p. 48, Ex. 4).
Let $u = (1,0)$ and $v = (0,1)$ be the elements of the canonical basis of the vector space $E = \mathbf{Q}^2$ over $\mathbf{Q}$. Recursively define a sequence of elements u_n of E by $u_0 = u$ and

$u_n = pu_{n+1} + v$ for $n \geq 0$. Let F be the A-submodule generated by v and the u_n ; show that F is not a free module, but contains no nonzero divisible submodule (cf. Ex. 6, *a*)).

¶ 8) Let A be a principal ideal domain which is not a field.
a) Show that the product A-module A^N admits no countable system of generators. (If A is countable, consider Card(A^N) ; otherwise notice that, if for all $x \in A$ we let v_x denote the element $(x^n)_{n \geq 0}$ of A^N, then the v_x form a free system).
b) Let π be an irreducible element of A. Let S denote the submodule of A^N consisting of those sequences $z = (z_n)_{n \in N}$ for which there exists a sequence $(k(n))_{n \in N}$ of integers tending to $+\infty$ (depending on z), such that $z_n \in A\pi^{k(n)}$ for all n. Show that the (A/π)-vector space $S/\pi S$ has a countable basis.
c) Deduce from *a*) and *b*) that A^N *is not a free* A-*module*. (Otherwise, S would be a free A-module with a countable basis ; but $(x_n) \mapsto (x_n \pi^n)$ is an injective A-linear map from A^N into S.)

9) Let A be a principal ideal domain and let M be the dual of the A-module A^N. Let $f : M \to A^N$ be the transpose map of the canonical injection $A^{(N)} \to A^N$ (where A^N is identified with the dual of $A^{(N)}$).
a) Let π be an irreducible element of A and let $\varphi : A^N/A^{(N)} \to A$ be a linear form ; show that if $u \in A^N/A^{(N)}$ is the class of an element $(u_n) \in A^N$ such that $u_n \in (\pi^n)$ for all n, then $\varphi(u) = 0$. If A contains two non-associate irreducible elements, deduce that f is injective. (Using the above and Bezout's identity, show that the dual of $A^N/A^{(N)}$ is zero.)
b) Give A the topology defined in Ex. 5 of VII, p. 60. Show that, if $f(M)$ is not contained in $A^{(N)}$, then A is complete. (Reduce the problem to showing that, if $\varphi : A^N \to A$ is a linear form such that $\varphi(e_n) \neq 0$ for every element e_n of the canonical basis of $A^{(N)}$, then $\mathrm{Im}(\varphi)$ contains the sum of any series of elements of A which tends to 0 sufficiently quickly.)
c) If A contains two non-associate irreducible elements, then the canonical map from $A^{(N)}$ (resp. A^N) into its double dual is bijective (use *a*) and *b*)).

10) Let M be a nonzero **Z**-submodule of the ring of p-adic integers $\mathbf{Z}_p$ such that M is a pure **Z**-submodule of the **Z**-module $\mathbf{Z}_p$ (VII, p. 55, Ex. 7). Show that $M + p\mathbf{Z}_p = \mathbf{Z}_p$; deduce that M is indecomposable (observe that $\mathbf{Z}_p/p\mathbf{Z}_p$ is isomorphic to $\mathbf{Z}/p\mathbf{Z}$). Give an example of a decomposable **Z**-submodule of $\mathbf{Z}_p$ (observe that $\mathbf{Z}_p$ has the cardinality of the continuum).

¶ 11) Let A and B be two infinite sets of rational prime numbers, with no common elements, and not containing the number 5. Let $(e_i)_{1 \leq i \leq 4}$ be the canonical basis of the **Q**-vector space $\mathbf{Q}^4$.
a) Let M be the **Z**-submodule of $\mathbf{Q}^4$ generated by the elements e_1/m for $m \in A$; let N be the **Z**-submodule of $\mathbf{Q}^4$ generated by the elements e_2/m with $m \in A$, e_3/n with $n \in B$, and $(e_2 + e_3)/5$. Then $M \cap N = 0$. Put $e_1' = 8e_1 + 3e_2$ and $e_2' = 5e_1 + 2e_2$. Let M′ be the **Z**-submodule of $\mathbf{Q}^4$ generated by the elements e_1'/m with $m \in A$; let N′ be the **Z**-submodule of $\mathbf{Q}^4$ generated by the elements e_2'/m with $m \in A$, e_4/n with $n \in B$, and $(3e_2' + e_3)/5$. Show that $M' \cap N' = 0$, that $M \oplus N = M' \oplus N'$; that M and M′ are

isomorphic and indecomposable ; and that N and N′ are indecomposable but not isomorphic (compare VII, p. 69, Ex. 23).

b) Let M be the **Z**-submodule of $\mathbf{Q}^4$ generated by the elements e_1/m with $m \in \mathrm{A}$, e_2/n with $n \in \mathrm{B}$, and $(e_1 + e_2)/5$; let N be the **Z**-submodule of $\mathbf{Q}^4$ generated by the elements e_3/m with $m \in \mathrm{A}$, e_4/n with $n \in \mathrm{B}$, and $(e_3 + e_4)/5$. Put $e_1' = 2e_1 + e_3$, $e_2' = e_2 + 3e_4$, $e_3' = 17e_1 + 9e_3$, and $e_4' = e_2 + 2e_4$. Let M′ be the **Z**-submodule of $\mathbf{Q}^4$ generated by the elements e_1'/m for $m \in \mathrm{A}$, e_2'/n for $n \in \mathrm{B}$, and $(e_1' + 2e_2')/5$; let N′ be the **Z**-submodule of $\mathbf{Q}^4$ generated by the elements e_3'/m for $m \in \mathrm{A}$, e_4'/n for $n \in \mathrm{B}$, and $(e_3' + 2e_4')/5$. Show that $\mathrm{M} \oplus \mathrm{N} = \mathrm{M}' \oplus \mathrm{N}'$, that M is isomorphic to N and M′ is isomorphic to N′, but M′ is not isomorphic to M.

c) Let M be the **Z**-submodule of $\mathbf{Q}^4$ generated by $e_1/5^n$ $(n \geqslant 0)$; let N be the submodule of $\mathbf{Q}^4$ generated by $e_2/5^n$, $e_3/2^n$, $e_4/3^n$ $(n \geqslant 0)$, $(e_2 + e_3)/3$ and $(e_2 + e_4)/2$. Put $e_1' = 3e_1 - e_2$ and $e_2' = 2e_1 - e_2$. Let M′ be the **Z**-submodule of $\mathbf{Q}^4$ generated by the elements $e_1'/5^n$, $e_3/2^n$ $(n \geqslant 0)$ and $(e_1' - e_3)/8$; let N′ be the **Z**-submodule of $\mathbf{Q}^4$ generated by the elements $e_2'/5^n$, $e_4/3^n$ $(n \geqslant 0)$ and $(e_2' - e_4)/2$. Show that $\mathrm{M} \oplus \mathrm{N} = \mathrm{M}' \oplus \mathrm{N}'$, that M′ and N′ are indecomposable of rank 2, and that M has rank 1.

§ 4

1) Let $U = (\alpha_{ij})$ be a matrix with m rows and n columns over the principal ideal domain A.

a) Suppose first of all that the α_{ij} are setwise coprime. Show that there exist two invertible square matrices P and Q with entries in A, such that one of the entries of the matrix PUQ is equal to 1. (For each $\alpha_{ij} \neq 0$, let $\mathrm{S}(\alpha_{ij})$ be the sum of the exponents in a decomposition of α_{ij} into irreducible elements, and let $s(U)$ be the least of the numbers $s(\alpha_{ij})$; if $s(U) > 0$, show that there exist two invertible square matrices R and S such that

$$s(RUS) < s(U) ;$$

use Bezout's identity, and restrict the matrices R and S considered, after permuting the rows and columns if necessary, to those of the form $\begin{pmatrix} T & 0 \\ 0 & I \end{pmatrix}$, where T is a 2×2 matrix and I an identity matrix, or to products of such matrices and permutation matrices.)

b) If δ_1 is a gcd of the entries of U, show that there exist two invertible matrices P_1, Q_1, such that

$$P_1 U Q_1 = \begin{pmatrix} \delta_1 & 0 \\ 0 & U_1 \end{pmatrix}$$

where all the entries of U_1 are divisible by δ_1 (use *a*)).

c) For $\mathrm{A} = \mathbf{Z}$, use *b*) to obtain a method of calculating the invariant factors of an explicit matrix with entries in **Z**. Apply this method to the matrix

$$\begin{pmatrix} 6 & 8 & 4 & 20 \\ 12 & 12 & 18 & 30 \\ 18 & 4 & 4 & 10 \end{pmatrix}.$$

2) Let A be a principal ideal domain. Then two submodules M and N of A^q are transforms of one another under automorphisms of A^q if and only if they have the same invariant factors with respect to A^q.

3) Let A be a principal ideal domain with field of fractions K, let E be a vector space over K, let M be a finitely generated A-module of rank n contained in E, and let N be a finitely generated A-module of rank $p \leqslant n$ contained in KM. Show that there exists a basis $(e_i)_{1 \leqslant i \leqslant n}$ of M and a basis of N consisting of vectors $\alpha_i e_i$ $(1 \leqslant i \leqslant p)$, where $\alpha_i \in K$ and α_i divides α_{i+1} (with respect to the ring A) for $1 \leqslant i \leqslant p-1$; show also that the fractional ideals $A\alpha_i$ are uniquely determined (*invariant factors of* N *with respect to* M).

4) Let G be a finite abelian group.

a) For every prime number p, show that the number of elements of order p in G is $p^N - 1$, where $N = \sum_{n=1}^{\infty} m(p^n)$ (in the notation of Prop. 9 of VII, p. 24).

b) Deduce that if the equation $x^p = 1$ has at most p solutions in G, for each prime number p, then G is cyclic.

5) Let G be a finite abelian group of order n. Show that for every integer q dividing n there exists a subgroup of order q in G (decompose G as a direct sum of indecomposable cyclic subgroups).

¶ 6) Let A be a commutative ring and let E be an A-module. Let $\mathscr{E}$ denote the ring $\mathrm{End}_Z(E)$ of endomorphisms of the abelian group E (without operators) ; for every subring $B \subset \mathscr{E}$, let B′ denote the centraliser of B in $\mathscr{E}$ (the subring of $\mathscr{E}$ consisting of elements which commute with every element of B).

a) Let A_0 be the subring of $\mathscr{E}$ consisting of the homotheties $x \mapsto \lambda x$ $(\lambda \in A)$; if $\mathfrak{a}$ is the annihilator of E, then A_0 is known to be isomorphic to $A/\mathfrak{a}$, and its centraliser A_0' in $\mathscr{E}$ is the ring $\mathrm{End}_A(E)$ of endomorphisms of E as an A-module. Show that the centraliser A_0'' of A_0' is commutative, and that its centraliser A_0''' is equal to A_0' (notice that $A_0 \subset A_0'$).

b) If the submodule F is a direct factor of E, show that $u(F) \subset F$ for all $u \in A_0''$ (consider the projection of E onto F obtained from the decomposition of E as a direct sum of F and another submodule).

c) Suppose that E is the direct sum of a family (F_ι) of cyclic submodules. If $u \in A_0''$ show that, for each index ι, there exists an element $\alpha_\iota \in A$ such that $u(x) = \alpha_\iota x$ for all $x \in F_\iota$ (use *b*)).

d) Suppose E is the direct sum of a sequence (E_n) (finite or infinite) of cyclic submodules such that, if $\mathfrak{b}_n$ denotes the annihilator of E_n, then the sequence $(\mathfrak{b}_n)$ is increasing. Show that $A_0'' = A_0$ (use *c*), and the fact that for each index $i > 1$ there exists an endomorphism $v_i \in A_0'$ which maps E_{i-1} into E_i).

Apply this to the case where A is a principal ideal domain and E is a finitely generated A-module.

7) Let A be a principal ideal domain with field of fractions K, and let E be a finitely generated, torsion-free A-module of rank n. Let E* be the dual of E ; then E and E* are each isomorphic to A^n.

a) Let M be a submodule of E ; show that, if M^0 is the submodule of E^* orthogonal to M, then the module E^*/M^0 is torsion-free, and consequently M^0 is a direct factor in E^* ; the submodule M^{00} of E orthogonal to M^0 is equal to $E \cap KM$ (considering E as being canonically embedded in an n-dimensional vector space over K). We can consider E^*/M^0 as being canonically embedded in the dual M^* of M ; show that the invariant factors of E^*/M^0 with respect to M^* are equal to the invariant factors of M with respect to E (use Th. 1 of VII, p. 18).
b) Let F be a second finitely generated, torsion-free A-module, and let u be a linear transformation from E into F. Show that the invariant factors of ${}^t u(F^*)$ with respect to E^* are equal to those of $u(E)$ with respect to F (proceed as in Prop. 5 of VII, p. 21).

8) Let G be a module of finite length over a principal ideal domain A. For any pair of free modules M and N such that $N \subset M$ and G is isomorphic to M/N, the invariant factors of N with respect to M which are distinct from A are identical to the invariant factors of G (VII, p. 20, Def. 2) ; their number is called the *rank* of G.
a) Show that the rank r of G is the least number of cyclic submodules of G of which G is the direct sum, and is equal to the greatest of the ranks of the π-primary components of G.
b) Show that the rank of a submodule or a quotient module of G is at most equal to the rank of G (consider G as a quotient of two free modules of rank r).
c) For each $\lambda \in A$, show that the rank of the submodule λG is equal to the number of invariant factors of G which do not divide λ (notice that, if $E = A/(A\alpha)$, then the module λE is isomorphic to $(A\lambda)/((A\alpha) \cap (A\lambda))$. Deduce that the k-th invariant factor of G (in *decreasing* order) is a gcd of those $\lambda \in A$ such that λG has rank $\leqslant r - k$.
d) Let $A\alpha_k$ $(1 \leqslant k \leqslant r)$ be the invariant factors of G arranged in decreasing order. Let H be a submodule of G of rank $r - q$, and let $A\beta_k$ $(1 \leqslant k \leqslant r - q)$ be its invariant factors, arranged in decreasing order. Deduce from *b*) and *c*) that β_k divides α_{k+q} for $1 \leqslant k \leqslant r - q$. Similarly show that, if G/H has rank $r - p$ and invariant factors $A\gamma_k$ $(1 \leqslant k \leqslant r - p)$ in decreasing order, then γ_k divides α_{k+p} for $1 \leqslant k \leqslant r - p$.
e) Conversely, let L be a torsion module of rank $r - q$, whose invariant factors $A\lambda_k$ $(1 \leqslant k \leqslant r - q)$ are such that λ_k divides α_{k+q} for $1 \leqslant k \leqslant r - q$. Show that there exist submodules M and N of G such that L is isomorphic to M and to G/N (decompose G into a direct sum of submodules isomorphic to $A/A\alpha_k$).

¶ 9) Let A be a principal ideal domain and K its field of fractions ; let E be a vector space over K, let M be a finitely generated A-module of rank n contained in E, let N be a finitely generated A-module of rank p contained in KM, and let P be a finitely generated A-module of rank q contained in KN.
a) Let $A\alpha_i$ $(1 \leqslant i \leqslant p)$ be the invariant factors of N with respect to M, arranged in decreasing order (Ex. 3). Show that, for $1 \leqslant k \leqslant p$, the element α_k is a gcd of those elements $\lambda \in K$ such that the rank of the quotient module $(N + \lambda(KN \cap M))/N$ is $\leqslant p - k$ (cf. Ex. 8, *d*)).
b) Let $A\beta_j$ $(1 \leqslant j \leqslant q)$ be the invariant factors of P with respect to M and $A\gamma_j$ $(1 \leqslant j \leqslant q)$ the invariant factors of P with respect to N, arranged in decreasing order. Show that $\alpha_1\gamma_j$ divides β_j for $1 \leqslant j \leqslant q$ (cf. Ex. 8).
c) Show that $\gamma_1\alpha_j$ divides β_j for $1 \leqslant j \leqslant q$. (First consider the case $n = p = q$, and apply Ex. 8. In general, show that we may always assume $p = n$, and consider a submodule Q of M such that $P + \rho Q$ has rank p for all $\rho \in A$; then choose the ideal $A\rho$ sufficiently small, and apply Prop. 4 of VII, p. 20.)

d) For any submodule H of N of rank $k \leqslant p$, let (μ_H) be the fractional ideal consisting of $\mu \in K$ such that $\mu(M \cap KH) \subset H$. Deduce from *c*) that for each $1 \leqslant k \leqslant p$ the element α_k is a gcd of the μ_H as H runs through all submodules of N of rank k.

¶ 10) Let A be a principal ideal domain and let M be a submodule of rank n in $E = A^n$; let $A\alpha_i$ $(1 \leqslant i \leqslant n)$ be the invariant factors of M with respect to E, arranged in decreasing order. Let N be a submodule of M admitting a complement in M (that is, such that $N = M \cap KN$, where K is the field of fractions of A).

a) Let P be a complement of N in M. Show that the submodule $(P + (E \cap KN))/M$ of E/M is isomorphic to $(E \cap KN)/N$. Deduce that, if N has rank p and $A\beta_k$ $(1 \leqslant k \leqslant p)$ are the invariant factors of N with respect to E, then β_k divides α_{k+n-p} for $1 \leqslant k \leqslant p$ (use Ex. 8, *d*)).

b) Give an example to show that the module E/M is not necessarily isomorphic to the product of the modules $(E \cap KN)/N$ and $E/(P + (E \cap KN))$ (take E/M to be indecomposable).

¶ 11) Let A be a principal ideal domain, let H be a submodule of $E = A^n$, and let H^0 be the submodule of the dual E^* of E orthogonal to H (Ex. 7). Let M be a submodule of rank p of E and let $A\alpha_i$ $(1 \leqslant i \leqslant p)$ be the invariant factors of M with respect to E, arranged in decreasing order. Let (ν_H) be the gcd of the ideals $f(M)$ of A as f runs through H^0 ; show that, for $1 \leqslant k \leqslant p$, the element α_k is a gcd of the ν_H for all submodules H of E of rank $k - 1$ which admit a complement in E (first consider the case $k = 1$; apply the result to the submodule $(M + H)/H$ of E/H, and use Ex. 8, *d*), applied to the quotient module $E/(M + H)$ of E/M).

12) Let G be a module of finite length over a ring A. Show that a submodule H of G can only be isomorphic to G if it is equal to G (cf. II, p. 212, Prop. 16).

13) Let A be a principal ideal domain with field of fractions K.

a) Show that, for every A-module M, the canonical homomorphism $c_M : M \to D(D(M))$ is injective (cf. II, p. 389, Ex. 13).

b) If $u : M \to N$ is a homomorphism of A-modules, define natural isomorphisms from $D(\mathrm{Im}(u))$ onto $\mathrm{Im}(D(u))$, from $D(\mathrm{Ker}(u))$ onto $\mathrm{Coker}(D(u))$, and from $D(\mathrm{Coker}(u))$ onto $\mathrm{Ker}(D(u))$.

c) Let M be an A-module and N a submodule of M. In the notation of Sect. 9, show that $N^{00} = N$ (use *a*) and *b*)). The module N^0 has finite length if and only if M/N has finite length. Conversely, for any submodule N′ of D(M) of finite length, the module N'^0 is a submodule of M such that M/N^0 has finite length, and $N' = N'^{00}$.

d) If M is a torsion A-module, and M_π denotes the π-primary component of M, show that D(M) is isomorphic to $\prod_{\pi \in P} \mathrm{Hom}(M_\pi, U_\pi)$, where P is a system of representatives of irreducible elements of A and U_π is the π-primary component of K/A for each $\pi \in P$ (VII, p. 54 Ex. 3).

e) Deduce from *d*) that D(K/A) is isomorphic to the product $\prod_{\pi \in P} \hat{A}_{(\pi)}$, where $\hat{A}_{(\pi)}$ denotes the completion of the ring $A_{(\pi)}$ defined in VII, p. 54, Ex. 2, with respect to the topology defined by taking the ideals $\pi^n A_{(\pi)}$ to be a basis of open neighbourhoods of 0 (this completion can also be identified with the inverse limit $\varprojlim (A/A\pi^n)$).

f) If M is an A-module of finite length, and $\mathfrak{a}$ is its annihilator, show that D(M) is isomorphic to the *dual* of the faithful $(A/\mathfrak{a})$-module associated to M.

g) Let M be an A-module of finite length. Show that if N is a submodule of M, then there exists a submodule P of M such that M/N is isomorphic to P and M/P is isomorphic to N (use VII, p. 26, Prop. 10 and VII, p. 27, Prop. 12) ; two submodules of M with these properties are called *reciprocal*. Show that, for all $\alpha \in A$, the submodule $M(\alpha)$ of M consisting of all $x \in M$ such that $\alpha x = 0$, and the submodule αM, are reciprocal.

¶ 14) Let M be a module of finite length over a principal ideal domain A with field of fractions K.
a) Show that if N is a submodule of M then the rank of M is at most equal to the sum of the ranks of N and M/N (let $M = E/H$, where E is a free module of rank n and H is a submodule of rank n of E ; if $N = L/H$, where $H \subset L \subset E$, notice that, if the rank of N is p, then there exists a submodule R of H, of rank $n - p$, admitting a complement S in L such that $S \cap H$ is a complement of R in H ; moreover, $(E \cap KR)/R$ has rank $\geqslant n - p$ and is isomorphic to a quotient module of E/L (Ex. 10, *a*)) ; complete the proof using Ex. 8, *b*)).
b) Let N be a submodule of M of rank p, and let q be the rank of M/N. Let $A\alpha_i$ $(1 \leqslant i \leqslant n)$ be the invariant factors of M and $A\beta_k$ $(1 \leqslant k \leqslant p)$ those of N, arranged in decreasing order. Show that β_k is a multiple of $\alpha_{k-(p+q-n)}$ (use *a*) and Ex. 8, *c*), noticing that $(\lambda M)/(\lambda N)$ is isomorphic to a quotient module of M/N).

¶ 15) *a*) Let M be a module of finite length and rank n over a principal ideal domain A, and let $A\alpha_i$ $(1 \leqslant i \leqslant n)$ be its invariant factors in decreasing order. Let (β_i) $(1 \leqslant i \leqslant n)$ be a sequence of elements of A such that β_i divides α_i for $1 \leqslant i \leqslant n$ and α_i divides β_{i+1} for $1 \leqslant i \leqslant n-1$. Let $(M_i)_{1 \leqslant i \leqslant n}$ be a sequence of submodules of M such that M is the direct sum of the M_i and each M_i is isomorphic to $A/A\alpha_i$; let a_i be a generator of M_i for $1 \leqslant i \leqslant n$. Put

$$b = \beta_1 a_1 + \frac{\beta_1\beta_2}{\alpha_1} a_2 + \frac{\beta_1\beta_2\beta_3}{\alpha_1\alpha_2} a_3 + \cdots + \frac{\beta_1\beta_2 \dots \beta_n}{\alpha_1\alpha_2 \dots \alpha_{n-1}} a_n .$$

Show that the quotient of M by the cyclic submodule Ab is isomorphic to the direct sum of the n modules $A/A\beta_i$ $(1 \leqslant i \leqslant n)$.
b) Let M and N be two modules of finite length over A ; let $A\alpha_i$ $(1 \leqslant i \leqslant n)$ and $A\beta_k$ $(1 \leqslant k \leqslant p)$ be the invariant factors of M and N respectively, arranged in decreasing order ; suppose that $p \leqslant n$, that β_k divides α_{k+n-p} for $1 \leqslant k \leqslant p$, and that α_k divides $\beta_{k+(p+q-n)}$ for $1 \leqslant k \leqslant n-q$, where q is an integer such that

$$q \leqslant n \leqslant p + q .$$

Show that there exists a submodule P of M, isomorphic to N, such that M/P has rank $\leqslant q$ (decompose each of the two modules M and N as a direct sum of q modules, in such a way as to reduce the problem to the case $q = 1$; then use *a*) and Ex. 13, *g*)).
c) For a module N over A to be isomorphic to a direct factor of a module M of finite length, it is necessary and sufficient that every elementary divisor of N be an elementary divisor of M, and that its multiplicity in N be at most equal to its multiplicity in M. Hence obtain an example of a module M of finite length and a submodule N of M such that the

rank of M is equal to the sum of the ranks of N and M/N, and such that N does not admit a complement in M (use *a*)).

16) Let M be a module over a commutative ring A. A submodule N of M is called *characteristic* if $u(\mathrm{N}) \subset \mathrm{N}$ for every endomorphism u of M. For all $\alpha \in \mathrm{A}$, the submodule $\mathrm{M}(\alpha)$ of elements annihilated by α, and the submodule $\alpha\mathrm{M}$, are characteristic.
a) If N is a characteristic submodule of M, and P and Q are two complementary submodules of M, show that N is the direct sum of $\mathrm{N} \cap \mathrm{P}$ and $\mathrm{N} \cap \mathrm{Q}$ (consider the projections from M onto P and Q).
b) Let M be a module of finite length over a principal ideal domain A; let $\mathrm{A}\alpha_i$ be the invariant factors of M, arranged in decreasing order $(1 \leqslant i \leqslant n)$; let $(\mathrm{M}_i)_{1 \leqslant i \leqslant n}$ be a sequence of submodules of M such that M is the direct sum of the M_i and each M_i is isomorphic to $\mathrm{A}/\mathrm{A}\alpha_i$. Let N be a characteristic submodule of M and let $\mathrm{A}\beta_i$ be the annihilator of $\mathrm{N} \cap \mathrm{M}_i$ $(1 \leqslant i \leqslant n)$; if $\alpha_i = \beta_i\gamma_i$, show that β_i divides β_{i+1} and γ_i divides γ_{i+1} for all $1 \leqslant i \leqslant n-1$ (notice that, for $1 \leqslant i \leqslant n-1$, there exists an endomorphism of M which maps M_{i+1} onto M_i, and an endomorphism of M which maps M_i onto a submodule of M_{i+1}).
c) Conversely, let N be a submodule of M which is a direct sum of submodules $\mathrm{N}_i \subset \mathrm{M}_i$; suppose the annihilators $\mathrm{A}\beta_i$ of the N_i $(1 \leqslant i \leqslant n)$ satisfy the above conditions. Then show that N is a characteristic submodule of M. Deduce that if two characteristic submodules of M have the same invariant factors then they are equal.

17) Let A be a principal ideal domain, let α be a nonzero element of A, and let E be the module $\mathrm{A}/\mathrm{A}\alpha$. Then the product E^n is isomorphic to $\mathrm{A}^n/(\alpha\mathrm{A}^n)$.
a) Let u be a linear map from E^n into E^m; show that u can be obtained from a linear map v from A^n into A^m on passing to quotients.
b) Let $\mathrm{A}\beta_i$ $(1 \leqslant i \leqslant p \leqslant \inf(m, n))$ be the invariant factors of $v(\mathrm{A}^n)$ with respect to A^m, arranged in decreasing order; let $q \leqslant p$ be the greatest of the indices i such that α does not divide β_i, and let δ_i be a gcd of α and β_i for $1 \leqslant i \leqslant q$. Show that the kernel $u^{-1}(0)$ of u is the direct sum of $n-q$ modules isomorphic to E and q cyclic modules isomorphic to $\mathrm{A}/\mathrm{A}\delta_i$ respectively $(1 \leqslant i \leqslant q)$. If $\alpha = \gamma_i\delta_i$ $(1 \leqslant i \leqslant q)$, show that the module $u(\mathrm{E}^n)$ has invariant factors $\mathrm{A}\gamma_i$.

18) Let C be a principal ideal domain. Consider a system of linear equations

$$\sum_{j=1}^{n} \alpha_{ij}\xi_j = \beta_i \quad (1 \leqslant i \leqslant m)$$

where the coefficients and the right hand sides belong to C, and let A denote the matrix (α_{ij}) with m rows and n columns, and B the matrix obtained by extending A by the $(n+1)$-st column (β_i). Show that the system admits at least one solution in C^n if and only if the following conditions are satisfied: 1) A and B have the same rank p; 2) a gcd of the minors of order p of A is equal to a gcd of the minors of order p of B (use Prop. 4 of VII, p. 20 and Ex. 9, *c*)).

19) Let C be a principal ideal domain. Consider a system of m « linear congruences » $\sum_{j=1}^{n} \alpha_{ij}\xi_j \equiv \beta_i \pmod{\alpha}$, where the coefficients, the right hand sides, and the element α all

belong to C, and where α is nonzero ; let A denote the matrix (α_{ij}), let B denote the matrix obtained by extending A by the column (β_i), and let δ_k (resp. δ'_k) be a gcd of the k-th order minors of A (resp. B). Then the system admits at least one solution in C^n if and only if :
a) If $m \leqslant n$, a gcd of α^m, $\alpha^{m-1}\delta_1$, ..., $\alpha\delta_{m-1}$, δ_m is equal to a gcd of α^m, $\alpha^{m-1}\delta'_1$, ..., $\alpha\delta'_{m-1}$, δ'_m ;
b) if $m > n$, a gcd of α^{n+1}, $\alpha^n\delta_1$, ..., $\alpha\delta_n$ is equal to a gcd of α^{n+1}, $\alpha^n\delta'_1$, ..., δ'_{n+1}.

(Reduce the system of linear congruences to a system of linear equations, and use Ex. 18.)

20) *a*) Let C be a principal ideal domain, and let

$$f(x_1, .., x_p) = \sum_{(i_k)} \alpha_{i_1 \dots i_p} \xi_{1,i_1} \cdots \xi_{p,i_p}$$

be a p-linear form on E^p, where E is the C-module C^n (and where ξ_{ij} denotes the j-th coordinate of x_i). Show that if δ is a gcd of the coefficients $\alpha_{i_1 \dots i_p}$ then there exists an element $(a_k) \in E^p$ such that $f(a_1, ..., a_p) = \delta$ (reduce to the case $\delta = 1$, and argue by induction on p, using Th. 1 of VII, p. 18).
b) Let $n > 1$ be an integer, and α_i $(1 \leqslant i \leqslant n)$ be n elements of C with gcd δ. Show that there exists a square matrix A of order n over C whose first column is (α_i) and whose determinant is δ (use *a*)).

21) Let C be a principal ideal domain, and let $A = (\alpha_{ij})$ be a square matrix of order n over C.
a) Show that there exists a square matrix U of order n over C, with determinant 1, such that the last column of the matrix UA has its first $(n-1)$ entries zero, and its last entry equal to the gcd of the entries in the n-th column of A (use Ex. 20, *b*)).
b) Deduce from *a*) that there exists a square matrix V of order n over C, with determinant 1, such that the matrix $VA = (\beta_{ij})$ is lower triangular (« Hermite reduced form of A »).
c) If W is a square matrix of order n over C, with determinant 1, such that $WA = (\gamma_{ij})$ is lower triangular, show that β_{ii} and γ_{ii} are associate for $1 \leqslant i \leqslant n$.

¶ 22) A module M over a principal ideal domain is said to have *finite rank* if there exists an integer $m \geqslant 0$ with the following property : every finitely generated submodule of M is contained in a submodule generated by m elements or fewer. The smallest integer m with this property is called the *rank* of M.
a) Show that if M has finite rank, then every quotient module has finite rank, at most equal to that of M.
b) For a torsion-free module M, the notion of rank defined above coincides with that defined in II, p. 315.
c) Show that if M has finite rank then every submodule of M has finite rank, at most equal to that of M.
d) Let M be a π-primary module such that $M(\pi) = M$ (in the notation of VII, p. 7). Show that if M has finite rank then it is finitely generated.
e) Let M be a π-primary module with no element of infinite height. Show that if M has finite rank then it is finitely generated, and the notion of rank coincides with that defined in

Ex. 8. (Notice that if $M(\pi)$ is finitely generated then so is M; by proving that the heights of elements of $M(\pi)$ in M are bounded ; use Ex. 5 of VII, p. 55 to show this.)
f) Let M be a π-primary module of finite rank r, and let M_0 denote the submodule consisting of all the elements of infinite height in M. Show that M_0 is a divisible module, and the direct sum of a finite number $p \leqslant r$ of indecomposable submodules U_π (VII, p. 54, Ex. 3), and that M is the direct sum of M_0 and a finitely generated submodule N of rank $r - p$. (Notice that M/M_0 is finitely generated, using *e*) ; then apply Ex. 3 of VII, p. 54.)
g) Deduce from the above that an A-module M has finite rank r if and only if the quotient module M/T of M by its torsion submodule T has finite rank $p \leqslant r$, and that each π-primary component of T has rank $\leqslant r - p$, with at least one of them having rank $r - p$.
h) Let (p_n) be the sequence of distinct prime numbers ; let E be the direct sum of the cyclic **Z**-modules $\mathbf{Z}/p_n^2\mathbf{Z}$, and let e_n denote the element whose coordinates all vanish except the n-th, which is equal to the residue class of 1 mod p_n^2. Let M be the submodule of the **Z**-module $E \times \mathbf{Q}$ generated by the elements $(e_n, 1/p_n)$. Show that M is a **Z**-module of rank 2, that the torsion submodule T of M is generated by the elements $(p_n e_n, 0)$, and that T is not a direct factor of M. (If $\bar{x}_n$ is the class of $(e_n, 1/p_n)$ mod T, show that for all $m \neq n$ there exists a class $\bar{y}_{nm}$ mod T such that $\bar{x}_n = p_m \bar{y}_{nm}$, but that there is no element x_n in the class $\bar{x}_n$ such that for all $m \neq n$ there exists $y_{nm} \in M$ with $x_n = p_m y_{nm}$.)
i) If M is a finitely generated A-module, show that the rank of M is equal to the smallest cardinal $\gamma(M)$ of any generating set for M.
j) Let M be a torsion A-module of finite rank. Show that if N is a proper submodule of M then N cannot be isomorphic to M (use *f*)).

¶ 23) Let A be a principal ideal domain and let M be a finitely generated A-module. Show that if N and P are two A-modules such that $M \oplus N$ and $M \oplus P$ are isomorphic, then N and P are isomorphic. (Reduce to the case where M is cyclic ; by identifying $M \oplus N$ with $M \oplus P$, we may consider an A-module E decomposed in two ways as a direct sum of submodules $F \oplus N$ and $G \oplus P$, where F and G are cyclic and isomorphic. In the case where F and G are isomorphic to A, show that if $N \neq P$ then N is the direct sum of $N \cap P$ and a submodule isomorphic to A. If F and G are isomorphic to $A/A\pi^n$, where π is an irreducible element of A, consider two generators a, b of F, G respectively, and examine the cases $\pi^{n-1}a \notin G$, $\pi^{n-1}b \notin F$, and finally the case where both $\pi^{n-1}a \in G$ and $\pi^{n-1}b \in F$; in this last case, consider the element $a + b$ of E and the submodule it generates.)

¶ 24) An abelian group G is isomorphic to a subgroup of the multiplicative group K^* of nonzero elements of some commutative field K if and only if the torsion subgroup T of G has rank $\leqslant 1$ (Ex. 22). (For the necessity of this condition, cf. V, p. 79, Prop. 2 ; use the same reference to show that the condition is sufficient for T to be isomorphic to a subgroup of E^*, for some field E. For all $\alpha \in G/T$ let g_α be a coset representative of α in G, then, identifying T with a subgroup of E^*, we may write $g_\alpha g_\beta = e_{\alpha\beta} g_{(\alpha\beta)}$ for all α, $\beta \in G/T$, where $e_{\alpha\beta} \in E$. Show that there exists an E-algebra B with basis $(b_\alpha)_{\alpha \in G/T}$ such that $b_\alpha b_\beta = e_{\alpha\beta} b_{(\alpha\beta)}$ for all α, β, and prove that B is an integrai domain using VI, p. 33, Ex. 20, *c*). Then the field of fractions K of B satisfies the desired conditions.)

25) Generalise Th. 1 of VII, p. 18 to left modules over a ring A which is not necessarily commutative, in which every left ideal and every right ideal is principal (an example of such

a ring is the tensor product $K[X] \otimes_K D$, where K is a commutative field and D is a K-algebra which is a field but not necessarily commutative).

§ 5

All fields in this section are commutative unless explicitly stated otherwise.

1) Let u be an endomorphism of a finite dimensional vector space E, and let V be a subspace of E invariant under u. Show that the characteristic polynomial of the restriction of u to V divides that of u. Hence obtain an elementary proof of the Cayley-Hamilton theorem (reduce to the case where E_u is cyclic, and use formula (2) of VII, p. 30).

2) *a*) Show that every square matrix over a field K is similar to its transpose (reduce to the case of a Jordan matrix).
b) Show that over the ring **Z** of rational integers, the matrix $\begin{pmatrix} 8 & 2 \\ 0 & 1 \end{pmatrix}$ is not similar to its transpose.
c) Let u be an endomorphism of a finite dimensional vector space over K. Then every eigenvalue λ of u is also an eigenvalue of the transpose endomorphism ${}^t u$ of E^*. Moreover, if λ and μ are two distinct eigenvalues of u, if x is an eigenvector of u corresponding to λ, and if y^* is an eigenvector of ${}^t u$ corresponding to μ, then $\langle x, y^* \rangle = 0$.

3) Let u be an endomorphism of a finite dimensional vector space E, let λ be an eigenvalue of u, and let v be the endomorphism $u - \lambda$. Show that E is the direct sum of $v^{-1}(0)$ and $v(E)$ if and only if λ is a simple root of the minimal polynomial of u.

4) Show, by adjoining a transcendental element to the base field, that formula (6) in Prop. 10 of VII, p. 36 can be deduced from formula (8). Prove the latter directly by decomposing the polynomial q into linear factors.

¶ 5) *a*) Let K be an algebraically closed field, and let A be a square matrix of order n over K, in the form of a diagonal block $\operatorname{diag}(A_i)$ of lower triangular matrices A_i, such that the diagonal entries of A_i are all equal to the same element α_i of K ($1 \leqslant i \leqslant r$), and $\alpha_i \neq \alpha_j$ for $i \neq j$. Show that every square matrix B which commutes with A is a diagonal block $\operatorname{diag}(B_i)$, where B_i is the same size as A_i for $1 \leqslant i \leqslant r$.
b) Let M be a set of commuting square matrices of order n over K. Show that by applying the same similarity to all the matrices in M, we can make them all into diagonal blocks $\operatorname{diag}(X_i)$ of lower triangular matrices, where the number r of X_i's and the order m_i of X_i for each i, are the same for each matrix in M, and where all the eigenvalues of X_i are equal. (Use *a*) to reduce to the case where $r = 1$, then Lemma 3 of VII, p. 41 to show that in this case all the matrices in M have nonzero common eigenvector, and complete the proof by induction on n.)

6) Two pairs (A_1, A_2) and (B_1, B_2) of square matrices of order n over a commutative field K are said to be ***equivalent*** if there exist two invertible square matrices P and Q over K such that $A_1 = PB_1Q$ and $A_2 = PB_2Q$. If A_1 and B_1 are invertible, show that the pairs are equivalent if and only if the matrices $XA_1 + A_2$ and $XB_1 + B_2$ are equivalent over the ring $K[X]$.

7) Let A be a matrix with m rows and n columns, of rank r, over a field K. If P is a square matrix of order m and Q is a square matrix of order n over K, such that $PAQ = A$, show that there exists a matrix P' similar to P and a matrix Q' similar to Q such that

$$P' = \begin{pmatrix} R & S \\ 0 & U \end{pmatrix}, \qquad Q' = \begin{pmatrix} R^{-1} & 0 \\ S' & U' \end{pmatrix}$$

where R is an invertible square matrix of order r, U (resp. U') is a square matrix of order $m - r$ (resp. $n - r$), and S (resp. S') is a matrix with r rows and $m - r$ columns (resp. $n - r$ rows and m columns).

8) Let E be a vector space of finite dimension n over a field K, let u be an endomorphism of E, let V be a subspace of E invariant under u, and let $u \mid V$ be the restriction of u to V. If f_i $(1 \leqslant i \leqslant n)$ (resp. g_j $(1 \leqslant j \leqslant m)$) are the similarity invariants of u (resp. $u \mid V$), show that g_j divides f_{j+n-m} for $1 \leqslant j \leqslant m$ (cf. VII, p. 64, Ex. 8, *d*)). State and prove the analogous property of the endomorphism of E/V induced by u.

¶ 9) Let A be a square matrix of order n, let B be a square matrix or order m, and let C be a matrix with m rows and n columns over a field K. Reduce the equation $XA = BX + C$, where X is an unknown matrix with m rows and n columns over K, to a system of linear congruences in the polynomial ring K[Y] (for all $f \in K[Y]$, notice that $Xf(A) = f(B)X + H(f)$ for some well defined matrix $H(f)$ with m rows and n columns ; let $\bigoplus_i E_i$ be the decomposition of the module E associated to A (Sect. 1) as a direct sum of cyclic modules, whose annihilators are the similarity invariants f_i of A which are distinct from 1 (Prop. 2), and let a_i be generators of the E_i ; in the same way define the submodules F_j, generated by b_j, and the polynomials g_j corresponding to B ; write $Xf_i(A)\,a_i = 0$; then decompose Xa_i and $H(f_i)\,a_i$ in terms of the F_j, and show that the conditions thus obtained from the equations $Xf_i(A)\,a_i = 0$ are necessary and sufficient for the existence of a solution).

10) a) Show that, over an arbitrary field K, the matrix

$$\begin{pmatrix} \lambda & \alpha & \beta_{13} & \beta_{14} & \dots & \beta_{1n} \\ 0 & \lambda & \alpha & \beta_{24} & \dots & \beta_{2n} \\ 0 & 0 & \lambda & \alpha & \dots & \beta_{3n} \\ \dots & \dots & \dots & \dots & \dots & \dots \\ 0 & 0 & 0 & 0 & \dots & \alpha \\ 0 & 0 & 0 & 0 & \dots & \lambda \end{pmatrix}$$

is similar to the Jordan matrix $U_{n,\lambda}$ if $\alpha \neq 0$.

b) If $\lambda \neq 0$ and K is of characteristic 0, show that $(U_{n,\lambda})^m$ is similar to the Jordan matrix U_{n,λ^m} for every integer m (positive or negative).

c) Show that $(U_{n,0})^m = 0$ for $m \geqslant n$. For $0 < m < n$, let $n - 1 = km + q$ with $0 \leqslant q < m$; show that $(U_{n,0})^m$ has $q + 1$ similarity invariants equal to X^{k+1} and $m - q - 1$ similarity invariants equal to X^k (arrange the canonical basis of K^n in a suitable order).

d) Similarly, determine the similarity invariants of $f(U_{n,\lambda})$ for $f \in K[X]$.

e) Discuss the same problems for K of characteristic $p \neq 0$.

11) Let K be an algebraically closed field, let A be an invertible matrix over K, and $m > 0$ an integer. Show that, if m is not a multiple of the characteristic of K, then there exists a polynomial $g(X) \in K[X]$ such that $(g(A))^m = A$ (use the Cayley-Hamilton theorem and Ex. 28 of VII, p. 54). Generalise to the equation $f(U) = A$ ($f \in K[X]$, and U an unknown $n \times n$ matrix over K).

¶ 12) *a*) Let A and B be two square matrices of order n over a commutative field K, and let f_i and g_i $(1 \leq i \leq n)$ be their respective similarity invariants. Show that the vector space (over K) of square matrices U, of order n, such that $UA = BU$, is isomorphic to the space $M_0 = M/N$ defined as follows : M is the space of square matrices $(u_{ij}(X))$ of order n over $K[X]$ such that $u_{ij}(X)\, f_j(X)$ is a multiple of $g_i(X)$ for each pair i, j of indices ; and N is the subspace of M consisting of those matrices for which $u_{ij}(X)$ is a multiple of $g_i(X)$ for each pair i, j of indices. (Consider U, A and B as the matrices of endomorphisms u, v and w respectively of $E = K^n$, and notice that u is a $K[X]$-linear map from E_v into E_w ; now express E_v and E_w as quotient modules.)
b) When $B = A$, the matrices U such that $UA = AU$ form a ring P ; then M is a ring of square matrices, and N a 2-sided ideal of M, such that P is isomorphic to the quotient ring M/N. Show that P is identical to the ring $K[A]$ generated by I_n and A if and only if the minimal polynomial of A is equal to its characteristic polynomial.
c) For arbitrary A and B, show that the dimension of M_0 is equal to $\sum_{i,j} m_{ij}$, where m_{ij} is the degree of the gcd of f_i and g_j (same method as *a*)).
d) Show that the greatest of the ranks of the matrices $U \in M_0$ is equal to $\sum_k d_k$, where d_k is the degree of the gcd of f_k and g_k (use Ex. 8).
e) Generalise the results of *a*), *c*) and *d*) to the case where A is a square matrix of order n, B is a square matrix of order m, and U is a matrix with m rows and n columns.

¶ 13) Let (Z_{ij}) $(1 \leq i, j \leq n)$ be n^2 indeterminates over a field K_0. Let A be the polynomial ring $K_0[Z_{11}, \ldots, Z_{nn}]$, and let $K = K_0(Z_{ij})$ be its field of fractions ; let χ_U be the characteristic polynomial of the matrix $U = (Z_{ij})$ over K.
a) Show that χ_U is a separable irreducible polynomial (show that, if χ_U were reducible over K, it would be the product of two polynomials in $K_0[X, Z_{ij}]$ of degree > 0 in X ; hence deduce that the characteristic polynomial of every square matrix of order n over K would be reducible ; then derive a contradiction from the example of a polynomial with algebraically independent coefficients over K_0 ; use the analogous method to prove separability).
b) Show that U is similar to a matrix in which every entry is equal to 0, 1 or a coefficient of the polynomial χ_U (cf. VII, p. 30, formula (2)). Deduce that, if K_0 is an infinite field, and p is a polynomial in A such that $p(u'_{ij}) = p(u''_{ij})$ whenever the matrices (u'_{ij}) and (u''_{ij}) over K_0 are similar, then p belongs to the subring of A generated by the coefficients of χ_U. State and prove the analogous result for rational functions $r \in K = K_0(Z_{ij})$.

14) Let K be an imperfect field of characteristic p, let a be an element of $K - K^p$, and let L be the field $K(a^{1/p})$. Consider the K-vector space $E = L \otimes_K L$; let u (resp. u') be the K-endomorphism of E given by multiplication by $a^{1/p} \otimes 1$ (resp. $1 \otimes a^{1/p}$) in the algebra E. Show that u and u' are semi-simple, that u and u' commute, and that $u - u'$ is nonzero nilpotent.

Historical Note

(Chapters VI and VII)

(N.B. — Roman numerals in parentheses refer to the bibliography at the end of this note.)

Elementary arithmetical operations, in particular operations with fractions, inevitably lead to a number of empirical observations about divisibility between integers. But neither the Babylonians (although they were such experts in algebra), nor the Egyptians (despite their acrobatic skill at manipulating fractions) seem to have been acquainted with the general rules governing these properties, and the initiative in this respect is due to the Greeks. Their arithmetical work, a masterly exposition of which can be found in Books VII and IX of Euclid (I), is in no way inferior to their most beautiful discoveries in other branches of Mathematics. The existence of the gcd of two integers is proved right at the beginning of Book VII by the procedure known as the « Euclidean algorithm » * ; it serves as the basis for all the subsequent developments (properties of prime numbers, existence and calculation of lcm, etc.), based on arguments not substantially different from those in Chap. VI, § 1 above ; and the crowning glory is formed by the two remarkable theorems proving the existence of infinity of prime numbers (Book IX, Prop. 20) and giving a procedure for constructing even perfect numbers (cf. VII, p. 53, Ex. 24 ; this procedure in fact gives all the even perfect numbers, as would be proved by Euler). It is only the existence and uniqueness of prime factorisation that is not proved in a general manner ; however Euclid does prove explicitly that every integer is divisible by a prime number (Book VII, Prop. 31), as well as the following two propositions (Book IX, Prop. 13 and 14) :

« *If as many numbers as we like are in a progression, beginning with* 1, *with constant ratio* [i.e. geometric], *and if the number following* 1 *is prime, then the*

* If a_1 and a_2 are two integers, such that $a_1 \geqslant a_2$, one defines a_n (for $n \geqslant 3$) recursively as the remainder on dividing a_{n-2} by a_{n-1} ; if m is the smallest index such that $a_m = 0$, then a_{m-1} is the gcd of a_1 and a_2. This is the transposition into the integers of the method of successive subtractions (sometimes also called ἀνθυφαίρεσις) for finding a common measure for two lengths. The latter no doubt goes back to the Pythagoreans, and seems to have been the basis of a pre-Eudoxean theory of irrational numbers.

largest number will not be divisible by any number except those appearing in the progression » (in other words, a power p^n of a prime number p can be divisible only by powers of p of exponent $\leqslant n$).

« *If a number is the smallest which is divisible by* [given] *prime numbers, it will not be divisible by any other prime number with the exception of those initially* [given as] *dividing it* » (in other words, a product of distinct primes $p_1, \dots, p_k$ has no prime factor other than $p_1, \dots, p_k$).

Thus it seems that if Euclid does not state the general theorem, it is only for want of an adequate terminology and notation for arbitrary powers of an integer. *

Although a careful study makes it seem likely that Euclid's text is made up of several successive layers, each corresponding to a stage in the development of Arithmetic **, it seems that this evolution took place entirely between the beginning of the 5th and the middle of the 4th centuries B.C., and one must admire the skill and the logical confidence which it represents : the next comparable progress made in Arithmetic would not occur for two millenia.

It was the so-called « indeterminate » or « Diophantine » problems which were at the root of subsequent developments in Number Theory. The term « Diophantine equations », as it is used today, is not entirely justified historically ; it is generally understood to mean polynomial equations (or systems of equations) with integer coefficients, for which only integer solutions are sought : a problem which is usually impossible if the equations are « determinate », that is have only finitely many (real or complex) solutions, but which on the other hand often admits solutions when there are more unknowns than equations. Now, while Diophantus does indeed seem to have been the first to consider « indeterminate » problems, he looks for integer solutions only in exceptional cases, and is usually content to find a single solution in the rational numbers (II). That was a type of problem which he was usually able to solve by algebraic calculations, where the

* In support of this thesis, one can also remark that the proof of the theorem on perfect numbers is basically just another special case of the unique prime factorisation theorem. Moreover, all the evidence shows that from this period onwards the factorisation of an explicit number into primes was well known and in current use ; but no complete proof of the factorisation theorem is to be found before that given by Gauss at the beginning of the *Disquisitiones* ((VIII) t. I, p. 15).

** Cf. B. L. van der WAERDEN, Die Arithmetik der Pythagoreer, *Math. Ann.*, vol. CXX (1947-49), p. 127. An example of a passage left over from a previous version is provided by Prop. 21 to 34 of Book IX, which deals with the most elementary properties of divisibility by 2, and no doubt goes back to a period when the general theory of prime numbers had not been developed. Moreover, it is known that the categories Even and Odd played a major rôle in the metaphysical speculations of the early Pythagoreans, to whom it is naturally tempting to attribute this segment.

arithmetical nature of the unknowns was not relevant * ; also the theory of divisibility plays only a minor rôle (the word for prime number is used only once ((II), Book V, problem 9, vol. 1, pp. 334-335), and the notion of coprime numbers is invoked only in connection with the theorem which asserts that the quotient of two coprime numbers can be a square only if each of them is a square) **.

The study of integer solutions of indeterminate equations only begins properly with the Chinese and Hindu mathematicians of the early middle ages. The former seem to have been led to considerations of this nature by the practical problems of drawing up calendars (where the determination of the common periods of various cycles of astronomical phenomena constitutes precisely a linear « Diophantine » problem) ; at any rate they were responsible (certainly between the 4th and the 7th century A.D.) for a rule for solving simultaneous linear congruences (cf. VI, p. 33, Ex. 25). As for the Hindus, whose Mathematics flourished between the 5th and the 13th centuries, not only did they know how to deal methodically (by applying the Euclidean algorithm) with systems of linear Diophantine equations in arbitrarily many unknowns ***, but they were the first to attack and solve quadratic problems, among them certain special cases of « Fermat's equation » $Nx^2 + 1 = y^2$ ((III), vol. II, pp. 87-307).

This is not the place for us to pursue the history of the theory of Diophantine equations of degree > 1, which, by way of the work of Fermat, Euler, Lagrange and Gauss, was to lead to the theory of algebraic integers in the 19th century. As we have already noted (cf. Historical Note to Chap. II and III), the study of linear systems, which no longer seemed to present problems worthy of interest, was rather neglected during this period : in particular there was no search for general existence conditions for an arbitrary system, nor for a description of the set of solutions. Nevertheless Hermite, in his Number-theoretical research in the 19th

* While Diophantus' work on indeterminate problems is always reduced to problems in a single unknown, *via* a numerical choice of the other unknowns in such a way as to make a solution to his final equation possible, it seems that the main reason for using this method was his notation, which did not allow him to calculate with several unknowns at once ; in any case he keeps track of the numerical substitutions he has made throughout the calculation, and modifies them later if need be, by writing down a compatibility condition for the substituted variables, and solving this auxiliary problem first. In other words, he handles these substituted numerical values like we handle parameters, so much so that what he actually does amounts to finding a rational parametric representation of a given algebraic variety, or of a subvariety of it (cf. (II *bis*)).

** Various indications, however, point to more advanced arithmetical knowledge in Diophantus : he knew for example that the equation $x^2 + y^2 = n$ has no rational solutions if n is an integer of the form $4k + 3$ (Book V, problem 9 and Book VI, problem 14 ((II), vol. 1, pp. 332-335 and p. 425 ; cf. also (II *bis*), pp. 105-110)).

*** Astronomical problems were also among those which led the Hindus to consider this type of equation (cf. (III), vol. II, pp. 100, 117 and 135).

century, was led to make use of several Lemmas on linear Diophantine equations, notably a « reduced form » for a linear transformation with integer coefficients ((XIII), pp. 164 and 265) ; finally, after Heger in 1858 had given the existence condition for a system whose rank is equal to the number of equations, H. J. Smith in 1861 defined the invariant factors of an integer matrix, and obtained the general theorem that such a matrix reduces to the « normal form » which we gave in VII, p. 22, Cor. 1 (XVII).

But meanwhile, following its introduction by Gauss (cf. Historical Notes to Chap. I, II and III), and the important part it played in the subsequent development of Number Theory, the notion of an abelian group was gradually being made precise. In his particularly deep study, presented in the *Disquisitiones*, of the finite abelian group of classes of quadratic forms of a given discriminant, Gauss soon realised that certain of these groups are not cyclic : « *in this case,* » he says, « *one basis* [that is to say, one generator] *cannot suffice, it is necessary to take two or a greater number which, through multiplication and composition* *, *can produce all the classes* » ((VIII), vol. I, pp. 374-375). It is not certain that Gauss intended these words to describe the decomposition of the group as a direct product of cyclic groups ; nevertheless, in the same article of the *Disquisitiones*, he proves that there exists an element in the group whose order is the lcm of the orders of all the elements — in other words he obtains the existence of the largest invariant factor of the group ((VIII), vol. 1, p. 373) ; and on the other hand, the notion of direct product was known to him, for, in a manuscript dating from 1801, but not published during his lifetime, he sketches a general proof of the decomposition of a finite abelian group into a direct product of p-groups ** ((VIII), vol. II, p. 226). In any case, in 1868 Schering, the editor of Gauss' collected works, inspired by these results (notably by this manuscript which he had just found), proved (still for the group of classes of quadratic forms) the general decomposition theorem (XVIII) by a method which, as repeated in abstract terms two years later by Kronecker (XX), is essentially that which we have used above (VII, p. 18, Th. 1). As for torsion-free abelian groups, we have already said (cf. Historical Note to Chap. II and III) how the theory of elliptic functions and abelian integrals, developed by Gauss, Abel and Jacobi, gradually led to attention being paid to their structure ; the first and best known example of a decomposition of an infinite group into a direct sum of cyclic groups was given in 1846 by Dirichlet in his paper on the units of an algebraic number field (XI). But it was not until 1879 that the connection between the theory of finitely generated abelian groups and Smith's theorem was recognised and explicitly used by Frobenius and Stickelberger ((XXIII), § 10).

* Gauss uses additive notation for the law of composition of classes, thus by « multiplication » he means the product of a class by an integer.

** Abel also proved this property *en passant* in his note on abelian equations ((IX), vol. I, pp. 494-497).

Around the same period the theory of similarity of matrices (with real or complex entries) was also reaching completion. The notion of an eigenvalue of a linear transformation appeared explicitly in the theory of systems of linear differential equations with constant coefficients, applied by Lagrange (VI*a*) to the theory of small oscillations and by Lagrange (VI*b*) and Laplace (VII*a*) to the « secular » perturbations of the planets. It is implicit in many other problems which were also attacked around the middle of the 18th century, such as that of finding the axes of a conic or a quadric (first solved by Euler (V*a*)), or the study (also developed by Euler (V*b*)) of the principal axes of inertia of a solid body (discovered by De Segner in 1755) ; we know now that it is also involved (in a more disguised form) in the beginnings of the theory of partial differential equations, in particular the equation of a vibrating string. But (leaving this last case aside) the relationship between these various problems was barely recognised before Cauchy (X). Moreover, as most of them involve the use of symmetric matrices, it was principally because of the latter that eigenvalues were initially studied ; we will return to this point in more detail in the Historical Notes following the chapters of this treatise devoted to hermitian operators ; let us simply note here that, as early as 1826, Cauchy proved the invariance of the eigenvalues of such matrices under similarity, and proved that they are real for a 3×3 symmetric matrix (X*a*), a result which he generalised three years later (X*b*) to arbitrary real symmetric matrices. * The general notion of a projection, introduced by Möbius in 1827, rapidly led to the problem of classifying such transformations (in 2 and 3 dimensions first of all), which is nothing other than the problem of classifying the corresponding matrices up to similarity ; but for a long time this question was treated only by the « synthetic » methods which were in vogue in the mid 19th century, and its (in any case rather slow) progress does not seem to have affected the theory of eigenvalues in any way. The same is not true of another geometric question, the classification of « pencils » of conics or quadrics, which from the modern point of view amounts to the study of the elementary divisors of the matrix $U + \lambda V$, where U and V are two symmetric matrices ; it was certainly in this spirit that Sylvester attacked this problem in 1851, carefully examining (in order to find « canonical forms » for the pencil under consideration) what happens to the minors of the matrix $U + \lambda V$ when a value for λ is substituted which annihilates the determinant (XIV). The purely algebraic aspect of the theory of eigenvalues was progressing simultaneously ; so it was that several authors (including Sylvester himself) proved around 1850 that the eigenvalues of U^n are the n-th powers of the eigenvalues of U, while Cayley

* An attempt to prove this result in the particular case of the « secular » perturbations of the planets had previously been made by Laplace (VII*b*) in 1784. As for the third degree equation giving the axes of a real quadric, Euler had stated without proof that it has real roots, and an attempted proof by Lagrange in 1773 (VI*e*) was incomplete ; this point was first proved rigourously by Hachette and Poisson in 1801 (*Journal de l'École Polytechnique*, cahier 11 (year X), pp. 170-172).

announced in 1858, in the paper in which he introduced matrix arithmetic (XVI), the « Cayley-Hamilton Theorem » for an arbitrary square matrix *, though he contented himself with a direct proof for 2×2 and 3×3 matrices. Finally Weierstrass in 1868, using Sylvester's methods, obtained « canonical forms » for a « pencil » $U + \lambda V$ where this time U and V are square matrices, not necessarily symmetric, subject only to the condition that $\det(U + \lambda V)$ not be identically zero ; he deduced from that the definition of the elementary divisors of an arbitrary square (complex) matrix, and proved that they characterise the latter up to similarity (XIX) ; these results, incidentally, were partially (and apparently independently) recovered by Jordan two years later ** (XXI). Here again it was Frobenius who showed in 1879 that Weierstrass' theorem can easily be deduced from Smith's theorem extended to polynomials ((XXII), § 13) ; his procedure is the basis of the proof of this theorem which we have given above (VII, p. 35).

We have just referred to the theory of divisibility for polynomials in one variable ; the question of division of polynomials must naturally have arisen in the earliest days of algebra, as the inverse operation to multiplication (the latter being known even to Diophantus, at least for polynomials of low degrees) ; but one can imagine that it is barely possible to attack this problem in a general fashion before a coherent notation for the various powers of the variable has been established. In fact we find very few examples of the « Euclidean » division procedure for polynomials, as we know it, before the middle of the 16th century *** ; and S. Stevin (essentially, using exponent notation) seems to have been the first to have had the idea of deducing the extension of the « Euclidean algorithm » to find the gcd of two polynomials ((IV, vol. I, pp. 54-56). Apart from that, the theory of divisibility had been restricted to the rational integers until the mid 18th century. It was Euler in 1770 who began a new chapter in Arithmetic by somewhat rashly extending the notion of divisibility to the integers of a quadratic extension : seeking to determine the divisors of a number of the form $x^2 + cy^2$ (x, y and c rational integers), he put

$$x + y\sqrt{-c} = (p + q\sqrt{-c})(r + s\sqrt{-c}) \quad (p, q, r, s \text{ rational integers})$$

and, taking norms of each side, he had no hesitation in asserting that all the divisors of $x^2 + cy^2$ are obtained in this way as $p^2 + cq^2$ (Vc). In other words,

* Hamilton, incidently, had proved this theorem for 3×3 matrices some years previously ((XV), pp. 566-567).

** Jordan did not mention the invariance of the normal form which he obtained. It is interesting to note as an aside that he dealt with the problem not for complex matrices, but for matrices over a finite field. On the other hand, let us point out that Grassmann had given a method for reducing a (complex) matrix to triangular form as early as 1862, and explicitly mentioned the connection between this reduction and the classification of projections (*Ges. Math. Werke,* vol. I_2, Leipzig (Teubner), 1896, pp. 249-254).

*** Cf. for example H. Bosmans, Sur le « libro del Algebra » de Pedro Nuñez, *Bibl. Math.* (3), vol. VIII (1907-1908), pp. 154-169.

Euler argued as if the ring $\mathbf{Z}[\sqrt{-c}]$ were a principal ideal domain ; a little later he used an analogous argument to apply the method of « infinite descent » to the equation $x^3 + y^3 = z^3$ (he reduced the problem to that of finding a cube root of $p^2 + 3q^2$, which he does by putting $p + q\sqrt{-3} = (r + s\sqrt{-3})^3$). But Lagrange proved as early as 1773 (VIc) that the divisors of numbers of the form $x^2 + cy^2$ are not all of this form, the first example of the fundamental difficulty which was to appear much more clearly in the studies of Gauss and his successors into divisibility in cyclotomic fields * ; it is not possible, in general, to extend the essential properties of divisibility of rational integers, such as the existence of gcd's and the uniqueness of prime factorisation, directly to these fields. This is not the place to describe how Kummer, for cyclotomic fields (XII)**, and then Dedekind and Kronecker for arbitrary algebraic number fields, succeeded in overcoming this formidable obstacle by the invention of ideal theory, one of the most decisive advances of modern algebra. But Dedekind, ever curious about the foundations of various Mathematical theories, was not content with this success ; and by analysing the mechanism of the divisibility relations he laid the foundations of the theory of lattice ordered groups, in a paper (unknown to his contemporaries, and lost in obscurity for 30 years) which is without doubt one of the earliest works of axiomatic Algebra (XXIV) ; modulo notation, his work was very close to the modern form of this theory, as we have presented it in Chap. VI, § 1.

*

* *

From the middle of the 18th century, the order of the day was a search for a « fundamental theorem of algebra » (cf. Historical Note to Chap. IV and V). We do not need to mention here the attempt of d'Alembert, who initiated the series of proofs which use infinitesimal calculus (cf. Gen. Top., Historical Note to Chap. VIII). But Euler attacked the problem in a completely different manner in 1749

* Gauss seems to have hoped at one time that the ring of integers in the field of n-th roots of unity would be a principal ideal domain ; in a manuscript unpublished in his lifetime ((VIII), vol. II, pp. 387-397), he proves the existence of a Euclidean division process in the field of cube roots of unity, and gives some indications of an analogous process in the field of 5th roots of unity ; he uses these results to prove by an « infinite descent » argument more correct than Euler's that the equation $x^3 + y^3 = z^3$ has no solution in the field of cube roots of unity, indicates that one can extend the method to the equation $x^5 + y^5 = z^5$, but stops short of the equation $x^7 + y^7 = z^7$, saying that here it is impossible to reject *a priori* the case where x, y and z are not divisible by 7.

** In his earliest work on « ideal numbers », Kummer explicitly indicated the possibility of applying his method, not only to cyclotomic fields, but also to quadratic fields, and hence of recovering Gauss' results on binary quadratic forms ((XII), pp. 324-325).

(V*d*) : for every polynomial f with real coefficients, he attempted to prove the existence of a factorisation $f = f_1 f_2$ into two (non constant) polynomials with *real* coefficients, which would give him a proof of the « fundamental theorem » by induction on the degree of f. It even suffices, as he noticed, to stop at the first odd degree factor, and consequently all the difficulties are reduced to the consideration of the case where the degree n of f is even. Euler then restricts himself to the case when the desired factors each have degree $n/2$, and he indicates that by a suitable process of elimination one can express the unknown coefficients of f_1 and f_2 as rational functions of a root of an equation with real coefficients, whose extreme terms have *opposite signs* and which consequently has at least one real root. But Euler's proof was only a sketch, and skipped over a number of essential points ; it was not till 1772 that Lagrange succeeded in resolving the difficulties raised by this proof (VI*d*) by means of an extremely long and minute analysis, in which he demonstrated a remarkable virtuosity in the use of the « Galois methods » which he had recently created (cf. Historical Note to Chap. IV and V).

All the same, Lagrange, like Euler and all his contemporaries, had no hesitation in arguing formally within a « field of roots » of a polynomial (that is, in his language, to consider « imaginary roots » of this polynomial) ; the Mathematics of his period had provided no justification for this type of argument. Gauss, who was from the outset resolutely hostile to the unrestricted formalism of the 18th century, came out strongly against this abuse in his dissertation ((VIII), vol. III, p. 3). But it would have been unlike him not to have sensed that it was a case of a superficially faulty presentation of an argument which was intrinsically correct. We find him also, a few years later ((VIII), vol. III, p. 33 ; cf. also (VIII*bis*)), taking up a simpler form of Euler's argument, which had been suggested in 1759 by Foncenex (who, however, had been unable to use it to any advantage), to obtain a new proof of the « fundamental theorem » in which he carefully avoids all mention of « imaginary » roots : the latter being replaced by skilful adjunction and specialisation of indeterminates. It is essentially this proof of Gauss that we have given in the text (VI, p. 26, Th. 3), with simplifications made possible by the use of algebraic extensions.

The rôle of Topology in the « fundamental theorem » was thus reduced to the single theorem that a polynomial with real coefficients cannot change sign in an interval without having a zero (Bolzano's Theorem for polynomials). This theorem is also at the root of all the criteria for separation of the real roots of a polynomial (with real coefficients), which was one of the favourite Algebraic topics in the 19th century *. In the course of this research it becomes obvious that it is the order structure of **R**, much more than its topology, which plays the

* On these topics, which we will not discuss in the present work, the reader may consult for example J.-A. SERRET, Cours d'Algèbre supérieure, 3rd ed., Paris (Gauthier-Villars), 1866, or B. L. VAN DER WAERDEN, Moderne Algebra, vol. I (1st ed.), Berlin (Springer) 1930, pp. 223-235.

essential rôle * ; for example Bolzano's Theorem for polynomials remains true in the field of all real algebraic numbers. This train of ideas resulted in the abstract theory of ordered fields, created by E. Artin and O. Schreier (XXV) ; one of its most remarkable results is certainly the discovery that the existence of an order relation on a field is related to purely algebraic properties of the field. This is the theory presented in § 2 of Chap. VI.

* The tendency to attach overwhelming importance to the order structure of the real numbers can also be seen in their definition by means of Dedekind « cuts », which is basically a process which can be applied to all ordered sets (cf. VI, p. 35, Ex. 30 onwards).

Bibliography

(I) *Euclidis Elementa*, 5 vol., ed. J. L. Heiberg, Lipsiae (Teubner), 1883-88.

(I *bis*) T. L. HEATH, *The thirteen books of Euclid's Elements...*, 3 vol., Cambridge, 1908.

(II) *Diophanti Alexandrini Opera Omnia...*, 2 vol., ed. P. Tannery, Lipsiae (Teubner), 1893-95.

(II *bis*) T. L. HEATH, *Diophantus of Alexandria*, 2nd ed., Cambridge, 1910.

(III) B. DATTA and A. N. SINGH, *History of Hindu Mathematics*, 2 vol., Lahore (Motilal Banarsi Das), 1935-38.

(IV) S. STEVIN, *Les œuvres mathématiques...*, ed. A. Girard, Leyde (Elsevier), 1634, vol. I.

(V) L. EULER : *a*) *Introductio in Analysin Infinitorum* (*Opera Omnia*, (1), t. IX, Zürich-Leipzig-Berlin (O. Füssli and B. G. Teubner), 1945, p. 384) ; *b*) *Theoria motus corporum solidorum seu rigidorum* (*Opera Omnia* (2), t. III, Zürich-Leipzig-Berlin (O. Füssli and B. G. Teubner), 1948, p. 200-201) ; *c*) *Vollständige Anleitung zur Algebra* (*Opera Omnia* (1), t. I, Leipzig-Berlin (Teubner), 1911, p. 422) ; *d*) Recherches sur les racines imaginaires des équations (*Opera Omnia* (1), t. VI, Leipzig-Berlin (Teubner), 1921, p. 78).

(VI) J.-L. LAGRANGE, *Œuvres*, Paris (Gauthier-Villars), 1867-1892 : *a*) Solutions de divers problèmes de Calcul intégral, t. I, p. 520 ; *b*) Recherches sur les équations séculaires du mouvement des nœuds, t. VI, p. 655-666 ; *c*) Recherches d'arithmétique, t. III, p. 695-795 ; *d*) Sur la forme des racines imaginaires des équations, t. III, p. 479 ; *e*) Nouvelle solution du problème de rotation d'un corps quelconque qui n'est animé par aucune force accélératrice, t. III, p. 579-616.

(VII) P. S. LAPLACE : *a*) Mémoire sur les solutions particulières des équations différentielles et sur les inégalités séculaires des planètes (*Œuvres*, t. VIII, Paris (Gauthier-Villars), 1891, p. 325-366) ; *b*) Mémoire sur les inégalités séculaires des planètes et des satellites (*Œuvres*, t. XI, Paris (Gauthier-Villars), 1895, p. 49-92).

(VIII) C. F. GAUSS, *Werke*, t. I (Göttingen, 1870), t. II (*ibid.*, 1876) et t. III (*ibid.*, 1876).

(VIII *bis*) *Die vier Gauss'schen Beweise für die Zerlegung ganzer algebraischer Functionen in reelle Factoren ersten oder zweiten Grades* (Ostwald's Klassiker, n° 14, Leipzig (Teubner), 1904).

(IX) N. H. ABEL, *Œuvres*, t. I, ed. Sylow and Lie, Christiania, 1881.

(X) A. L. CAUCHY : *a*) *Leçons sur les applications du Calcul infinitésimal à la Géométrie* (*Œuvres complètes* (2), t. V, Paris (Gauthier-Villars), 1903, p. 248) ; *b*) Sur l'équation à l'aide de laquelle on détermine les inégalités séculaires des planètes (*Œuvres complètes* (2), t. IX, Paris (Gauthier-Villars), 1891, p. 174).

(XI) P. G. LEJEUNE-DIRICHLET, *Werke*, t. I, Berlin (G. Reimer), 1889, p. 619-644.

(XII) E. KUMMER, Zur Theorie der complexen Zahlen, *J. de Crelle*, t. XLIII (1847), p. 319 (Collected papers, vol. I, Heidelberg (Springer V.), 1975, p. 203).

(XIII) Ch. HERMITE, *Œuvres*, t. I, Paris (Gauthier-Villars), 1905.

(XIV) J. J. SYLVESTER, *Collected Mathematical Papers*, vol. I, Cambridge, 1904 : An enumeration of the contacts of lines and surfaces of the second order, p. 219 (= *Phil. Mag.*, 1851).

(XV) W. R. HAMILTON, *Lectures on Quaternions*, Dublin, 1853.

(XVI) A. CAYLEY, *Collected Mathematical Papers*, Cambridge, 1889-1898 : A memoir on the theory of matrices, t. II, p. 475-496 (= *Phil. Trans.*, 1858).

(XVII) H. J. Smith, *Collected Mathematical Papers*, vol. I, Oxford, 1894 ; On systems of linear indeterminate equations and congruences, p. 367 (*Phil. Trans.*, 1861).

(XVIII) E. Schering, Die fundamental Classen der zusammensetzbaren arithmetischen Formen, *Abh. Ges. Göttingen*, t. XIV (1868-69), p. 13.

(XIX) K. Weierstrass, *Mathematische Werke*, Bd. II, Berlin (Mayer und Müller), 1895 : Zur Theorie der bilinearen und quadratischen Formen, p. 19.

(XX) L. Kronecker, Auseinandersetzungen einiger Eigenschaften der Klassenanzahl idealer complexer Zahlen, *Monats. Abhandl. Berlin* (1870), p. 881 (= *Werke*, t. I, Leipzig (Teubner), 1895, p. 273).

(XXI) C. Jordan, *Traité des substitutions et des équations algébriques*. Paris (Gauthier-Villars), 1870, p. 114-125.

(XXII) G. Frobenius, Theorie der linearen Formen mit ganzen Coefficienten, *Gesammelte Abhandlungen*, vol. I, Heidelberg (Springer V.), 1968, p. 482 (= J. de Crelle, 1879).

(XXIII) G. Frobenius und L. Stickelberger, Ueber Gruppen von vertauschbaren Elementen, *J. de Crelle*, t. LXXXVI (1879), p. 217 (= Frobenius, Ges. Abh., vol. I, p. 545).

(XXIV) R. Dedekind, *Gesammelte mathematische Werke*, t. II, Braunschweig (Vieweg), 1932 : Ueber Zerlegungen von Zahlen durch ihre grössten gemeinsamen Teiler, p. 103.

(XXV) *a*) E. Artin und O. Schreier, Algebraische Konstruktion reeler Körper, *Abh. Math. Sem. Univ. Hamburg*, t. V (1927), p. 83 ; *b*) E. Artin, Ueber die Zerlegung definiter Funktionen in Quadrate (*ibid.*, p. 100) ; *c*) E. Artin und O. Schreier, Eine Kennzeichnung der reell abgeschlossenen Körper (*ibid.*, p. 225).

Index of notation

Index of terminology

Table of contents

Printing and Binding: Strauss GmbH, Mörlenbach